Sixth Edition

Measurement and Evaluation

—in—

Human Performance

James R. Morrow, Jr., PhD
University of North Texas, Regents Professor Emeritus

Dale P. Mood, PhD
University of Colorado, Professor Emeritus

Weimo Zhu, PhD
University of Illinois at Urbana–Champaign

Minsoo Kang, PhD
University of Mississippi

HUMAN KINETICS

Library of Congress Cataloging-in-Publication Data

Names: Morrow, James R., Jr., 1947- author. | Mood, Dale, author. | Zhu,
  Weimo, 1955- author. | Kang, Minsoo (Physical education teacher),
  author.
Title: Measurement and evaluation in human performance / James R. Morrow,
  Jr., Dale P. Mood, Weimo Zhu, Minsoo Kang.
Description: Sixth edition. | Champaign, IL : Human Kinetics, [2023] |
  Includes bibliographical references and index.
Identifiers: LCCN 2022014957 (print) | LCCN 2022014958 (ebook) | ISBN
  9781492599586 (paperback) | ISBN 9781492599593 (epub) | ISBN
  9781492599616 (pdf)
Subjects: MESH: Physical Education and Training | Physical Fitness |
  Monitoring, Physiologic--methods | Athletic Performance
Classification: LCC QP301 (print) | LCC QP301 (ebook) | NLM QT 255 | DDC
  613.7028/7--dc23/eng/20220801
LC record available at https://lccn.loc.gov/2022014957
LC ebook record available at https://lccn.loc.gov/2022014958

ISBN: 978-1-4925-9958-6 (print)

Excel screen captures in chapter 2 and the appendix are used with permission of Microsoft.

The web addresses cited in this text were current as of April 2022, unless otherwise noted.

Acquisitions Editor: Diana Vincer; **Copyeditor:** Heather Gauen Hutches; **Proofreader:** Erin Cler; **Indexer:** Rebecca L. McCorkle; **Permissions Manager:** Laurel Mitchell; **Senior Graphic Designer:** Sean Roosevelt; **Cover Designer:** Dawn Sills; **Art Director:** Keri Evans; **Cover Design Specialist:** Susan Rothermel Allen; **Photograph (cover):** vm /E+/Getty Images; **Photo Asset Manager:** Laura Fitch; **Photo Production Manager:** Jason Allen; **Senior Art Manager:** Kelly Hendren; **Illustrations:** © Human Kinetics **Printer:** Walsworth

Printed in the United States of America   10 9 8 7 6 5 4 3 2 1

The paper in this book was manufactured using responsible forestry methods.

Human Kinetics
1607 N. Market Street
Champaign, IL 61820
USA

United States and International
Website: **US.HumanKinetics.com**
Email: info@hkusa.com
Phone: 1-800-747-4457

Canada
Website: **Canada.HumanKinetics.com**
Email: info@hkcanada.com

E8131

Tell us what you think!
Human Kinetics would love to hear what we
can do to improve the customer experience.
Use this QR code to take our brief survey.

We dedicate the sixth edition of *Measurement and Evaluation in Human Performance* to our friend, colleague, and coauthor, Allen W. Jackson, who passed away in August 2020. Allen was a coauthor on the first four editions of this textbook and a major contributor to the fifth edition. This book would never have had a first edition if not for Allen. Allen lived what this textbook is all about. He was a scientist, researcher, and teacher who contributed much to the field and approached his responsibilities in a timely and professional manner. This textbook and the entire field of measurement and evaluation in human performance are better because of Allen Jackson. He is greatly missed.

JRM, DPM, WZ, MK

Contents

Preface

In *Measurement and Evaluation in Human Performance, Sixth Edition,* our objective remains the same as when we developed the original text nearly three decades ago: We want to provide an effective, interactive textbook that is accessible to undergraduate students in human performance, kinesiology, exercise science, or physical education.

You are about to study concepts that we believe are among the most important you will learn as an undergraduate. We are confident that you want to make good decisions in all aspects of your professional and personal life. In the truest sense, this means you want to make evidence-based decisions that are accurate, trustworthy, truthful, and relevant to the issue. That is exactly what this textbook and the course in which you are enrolled are all about. The concepts of reliability, objectivity, and validity that serve as the focus of this textbook can be used in all of your courses and throughout all phases of life.

Whether your desired career path is physical therapy, teaching, athletic training, general kinesiology or exercise science, health and fitness, sport studies, or sport management, in this textbook you will find concepts important to your daily work. The ability to measure physical activity, physical fitness, and physical skills and understand and use sound measurement procedures will be vital to your work, whether testing in the psychomotor, cognitive, or affective domains.

When you obtain data as a result of some type of testing or survey process, you want to be confident that the data you have for decision making are trustworthy—that is, can you count on these data? Are they truthful? Whether you are testing the fitness level of a student or client or receiving test results from your personal physician, you want the results to be valid. In this text we will help you learn how to gather data, analyze them, interpret the results, and feel confident that what you have learned is valuable enough to make good decisions.

We have been university professors for more than 130 combined years. We have taught hundreds of sections and thousands of students and received much feedback and commentary. We have attempted to incorporate all of this valuable information into this textbook so that you will find the materials engaging, interesting, entertaining, informative, and useful. Although many students may be intimidated by the content presented in a measurement and evaluation course, the level of mathematics involved in this textbook is no more than that in an introductory algebra course. We combine these introductory concepts with reliability and validity theory in a way that can help you appreciate and learn this material. That does not mean that the course is necessarily easy—it does require study, review, and logical thinking to help you put all of the concepts together. We believe we do this well in the textbook, and we hope that when you have a decision facing you, you will immediately think about how you might gather data, analyze them, and use your analyses to make a good decision.

UPDATES TO THE SIXTH EDITION

As we present the sixth edition of our text, we have maintained elements that have worked well, added some new points, and reorganized information to permit instructors and students to have a better understanding of the textbook and apply the information to their specific courses and career goals. The key changes in the sixth edition are the following:

• The focus in our text is new developments in measurement concepts and methods, data collection, and analysis so that you can make appropriate decisions based on evidence. The concept of evidence-based decision making is important regardless of your chosen career path.

• As in the fifth edition, we include procedures to use both SPSS (IBM SPSS Statistics) and Microsoft Excel for calculating statistics, with examples in an appendix that includes instructions and screen captures.

• New material has been added to all chapters, including the latest views on validity evidence (chapter 6), administration and interpretation of the Functional Movement Screen (chapter 11), and information on video analysis apps and employment-related testing (chapter 12), to cite a few.

• In addition, the student and instructor resources on HK*Propel* have been updated for this edition. We continue to use large datasets in each chapter in order for you to apply previously learned concepts and methods as well as practice valuable computer skills for measurement and evaluation. Although students often overlook or ignore these ancillaries, they are presented to help you learn and review content—you should definitely take advantage of these many resources!

TEXTBOOK ORGANIZATION

The textbook is presented in four parts. Part I, Introduction to Tests and Measurements in Human Performance, consists of two chapters. Chapter 1 introduces you to concepts in tests and measurement in human performance and the domains of interest in which you will use these concepts and tools. We use the information in this textbook every day of our lives. However, to do all of this work by hand would be tedious and susceptible to errors. Therefore, chapter 2, Using Technology in Measurement and Evaluation, introduces you to SPSS and Excel to help you complete analyses. The use of technology and statistical packages may not eliminate all errors, but it does save you a great deal of time. Technology applications and statistical analysis examples are then used throughout the remainder of the textbook.

Part II, Basic Statistical Concepts, consists of three chapters and provides the statistical background for many of the decisions and interpretations that you will encounter in the remainder of the textbook. The concepts that you learn in part II are the reason some students refer to this course content as statistics. Although the basic statistics presented are important, we prefer to consider these as foundational elements for reliability and validity decisions. Importantly, we are not mathematicians and do not expect you to be. Although having a stronger mathematics background can be an advantage, we present material in such a way that minimal mathematical expertise is expected or required. Moreover, continuing to practice and use SPSS and Excel will help you complete time-consuming and difficult tasks with large numbers of observations quickly and accurately. Chapter 3, Descriptive Statistics and the Normal Distribution, illustrates how to describe test results and interpret them with graphs and charts. Chapter 4, Correlation and Prediction, helps you understand relations between variables and how knowledge about one variable tells you something about other variables. The content of chapter 5, Inferential Statistics, is used daily by human performance researchers to help interpret the results from research studies and determine how these results can be generalized to make decisions. For example, it is common knowledge that physical activity, physical fitness, and energy expenditure are related to quality of life, health, and disease and death risk. Much of what we know about the relations between these variables has been generated as a result of the types of analyses you will learn in these chapters. After completing part II, you will be able to read, interpret, understand, and use the scientific research literature related to your profession.

Part III, Reliability and Validity Theory, presents the most important concepts of the textbook. Everything that is done throughout the textbook can be directed toward or emanates from the concepts of reliability and validity. After the basic concepts of reliability and validity are introduced, how to connect scientific evidence and make a correct judgment based on the evidence of a test or measure is then described in detail. Chapter 6 presents Reliability and Validity from a theoretical and comprehensive perspective, whereas chapter 7 does so from a criterion-referenced perspective, including how to set standards and cut scores. Consider the most recent test you completed: How do you know the results were reliable and accurate? You will be better able to interpret these results once you have completed part III.

Part IV, Human Performance Applications, consists of eight chapters that illustrate practical settings where you will use the kinds of knowledge you have gained in this textbook. To complete an assessment, you have to make a judgment on or evaluate the number you got from a valid and reliable measure. Is it fair, good, or excellent; fit or not fit; passed or failed? To do so, you have to put the number into a reference framework. In Chapter 8, Evaluation: Theory and Practice, you will learn two commonly used frameworks for evaluation (norm- and criterion-referenced) and how to apply them in both school and nonschool settings. Other critical issues related to evaluation, such as fairness and objectivity, will also be introduced. Even students whose career options are not directed toward teaching can then apply this information to the classes they are currently enrolled in or may take in the future. For example, are you well informed about the grading procedures used in your classes? What might you encourage the professor to do to make grading fairer? This chapter will help you better answer these questions.

Although it is difficult to discriminate among levels of knowledge, it is important that you are able to do so in order to gather accurate data. Chapter 9, Developing Written Tests and Surveys, instructs you how to create tests that accurately discriminate among those with differing levels of knowledge. Another important ability in this area is the development of survey and questionnaire materials that accurately reflect the knowledge or attitudes of an individual or a group.

Chapters 10 and 11 are devoted to the assessment of fitness, one of the most important areas in human performance. Chapter 10 focuses on health-related fitness, including its definition, brief history, major components, laboratory and field tests, and commonly used fitness batteries for children, adults, older adults, and special populations. Chapter 11 focuses on performance-related fitness, including its components and commonly used tests, as well as critical issues related to selection, administration, and use of performance-related tests. Upon the completion of these two chapters, you should have adequate knowledge and skills to select and administer a number of fitness-related tests.

Chapter 12 illustrates reliability, validity, and practical issues when assessing motor abilities, skills, and performance. Many textbooks use this type of chapter to list tests that can or should be used to assess these particular elements; however, we prefer to identify the important concepts to be considered when choosing a test and then let you decide if a particular test appropriately meets your needs. Assume that you want to measure the physical skills or abilities of students, clients, or athletes with whom you work. How might you best assess their abilities and skills, and how might you interpret and use these results? You will be able to use measurement techniques, including sport analytics and new video analysis technology, to answer many of these questions after completing chapter 12.

Chapter 13, a new chapter, is devoted specifically to assessing physical activity and energy expenditure, another very important area in human performance. After an introduction to the concept of physical activity, historical efforts made to assess physical activity, and its relationship with fitness and energy expenditure, the methods of assessing of physical activity are described in detail, including their advantages and disadvantages.

Chapter 14, Psychological Measurements in Sport and Exercise, presents the latest developments in sport and exercise psychology, including the effects of physical activity on academic performance and using exercise for the management of symptoms of chronic diseases. The newest technology applications in psychology—using magnetic resonance imaging (MRI) and electroencephalography (EEG) techniques to quantify changes in brain structure and cognitive functions—are also discussed. Assessment principles and issues in the affective domain (e.g., attitudes and beliefs) are introduced. A set of commonly used psychological scales are also included in HKPropel. You will be better able to interpret results obtained from the affective domain after completing chapter 14.

Chapter 15 introduces a set of alternative ways to assess student learning, namely performance-based assessment. Although they are most appropriate for those with career goals in public or private school instruction, the principles and formats learned can be applied in other areas, such as coaching or clinical exercise interventions, as well as to your own professional development (e.g., you may show your competency and job skills with a portfolio).

Appendix: Microsoft Excel Applications provides support for those without access to SPSS by presenting directions for calculating measurement and evaluation statistics with Excel. The results are similar to those obtained with SPSS. Templates are provided in HKPropel in some cases rather than listing the many necessary steps required with Excel. Some Excel resources are also available on HKPropel.

STUDENT RESOURCES

Key to the sixth edition of our textbook is the variety of student resources that help you learn the material and apply it to your career in human performance and everyday decision making. In each chapter you'll find many items that will help you understand and retain information:

• Outlines are presented at the start of each chapter, introducing you to the organization of the chapter and helping you locate key information.

• Objectives indicate the main points you should take away after you have finished each chapter.

• Key terms are highlighted throughout the text, with accompanying definitions provided for you in the glossary.

• A Measurement and Evaluation Challenge introduces and closes each chapter. The opening scenario presents a person faced with a measurement or evaluation issue; the closing shows how the person can solve the issue presented in the opening scenario using concepts covered in the chapter.

• Mastery Items test whether you have mastered a particular concept. These items include problems and student activities that will help you confirm that you have mastered the information at certain points throughout a chapter. Some Mastery Items require you to complete the task on HKPropel.

• Dataset Applications give you an opportunity to practice many of the techniques presented in chapters 2 through 15. Large datasets are available in each chapter's section of HKPropel (described in the next section). You will gain valuable experience in using statistical software by using these datasets.

We encourage you to use all of the resources provided with this textbook. They will help you better understand the concepts to prepare you for examinations as well as apply them to your career.

STUDENT RESOURCES IN HK*Propel*

New to this edition are several additional resources delivered online through HK*Propel*. These help bring the content to life and further student comprehension. The resources include a printable outline of each chapter, an interactive chapter quiz, homework problems with answers to the odd-numbered problems, student activities, answers to the Mastery Items, large datasets that are used to complete the Dataset Application activities in the chapters, and dozens videos that provide brief reviews of chapter content.

WWW. There are callouts throughout the chapters that guide you to HK*Propel*. When you see one of these icons in the margin, go to the appropriate location in HK*Propel* and download the information or complete the activity.

Students who don't use the many resources provided in HK*Propel* miss great opportunities to better understand the course content. Comprehension and application are what you will be tested on. If you want to do well in the course, apply yourself and use the many resources available to you.

To access the HK*Propel* resources, see the card at the front of the print book for your unique access code. For e-book users, reference the HK*Propel* access code instructions on the page immediately following the book cover.

HELPFUL STUDY HINTS

Here are a few hints that should help you learn and use the contents of your measurement and evaluation textbook. Frankly, most of these suggestions apply to any course in which you are enrolled and are common knowledge for most students. The hard part is actually doing what you know you need to do. These are our suggestions:

1. Download the chapter outlines from HK*Propel*.
2. As you read the chapter, highlight key points that pique your interest.
3. Attend class every day. Do not sit near your friends. Take notes in class. Ask questions.
4. Study together in groups.
5. After reading the chapter, return to the Measurement and Evaluation Challenge to see if you can determine how the chapter information helped you address the issue.
6. After you conduct the analyses as directed for each large dataset, conduct additional analyses of your own.
7. If you have questions, consider how you might learn the answer yourself. Look through the chapter, review your notes, look at the homework, and go to HK*Propel*. Some instructors suggest, "Ask three and then me." The idea is for you to consult three classmates or other sources about your questions. The interaction will help all of you learn the material better. If you still have questions after that, go ask the instructor.

You are encouraged to use this textbook and the various resources to their fullest extent. We have learned that students who apply themselves, spend the time necessary for learning, attend class, are prepared, and follow the preceding suggestions have a greater appreciation for reliability and validity theory, understand the concepts better, and apply the concepts more often in their careers than students who do not use these techniques, methods, and strategies. We believe that we have prepared a textbook that is understandable, interesting, informative, and easy to read that will help you use measurement and evaluation concepts throughout your academic, personal, and professional lives.

INSTRUCTOR RESOURCES IN HK*Propel*

A variety of instructor resources are available online within the instructor pack in HK*Propel*:

- *Presentation package.* The presentation package includes nearly 600 slides that cover the key points from the text. Instructors can easily add new slides to the presentation package to suit their needs.

- *Instructor guide.* The instructor guide provides sample course syllabi, chapter review questions and answers, homework problems and answers, and answers to all of the Mastery Items and student activities. Although we hope instructors will use these materials, we also encourage instructors to create their own additions and extensions of these materials so that the course reflects the instructor personally and addresses current important topics.

- *Test package.* The test package contains a bank of more than 900 questions in multiple-choice format. The files may be downloaded for integration with a learning management system or printed as paper-based tests. We encourage instructors to use and modify these items so that they reflect current information from the scientific literature. The literature about measurement and evaluation in human performance is expanding daily, and instructors should develop test items around these scientific advances. Instructors may also create their own customized quizzes or tests from the test bank questions to assign to students directly through HK*Propel*. Multiple-choice questions are automatically graded, and student scores can be easily reviewed by instructors in the platform.

- *Chapter quizzes.* Ready-made chapter quizzes allow instructors to assess student comprehension of the most important concepts in each chapter. Each quiz contains 10 multiple-choice questions. Each quiz may be downloaded or assigned to students within HK*Propel*. The chapter quizzes are automatically graded with scores available for review in the platform.

- *Image bank.* The image bank includes most of the figures, content photos, and tables from the text. You can use these items to add to the presentation package or to create your own PowerPoint presentation covering the textbook material. These items can also be used in handouts for students.

- *Videos.* There are dozens of brief videos that summarize chapter topics. You can use the videos for review, for class preparation, or as supplemental material during instruction. Students can use the videos for further clarification on a topic and in preparation for chapter examinations.

Instructor ancillaries are free to adopting instructors, including an e-book version of the text that allows instructors to add highlights, annotations, and bookmarks. Please contact your sales manager for details about how to access instructor resources in HK*Propel*.

Part I

Introduction to Tests and Measurements in Human Performance

We all want to make good decisions. In part I, we introduce you to the concepts of measurement and evaluation and their importance in decision making. These concepts are the foundation for your study of measurement in human performance throughout the remainder of the book. Chapter 1 presents an overview of the scope and use of measurement in human performance. Chapter 2 describes the latest technology applications in human performance, with specific attention to applications of measurement, testing, and evaluation; this chapter introduces advanced technology for conducting many of the exercises in the remainder of the book. Specifically, you will use the Internet and statistical software (IBM SPSS Statistics and Microsoft Excel) to help solve measurement and evaluation problems.

Part I will provide you with much of the background, knowledge, and computer skills necessary to make valid measurement decisions. For example, you'll learn to create data tables in SPSS, read Excel files in SPSS, and analyze a dataset using the appropriate procedures. You will also learn to select and use apps to solve various assessment problems. These procedures will be used throughout the remainder of the text in Mastery Items and in other activities available to you through your textbook and through HK*Propel*. On the HK*Propel* site, you will find many helpful resources as you learn more about measurement and evaluation.

Concepts in Tests and Measurements

OBJECTIVES

After studying this chapter, you will be able to

- define the terms *test*, *measurement*, and *evaluation*;
- differentiate between norm-referenced and criterion-referenced standards;
- differentiate between formative and summative evaluation;
- discuss the importance of measurement and evaluation processes;
- identify the purposes of measurement and evaluation;
- identify the importance of objectives in the decision-making process; and
- differentiate among the cognitive, affective, and psychomotor domains as they relate to human performance.

WWW The lecture outline in HK*Propel* will help you identify the major concepts of the chapter.

MEASUREMENT AND EVALUATION CHALLENGE

This first measurement and evaluation challenge presents a scenario that relates to most of the chapters and concepts you will study throughout this textbook. We first describe a scenario, and then at the end of the chapter we explain how you might answer the questions that arise in the scenario as you learn about measurement and evaluation processes in human performance.

Imagine that your father talks with you about his recent physical examination. It has been a number of years since he had a medical examination. After conducting a battery of tests and asking about his lifestyle, the physician told your father that his weight, blood pressure, physical activity level, cholesterol, nutritional habits, and stress levels have increased his chances of developing cardiovascular disease. Your father is concerned because he doesn't know what the reported numbers mean, and he is concerned about the accuracy of these measurements. Your father tells you that he feels great, was physically active throughout high school and college, looks better than most people his age, and cannot imagine that he is truly at an elevated risk. To help your father interpret these results and encourage him to make the necessary lifestyle changes to reduce his cardiovascular risk, answer the following questions:

1. How does one know if the measures taken are accurate?

2. What evidence suggests that these characteristics are truly related to developing cardiovascular disease?

3. How likely is it that the physician's evaluation of the tests is correct?

4. What aspect of the obtained values places one at increased risk? For example, how was a systolic blood pressure of 140 mmHg originally identified as the point at which one is at increased risk? Why not 130 mmHg or 150 mmHg? Why has the blood pressure associated with risk been decreased from 140 to 130 and even 120 mmHg?

5. Your father reports being physically active, but what does that mean? Is he engaging in sufficient physical activity to be at increased health or reduced risk for negative health outcomes? Similar questions could be asked about each of the measurements obtained: What evidence exists that changing any of these factors will reduce risk?

Interpreting measurement results and determining the quality of the information one receives are what this course is all about. Information gained from this course will help you make informed decisions about the accuracy and veracity of obtained measures as well as other decisions based on human performance measurements. In general, good measurement and subsequent evaluation should lead to good decisions, such as changing one's lifestyle to improve health. We focus on measurements obtained in the cognitive, psychomotor, and affective domains.

Why is testing important? Is it really necessary to know about statistical concepts and how to apply them? What decisions are involved in the measurement process? How can one determine the influence of measurement errors on decisions? How you answer these questions is important to your development as a competent professional in the field of human performance.

We all gather data before making decisions, whether the decision-making process occurs in research, in education, or in our personal lives. For example, researchers gather data on fitness characteristics to examine the relationships among fitness, physical activity, mortality, morbidity, and quality of life. Examples of variables measured might include blood pressure, cholesterol levels, and the amount and type of physical activity completed. Weight loss and weight control are major health concerns, so you might be interested in measuring energy expenditure to estimate caloric balance. You might gather data about the weather before venturing out for a morning run and adjust your behavior based on the data you obtain. Likewise, before purchasing a stock for investment, you might gather data on the company's history, leadership, earnings, and goals. All of these are examples of testing and measuring. In each case, making the best possible decision is based on collecting relevant data and using them to make an accurate decision: The quality of the data collected will affect the quality of one's decisions. The statistical and measurement concepts presented in this text provide a framework for making accurate decisions based on sound data.

The course you are embarking on was historically called "Tests and Measurements." Although some students might refer to this as a statistics book, that does not accurately describe this text, nor most of the courses that use it. Some basic statistical concepts are presented in part II (chapters 3, 4, and 5); however, the statistical and mathematical knowledge necessary for testing and measurement is not extensive, requiring only basic algebraic concepts. On the other hand, every chapter in this text focuses in some way on the important issues of reliability and validity. To make good decisions, you must measure and evaluate accurately. Making effective decisions depends on first obtaining relevant information.

In kinesiology (a subdomain of human movement and performance), the measurement and evaluation of movement is now becoming known as *kinesmetrics* (Mahar and Rowe, 2014; Zhu, 2010). Consider youth physical activity levels in the United States. How is physical activity measured, reported, and tracked? The U.S. Department of Health and Human Services' *Physical Activity Guidelines for Americans* (USDHHS, 2018) recommends 60 minutes of moderate-to-vigorous physical activity (MVPA) daily for children and youth. However, the Centers for Disease Control and Prevention's Youth Risk Behavior Surveillance System (YRBSS) reported that in 2020, less than half of high school youth had been physically active for at least 60 minutes each day (23.2%), had exercised to strengthen or tone their muscles 3 days per week (49.5%), had met both aerobic and muscle-strengthening physical activity guidelines (16.5%), or had attended physical education classes on all 5 days in an average school week (25.9%) (Merlo et al., 2020). Are these reported values reliable and valid? Can intervention intended to affect lifestyle behavior changes be based on these data? This is where testing and measurement (and reliability and validity) enter the picture.

NATURE OF MEASUREMENT AND EVALUATION

The terms *measurement*, *test*, and *evaluation* refer to specific elements of the decision-making process. Although the three terms are related, each has a distinct meaning and should be used correctly. **Measurement** is the act of assessing, which usually results in assigning a number to quantify the amount of the characteristic that is being assessed. For example, people might be asked to self-report the number of days per week they engage in MVPA (as was done with the YRBSS results presented earlier) or asked to run 1 mile (1.61 km) so their recorded time can be used to estimate their aerobic capacity.

A **test** is an instrument or tool used to make the particular measurement. This tool can be written, oral, physiological, or psychological, or it can be a mechanical device (such as a treadmill). To determine the amount of MVPA conducted in a week, one might use self-report, direct observation, pedometers, or a motion sensor (i.e., pedometer or accelerometer).

Finally, **evaluation** is a judgment and statement of quality, goodness, merit, value, or worthiness about what has been assessed, made by comparing a measured value (obtained with a test) to some criterion or other measure. For example, once we determine how physically active a person is, we can compare that result to national standards based on the U.S. Health and Human Services Guidelines for Physical Activity (USDHHS, 2018) to see if the person is sufficiently active to accrue health benefits. You could evaluate someone's MVPA per week and compare it to the national adult recommendation of 150 minutes per week. Alternatively, you may report how many steps the person averages per day and compare that to how many steps people of that age generally take per day. Evaluation therefore enters the discussion when you tell the person whether he or she meets a set standard.

Consider another example: Suppose you want to measure cardiorespiratory fitness, often reported as $\dot{V}O_2$max, a measure of aerobic capacity. You can measure a person's maximal oxygen consumption or uptake in several ways. You might have someone perform a maximal run on a treadmill while you collect and analyze expired gases. You might collect expired gases from a maximal cycle ergometer protocol. You might have a participant perform either a submaximal treadmill exercise or a cycle exercise, and then predict $\dot{V}O_2$max from heart rate or workload. You might measure the distance a person runs in 12 minutes or the time it takes to complete a 1.5-mile (2.4 km) run. Each of these tools results in a number, such as percent O_2 and CO_2, heart rate, minutes, or yards. Having assessed $\dot{V}O_2$max with one of these tools does not mean that you have evaluated it. Obtaining and reporting data mean little unless you reference the data to something. This is where evaluation enters the process.

Assume that you test someone's $\dot{V}O_2$max. Furthermore, assume that this participant has no knowledge of what the $\dot{V}O_2$max value means. Certainly, the participant might be aware that the treadmill test is used to measure cardiorespiratory fitness. However, the first question most people ask after completing some measurement is *How did I do?* or *How does it look?* To simply report, "Your $\dot{V}O_2$max was 30 ml · kg^{-1} · min^{-1}" says little. You need to provide an evaluation of the performance, which is often based on normative data. For example, a $\dot{V}O_2$max of 30 might be considered quite good for a 70-year-old female but quite poor for a healthy 25-year-old male.

WWW Go to HK*Propel* to complete Student Activity 1.1.

Norm-Referenced and Criterion-Referenced Standards

To make an evaluative decision, you must have a reference perspective. You can make evaluative decisions from either norm-referenced (normative) or criterion-referenced standards. Using a **norm-referenced standard** means comparing a performance with that of others (perhaps those of the same sex, age, or class).* Thus, as in the earlier example, you might report that a $\dot{V}O_2$max of 30 is relatively good or poor for someone's particular age and sex.

Conversely, you might not be interested in how someone compares with others; rather the comparison is with a standard, or criterion, you would like him or her to achieve. Assume that the $\dot{V}O_2$max of 30 was measured on someone who previously had a heart attack. A physician may be interested in whether the patient achieved a $\dot{V}O_2$max of at least 25 ml · kg^{-1} · min^{-1}, which would indicate that the patient had achieved a functional level of cardiorespiratory fitness. This is an example of a **criterion-referenced standard**. The criterion often is initially based on norm-referenced data and the best judgment of experts in the content area.

*The issue of "sex" versus "gender" is a complicated topic that continues to evolve in terms of how data are collected. Some normative data are based on differences that correspond to biological sex, yet subjects completing paperwork may be providing gender information rather than biological sex. Throughout this text, you will see sample data and data analyses that include "gender" as a variable. That variable might reflect the data you have gathered, depending on the question asked in your survey or paperwork. Or the data you have collected may instead reflect subjects' biological sex. In such a case, you will change the terminology in your datasets accordingly.

Consider assessing physical activity behaviors with a pedometer. The number of steps taken could indicate general physical activity behavior and be used to determine if one is sufficiently active to accrue health benefits. One might be interested in how many steps he or she takes compared to others (a norm-referenced comparison), but the more important measure might be if the person is taking sufficient steps for health benefits (a criterion-referenced comparison)—the fact that one person takes more steps than another does not mean that the more active person is sufficiently active. Tudor-Locke and her colleagues (Tudor-Locke and Bassett, 2004; Tudor-Locke, Hatano, Pangrazi, and Kang, 2008) suggested the following pedometer criteria for adult public health:

Steps per day	Public health index
<5000	Sedentary lifestyle
5000-7499	Low active
7500-9999	Somewhat active
10,000-12,499	Active
≥12,500	Highly active

No national criterion has been developed for steps per day to be healthy. Paluch et al. (2021) published work on step counts in young adults and found that those taking at least 7000 steps per day had a 50% to 70% lower risk of mortality. Additionally, they reported that there was no association between step intensity and mortality. Setting such a criterion is a difficult challenge for measurement and evaluation experts. You will learn much more about setting standards and the validity of standards in chapter 7.

Changes in youth fitness evaluation processes over the last 40+ years provide a good comparison of norm-referenced versus criterion-referenced standards. Fitness scores used to be evaluated using a norm-referenced system—that is, a child completed a fitness test and then was informed what percentage of children he or she did better than (a norm-referenced comparison). However, many youth fitness tests are now criterion referenced, such as the FitnessGram, which uses criterion-referenced standards (called Healthy Fitness Zones) for assessing **health-related physical fitness**. Consider a 12-year-old boy who completes a 1-mile (2.4 km) run and his predicted $\dot{V}O_2$max is 37.5. This does not meet the minimum criterion for the FitnessGram Healthy Fitness Zone (40.3). It is not important how one compares with peers on the FitnessGram; what is important is whether one achieves the Healthy Fitness Zone.

Mastery Item 1.1

Are the following measures usually evaluated from a norm-referenced or a criterion-referenced perspective?

- Blood pressure
- Fitness level
- Blood cholesterol
- A written driver's license examination
- Performance in a college class

WWW! **Go to HK*Propel* to complete Student Activity 1.2.**

Formative and Summative Evaluation

Evaluations may be either formative or summative. **Formative evaluations** are initial or intermediate evaluations, such as the administration and evaluation of a pretest. Formative evaluation should occur throughout an instructional, training, or research process. For example, after shoulder surgery, your goal may be to regain range of motion (ROM) in the shoulder joint. Your physical therapist could assess your ROM and suggest activities to improve it. These ongoing evaluations need not involve formal testing; simple observation and feedback sequences between the therapist and the patient are fine. Formative evaluation processes, including ongoing measurement and feedback, are essential to track changes and measure achievement toward goals (in this course, for example). **Summative evaluations**, on the other hand, are final evaluations that typically come at the end of an instructional or training unit. You, as a student in this course, are interested in the summative evaluation—the grade—that you will receive at the end of the semester.

The difference between formative and summative evaluations might seem to be merely the difference in timing of data collection; however, the actual use of the data collected distinguishes the evaluation as formative or summative. Thus, in some situations the same data can be used for formative and summative evaluations.

A weight-loss program provides a simple and useful example for applying both formative and summative evaluations. Assume that you have measured a participant's body weight and body fat percentage. Your formative evaluation indicates that this participant has 30% body fat and needs to lose 10 pounds (4.5 kg) to achieve the desired 25% body fat. You establish a diet and exercise program designed to produce a weight loss of 1 pound (0.45 kg) per week for 10 weeks. Each week you weigh the participant, measure body fat percent, and give the person feedback on these numbers so that the participant knows the amount of progress (or lack of progress) that is occurring each week. At the end of the 10-week program, you measure body weight and body fat percent again and conduct a simple summative evaluation. Were the goals achieved at the end of the program?

Consider pedometer steps again. One might have a long-term goal of 10,000 steps per day but a short-term goal of increasing the steps taken by 500 per day over a period of a few weeks. This goal can then be adjusted as one becomes increasingly physically active. The short-term goals can be viewed as formative, and the final long-term goal can be viewed as the summative evaluation. Can you see how you might use formative and summative evaluations in many of your daily decisions, regardless of your career goal?

WWW **Go to HK*Propel* to complete Student Activity 1.3.**

PURPOSES OF MEASUREMENT, TESTING, AND EVALUATION

Prospective professionals in kinesiology, human performance, physical activity, health promotion, and the fitness industry must understand measurement, testing, and evaluation because they will be making evaluative decisions daily. Regardless of your specific area of interest, the best tools to use and how to interpret data may be the most important concepts that you will study. Related evaluation concepts are objectivity (rater consistency), reliability (consistency), relevance (relatedness), and validity (truthfulness). These terms are discussed in greater detail in chapters 6 and 7.

The evaluative process in human performance may be used in many ways. For instance, your employer might hold you accountable for a project; that is, you might be responsible for obtaining a particular outcome. Tests, measurement, and evaluation processes are used to show whether you have met the goals. Obviously, you want the evaluation to accurately reflect the results of your work—assuming that you did a good job! Certainly, if you enter the teaching profession, you will hold your students accountable for learning and retaining the content of the courses you teach. Likewise, your students should hold you accountable for preparing the best possible tests to evaluate their performance.

WWW! **Go to HK*Propel* to complete Student Activity 1.4.**

As you will discover during your course of study, you need considerable knowledge and skill to conduct effective measurement and evaluation. Having a thorough understanding of the purposes of executing a measurement and evaluation process is critical. There are six general purposes of measurement and evaluation: placement, diagnosis, prediction, motivation, achievement, and program evaluation.

Placement

Placement is an initial test and evaluation allowing a professional to group participants according to their needs or abilities. All participants in a group can then have a similar starting point and can improve at a fairly consistent rate. It is difficult to teach a swimming class if half the students are beginners and the others are members of the swim team, but even less extreme differences can affect learning.

Diagnosis

Evaluation of test results is often used to determine weaknesses or deficiencies in students, patients, athletes, and fitness program participants. For example, cardiologists may administer treadmill stress tests to obtain exercise electrocardiograms of cardiac patients to diagnose the possible presence and magnitude of cardiovascular disease. Recall the measurement and evaluation challenge highlighted at the beginning of this chapter. The doctor made a diagnosis based on a number of physiological and behavioral measures. This was possible because of the known relationships between the measures and the incidence of heart disease. Although there is currently much interest in MVPA, there is also much interest in sedentary behaviors—the amount of time in which one engages in each can be diagnostic of physically active lifestyle behaviors.

© Human Kinetics

Athletic teams are created by placing players of similar skill together, and those players are motivated to reach a certain level of achievement. *Placement*, *motivation*, and *achievement* are all important factors in measuring and evaluating human performance.

Prediction

One of the goals of scientific research is to predict future events or results from present or past data. This is also one of the most difficult research goals to achieve. You probably took the Scholastic Aptitude Test (SAT) or the American College Test (ACT) during your junior or senior year of high school. Your test scores were likely part of the admissions process used by your college or university and may be viewed as predictors of your future success in college. The exercise epidemiologist may use physical activity patterns, cardiovascular endurance measures, blood pressure, body fat, or other factors to predict your risk of developing cardiovascular disease.

Motivation

The measurement and evaluation process is necessary for motivating your students and program participants. People need the challenge and stimulation they get from an evaluation of their achievement: There would not be any athletes if there were only practices and no games or competitions! What would motivate you to study and learn the material for this course if you knew you would not be tested? Consider how simply measuring your weight can be motivational or how knowing the number of steps you take per day might provide impetus to increase physical activity behaviors.

Achievement

In a program of instruction or training, a set of objectives must be established by which participants' achievement levels can be evaluated. For instance, in this course, your final achievement level will be evaluated and a grade will be assigned on the basis of how well you met objectives set forth by the instructor, which will be introduced in chapter 8. Developing the knowledge, fitness, skills, and physical activity needed for proper grading is one objective of this book; chapters 8-13 are devoted to these topics. Improvement in human performance is an important goal in instruction and training programs, but it is difficult to evaluate fairly and accurately. Is the final achievement level going to be judged with criterion-referenced standards on a pass/fail basis or with norm-referenced standards and grades (i.e., grading on the curve)? Assessment of achievement is a summative evaluation task.

Program Evaluation

The goal of program evaluation is to demonstrate (with sound evidence) the successful achievement of program objectives. You may have to conduct such evaluations in the future to justify your treatment, instruction, and training programs. Perhaps you have a program objective of increasing MVPA in your community, which you will measure with self-reported MVPA and pedometer steps. You would then make decisions based on the data you obtain. Alternatively, if you are a physical education teacher, you may be asked to demonstrate that your students are receiving appropriate physical activity experiences. You might compare your students' fitness test results with the test results of students in your school district or with national test results, or you might gather student and parent evaluations of your program. Professionals in community, corporate, or commercial fitness centers can evaluate their programs in terms of membership levels and participation, participant test results, participant evaluations, and physiological assessments. Your job and professional future could depend on your being able to conduct a comprehensive and effective program evaluation.

WWW Go to HK*Propel* to complete Student Activity 1.5.

RELIABILITY AND VALIDITY

The terms *reliability* and *validity* are the most important concepts in this textbook and will appear frequently throughout. **Reliability** refers to consistency of measurement. **Validity** refers to truthfulness of measurement. Regardless of the purpose for which you are using measurement, you want your results to be reliable and valid.

If a test process is reliable and it is relevant to what is being measured, then it is valid. Consider the conundrum of a valid test that is not reliable: How can this be? How can a test truthfully measure something but not do so consistently? It can't! Testing processes must first be reliable before they can be valid. When obtaining any measure for any purpose, your first thoughts should be about reliability and validity. Essentially, reliability + relevance → validity. A word closely associated with reliability is **objectivity**, which refers to reliability between raters (interrater reliability). Objectivity, reliability, and validity are presented in detail in chapters 6 and 7; these concepts are key to measurement in human performance.

DOMAINS OF HUMAN PERFORMANCE

The purposes of measurement, testing, and evaluation we've just discussed are related to the objectives of your program, or the specific outcomes that you hope to achieve. To be accurately measured and truthfully evaluated, these outcomes need to be measurable. Objectives in the area of human performance fall into three areas: the cognitive domain, the affective domain, and the psychomotor domain. Students of measurement and evaluation in education or psychology are concerned with objectives in the first two areas. For students of human performance, the distinctive objectives are those in the psychomotor domain.

Bloom (1956) presented a **taxonomy** (classification system) of objectives in the **cognitive domain**; these deal with knowledge-based information (see table 1.1). Bloom's levels include knowledge, comprehension, application, analysis, synthesis, and evaluation. Anderson and Krathwohl (2001) later modified Bloom's original taxonomy to include *creating* as the highest level of cognitive endeavors.

Objectives in the **affective domain** concern psychological and emotional attributes. Affective objectives—which concern, for example, how people feel about their performance—are important but often difficult to measure and are not normally used for grading purposes. Krathwohl, Bloom, and Masia (1964) list affective objectives as receiving, responding, valuing, organizing, and characterizing by a value complex.

The third domain of performance is the **psychomotor domain** (Harrow, 1972); these objectives are reflexive movements, fundamental movements, perceptual abilities, physical abilities, skilled movements, and nondiscursive communication. The measurement techniques and concepts associated with the psychomotor domain differentiate human performance students from students in other areas. Note that other taxonomies may be used for the cognitive, affective, and psychomotor domains; those in table 1.1 are only examples.

When you are evaluating participants, you must take into account the level of the domain your participants have achieved. Each taxonomy is a hierarchy; each level is based on the earlier levels having been achieved. For example, it would be inappropriate for you to attempt to measure complex motor skills in 7-year-old children because most of them have not achieved prior levels of the taxonomic structure. Likewise, it is difficult, if not impossible, for younger participants to achieve the higher-level cognitive objectives of a written test.

Table 1.1 Taxonomies in Domains of Human Performance

Taxonomy of the cognitive domain (Bloom, 1956)	Taxonomy of the affective domain (Krathwohl, Bloom, and Masia, 1964)	Taxonomy of the psychomotor domain (Harrow, 1972)
Knowledge • Of specifics • Of ways and means of dealing with specifics • Of the universals and abstractions in a field **Comprehension** • Translation • Interpretation • Extrapolation **Application Analysis** • Of elements • Of relationships • Of organizational principles **Synthesis** • Production of unique communications • Production of a plan for operations • Derivation of a set of abstract relations **Evaluation** • Judgments in terms of internal evidence • Judgments in terms of external evidence	**Receiving** • Awareness • Willingness to receive • Controlled or selected attention **Responding** • Acquiescence in responding • Willingness to respond • Satisfaction in response **Valuing** • Acceptance of a value • Preference for a value • Commitment **Organizing** • Conceptualization of a value • Organization of a value system **Characterizing by a value complex** • Generalized set • Characterization	**Reflexive movements** • Segmental reflexes • Intersegmental reflexes • Suprasegmental reflexes **Fundamental movements** • Locomotor movement • Nonlocomotor movement • Manipulative movement **Perceptual abilities** • Kinesthetic discrimination • Visual discrimination • Auditory discrimination • Tactile discrimination • Coordinated discrimination **Physical abilities** • Endurance • Strength • Flexibility • Agility **Skilled movements** • Simple adaptive skill • Compound adaptive skill • Complex adaptive skill **Nondiscursive communication** • Expressive movement • Interpretive movement

It is important to define the difference between physical activity and physical fitness. **Physical activity** is a behavior defined as bodily movement. Physical activities can range from those that substantially increase energy expenditure to light activities, strengthening activities, and even to physical inactivity (i.e., sedentary behaviors). These behaviors can take place during leisure, transportation, occupational, and household activities. **Physical fitness**, on the other hand, is a set of attributes that people have or achieve that relates to the ability to perform physical activity. In other words, physical activity is something people do, whereas physical fitness is something people have or achieve. Heredity plays an important role in both factors but is probably more important in physical fitness. Physical activity is more difficult to measure reliably and validly than physical fitness, because measuring a behavior is generally more difficult than measuring an attribute.

Pettee Gabriel, Morrow, and Woolsey's (2012) diagram (figure 1.1) illustrates the differences between physical activity (the behavior) and physical fitness (the attribute). Measurement considerations are key to assessing physical activity behaviors and physical attributes.

Just as there are a variety of physical activity surveys (e.g., BRFSS, YRBSS), fitness tests, and test protocols for measuring specific fitness attributes, there are a variety of techniques to measure physical activity and fitness. These techniques include use of motion sensors, written recalls, self-reporting, and heart rate biotelemetry. An increasing body of scientific literature is appearing about the reliability and validity of physical activity measurement in a variety of situations and populations. In chapter 13, we explore the measurement and evaluation of physical activity.

www **Go to HK*Propel* to complete Student Activity 1.6.**

PSYCHOMOTOR DOMAIN: PHYSICAL ACTIVITY AND PHYSICAL FITNESS

For many years, physical educators, exercise scientists, personal trainers, athletic coaches, and public health leaders have been concerned about the definition, the reliable and valid measurement, and the evaluation of physical fitness in people of all ages. This concern has led to a growing number of fitness tests and protocols for both mass and individual testing. For example, though each consists of different test items, the Cooper Institute's FitnessGram, the President's Council on Fitness, Sports & Nutrition, the President's Challenge, and the European test Eurofit are all test batteries used to assess levels of youth physical fitness. A large number of research studies have been conducted to demonstrate the feasibility, reliability, and validity of such fitness tests, including normative surveys to establish the fitness levels of various populations. You'll learn more about these test batteries and other physical fitness testing in chapters 10 and 11.

In the latter part of the 20th century, the health-related aspects of physical activity became a dominant concern of public health officials. The culmination of this concern was presented in the release of *Physical Activity and Health: A Report of the Surgeon General* (USDHHS, 1996). This publication, led by senior scientific editor Steven N. Blair, presented a detailed case for the health benefits of a lifestyle that includes regular and consistent participation in moderate-to-vigorous levels of physical activity. Based on the 2011 data from the Centers for Disease Control and Prevention's Behavioral Risk Factor Surveillance System (BRFSS), Harris et al. (2013) reported that that only about half of the adult U.S. population engaged in the recommended 150 minutes of MVPA per week and less than 30% met the muscle-strengthening guideline of 2 or more days per week. Unfortunately, these numbers currently remain approximately the same, according to the latest CDC data (www.cdc.gov/nchs/fastats/exercise.htm).

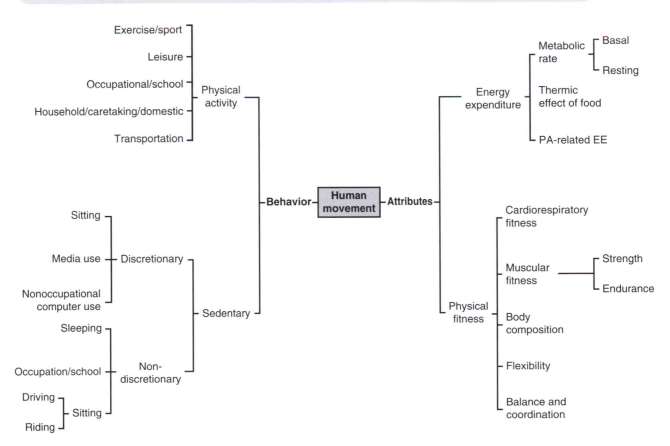

Figure 1.1 A framework for physical activity.

Adapted from Pettee Gabriel, Morrow, and Woolsey (2012).

MEASUREMENT AND EVALUATION CHALLENGE

The issues just presented about standards, evaluation, purposes, and domains of human performance directly relate to the questions you have about your father's measurements. How do you know they were accurate? How do you know they are truly predictive? How do you know if he is really at increased risk? How do you know if specific interventions help reduce risk? Throughout the remainder of this course, you will learn about the tools available to answer these questions and more.

In chapter 2, you will learn about accessing the Internet to obtain information and calculate health risks. You will also be introduced to powerful computer programs to help you analyze data and make decisions; you will use these computer programs throughout the remainder of the book. Chapters 3 through 5 will help you understand the statistical procedures necessary for making evidence-based decisions. Chapter 3 presents information about descriptive statistics and measurement distributions. Chapter 4 presents information about quantifying the relationship of one variable to another (e.g., relating decreased physical activity to increased cardiovascular risk). Chapter 5 presents an overview of research methods to help you decide if an intervention makes a significant difference in a specific outcome of interest (e.g., does moderate exercise reduce body fat?). Chapters 6 and 7 illustrate how to determine the reliability and validity of measurement. You will learn how to determine the best measurement to use, how to interpret the result of the measurement, what influences measurement errors, and how to collect and evaluate different types of evidence to support the validity. Chapter 7 also illustrates how specific health standards are set, whether these standards affect the risk of developing a specific disease, and related reliability and validity issues. Chapter 8, although primarily concerned with grading, contains important information on evaluation, including how to properly add various measurements together to obtain a composite score. Chapter 9 provides tools with which to make an accurate assessment of student knowledge and achievement. Because knowledge is required but not sufficient for behavior change, you will learn how to determine if a knowledge test is accurate and truly reflects learning. Chapter 10 illustrates how to measure risk factors associated with cardiovascular disease. As an example, you will learn about some simple tests of health-related physical fitness, including aerobic capacity, body composition, muscular strength and endurance, flexibility, and bone density. Chapter 11 illustrates how to assess performance-related and sports-related fitness and how to use some simple tests. Chapter 12 presents information about the assessment of motor abilities, sport skills and performance, and other topics of particular interest to coaches and teachers. Chapter 13 is primarily concerned with the measurement of physical activity and sedentary behavior. Chapter 14 illustrates methods of measuring psychological stress levels, which might be pertinent in your father's case. Finally, chapter 15 provides examples and information about how to obtain alternative measures. For example, one could measure physical activity by asking for a self-report, by using a pedometer, or by directly observing a participant's daily behaviors.

SUMMARY

As a student, you are likely aware that nearly all educational decisions rely greatly on the processes of measurement and evaluation. As a human performance professional, you will have to make a variety of decisions regarding the methods of collecting and interpreting data in the measurement process. Figure 1.2 illustrates the relationships among testing, measurement, and evaluation. A wide range of instruments (tests) are used to assess abilities in the cognitive, psychomotor, and affective domains. You will have to develop specific objectives according to the domain and select tests that produce objective, reliable, relevant, and valid measurements of those objectives. Once you collect the data, you will make evaluative decisions based on norm-referenced or criterion-referenced standards. In norm-referenced comparisons, a participant's performance is compared with that of others who were tested. Criterion-referenced standards are used to compare a participant's performance with a predetermined standard. Evaluations can also be formative (judged during the program or at certain intervals) or summative (judged at the end of the program).

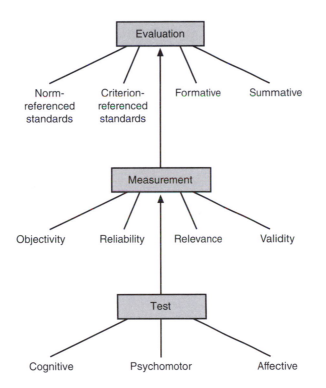

Figure 1.2   Relationships among testing, measurement, and evaluation.

www **Go to HK*Propel* for videos, homework assignments, and quizzes that will help you master this chapter's content.**

CHAPTER 2

Using Technology in Measurement and Evaluation

OBJECTIVES

After studying this chapter, you will be able to

- identify the potential uses of computers in your field;
- identify sources of computer software and hardware for use in exercise science and physical education;
- present examples of technology and computer use in exercise science, physical education, kinesiology, and clinical health and fitness settings and describe how various testing procedures can be facilitated with computers;
- use SPSS to create and save data files;
- use Microsoft Excel to create a data file to be used in SPSS; and
- use SPSS and Excel to analyze data and set the groundwork for reliability and validity analyses that will follow in subsequent chapters.

www The lecture outline in *HKPropel* will help you identify the major concepts of the chapter.

MEASUREMENT AND EVALUATION CHALLENGE

Jamiyah works for a research company that conducts interventions intended to increase the amount of moderate-to-vigorous physical activity (MVPA) in which people engage. The company's intent is to encourage people to engage in sufficient MVPA to meet the *Physical Activity Guidelines for Americans* (USDHHS, 2018). Their current project involves strategies to get people to walk more, and the study participants have been asked to wear motion sensors in order to collect data. The question facing the study team is *How many consecutive days of monitoring are necessary to obtain a reliable measure of MVPA?* Additionally, they wonder if a weekend day or two is necessary. Their study participants come from throughout a large metropolitan area and represent the public at large. Handling such data by hand would be cumbersome and time consuming. The study team plans to store their data on a computer and have them readily available for analyses throughout their study.

There is no question that technology has changed our lives. We watch TV or sit in front of computers for many hours, we google, we chat online, we post photos and videos with smartphones, we make friends through social media, and the list goes on and on. As a result of these and other changes, many of us now live increasingly sedentary lifestyles. From a physical activity and public health point of the view, this shift has been largely negative. Fortunately, technological advances have also provided us with many new means to examine human performance and help us improve our instruction in physical education and promote physical activity. Simply being able to examine a performance, however, will not help us understand the performance. To do so, we must analyze it and compare it to a relative standard (e.g., compare with peers of the same age and sex) or with an absolute standard (e.g., determine if a certain skill competency has been achieved). This is often accomplished through the use of a computer, specifically with statistical analysis software.

In this chapter, we will introduce how technology can help us measure and evaluate human performance. After briefly discussing some key developments in technology, we will describe some examples of how technology can be used to measure human performance. Thereafter, we will demonstrate how to use SPSS, a statistical software program used to analyze data.

PRINCIPLES AND PRACTICE OF USING TECHNOLOGY IN ASSESSMENT

Movement is the key characteristic of human performance. To move, muscles contract, heart rate increases, blood circulation speeds up, the amount and ratio of the air breathed in and out change, body temperature raises, and so on. As a result, force, speed, distance, and other factors of movement are generated. With the help of technology, we are better able to record and quantify these changes. Some types of sensors are usually required to convert the "movement physics" (e.g., speed, pressure, light) to electronic signals, which are then transferred to a data acquisition system to further convert raw signals to desirable and meaningful parameters.

Early Efforts

Technology is not new to the measurement and evaluation of human performance. In fact, it has been around for many years. Electronic gauges and the telephone are just two familiar early examples.

Electronic gauges are devices that monitor performance either by detecting and processing mechanical movement of the human body or through bioelectronic signals associated with movement. The most familiar electronic gauges used in practice are pedometers, accelerometers, heart rate (HR) monitors, and telephones.

The pedometer, known also as a step counter, is a device to record the number of walking steps taken. The pedometer is the oldest and most popular device for measuring, recording, and promoting physical activity (PA). The first nonelectronic pedometer was developed in the 15th century by Leonardo da Vinci and used to measure plots of land by counting steps (Gibbs-Smith and Rees, 1978; Montoye et al., 1996). The real development and usage of pedometers to measure and promote PA started in Japan in the 1960s. In 1965, the Japan Walking Association invented the "10,000 steps/day" slogan and Yamasa, a Japanese company, produced a nonelectronic pedometer called "Manpo-Kei," meaning "10,000-step meter" in Japanese (Hatano, 1993). Eventually, the device was computerized. The relationship between steps and energy consumption was soon established and integrated into pedometers. Yamasa developed and released the first calorie-meter type of pedometer in the 1980s. Pedometers were used by U.S. researchers in the late 1980s and became popular in PA promotion in the 1990s.

Similar to pedometers, accelerometers are small, portable electronic devices that can measure PA by recording minute-by-minute data of body acceleration and reporting the data as activity counts per minute. Differing from the pedometer, which can only measure steps taken, accelerometers can record most movements and provide detailed information on frequency, duration, and intensity of activities. In addition, it can store data for long periods with flexible recording intervals. Accelerometers have been used in many research studies, have shown acceptable validity and good reliability (Plasqui and Westerterp, 2007), and have been accepted as an accurate and objective field measure for PA (Freedson and Miller, 2000). However, accelerometers cannot measure certain activities, such as movement involving only the upper or lower body or water-related activities (Dale et al., 2002).

Between 2003 and 2006, ActiGraph, a well-studied accelerometer, was included in the National Health and Nutrition Examination Survey (NHANES) to monitor PA among the U.S. population. As a result of this effort, several research reports (e.g., Matthews et al., 2008; Troiano et al., 2008) were generated, providing an estimation of PA participation in the United States based on an objective measure for the first time. With the popularity of wearable devices, such as Fitbit, accelerometer-based devices are now widely available for large-scale PA monitoring and interventions.

The heart rate (HR) monitor is another early (1907) instrument used to assess physiological responses during exercise (Montoye et al., 1996). HR has also been used to estimate energy expenditure (EE) because of its linear relationship with VO_2 (Wilmore and Haskell, 1971). However, HR is sensitive to age, sex, temperature, humidity, fatigue, hydration, training status, and emotional stress; thus, EE estimated in this way can be biased by these factors (Rennie et al., 2000). In addition, the HR monitor itself could possibly be interrupted or delayed in recording each heart beat (Montoye et al., 1996). Overall, moderate to high correlations ($r = 0.53$ to 0.73) have been reported between total energy expenditure (TEE) and HR (Schulz, Westerterp, and Brück, 1989).

Finally, the telephone played an early critical role in monitoring the population's level of PA participation. In the past, major national surveillance systems, such as the National

Health and Nutritional Examination Survey (NHANES) and the Behavioral Risk Factor Surveillance System (BRFSS), used the telephone to survey the U.S. population concerning their PA participation.

A Few Significant Developments

The development of the computer and associated information technologies was among the most exciting and significant developments of the 20th century. They, along with a number of other emerging technologies (interactive technology, cell phone and wireless technology, GPS and GIS, voice recognition, etc.), can be used to assess human performance.

Computers

Undoubtedly, the development of computers is one of the most important technological advances of our time. Computers have become commonplace in our lives, and tasks that were once tedious and time consuming are now completed in a matter of seconds by computers that can easily fit in a briefcase or even in your hand. Wireless communication is available nearly everywhere, which permits those gathering data to record measures in nearly any location for real-time analysis. Clearly, your ability to use a computer will affect you in your career as well as your daily activities.

Steve Jobs, an inventor, designer, entrepreneur, and the cofounder of Apple, famously compared computers to "bicycles for the mind":

> I think one of the things that really separates us from the high primates is that we're tool builders. I read a study that measured the efficiency of locomotion for various species on the planet. The condor used the least energy to move a kilometer. And, humans came in with a rather unimpressive showing, about a third of the way down the list. It was not too proud a showing for the crown of creation. So, that didn't look so good. But, then somebody at Scientific American had the insight to test the efficiency of locomotion for a man on a bicycle. And, a man on a bicycle, a human on a bicycle, blew the condor away, completely off the top of the charts.
>
> And that's what a computer is to me. What a computer is to me is it's the most remarkable tool that we've ever come up with, and it's the equivalent of a bicycle for our minds. (Jobs, 1990)

Mastery Item 2.1

Listen to Steve Jobs' famous talk on "bicycles for the mind" at the following link: www.youtube.com/watch?v=ob_GX50Za6c

Originally, mainframe computers were large machines that took up entire rooms (or even floors) in buildings. Users typically had to be connected via telephone or other electronic line to the mainframe, and mainframes were inaccessible to most people. However, the development of the microprocessor has resulted in small, powerful, relatively inexpensive personal computers. Indeed, it has been suggested that if the change occurring in computer technology over the past several decades had also occurred in the automobile industry, cars would now be able to travel much farther on a gallon or liter of gas! However, even though computers are now widely available, many students and professionals who measure and evaluate human performance have not taken full advantage of the power that computers can provide.

Information Technology (IT)

IT refers to the study, design, development, implementation, support, or management of computer-based information systems, particularly software applications and computer hardware. IT deals with the use of computers and computer software to convert, store, protect, process, transmit, and securely retrieve information. Among countless IT inventions and applications that have changed our lives in the information age, the Internet and World Wide Web are perhaps the most significant. The Internet is a worldwide system of computer networks—a network of networks in which users at any one computer can, with permission, get information from any other computer. It was conceived by the Advanced Research Projects Agency of the U.S. government in 1969 to create a network that would allow a research computer at one university to be able to "talk to" research computers at other universities. With the development of desktop computers and easy-access software, the Internet has become a public, cooperative, and self-sustaining facility accessible to hundreds of millions of people worldwide.

Additionally, the development and worldwide use of the Internet have had a significant impact on how people obtain information and communicate with one another. According to the Pew Research Center (www.pewresearch.org/internet/fact-sheet/internet-broadband), about half of U.S. adults used the Internet in 2000—a number that increased to 95% by 2021. Young adults also use the Internet more than older adults: 99% of those ages 18 to 29, 98% of ages 30 to 49, 96% of ages 50 to 64, and 75% of ages 65 or older. A positive relationship exists between Internet usage and household income and amount of education. It is now nearly impossible to think of a career in human performance that does not require the use of computers to analyze or report data. Clearly, regardless of your age or profession, you will have Internet resources and computer applications that help you analyze and report data in meaningful, interesting ways.

Many Internet applications can be directly or indirectly used for human performance assessments, including e-mail, social networking services (e.g., Facebook and Twitter), and video publishing platforms (e.g., YouTube). Numerous professional organization websites are of specific interest and value to the human performance student, including the American College of Sports Medicine (ACSM; www.acsm.org), the Society of Health and Physical Educators (SHAPE America; www.shapeamerica.org), the American Kinesiology Association (AKA; www.americankinesiology.org) and the American Heart Association (AHA; www.heart.org). Government agencies such as the U.S. Centers for Disease Control and Prevention (www.cdc.gov) and the National Heart, Lung, and Blood Institute (www.nhlbi.nih.gov) also provide important health information. The Behavioral Risk Factor Surveillance System (BRFSS) (www.cdc.gov/brfss) and the Youth Risk Behavior Surveillance System (YRBSS) (www.cdc.gov/HealthyYouth/yrbs) introduced in chapter 1 have excellent resources for reporting data. Many other websites are available that provide useful scientific and content-based information.

Interactive Technology

Interactive technology refers to technology that allows the user and device to interact through an interface, such as a computer or video game. The Nintendo Wii (pronounced "we") is a good example of interactive technology with applications for human performance. Exercise and physical therapy communities have adopted the Wii for use in rehab programs and, in general, responses have been very positive. For example, virtual exercises with the Wii have been demonstrated as useful adjunctive therapy to traditional treatment to improve static and dynamic balance in stroke patients (Karasu et al., 2018).

Mobile Phones and Apps

Like the computer and Internet, the cell phone is one of the greatest inventions of the information age. Technically, a cell phone functions like an extremely sophisticated wireless radio, an invention that can trace its roots to Nikola Tesla in the 1880s (formally presented in 1894 by a young Italian named Guglielmo Marconi). The concept of using hexagonal cells for mobile phone base stations was invented in 1947 by Bell Labs engineers at AT&T and further developed by Bell Labs during the 1960s. In 1970, Amos Joel of Bell Labs invented "call handoff," which allowed a mobile phone user to travel through several cells during the same conversation. Martin Cooper of Motorola is widely considered to be the inventor of the first practical handheld cell phone used in a nonvehicle setting. Using a modern (if somewhat heavy) portable handset, Cooper made the first call on a handheld cell phone in 1973. Just a few years ago, the average cell phone user was most likely to be wealthy, but today's cell phone users encompass all ages, races, and economic classes.

The mobile application, known also as a *mobile app* or simply an *app*, is a computer program or software application designed to run on mobile devices, such as phones, tablets, or smartwatches. As of the first quarter of 2021, according to a report by Statista Research Development, there were 3.48 million apps in the Google Play store for Android users and 2.22 million apps in the App Store for Apple users, including some developed for assessing human performance (some examples are listed in chapter 12).

Technology Application Examples

It is a great challenge to keep up with the fast-changing pace of new technology development and applications. Therefore, only a few examples of technology applications related to assessment will be described below.

Measuring and Tracking Athletes' Performance

Technology is used to assess athletes' performances at all stages by coaches, medics, technicians, and athletes themselves. Some examples include

- wearable devices to track cardiovascular rates;
- performance analysis software to record body movements in slow motion;
- game analysis software to track team, ball, and opposition movements;
- sensors placed on the body or in "smart clothing" with sensing fibers woven in to measure and record breathing, heart rate, hydration, and temperature; and
- devices to track an athlete's sleep, including patterns, amount, and quality of sleep.

Using Scanning Technology

With the development of scanning technology, it became possible to reveal an individual's ratio of fat to muscle, identify weak areas that could increase the risk of injury in the future, and understand the impact of exercise on brain structure. For example, with dual-energy X-ray absorptiometry (DXA), it is now possible to measure body composition and bone density accurately. With thermal imaging cameras, we can monitor physiological functions related to skin temperature, detect live muscles and tissues (whereas previously they had to be cut), and prevent possible injury. Finally, with functional magnetic resonance imaging (fMRI), which measures and maps the brain's activity, we can clearly see how and where exercise affects the brain.

Analyzing Individual and Team Data

With a combination of new technology and advanced statistical and artificial intelligence (AI) procedures, accurate and personalized data analysis has become possible. For example, sporting events like golf, baseball, and tennis rely heavily on swing analysis. With swing technologies, a range of easy-to-use motion capture and analysis programs, swing action now can be analyzed extensively. This allows players to try different techniques and postures to improve their swing.

Team performance analysis is another good example. With the tools of team performance analysis, information such as position of the ball, player movement and involvement, fatigue, work rate, time of a particular action, and the outcome of such action becomes available to both coach and players. This can result in targeted training to prepare both individual players and the team as a whole.

© Human Kinetics

This runner's running watch provides precise speed, distance, and pace data. Accompanying software allows the runner to download these data for detailed analyses. You can use similar technology to track and analyze your research data.

Mastery Item 2.2

Go to www.pubmed.gov (the U.S. National Library of Medicine) and search for "pedometer reliability validity" to see some of the research that has been conducted with pedometers.

USING COMPUTERS TO ANALYZE DATA

Computer technology is now pervasive in schools and businesses, and many require their students and employees to be computer literate. Some universities require students to have personal computers, whereas others provide them to students when they pay tuition. Computers have such a big influence in our daily lives (they're involved in everything from grocery shopping and banking to using the telephone) that computer literacy is a necessary skill.

Hardware and software must be differentiated here. Hardware consists of the physical machines that make up your computer and its accessories. Software is the computer code, generated by a programmer, through which you interact with the computer as you enter data and conduct analyses, create text, create graphs, and so on. Many inexpensive (or free!) apps provide the means for completing sophisticated analyses and data collection and reporting strategies. You don't have to be a software programmer to be a competent computer user—the vast majority of expert computer users do not do programming.

Mastery Item 2.3

Run an Internet search for a kinesiology topic of your choice (e.g., PA prevalence, obesity changes). See how many sources you can identify related to this topic. Consider how you might use this information in this class or in your career.

You must familiarize yourself with the computer features and uses specific to the field of human performance so that you can understand and use the concepts presented in this text. For example, the most important characteristics of any test are its reliability and validity. As you will learn in chapters 3, 4, 5, 6, and 7, computers can generate data related to reliability and validity in a matter of seconds. Many of the decisions that you will make in your field also require data analysis. Thus, we will introduce you to **SPSS (IBM SPSS Statistics)**, a powerful data analysis program that will help you save, retrieve, and analyze much of the measurement and evaluation data that you will encounter daily. SPSS makes number crunching fast, efficient, and almost painless. Chapters 3 through 15 provide you with many opportunities to practice using SPSS in scenarios similar to those you will encounter in your profession.

Additionally, we present information about how to create databases with **Microsoft Excel**. The benefit of creating your database with Excel is that it is readily available on most computers and it can be easily read with SPSS. Thus, you could create your database while at home and then conduct the analysis with SPSS. We provide more information for Excel users in the appendix.

Uses for a computer in human performance, kinesiology, and physical education include the following:

• *Accessing the Internet to obtain information relative to your specific job responsibilities.* Whether seeking normative strength measures for your personal trainer clients, researching reports regarding the most effective modality for treating patients in your PT clinic, or accessing health and fitness data from large-scale populations, you will use the Internet on a nearly daily basis once you have completed your training and entered your professional career.

• *Determining test reliability and validity.* Statistics presented in chapter 6 can be used to estimate the reliability (consistency) and validity (truthfulness) of test results in the cognitive, affective, and psychomotor domains. SPSS examples are provided in chapters 3 through 15.

• *Evaluating and reporting test results.* Computers can help evaluate and report individual test results. Likewise, you can quickly retrieve, analyze, and return test results to study participants. Consider the physical therapist who wants to track patient improvement: Using the computer to track and display data serves as an excellent source of formative and summative evaluation.

• *Conducting program evaluations.* Computers can calculate changes in overall student performance and learning across teaching units or track individual changes in a student's performance.

• *Conducting research activities.* You can compare an experimental group of study participants with a control group to determine if your new intervention program has a significant effect on cognitive or physiological performance.

• *Developing presentations.* Specialized software can be used to create powerful presentations for students, potential clients, patients, and professional peers. The presentations can include text, pictures, video, graphics, animations, and sound to effectively present your message. Perhaps your instructor is using the presentation package that accompanies this textbook to illustrate specific points.

- *Assessing student performance.* Students and clients are always interested in how they perform on tests, including what their individual scores are, how they are interpreted, what they mean, and what effect they have. Computers make it easy to provide the answers to all these questions.

- *Storing test items.* Programs that permit entry and manipulation of data records are called spreadsheets—essentially computer versions of a data matrix with rows and columns of information. Spreadsheets are an easy way for teachers to keep records of student grades, with students' names in the first column and data from course assignments in the remaining columns. If the instructor keeps a daily record of class grades, then average grades, final grades, and printed reports can be generated with a few computer keystrokes. Likewise, health and fitness professionals can keep records of workouts and changes in weight, strength, aerobic capacity, and so forth.

- *Creating written tests.* Rather than having to develop a new test each time you teach a unit, you can store test items on your computer and generate a different test each time you teach the unit. Test banks can also be built using word-processing or test-development software. Some test-development programs are quite sophisticated and permit you to choose items not only by content area but also by type of item, degree of difficulty, or date created.

- *Calculating statistics.* Physiological measurements often involve equations for estimating values. For example, skinfolds are used to estimate body fat percentage, and distance runs and heart rate measurements are used to estimate maximal oxygen consumption. Rather than manually completing these calculations, you can enter the formula into the computer once and automatically calculate the desired value for each person. For example, you may go to the National Heart, Lung, and Blood Institute (NHLBI) of the National Institutes of Health (NIH) website (www.nhlbi.nih.gov/health/educational/lose_wt/BMI/bmicalc.htm) and calculate your body mass index (BMI).

WWW **Go to HK*Propel* to complete Student Activity 2.1.**

Mastery Item 2.4

Think of some time-consuming tasks that you need to do regularly. How would a computer help you complete them more efficiently? What kinds of tasks can kinesiology, exercise science, or physical education majors complete using computers?

Physical fitness testing is an important component in most physical education, kinesiology, and exercise science programs. Not too many years ago, fitness test results were reported orally, on poorly prepared reports, or on mimeographed copies. Today, specialized software programs can analyze student results. The Cooper Institute's FitnessGram is an excellent example of one such program. Figure 2.1 illustrates the type of report that the teacher can give students and their parents to better inform them of their physical fitness status or progress.

WWW **Go to HK*Propel* to complete Student Activity 2.2.**

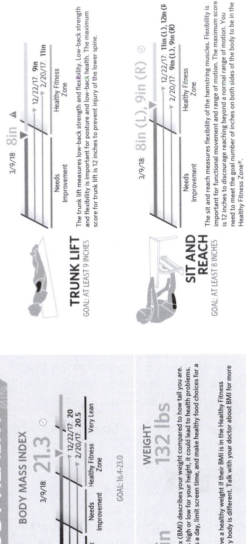

Figure 2.1 FitnessGram reports such as this one can be given to parents and students to illustrate a student's improvement.

Reprinted by permission from The Cooper Institute and GreenLight Fitness.

USING SPSS

Many of the decisions that you will have to make about reliability and validity, whether you are a kinesiologist, physical therapist, clinician, coach, instructor, or educator, are based on statistical evidence. Now, don't get scared! This is not a statistics text. However, the statistics presented in chapters 3, 4, and 5 provide the framework for much of your decision making when measuring and evaluating human performance. Although we will use SPSS (a sophisticated statistical package that is widely available on many university campuses) and Microsoft Excel throughout this text to assist you in calculating statistics used in reliability and validity decisions, your instructor may choose another type of software for these analyses. Regardless, the calculations will be nearly identical (within rounding error) and the interpretation of the results will be exactly the same, regardless of the particular statistical package you use. Nearly all of what we present in this text is available in both the full and student versions of SPSS. However, it is important to note that SPSS is continually updated and the specific method that you use to access SPSS at your location may not be the same as at other locations.

SPSS is software developed to analyze numbers; it permits you to enter and manipulate data and conduct analyses that result in a variety of numbers, charts, and graphs. However, SPSS must have a database on which to conduct the analysis. Thus, each analysis conducted by SPSS is run on a set of data created and saved through the SPSS data editor. The SPSS data editor lets you create a database (also called a *data matrix*) consisting of N rows of people with p columns of variables. Table 2.1 illustrates a data matrix consisting of 10 rows of people (represented by "id" numbers 1 through 10) with five variables ("gender," "age," "weightkg," "heightcm," "stepsperday"). Weight is measured in kilograms (kg) and height in centimeters (cm).

Table 2.1 Sample Database (Data Matrix)

id	gender	age	weightkg	heightcm	stepsperday
1	0	20	50	165	5000
2	0	24	51	160	6000
3	0	21	62	173	7000
4	0	19	59	178	6500
5	0	23	43	145	4500
6	1	22	86	193	4800
7	1	25	65	183	4000
8	1	24	61	178	4200
9	1	28	75	173	3900
10	1	20	70	178	3500

Each of the data tables used in your textbook is located on the website for each chapter. You can download them in SPSS or Excel formats. You will learn more about this in the following paragraphs.

MICROSOFT EXCEL PROCEDURES

Students without access to SPSS can use Excel to conduct the statistical procedures. The appendix contains steps to use Excel for each of the procedures illustrated in this chapter. SPSS will be used and illustrated throughout the remainder of the text. However, each of the procedures is also illustrated in the appendix by chapter. We have provided templates in HK*Propel* for some of the Excel procedures (particularly for chapters 5 and 7) because the number of steps necessary for calculating these statistics is large and the process is rather cumbersome. Excel users will be well served by reviewing the appendix at this time and then throughout the remainder of the textbook when conducting analyses. Each time SPSS is presented in a chapter, Excel users should turn to the appendix and use the appropriate steps. You should be aware that the statistical procedures learned in early chapters generalize to later chapters, so the same procedures are used repeatedly.

Getting Started

Locate and double-click on the icon for SPSS on your computer's desktop. Alternatively, you may have to go through the Start menu on the bottom left of your computer to locate SPSS. It will be different depending on your university or computer. Once you locate and begin SPSS, note that a blank data matrix appears. The upper left corner has the name Untitled1 [DataSet0]—SPSS Statistics Data Editor. The data editor lets you define and enter data (figure 2.2).

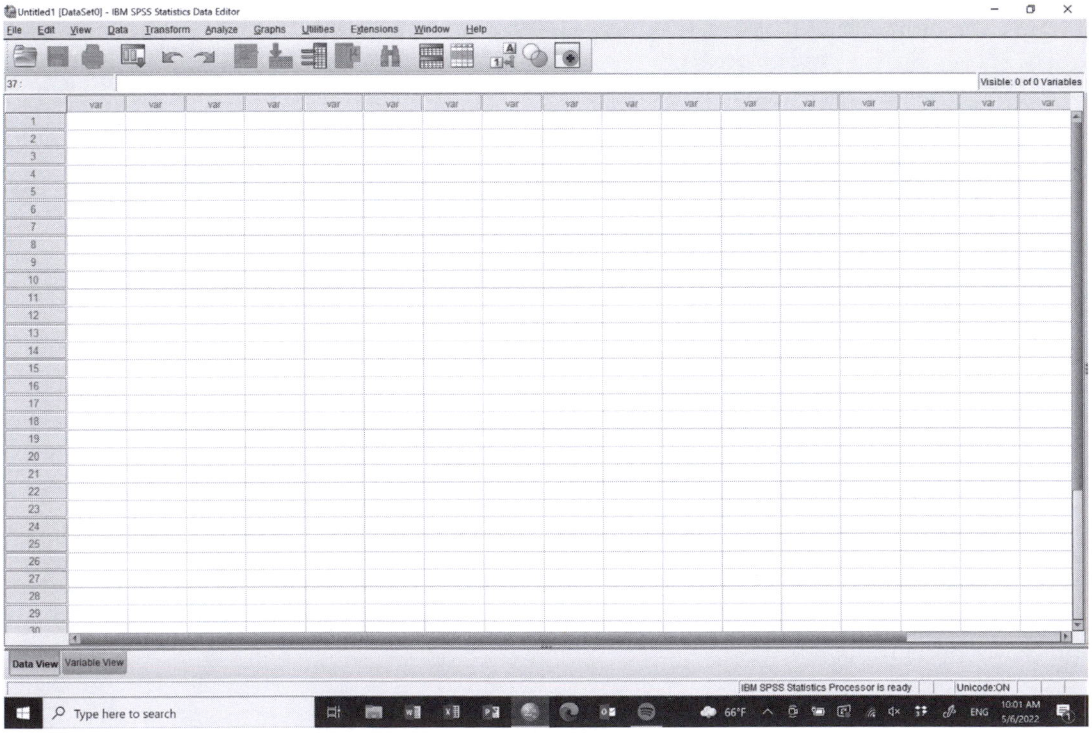

Figure 2.2 Screen capture of SPSS Data Editor.

Reprint Courtesy of IBM Corporation ©

Note there are two tabs near the bottom left of the SPSS window. One of them says Data View and the other says Variable View. The Data View window presents the data you have entered or provides a spreadsheet that permits you to enter data. The Variable View window permits you to define and name the variables themselves. It also permits you to identify variable labels, value labels, and missing values. These are illustrated in subsequent paragraphs. Note also that there are several pull-down menus across the top of the data matrix. These generally provide you with the following functions:

- *File.* Permits you to create a new data matrix, open a data matrix that was previously saved, save the current data matrix, print the current data matrix or analysis results, and exit the program.

- *Edit.* Permits you to undo a previous command; cut, copy, or paste something from the window; insert variables or cases; or find a specific piece of data.

- *View.* Permits you to change the font in which your data appear, change the appearance of the data matrix window, and so on.

- *Data.* Permits you to sort the data and select specific cases, among other functions.

- *Transform.* Permits you to modify your variables in a number of ways. You will use the Compute function under the transform tab often.

- *Analyze.* Lists the statistical procedures that you will use. Note that each of the options in this menu has an arrow next to it, which indicates that additional submenus are available to you under this particular statistical procedure. You will become familiar with many of these options and submenus as you work through the textbook.

- *Graphs.* Lists the various types of graphs you might use to present your data. We will use a limited number of these options with your textbook.

- *Utilities.* Permits you to modify your data matrix in a number of ways. We will not be using this menu in your textbook.

- *Extension.* Permits you to search for, download, and install extensions from the IBM SPSS Predictive Analytics collection on GitHub using an interface. We will not be using this menu in your textbook.

- *Window.* Allows you to split the Data View into four quadrants. This is helpful when you have a large number of cases or variables and you want to look through the data. The Window tab also lets you hide the data window when you are running several programs at a time and switch from the data matrix window to the SPSS output window once you have conducted an analysis. You can split the data window to facilitate ease of reading columns.

- *Help.* Provides you with a variety of help resources when you are running SPSS and have a question. You might find the Topics (and then Index) submenus helpful.

Each of the pull-down menus has additional functions that you might enjoy investigating, but we provide you with sufficient information to conduct the SPSS processes you will use with your textbook. You are encouraged to investigate these various menus as you learn SPSS and use the Help windows provided within SPSS. The more you interact with SPSS, the better you will understand its capabilities and use it to make your work much easier. SPSS instructions are based on version 25.0. These instructions may change slightly as new SPSS software versions are updated. We have provided SPSS sample datasets in HK*Propel* for each of the remaining chapters of your textbook.

Creating and Saving Data Files

Use the steps in Mastery Item 2.5 to create and save your first SPSS data file. Note that you will need to save the data matrix as table 2-1 (with a hyphen and not a period). This is because the computer could interpret the period as a file extension, which may cause you difficulty when you attempt to access the table at some later point. Thus, when you create and save your tables, name them with the following style: chapter number-table number. For example, in chapter 2, the second table would be table 2-2, the third table would be 2-3, and so on. Once you have the idea, you will be able to go through these steps quite rapidly.

SPSS variable names must begin with a letter and be no longer than 64 characters in length. You cannot use spaces in the name, and you should avoid special characters when naming a variable. We have used a mnemonic name to identify our variables. Thus, "weightkg" is actually weight in kilograms (kg). The mnemonic helps you remember exactly what the variable is and the units in which it was measured (figure 2.3).

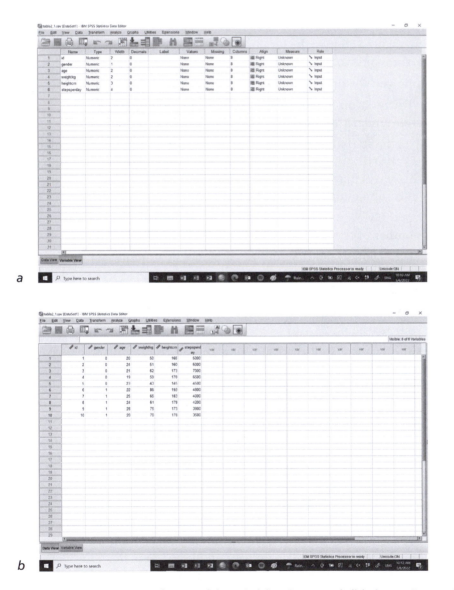

a

b

Figure 2.3 Screen capture of SPSS *(a)* variable view and *(b)* data view windows.

Reprint Courtesy of IBM Corporation ©

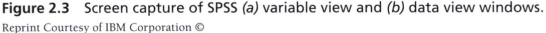 **Go to HK*Propel* to complete Student Activity 2.3.**

Mastery Item 2.5

Follow these step-by-step procedures to create an SPSS file named table 2-1 and save it to a data storage device (DSD) (e.g., flashdrive, disc, the cloud, etc.). Excel users should turn to the appendix for instructions.

Create and Save an SPSS Data File

1. Be certain to have a DSD with you before you start this assignment. In some systems you may have to save your data to an electronic account.

2. Place the DSD in the machine and note the drive location.

3. Locate the SPSS icon and click on it. (Alternatively, you might have to go to the Start button on the bottom left of your computer and locate SPSS among the programs listed in the Start menu.)

4. First you will name the variables, define the variables, and essentially build a codebook that helps you remember what the variables are.

5. Click on the Variable View tab on the bottom left and note that the window now looks like that illustrated in figure 2.3a, but without the information in it.

6. Name each of the variables in the first column. Note the variable name must start with a letter, have no spaces, contain no special characters, and be no more than 64 characters in length.

7. For the time being, skip over the Type, Width, and Decimals columns.

8. You can expand on the variable names in the Label column.

9. Click on the right edge of the Values column for the second variable (i.e., gender). Notice that you get a box that helps you define the values associated with numbers for gender. In our case, we have coded females as 0 and males as 1. Enter these values, click on the Add button each time, and then click on OK. You have defined your variables and you are ready to begin entering data.

10. Click on the Data View tab to get to the Data View window.

11. Enter the data from table 2.1 into SPSS. Your results should look like those in figure 2.3b.

12. You are now ready to save the data on your DSD. Be certain the DSD is inserted properly. Go to the File menu and scroll down to Save As (see figure 2.4).

13. In the File Name box, enter "table 2-1" (without the quotes).

14. Go to the Save In box at the top of the screen and click on the downward pointing arrow. Scroll down to the location where you just placed your DSD and click.

15. Now click Save. Your table 2.1 data are now saved to your DSD.

16. Go to the File menu, scroll down to Exit SPSS, and click. Doing so will exit you from SPSS.

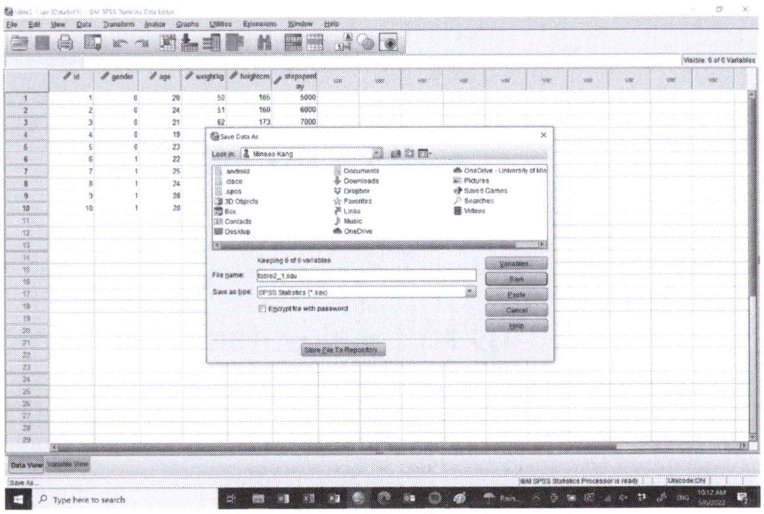

Figure 2.4 Screen capture of SPSS showing how to save a file.
Reprint Courtesy of IBM Corporation ©

Mastery Item 2.6

You may not always have SPSS available on the computer with which you are working. Thus, you can use Excel to enter your data and then have SPSS read the Excel data file. We will give you an example of how to do this with the data from table 2.1. Follow these steps to create an Excel database of table 2.1 and read the data into SPSS:

Create an Excel Database

1. Open Excel on your computer. You will see a blank data sheet.
2. Enter the variable names in the first row. Continue to use the SPSS restrictions on variable names. (Each variable name must start with a letter, have no spaces, contain no special characters, and be no longer than 64 characters.)
3. Place your cursor in cell a2 and begin to enter the data from table 2.1. Once you have entered all of the data, your Excel data file should look like that presented in figure 2.5.
4. Go to the File menu, scroll down to Save As, and save the Excel version of table 2.1 (remember to name it table 2-1) to your DSD just as you were instructed to do with the SPSS version of table 2.1.

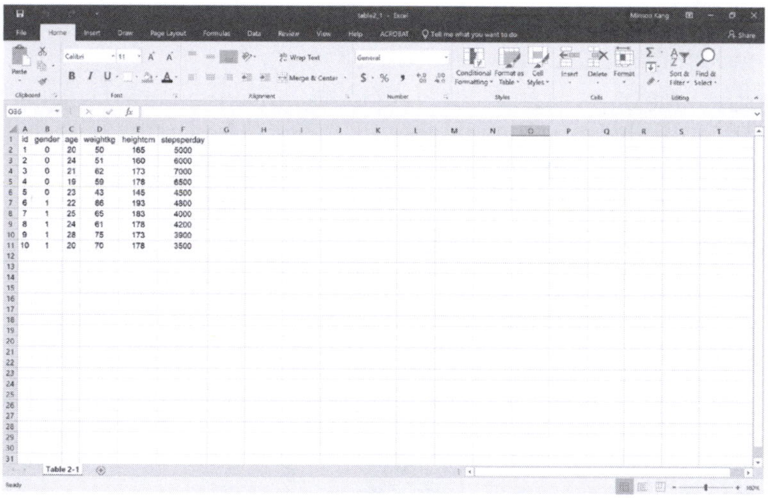

Figure 2.5   Excel data file with data from table 2.1 inserted.
Reprint Courtesy of IBM Corporation ©

You are now ready to access your Excel data with SPSS.

Access an Excel Database With SPSS

1. Open SPSS as you have previously been instructed.
2. Go to the File menu and scroll down to Open and over to Data . . . ; you will see the screen presented in figure 2.6.
3. You will be presented with an Open File window. Go to Files of type near the bottom of the window and click on the downward arrow so that you can then highlight the Excel (*.xls, *.xlsx, and *.xlsm) indicator. This will indicate that you want to import an Excel data file into SPSS. This is illustrated in figure 2.7.
4. Locate the Excel file that you want to read into SPSS. Click on the file name and it will then appear in the file name box. Click on Open.

5. You will see an Opening Excel Data Source window. Click OK. This screen is presented in figure 2.8.

6. Your data will automatically be placed into SPSS. Compare the results that you have just imported with those presented in figure 2.3*b*, where you entered the data directly into SPSS.

7. Note that only the variable names and data have been imported into SPSS. You will need to go to the Variable View tab and enter labels and values as you did originally with SPSS. That is one disadvantage of reading your data from an Excel file—you lose the codebook information and must create it.

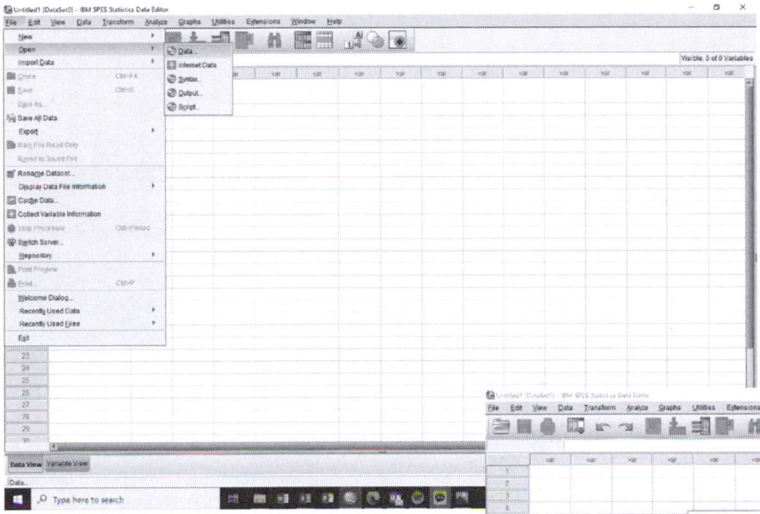

Figure 2.6   Opening an Excel data file in SPSS.
Reprint Courtesy of IBM Corporation ©

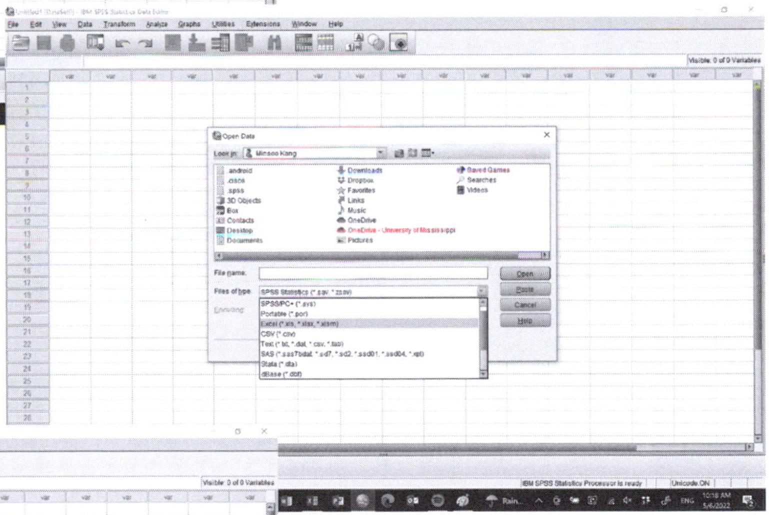

Figure 2.7   Accessing an Excel database to read into SPSS.
Reprint Courtesy of IBM Corporation ©

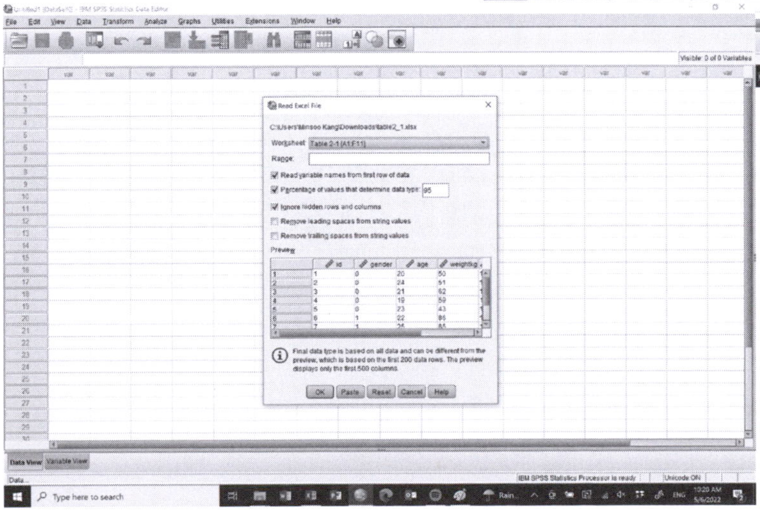

Figure 2.8   Opening an Excel file into SPSS.
Reprint Courtesy of IBM Corporation ©

Recalling and Analyzing Data

Now that you have created and saved the data, let's recall them and conduct an analysis with them using the procedures in Mastery Item 2.6. Mastery Item 2.7 will demonstrate one of the most powerful functions of SPSS—the ability to manipulate data easily. Body mass index (BMI) will be discussed further in chapter 10; we use it here because it provides an excellent example of SPSS data modification.

Mastery Item 2.7

Use table 2-1 that you created with SPSS to obtain some descriptive statistics on the data.

Recall and Analyze Data

1. First, go to your DSD and locate table 2-1. Double-click on it to begin SPSS.
2. Go to the Analyze menu.
3. Scroll down to Descriptive Statistics, over to Descriptives, and click.
4. When the Descriptives window appears, use the arrow to move "age," "weightkg," "heightcm," and "stepsperday" into the Variable(s) box.
5. Click OK and then compare your results with those presented in table 2.2.
6. If your results are different, go back to table 2.1 and compare the data in the table with what you entered in the SPSS Data Editor.

Table 2.2 Descriptive Statistics

	N	Minimum	Maximum	Mean	Standard deviation
Age in years	10	19	28	22.60	2.757
Weight in kilograms	10	43	86	62.20	12.709
Height in centimeters	10	145	193	172.60	13.293
Steps per day	10	3500	7000	4940.00	1183.404
Valid N (list-wise)	10				

Mastery Item 2.8

We'll use the following instructions to create some new variables for the 10 participants listed in table 2.1. We'll use weight and height to calculate each person's BMI.

Use Compute Statements in SPSS

1. Access your table 2-1 data as you did in Mastery Item 2.7.
2. Go to the Transform menu, scroll down to Compute Variable, and click—a new Compute Variable window will appear.
3. Type "weightlb" in the Target Variable box.
4. Put "weightkg" in the Numeric Expression box by using the arrow to move it.
5. Go to the keypad in the window and click on the "*" for multiplication.

6. Place the cursor next to the "*" in the Numeric Expression box and enter 2.2 (to calculate weight in pounds from weight in kilograms, you simply multiply weight in kilograms by 2.2).

7. Click on OK.

8. Note that a new variable, "weightlb," has been created and added in a column to the right of "stepsperday."

9. Follow the same steps to change height in centimeters to height in inches ("heightin"). Note that you divide height in centimeters by 2.54 to obtain height in inches.

Calculating BMI is a bit more involved. BMI is weight in kilograms divided by height in meters squared. BMI can be calculated from kilograms and centimeters or from inches and pounds. We'll take you step by step.

Calculate BMI in SPSS

1. Use the Compute Variable submenu (under Transform) to create a variable called BMI.

2. Use the Compute Variable statement to create BMI from "weightlb" and "heightin." The formula is weightlb/(heightin * heightin) * 703. Put this formula in the Numeric Expression box on the right and then click OK.

3. Save the revised version of table 2-1 to your DSD with the Save command under the File menu (see figure 2.9).

4. Calculate the mean for the BMI you just created and confirm that the mean value is 20.6524.

5. If you do not get this number, check the original numbers you entered and recheck how you created the variables at the various steps.

6. If you find a variable that you created incorrectly, simply highlight the column for this variable and press Delete. The column will be removed from the dataset and you can re-create it.

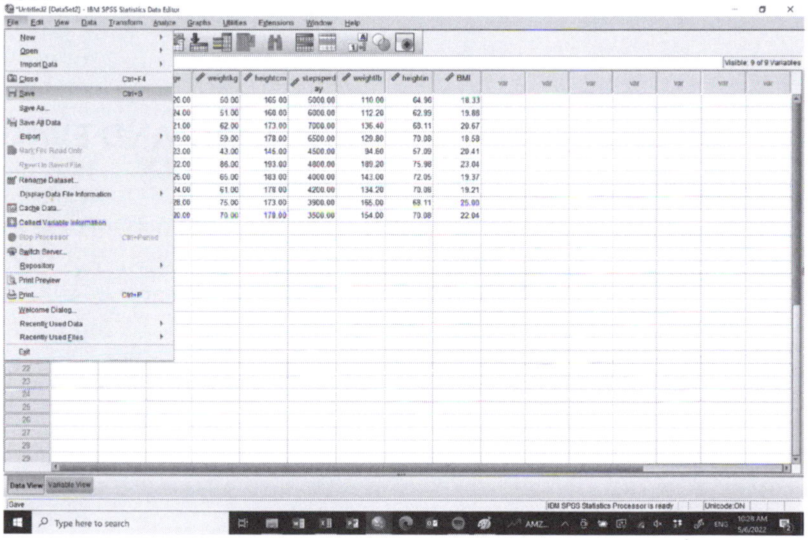

Figure 2.9   Saving an SPSS table 2-1 file with a new variable, BMI.
Reprint Courtesy of IBM Corporation ©

DOWNLOADING DATA MATRICES

As previously indicated, selected data tables from many of the chapters in this textbook are available in HK*Propel*. We will now illustrate how you can download the data for use in class assignments, practice, and learning. Open chapter 2 in HK*Propel*. Once you find the data matrices, you will notice that there are two columns with essentially the same names, but different file extensions—one contains data in the SPSS format, and the other contains the same data in Excel. To download a file, simply click on the file name. Depending on the settings on your computer, either the file will be automatically opened for you in the specific file format (i.e., SPSS or Excel) or you will be able to save it to your DSD.

- *SPSS download.* When downloading an SPSS file, you will get all of the data in the table as well as everything in the Variable View window that defines and describes the variables. If the computer that you are using does not have SPSS on it, you will not be able to view the SPSS data matrix. Don't worry—simply take your DSD to a computer that does have SPSS on it and then double-click on the SPSS data file to open it.

- *Excel download.* When you are downloading Excel files, the process is the same as that for SPSS. However, recall that the Excel file only includes the variable names and data, not the labels and values that represent the codebook found in SPSS's Variable View.

OPEN SOURCE STATISTICAL SOFTWARE

In contrast to commercial statistical packages, such as SPSS or SAS, free open source statistical software is also available. The R Project for Statistical Computing, for example, provides open source software for conducting statistical analysis (www.r-project.org). Although it provides a different look than SPSS, it permits completion of similar statistical analyses. Difficulty in learning and lack of technical support, however, are sometimes limitations of open source software.

MEASUREMENT AND EVALUATION CHALLENGE

Jamiyah can use SPSS, Excel, or some other statistical package to help her analyze how much MVPA the study participants are doing. Use the chapter 2 large dataset available in HK*Propel*. Assume these are pedometer step counts for a small sample (*n* = 100) of Jamiyah's study participants. Use SPSS to determine the average number of steps that are recorded on Monday, Wednesday, Friday, and Saturday. Go to Analyze → Descriptive Statistics → Descriptives, move the days to the right, and then click OK. How might these results influence Jamiyah's decision about how many days of steps should be recorded? Imagine that Jamiyah needs to analyze her data from the full study. Statistical packages will make her tasks much easier.

SUMMARY

Newer, faster, and more capable computers are continuing to change every aspect of our lives. Whether used for research, testing, evaluating, teaching, or grading, computers and specialized software can greatly help users to develop and record data with which to make decisions, develop written tests, make assessments, and more. Development of the Internet has also had implications for the gathering and transmission of knowledge that affects every educator, allied health professional, and fitness instructor. Although perhaps difficult to learn at first, computing skills are some of the most valuable skills any professional can have.

An excellent resource to help you learn more about the statistical methods that you will be studying in chapters 3, 4, and 5 can be found at the Rice Virtual Lab in Statistics (http://onlinestatbook.com/rvls.html). The various examples and simulations available at this site are excellent learning tools.

WWW Go to HK*Propel* for videos, homework assignments, and quizzes that will help you master this chapter's content.

Part II

Basic Statistical Concepts

As we have stressed, it is important to make good decisions based on accurate and valid measurements. In this section we emphasize the importance of statistics as a tool for helping to make these decisions. In general, statistics help us to determine the probability of an event's occurrence, which can be an important factor in decision making. In chapter 3, you will learn how to use measures of central tendency and variability to describe distributions, such as the distribution of physical activity behaviors (e.g., minutes of moderate-to-vigorous physical activity [MVPA] per week). You will also use the normal distribution to describe measured variables and calculate various numbers based on the normal distribution (e.g., percentiles). In chapter 4, you will learn about the relations between variables and the possibility of predicting one variable from one or more variables, such as how different measures of physical activity might be related (e.g., are weekly self-reported MVPA minutes related to steps taken per week?). In chapter 5, you will learn how to test scientific hypotheses of differences between groups, such as differences in physical activity behaviors following an intervention developed to increase physical activity. For example, what is the probability that the mean weekly minutes of MVPA involvement for two groups are *true* differences—that is, how likely is it that this difference would occur simply by chance? If this probability is extremely small, it might lead us to conclude that the first group was involved in some form of activity that resulted in increased MVPA minutes per week.

Although statistics can be quite involved, the level of mathematical skills necessary for these chapters is no higher than college algebra. The key to using statistics for measurement decisions is understanding the underlying reasoning and concepts and then applying the appropriate statistical procedures. Your research question and the scale of measurement of the variables involved will lead you to the appropriate statistical technique to analyze your data. Thus, careful reading and practicing of the concepts presented in part II, including the use of SPSS and Excel, will help you throughout the remainder of the textbook.

CHAPTER 3

Descriptive Statistics and the Normal Distribution

OUTLINE

OBJECTIVES

After studying this chapter, you will be able to

- illustrate types of data and associated measurement scales;
- calculate descriptive statistics on data;
- graph and depict data; and
- use SPSS computer software in data analysis.

WWW The lecture outline in HK*Propel* will help you identify the major concepts of the chapter.

MEASUREMENT AND EVALUATION CHALLENGE

Maaz, a college student, recently had a complete health and fitness evaluation conducted at the Cooper Clinic in Dallas, Texas. Part of the evaluation required him to run to exhaustion on a treadmill. Maaz ran for 24 minutes and 15 seconds. Using his treadmill time, the technician estimated Maaz's $\dot{V}O_2$max to be 50 ml · kg^{-1} · min^{-1}. How does Maaz interpret this value? Is this result high, average, or low? How does it compare with the results of others his age and sex? The concepts in this chapter will help Maaz better interpret statistical results from any type of test he might complete.

Researchers and teachers often work with large amounts of data. Data can consist of ordinary alphabetical characters (such as student names or sex), but data are typically numerical. In this chapter we cover the basics of data analysis to help you develop the skills necessary for measurement and evaluation. Understanding fundamental statistical analysis is required to accomplish this goal. If you can add, subtract, multiply, divide, and (with a calculator or computer) find a square root, you have the mathematical skills necessary for completing much of this work. In fact, with the SPSS software introduced in chapter 2, the computer does most of the work for you. However, you must understand the concepts of statistical analysis, when to use them, and how to interpret the results.

Descriptive statistics provide you with mathematical summaries of performance (e.g., the best score) and performance characteristics (e.g., central tendency, variability). They can also describe characteristics of the distributions, such as symmetry or amplitude.

SCALES OF MEASUREMENT

Taking a measurement often results in assigning a number to represent the measurement, such as the weight, height, distance, time, or amount of moderate-to-vigorous physical activity (MVPA). However, not all numbers are the same. Some types of numbers can be added and subtracted with meaningful results; others will result in little or no meaning. One method of classifying numbers is using scales of measurement, as presented here. Note that scales—or levels—of measurement represent a taxonomy from lowest level (nominal) to highest (ratio).

• *Nominal.* Naming or classifying, such as a football position (quarterback or tight end), sex (male or female), type of car (sports car, truck, SUV), or group into which you belong (experimental or control). A nominal scale is categorical in nature, simply identifying mutually exclusive things on some characteristic. There is no notion of order, magnitude, or size. Everyone within the group is assumed to have the same degree of the trait that determines their group. You are either in a group or not.

• *Ordinal.* Ranking, such as the finishing place in a race. Although 1 is ranked higher than 2, 3 is higher than 4, and so on, the differences between ranked positions are not comparable. For example, the person finishing first might be 0.2 seconds in front of second place, but third place might be 2 seconds later, and then fourth place 6 seconds later.

• *Continuous.* Numbers are said to be continuous in nature if they can be added, subtracted, multiplied, or divided and the results have meaning. The numbers occupy a distinct place on a number line. Continuous numbers can be either interval or ratio in form.

1. ***Interval.*** Using an equal or common unit of measurement, such as temperature (°F or °C) or IQ. The zero point is arbitrarily chosen; it does not mean that something doesn't exist. A value of zero simply represents a point on a number line. For example, in the centigrade temperature scale, 0 °C does not indicate the absence of heat but rather the temperature at which water freezes. It is possible to have temperatures below zero.

2. ***Ratio.*** Same as interval, except having an absolute (true) zero, such as temperature on the Kelvin scale. Weight, times, or shot-put distance are examples of ratio measures. With a true zero, ratios are possible. For example, if one person is 6 feet (1.82 m) tall and another is 3 feet (0.91 m) tall, the first person is twice as tall as the second.

One simple way to distinguish these scales is to determine if a scale has order, distance, or absolute zero (origin) features. *Order* means that a larger number represents a greater amount of an attribute or ability being measured; *distance* means that there are equal distances between measurement units; and *origin* means that the scale has an absolute zero, which represents a total absence of the attribute being measured. A nominal scale has none of these features; an ordinal scale has the only "order" feature; an interval score has both "order" and "distance" features; and finally, a ratio scale has all three features.

To help put these scales into perspective, consider physical activity. You might be interested in steps taken by men and women (sex is a nominal variable). You might be interested in moderate versus vigorous physical activity (an ordinal variable, because vigorous activity is higher intensity than moderate, but not all moderate or vigorous physical activity is the same). You might be interested in steps taken per day (a ratio variable, because if you take 5000 steps per day and your sister takes 10,000 steps, then she has taken twice as many steps as you).

© Human Kinetics

What nominal, ordinal, and continuous scales of measurement can you identify in this medal ceremony?

An important concept to remember is that certain characteristics must exist before mathematical operations can be conducted. The scales of measurement are hierarchical; each builds on the previous level or levels. That is, if a number is ordinal, it can also be categorized (nominal); if the number is interval, it also conveys ordinal and nominal information; and if the number is ratio in nature, it also conveys all three lower levels of information—nominal, ordinal, interval. Only interval and ratio numbers can be subjected to mathematical operations (e.g., added, divided). People sometimes use lower scales of measurement—ordinal and nominal—as if they were the higher-scale interval or ratio. Consider data in which you have males coded as 0 and females coded as 1. Does it make any sense to talk about the average sex being 0.40? Of course not! Sex is a nominal variable. It is similarly inappropriate to calculate an average from ordinal data, such as Likert scales. What is the average of "Agree" and "Disagree"? The average will always be the middle rank; it has no meaning. This is another reason it is important to distinguish the level of measurement of the data before applying statistical tests.

Mastery Item 3.1

Diving and gymnastic scores use what scale of measurement?

SUMMATION NOTATION

To represent what they want to accomplish mathematically, mathematicians have developed a shorthand system called *summation notation*. Although summation notation can become quite complex, for present purposes you need learn only a few concepts. Three points are important for you to remember: N is the number of cases, X is any observed variable that you might measure (e.g., height, weight, distance, MVPA minutes), and Σ (the capital Greek letter sigma) means "the sum of." In summation notation, $\Sigma X = (X_1 + X_2 + \ldots X_n)$, where n represents the nth (or last) observation. This reads, "The sum of all X values is equal to $(X_1$, plus X_2, plus $\ldots X_n)$."

Be certain to recall the order of operations when using summation notation, particularly parentheses and exponents. Recall that you do all operations within parentheses before moving outside them. If there are no parentheses, the precedence rules of mathematical operations hold: First, conduct any exponentiation, followed by multiplication and division and then addition and subtraction. For example, ΣX^2 is read as "the sum of the squared X scores," whereas $(\Sigma X)^2$ is "the sum of X, the quantity squared." The distinction is important because the two terms represent different values. With ΣX^2, one squares each X and then sums them up. With $(\Sigma X)^2$, one sums all of the X values and then squares that sum. You can confirm that these values will not be the same with the following values of X: 1, 2, 2, 3, 3. $\Sigma X^2 = 27$ and $(\Sigma X)^2 = 121$.

Mastery Item 3.2

Use the following numbers to calculate the summation notation values indicated: 3, 1, 2, 2, 4, 5, 1, 4, 3, 5.

Confirm that ΣX is equal to 30.
Confirm that ΣX^2 is equal to 110.
Confirm that the following equation resolves to 2.22.

$$\frac{\Sigma X^2 - \dfrac{(\Sigma X)^2}{n}}{n-1}$$

REPORTING DATA

You might be interested in how many steps you take per week or per day. Or, after you measure your students on a variable, they may want to know how they performed. If you only tell your students their scores, however, they would know little about how well they performed. Similarly, if you only learned the number of steps you take per week or day but had no comparison (either norm referenced or criterion referenced), the number alone would be rather meaningless. You need some additional information.

People often want to compare themselves with others who have completed a similar test or activity. As described in chapter 1, norm-referenced measurement allows this. For fitness and sports performance testing, norms are often provided for the parameters of sex and age.

One method for deciding how your performance compares with others is to develop a **frequency distribution** of the test results. A frequency distribution is a method of organizing data that involves noting the frequency with which various scores occur. The results from $\dot{V}O_2$max testing for 65 students are presented in table 3.1. Look at them and assume your measured $\dot{V}O_2$max was 46 ml · kg^{-1} · min^{-1}. How well did you do? This is difficult to determine when numbers are presented as they are in the table. But a frequency distribution can clarify how your score compares to those of others (i.e., a norm-referenced comparison). Alternatively, you might want to make a criterion classification based on your $\dot{V}O_2$max. You will learn more about these interpretations later.

Table 3.1 65 $\dot{V}O_2$max Values

48	45	50	49	46	47	47	49	50	50
45	51	51	48	49	46	44	44	52	53
48	43	48	41	48	49	47	49	51	54
51	43	53	45	48	47	51	46	49	50
48	48	45	46	49	48	46	48	52	54
52	50	51	47	45	47	43	47	49	50
44	55	48	50	53					

Mastery Item 3.3

Use SPSS to obtain a frequency distribution and percentiles for the 65 values presented in table 3.1. Confirm your analysis with the results presented in figure 3.1. SPSS commands for obtaining a frequency distribution and percentiles follow:

1. Start SPSS.
2. Open table 3.1 data from HK*Propel* (or enter the values from table 3.1).
3. Click on the Analyze menu.
4. Scroll to Descriptive Statistics and across to Frequencies and click.
5. Highlight "$\dot{V}O_2$max" and place it in the Variable(s) box by clicking the arrow key.
6. Click Statistics.
7. Click Percentile(s).
8. Type "25" into the blank box, then click Add.
9. Type "50" into the blank box, then click Add.
10. Type "75" into the blank box, then click Add.

11. Click Continue.
12. Click OK.

Statistics

$\dot{V}O_2$max

N	Valid	65
	Missing	0
Percentiles	25	46.0000
	50	48.0000
	75	50.0000

$\dot{V}O_2$max

		Frequency	Percent	Valid percent	Cumulative percent
Valid	41	1	1.5	1.5	1.5
	43	3	4.6	4.6	6.2
	44	3	4.6	4.6	10.8
	45	5	7.7	7.7	18.5
	46	5	7.7	7.7	26.2
	47	7	10.8	10.8	36.9
	48	11	16.9	16.9	53.8
	49	8	12.3	12.3	66.2
	50	7	10.8	10.8	76.9
	51	6	9.2	9.2	86.2
	52	3	4.6	4.6	90.8
	53	3	4.6	4.6	95.4
	54	2	3.1	3.1	98.5
	55	1	1.5	1.5	100.0
	Total	65	100.0	100.0	

Figure 3.1 $\dot{V}O_2$max for 65 students.

You may be asking yourself, *Why do this?* Recall that we were interested in determining how well you performed relative to the rest of the class. Your $\dot{V}O_2$max of 46 appears in the lower half of the distribution. The fifth column of the frequency distribution, Cumulative Percent, is also known as **percentile rank (PR)**. The PR is obtained by summing the percentage (third column, Percent) of scores that fall at and below the PR you are calculating. A PR represents the percent of observations at or below a given score. This is a norm-referenced comparison. This concept is extremely important because test results, such as those from standardized college admissions tests, are often reported in PR. For example, if you achieve at PR 90 (PR_{90}), this simply means that you have a score higher than 90% of the people who were tested. Conversely, if you scored at PR_{10}, 90% of the people taking the test had a higher score than you did. Your $\dot{V}O_2$max of 46 is at $PR_{26.2}$—that is, 26.2% of the people achieved at the same level or lower than you did, and thus 73.8% had a higher $\dot{V}O_2$max.

You might also be interested in determining whether your $\dot{V}O_2$max is high enough to put you at reduced risk for cardiovascular disease—a criterion-referenced standard. Research suggests that men with a $\dot{V}O_2$max ≥35 are sufficiently fit for health benefits. The $\dot{V}O_2$max of 46 exceeds this criterion.

Because there are five individuals who achieved a $\dot{V}O_2$max of 46, you may wonder what are the exact PRs for these scores. Fortunately, this can easily be computed by SPSS:

1. Go back to the SPSS file containing table 3.1 data.
2. Click on the Transform menu.
3. Scroll to Rank Cases and click.
4. Highlight "$\dot{V}O_2$max" and place it in the Variable(s) box by clicking the arrow key.
5. Click on Rank Types.
6. Uncheck the box labeled Rank and check the box labeled Fractional Rank as %.
7. Click Continue.
8. Click OK.

The scores are converted to PR, and a new column of PR ($P\dot{V}O_2$max) is added to your data file.

Determine Percentile Rank When Small Scores Mean Better Performance

The presentation in figure 3.1 assumes that a higher score is better than a lower score. However, this is not always the case. If lower scores are better—for example, in golf or a timed event where a faster time is better—you need to reflect this in your interpretation of the scores. For example, if the data presented in table 3.1 were golf scores for nine holes, your lower score of 41 represents the 98.46th percentile ($100 - 1.54 = 98.46$). Work from the new table 3.1 you have created with $P\dot{V}O_2$max to see the impact of reversing the scores when a lower score is a better score. For individual scores' PR convention, this can also be done using the Compute Variable function in SPSS:

1. Click on the Transform menu.
2. Scroll to Compute Variable and click.
3. Type "LOWPGOLF" in the box Target Variable.
4. Type "100 –" and then Highlight "$P\dot{V}O_2$max" and then click OK.

The scores are converted to new PR, and a new column of LOWPGOLF is added to your data file.

Percentile

You may be wondering about the meaning of percentiles in figure 3.1. The **percentile** can be defined simply as a score value for a specified percentage of cases in a distribution of scores. For example, the 50th percentile, known as the *median* and expressed as $X_{.50}$, is the score (X) that divides the distribution so that 50% (.50) of the scores fall above this point and 50% fall below it. Other percentiles also provide useful information, especially when standards of performance are needed to interpret a set of scores in norms, which are the estimates of some characteristics of a distribution of test scores for a specified population (e.g., the number of push-ups for 14-year-old U.S. boys). The commonly used percentiles include quartiles ($X_{.25}$, $X_{.50}$, and $X_{.75}$) and deciles ($X_{.10}$, $X_{.20}$, ..., $X_{.90}$).

If the percentile information of a test or measurement is available, a teacher or a test administrator will be able to answer the question, "What is the nth percentile?" and use the information to evaluate the performance of a student or the effectiveness of a program. For example, a teacher may wish to provide summary information on students' test performances to determine the effectiveness of a fitness program, or the manager of a fitness club may ask for information on specific percentiles (say 75th percentile) for certain measures of fitness administered to clients.

For an ordered dataset, proportions of the data that should fall above and below a given value can be estimated. The percentile is thus a specific value (e.g., a $\dot{V}O_2$max or push-ups score), whereas PR is the specific percent associated with that value. In this way, percentile and PR can be considered two sides of the same coin.

Dataset Application

Now it is time to try this with a larger dataset. Go to HK*Propel* and use SPSS to calculate percentiles for the mean number of steps taken per week by the 250 people in the chapter 3 large dataset. When you run the Frequencies you will see that the output is much longer (and unwieldy for you). To reduce this, click on the Statistics button in the Frequencies window; then click on the box next to Cutpoints and you will get 10 equal groups on your output. Click on the Continue button and then OK. Notice that the first part of your output is in deciles—10 groups. You'll see the steps per week of 10th, 20th, 30th percentiles and so on.

CENTRAL TENDENCY

Now you know where you fit into the distribution. You are somewhere near the lower 30% of the people tested. Let's further consider how you can interpret your score. One way is to determine how you compare with the so-called typical person who took the test. To do so, you look at where the scores tend to center—their **central tendency**. We will briefly describe three measures of central tendency:

• *Mean.* The arithmetic average; the sum of the scores divided by the number of scores. From summation notation, this definition may be represented as

$$M = \frac{\Sigma X}{n} \tag{3.1}$$

where M is the mean, X is the value of each observation, and n is the number of observations. Every score in the data is used to determine the mean—it is the most stable measure of central tendency. Using the four scores (4, 3, 2, 5), the mean is (4 + 3 + 2 + 5) / 4 = 3.5.

• *Median.* The middle score; the 50th percentile. To obtain the median, order the scores from high to low and find the middle one. The median value for the data presented in table 3.1 is 48, which is the middle score. Note that the median is a specific percentile: $P_{.50}$. The median is the most typical score in the distribution: Half of the scores are higher and half are lower.

• *Mode.* The most frequently observed score. The mode is the most unstable measure of central tendency but the most easily obtained one. You should confirm from figure 3.1 that the mode is 48 (occurring 11 times).

Dataset Application

Use the chapter 3 large dataset available with HK*Propel* and learn what the mean, median, and mode are for each of the variables in the dataset. Hint: Use Analyze → Descriptive Statistics → Frequencies → Statistics and click on the buttons in the Central Tendency box.

DISTRIBUTION SHAPES

Simply knowing a distribution's measures of central tendency does not tell you everything about the scores. Not all distributions have the same shape. Figure 3.2 illustrates several shapes that distributions can assume. The statistical term for the shape (or symmetry) of a distribution is **skewness**. A positively skewed distribution has a tail toward the positive end (right) of the number line, and a negatively skewed distribution has a tail toward the negative end (left) of the number line. Distributions with little skewness are characterized by values close to 0; in contrast, if skewness value is less than –1 or greater than 1, the distribution is highly skewed; if the value is between –1 and –0.5 or between 0.5 and 1, the distribution is moderately skewed; finally, if the value is between –0.5 and 0.5, the distribution is considered symmetric or normally distributed. If you have a skewed distribution, the measures of central tendency generally fall in alphabetical order from the tail to the highest point of the frequency distribution—thus, the mean is generally closest to the tail, the median typically in the middle, and the mode at the highest point when the distribution is skewed. Notice how the mode is the most easily identified measure of central tendency.

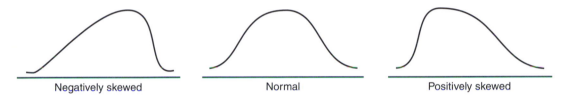

Negatively skewed　　　　　Normal　　　　　Positively skewed

Figure 3.2 Distribution shapes.

Another property of the distribution is associated with its shape. In figure 3.3, all three distributions are symmetrical curves with identical means, medians, and modes. However, there is an obvious difference in the amplitudes (or peakedness) of the distributions. The peakedness of a curve is referred to as its **kurtosis**. The top curve is said to be mesokurtic (i.e., curved an average amount), the middle curve is platykurtic (flat), and the bottom curve is leptokurtic (peaked). Leptokurtic curves have a positive kurtosis value and platykurtic curves have a negative kurtosis value. A mesokurtic curve is a normal distribution with no skewness.

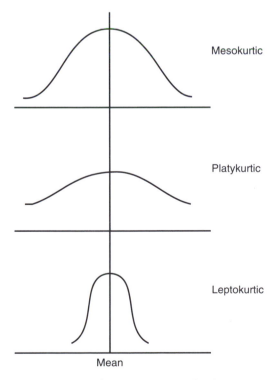

Mesokurtic

Platykurtic

Leptokurtic

Mean

Figure 3.3 Three symmetrical curves.

Dataset Application

Use the chapter 3 large dataset available with HK*Propel* and learn what the skewness and kurtosis are for each of the variables in the dataset. (Hint: Use Analyze → Descriptive Statistics → Frequencies → Statistics and click on the buttons in the Characterize Posterior Distribution box for Skewness and Kurtosis.) Be certain to move the variables from the left window to the right window under the Frequencies procedure. Can you identify which is the most positively skewed, most leptokurtic, and so on?

A good way to determine the shape of the distribution you are working with is to develop a histogram. A **histogram** is a graph that consists of columns to represent the frequencies with which various scores are observed in the data. It lists the score values on the horizontal axis and the frequency on the vertical axis.

WWW Go to HK*Propel* to complete Student Activity 3.1.

Mastery Item 3.4

Use the data in table 3.1 to create a histogram of the 65 $\dot{V}O_2$max values (see figure 3.4). SPSS commands for obtaining a histogram are the following:

1. Start SPSS.
2. Open table 3.1 data.
3. Click on the Analyze menu.
4. Scroll to Descriptive Statistics and across to Frequencies and click.
5. Highlight "$\dot{V}O_2$max" and place it in the Variable(s) box by clicking the arrow key.
6. Click the Charts button.
7. Click the Histograms button.
8. Click Show normal curve on histogram.
9. Click Continue.
10. Click OK.

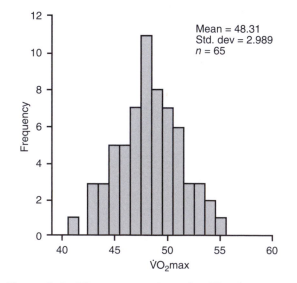

Figure 3.4   Histogram output for 65 values.

Note that you could also obtain a histogram directly by going to Graphs → Legacy Dialog → Histogram and placing the variable of interest in the variable box on the right. You can also have the normal curve displayed by clicking the box Display normal curve.

Dataset Application

Earlier you used SPSS to determine which variables in the chapter 3 large dataset were most skewed. Create some histograms to visually confirm what these distributions look like. In SPSS go to Graphs → Legacy Dialogs → Histogram and place the variable of interest in the Variables box. Then click on the Display Normal Curve box. Can you confirm your earlier results with these histograms?

VARIABILITY

A curve's kurtosis leads directly to the next important descriptive measure of a set of scores. The platykurtic curve in figure 3.3 is said to contain more heterogeneous (dissimilar) data, whereas the leptokurtic curve is said to have observations that are more homogeneous (similar). The spread of a distribution of scores is reflected in various measures of **variability**. Is a large variability bad? Not necessarily—the world is colorful, different, and diverse! Variability is also a common phenomenon in the field of human performance, whether in physical traits (e.g., body fat percent) or responses to stimulation (e.g., losing or not losing weight when completing the same intervention). In practice, we are often interested in two types of variability: between- and within-individual variability. Between-individual variability is the difference among individuals (e.g., height). Within-individual variability refers to the variability of an individual at different times (e.g., weight, performance, or mood) (Li, Chen, and Zhu, 2019).

In this chapter, we present three commonly used measures of variability: range, variance, and standard deviation.

Range

The **range** is the highest score minus the lowest score. It is the least stable measure of variability because it depends on only two scores and does not reflect how the remaining scores are distributed.

Variance

The **variance** (s^2) is a measure of the spread of a set of scores based on the squared deviation of each score from the mean. The variance is the most stable measure of variability and is used much more frequently than the range in reporting the heterogeneity of scores. Two sets of scores that have vastly different spreads will have vastly different variances. Many types of variance (such as observed variance, true variance, error variance, sample variance, between-subject variance, and within-subject variance) will become important as measurement and evaluation issues are presented throughout this textbook.

As an illustration of how to calculate the variance, consider the following scores: 3, 1, 2, 2, 4, 5, 1, 4, 3, 5. The steps to calculate the variance by hand are as follows:

1. Calculate the mean.
2. Subtract the mean from each score.

3. Take each difference (deviation) and square it.
4. Add the results together and divide by the number of scores minus 1.

Equation 3.2 (the didactic formula) is used to illustrate the variance.

$$s^2 = \frac{\Sigma(X-M)^2}{n-1} = \frac{\Sigma x^2}{n-1} = \frac{20}{9} = 2.22 \tag{3.2}$$

The variance, therefore, is the average of the squared deviations from the mean (hence, the term *mean square* is sometimes used for a variance). Note that you divide by $n-1$ and not n, so it isn't exactly the mean. You should use the following calculation formula to obtain the variance from a set of scores because it is generally easier to use.

$$s^2 = \frac{\Sigma X^2 - \frac{(\Sigma X)^2}{n}}{n-1} \tag{3.3}$$

Confirm that using the calculation formula will result in the same value for the variance as the didactic formula. (Hint: $\Sigma X^2 = 110$; $\Sigma X = 30$.) Note that when all of the scores are identical, the variance is zero. Typically, this is not something you want in measurement. Variation among scores is preferable. The reasons for this will be illustrated throughout this textbook.

Turn to table 3.2 and see that you have already learned to calculate the variance from the data that were provided there. You have simply used summation notation.

Table 3.2 Calculating the Variance

X (observed score)	–	M (mean)	=	x (deviation score)	x2 (squared deviation score)
3	–	3	=	0	0
1	–	3	=	–2	4
2	–	3	=	–1	1
2	–	3	=	–1	1
4	–	3	=	1	1
5	–	3	=	2	4
1	–	3	=	–2	4
4	–	3	=	1	1
3	–	3	=	0	0
5	–	3	=	2	4
Total				0	20

The types of variance will become increasingly important throughout the remainder of this textbook, but it is important that you understand variance at this point. Variance in general can be illustrated with a square, as in figure 3.5. Not everyone scored the same, so there is total or **observed score** variance. Not everyone has the same true knowledge about the content of the test, so there is **true score** variance. Last, there is some error in the test, but not everyone has the same amount of error reflected in their scores, so there is **error score** variance (error could result from several sources, such as student guessing or incorrect scoring). These important concepts will be further discussed in chapter 6.

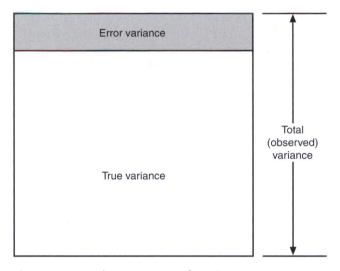

Figure 3.5 Three types of variance.

Although it is a calculated *number*, we will often speak of variance in conceptual terms (i.e., variability) throughout the text. It may be helpful for you to remember the square in figure 3.5 and the fact that not all scores are identical.

Standard Deviation

Whereas the variance is important, a related number—the **standard deviation**—is often used in descriptive statistics to illustrate the variability of a set of scores. The standard deviation (*s*), which is sometimes expressed as *S* or SD in the literature, is the square root of the variance. It is helpful to think of the standard deviation as a linear measure of variability. Consider the square used to illustrate variance (figure 3.5). When you take the square root of a square, you get a linear measure. The same concept holds true for the standard deviation. The standard deviation is important because it is used as a measure of linear variability for a set of scores and can tell us a great deal about their heterogeneity or homogeneity.

You are encouraged to use the calculation formula (equation 3.3) to calculate the standard deviation. Simply calculate s^2 (the variance) and take its square root. You should confirm that the standard deviation from the data in table 3.2 is 1.49 (i.e., $\sqrt{2.22}$).

Dataset Application

Use the chapter 3 large dataset available with HK*Propel* and determine the variance and standard deviation for each of the variables in the dataset. Hint: Use Analyze → Descriptive Statistics → Frequencies → Statistics and click on the buttons in the Dispersion box. Can you identify which are the most heterogeneous and homogeneous variables? Create histograms to visualize the variabilities.

Why is the standard deviation so important, and what exactly does it mean? For example, in a **normal distribution**, a bell-shaped, symmetric probability distribution (figure 3.6), knowing the standard deviation tells you a great deal—if you add to and subtract from the mean the value of 1 standard deviation, you will obtain a range that encompasses approximately 68.26% of the observations. If you add and subtract the value of 2 standard deviations, you will obtain a range that encompasses approximately 95.44% of the observations, and adding and subtracting the value of 3 standard deviations will give a range encompassing 99.74% of the observations. This is true regardless of what the mean and standard deviation are, as long as the distribution is normal. To summarize:

$$M \pm 1s \rightarrow 68.26\% \text{ of observations (where } s = \text{standard deviation)}$$

$$M \pm 2s \rightarrow 95.44\% \text{ of observations}$$

$$M \pm 3s \rightarrow 99.74\% \text{ of observations}$$

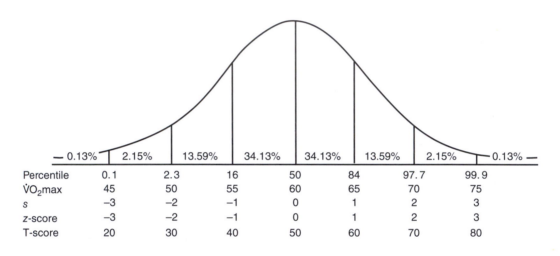

	–0.13%	2.15%	13.59%	34.13%	34.13%	13.59%	2.15%	0.13%	
Percentile		0.1	2.3	16	50	84	97.7	99.9	
V̇O₂max		45	50	55	60	65	70	75	
s		–3	–2	–1	0	1	2	3	
z-score		–3	–2	–1	0	1	2	3	
T-score		20	30	40	50	60	70	80	

Figure 3.6 Normal distribution of scores.

Using the information obtained about the mean and standard deviation for a set of observations, you can approximate the percentile for any observation. Looking again at figure 3.6, assume that it illustrates the observations on a recent maximal oxygen consumption ($\dot{V}O_2$max) test that had a mean of 60 ml · kg⁻¹ · min⁻¹ and a standard deviation of 5 ml · kg⁻¹ · min⁻¹. Use the figure to approximate percentiles for $\dot{V}O_2$max values of 50, 55, 60, 65, and 70. How likely would it be to obtain a $\dot{V}O_2$max in excess of 70 ml · kg⁻¹ · min⁻¹ for similar subjects?

The mean should be used as the measure of central tendency and the standard deviation as the measure of variability as long as the data are normally distributed. However, if the data distribution is either positively or negatively skewed, the median should be used as the measure of central tendency because it is always the middle value, whereas the mean is often seriously affected by extremely large or small values in a dataset, known as the outliers. Whenever the median is used as the measure of central tendency, the interpercentile range should be used as the variability measure. The most popular interpercentile range used in practice is the interquartile range, which can be computed as ($P_{.75} - P_{.25}$), representing the variability among the middle 50% of data.

STANDARD SCORES

Knowing the mean and standard deviation makes it easy to calculate a **standard score**, or a set of observations that have been standardized around a given mean and standard deviation. The most fundamental standard score is the z-score. It is calculated as follows:

$$z = \frac{X - M}{s} \tag{3.4}$$

In other words, you obtain the z-score for any observation by subtracting the mean, M, from the observed score, X, and dividing this difference by the standard deviation, s. If you do this for all observations in a set of data, you get a set of scores that themselves have a mean of 0 and a standard deviation of 1 (i.e., they have been standardized). Note that we have included z-scores in figure 3.6 (again, z-scores always have a mean of 0 and a standard deviation of 1).

Another commonly used standard score is the T-score, which is calculated as follows:

$$T = 50 + \frac{10(X - M)}{s} \quad \text{or} \quad T = 50 + 10z \tag{3.5}$$

The mean of the T-scores for all the observations in a set of data is always 50, and the standard deviation is always 10. Note that in a normal distribution, 99.74% of the scores will fall between a z-score of –3 and +3 and between T-scores of 20 and 80. Can you see why this is the case? (Hint: See figure 3.6.)

You are probably asking yourself, *What is the point of standard scores?* To answer this, let's assume you are a physical education teacher teaching a basketball unit in which you are going to grade students on only two skill items (we're not recommending this!). You assume dribbling and shooting are equally important in basketball (we probably don't agree with you!), so you wish to weight them equally. For shooting, you plan to measure the number of baskets a student makes in 1 minute; for dribbling, you measure the amount of time a student takes to dribble through an obstacle course. A quick look at the scores illustrates that you cannot simply add these two scores together for each student to determine who is the best basketball player. High scores are better on the shooting test, but low scores are better on the dribbling test. In addition, each score is measured in different units (i.e., number of baskets made and number of seconds in which dribbling was completed).

This is exactly where standard scores can help you. Generally, the set of scores with the larger variance would be weighted more heavily in the total if you simply added two raw scores together for each student. To weight these two skills equally on this test, you must first convert each person's scores to a standard form, such as z-scores or T-scores. Because the dribble test is timed, you must correct the obtained z-score so that a faster time results in a higher z-score. You do this by either reversing the order of X and M (M – X) in equation 3.4, or by simply changing the sign of the z-score (2 becomes –2, –1.5 becomes 1.5, etc.). Now that each test has the same weight, you can add each student's two z-scores together and obtain a total z-score. You could not do this using the original raw scores, nor could you do it with any sets of scores having unequal variances. But if you convert them to a standard score, they have the same weight because they have the same standard deviation (remember that z-scores always have a standard deviation equal to 1.0).

You can also take this example one more step and decide to give the shooting test twice the weight of the dribbling test. All you need to do is multiply the z-score for the shooting test by 2 and add it to the z-score for the dribbling test. The key concept is that whichever test is most variable will carry the greater weight. In fact, if there is no variability on an examination (i.e., if everyone scores an 80, therefore, mean = 80, s = 0), the examination contributes absolutely nothing toward differentiating student performance. This concept is further illustrated for you in chapter 8 on evaluation.

Table 3.3 summarizes important information about the mean and standard deviation of z- and T-scores. Note the means and standard deviations of these standard scores never change. You can also see this by looking at figure 3.6.

Table 3.3 Standard Score Means and Standard Deviations

	Mean	Standard deviation
z-score	0	1
T-score	50	10

NORMAL-CURVE AREAS (z-TABLE)

If you examine equation 3.4, you'll see that converting a score to a z-score actually expresses the distance a score is from its own mean in standard deviation units. That is, a z-score indicates the number of standard deviations that a score lies below or above the mean. In addition to weighting performance for grades, z-scores can be used for a number of other purposes, mainly determination of (a) percentiles and (b) the percentage of observations that fall within a particular area under the normal distribution. Consider again the normal distribution presented in figure 3.6. The area under the distribution is represented as a total of 100%. A student who scores at the mean has achieved the 50th percentile (i.e., has scored better than 50% of the group) and has a z-score of 0 (and a T-score of 50).

Observed scores alone may tell you little about performance (and, in fact, tell you nothing about relative performance). However, a positive z-score tells you that you have achieved better than average; a negative z-score indicates that you scored below the mean. Using the z-score and the table of normal-curve areas (table 3.4), you can determine the percentile associated with any z-score for data that are normally distributed.

The numbers in the body of table 3.4 are percentages of observations that lie between the mean and any given standard deviation distance from the mean. The values down the left side of the table are z-scores in whole numbers and tenths, and the numbers across the top represent z-scores to the hundredths place. (For example, for a z-score of 1.53, you would find 1.5 on the left and read across to the 0.03 column to locate the entry 43.70.) Verify that 34.13% of the scores lie between the mean and 1 standard deviation above the mean (i.e., a z-score of 1.00). Note that no negative z-scores are shown in the table: They are not necessary because the normal distribution is symmetrical. Therefore, 34.13% of the observations also fall between the mean and 1 standard deviation below the mean (i.e., a z-score of –1.00). Note that ±1 standard deviation is 68.26% (34.13 + 34.13). See if you can determine the percentage of scores that fall between 1 and 1.5 standard deviations above the mean. Use table 3.4 and the illustration in figure 3.7 to help you determine the answer (43.32 – 34.13 = 9.19%).

Table 3.4 Normal-Curve Areas

z	0.00	0.01	0.02	0.03	0.04	0.05	0.06	0.07	0.08	0.09
0.0	00.00	00.40	00.80	01.20	01.60	01.99	02.39	02.79	03.19	03.59
0.1	03.98	04.38	04.78	05.17	05.57	05.96	06.36	06.75	07.14	07.53
0.2	07.93	08.32	08.71	09.10	09.48	09.87	10.26	10.64	11.03	11.41
0.3	11.79	12.17	12.55	12.95	13.31	13.68	14.06	14.43	14.80	15.17
0.4	15.54	15.91	16.28	16.64	17.00	17.36	17.72	18.08	18.44	18.79
0.5	19.15	19.50	19.85	20.19	20.54	20.88	21.23	21.57	21.90	22.24
0.6	22.57	22.91	23.24	23.57	23.89	24.22	24.54	24.86	25.17	25.49
0.7	25.80	26.11	26.42	26.73	27.04	27.34	27.64	27.94	28.23	28.52
0.8	28.81	29.10	29.39	29.67	29.95	30.23	30.51	30.78	31.06	31.33
0.9	31.59	31.86	32.12	32.38	32.64	32.90	33.15	33.40	33.65	33.89
1.0	34.13	34.38	34.61	34.85	35.08	35.31	35.54	35.77	35.99	36.21
1.1	36.43	36.65	36.86	37.08	37.29	37.49	37.70	37.90	38.10	38.30
1.2	38.49	38.69	38.88	39.07	39.25	39.44	39.62	39.80	39.97	40.15
1.3	40.32	40.49	40.60	40.82	40.99	41.15	41.31	41.47	41.62	41.77
1.4	41.92	42.07	42.22	42.36	42.51	42.65	42.79	42.92	43.06	43.19
1.5	43.32	43.45	43.57	43.70	43.83	43.94	44.06	44.18	44.29	44.41
1.6	44.52	44.63	44.74	44.84	44.95	45.05	45.15	45.25	45.35	45.45
1.7	45.54	45.64	45.73	45.82	45.91	45.99	46.08	46.16	46.25	46.33
1.8	46.41	46.49	46.56	46.64	46.71	46.78	46.86	46.93	46.99	47.06
1.9	47.13	47.19	47.26	47.32	47.38	47.44	47.50	47.56	47.61	47.67
2.0	47.72	47.78	47.83	47.88	47.93	47.98	48.03	48.08	48.12	48.17
2.1	48.21	48.26	48.30	48.34	48.38	48.42	48.46	48.50	48.54	48.57
2.2	48.61	48.64	48.68	48.71	48.75	48.78	48.81	48.84	48.87	48.90
2.3	48.93	48.96	48.98	49.01	49.04	49.06	49.09	49.11	49.13	49.16
2.4	49.18	49.20	49.22	49.25	49.27	49.29	49.31	49.32	49.34	49.36
2.5	49.38	49.40	49.41	49.43	49.45	49.46	49.48	49.49	49.51	49.52
2.6	49.53	49.55	49.56	49.57	49.59	49.60	49.61	49.62	49.63	49.64
2.7	49.65	49.66	49.67	49.68	49.69	49.70	49.71	49.72	49.73	49.74
2.8	49.74	49.75	49.76	49.77	49.77	49.78	49.79	49.79	49.80	49.81
2.9	49.81	49.82	49.82	49.83	49.84	49.84	49.85	49.85	49.86	49.86
3.5	49.98									
4.0	49.997									
5.0	49.99997									

Adapted from Lindquist (1942).

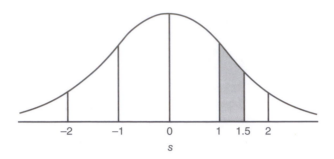

Figure 3.7 Area of normal curve between 1 and 1.5 standard deviations above the mean.

In summary, the important points to remember when using the table of normal-curve areas are that

- the reference point is the mean,
- z-scores are presented to the nearest one-hundredth, and
- the numbers in the body of the table are percentages.

Use table 3.4 to confirm that a z-score of 1.23 is (approximately) the 89th percentile. Remember that the reference point is the mean and that a percentile represents all people who score at or below a given point. (Hint: This includes those scoring below the mean if your observed score is above the mean, the percentile is greater than 50, or the z-score is greater than 0.00.)

Use table 3.4 to confirm that a z-score of –1.23 represents (approximately) the 11th percentile. You can do this using the following method:

1. Find 1.2 in the left column of the table and 0.03 at the top. They intersect at the percent value of 39.07.
2. Recall what this number means: 39.07% of the scores lie between the mean and a z-score of either +1.23 or –1.23.
3. Recall that 50% of the scores fall below the mean.
4. Thus, 50 – 39.07 = 10.93 (or approximately the 11th percentile).

Use table 3.4 to confirm that 14.98% of a set of normally distributed observations fall between a z-score of 0.50 and 1.00. (Draw a picture to help you see this.) Some hints for completing these types of items include the following:

- Draw a picture of the question (starting with the normal distribution).
- Consider whether the z-score is helpful. (It *often* is.)
- Can the z-table (table 3.4) help you? (It *often* can.)

If a score is reported in T-score form, simply transform it to z-score form and then estimate the percentile. To change from a T-score to a z-score, substitute the T-score mean (50) and standard deviation (10) into the z-score formula. For example, if your T-score is 30, your z-score would be

$$(30 - 50) / 10 = -2.00$$

Although you can use table 3.4 to convert from z-scores to percentiles, you can also use the table to convert from percentiles to z-scores for normally distributed data. This might be helpful if you wanted to create a summation of z-scores to determine the best overall performance. Assume you were told that you scored at the 69th percentile on a test. You can determine your z-score on the test as follows:

1. Calculate 69% – 50% = 19% to get the area between your percentile and the mean.
2. Find 19 in the body of the table—the closest value is 19.15.
3. Find what z-score that 19.15 corresponds to by going over to the left margin and up to the top row. The z-score is 0.50. (It is positive, because you were above the mean.)

If you wanted to confirm that this is a T-score of 55, you would simply substitute 0.50 into the T-score formula (equation 3.5).

www **Go to HK*Propel* to complete Student Activity 3.2.**

Mastery Item 3.5

Use table 3.4 to confirm that if the cutoff to achieve an A in this course is a z-score of 1.35, 8.85% of students would be expected to earn a grade of A.

Mastery Item 3.6

Verify that the percentile associated with a T-score of 68 is 96.41.

Mastery Item 3.7

Use SPSS to calculate z-scores for the data in table 3.1.

1. Start SPSS.
2. Open table 3.1 data.
3. Click on the Analyze menu.
4. Scroll to Descriptive Statistics and across to Descriptives and click.
5. Highlight "VO_2max" and place it in the Variables box by clicking the arrow key.
6. Click on the Save Standardized Values as Variables box.
7. Click OK.

The scores are converted to z-scores, and a new column of z-scores (ZVO_2max) is added to your data file. Be certain to save the modified data file because you will need the z-scores in the next mastery item.

Mastery Item 3.8

Use SPSS to calculate T-scores.

1. Start SPSS.
2. Open table 3.1 data.
3. Click on the Transform menu.
4. Click on Compute.
5. In the Target Variable box, type "Tscore."
6. In the Numeric Expression box, type "50 + 10*ZVO2max."
7. Click OK.

Now the T-scores have been added to your data file. Save the data file.

Mastery Item 3.9

Use the data that you just created and SPSS to verify that the mean and standard deviation for the *z*-scores and T-scores are 0 and 1 and 50 and 10, respectively.

MEASUREMENT AND EVALUATION CHALLENGE

What has Maaz learned in this chapter that will help him interpret the results from his treadmill test? Maaz can compare his $\dot{V}O_2$max to that of the average male of his age. He can determine his percentile based on his score, the mean score, and the variability in $\dot{V}O_2$max. Maaz has learned how to develop and interpret scores based on score distributions and norms. Assuming the average $\dot{V}O_2$max for a male of his age is 45 ml · kg^{-1} · min^{-1} with a standard deviation of 5 ml · kg^{-1} · min^{-1}, Maaz can easily determine that his performance is at the 84th percentile. Only 16% of people his age and sex have $\dot{V}O_2$max values that exceed 50 ml · kg^{-1} · min^{-1}.

SUMMARY

The descriptive statistics presented in this chapter are foundational to the remainder of this textbook. Of particular importance is your ability to understand and use the concepts of central tendency and variability and to use and interpret the normal curve and areas under sections of the normal distribution. If you are interested in more on statistical methods, see Mood, Morrow, and McQueen (2020). Thomas, Nelson, and Silverman (2015) also provide excellent examples of research and statistical applications in human performance.

At this point you should be able to accomplish the following tasks:

- Differentiate among the four levels of measurement and provide examples of each
- Calculate and interpret descriptive statistics
- Calculate and interpret standard scores
- Use a table of normal-curve areas (*z*-table) to estimate percentiles
- Use SPSS to enter data and to generate and interpret the following:
  - Frequency distributions and the percentiles associated with observed scores
  - Histograms for observed scores
  - Descriptive statistics (mean, standard deviation) on variables
  - *z*- and T-scores

www. **Go to HK*Propel* for videos, homework assignments, and quizzes that will help you master this chapter's content.**

CHAPTER 4

Correlation and Prediction

OBJECTIVES

After studying this chapter, you will be able to

- calculate statistics to determine relationships between variables;
- calculate and interpret the Pearson product-moment correlation coefficient;
- calculate and interpret the standard error of estimate (SEE);
- use scatterplots to interpret the relationship between variables;
- calculate and interpret Spearman's rank correlation coefficient;
- differentiate between simple correlation and multiple correlation; and
- use SPSS and Excel computer software in data analysis for correlational and regression analyses.

www The lecture outline in *HKPropel* will help you identify the major concepts of the chapter.

MEASUREMENT AND EVALUATION CHALLENGE

Now that you know how to calculate descriptive statistics, you are ready to learn other important statistical procedures. In chapter 3, a technician reported Maaz's $\dot{V}O_2$max as 50 ml · kg^{-1} · min^{-1}. Maaz had recently read that $\dot{V}O_2$max should actually be measured by collecting expired air and analyzing it for oxygen and carbon dioxide content, but the technician had not conducted any of these measurements during his treadmill run. Maaz now realizes that his treadmill time was related to his actual $\dot{V}O_2$max. Correlational procedures are used to determine the relationship between and among variables. Indeed, predictions and estimations of one variable from one or more other variables are common in kinesiology, human performance, and sport and exercise science. For example, in Maaz's case, his treadmill time was used to estimate his $\dot{V}O_2$max without having to actually collect expired oxygen and carbon dioxide.

How are two variables related? As performance on one variable increases, does performance on the other variable change? The relationship between variables is examined statistically by the **correlation coefficient**. **Correlations** help you describe relations and, in some cases, predict outcomes; they are useful in measurement theory.

CORRELATION COEFFICIENT

In chapter 3, you learned to describe data using measures of central tendency and variability, and we discussed one variable or measure at a time. However, teachers, clinicians, and researchers often measure more than one variable. They are then interested in describing and reporting the **relationship**, or the statistical association, between these variables. This is a fundamental statistical task that data analysts, kinesiologists, teachers, clinicians, and researchers should be able to perform. A few examples include the following:

- What is the relationship between the bench press and the leg press strength tests? That is, do these strength measures have anything in common?

- Will a child who has better gross motor function be more physically active? In other words, what makes a child more active?

- What is the relationship between sitting time and lower-back pain? Can more breaks for movement reduce or even eliminate lower-back pain?

- What are the key factors that led to the sudden increase in childhood obesity?

To answer these relationship-related questions, we would calculate the correlation coefficient, specifically the **Pearson product-moment (PPM) correlation coefficient**, symbolized by *r*. The correlation coefficient is an index of the **linear relationship** between two variables, an association that can best be depicted by a straight line. It indicates the magnitude (e.g., amount of relationship) as well as the direction of the relationship. Figure 4.1 illustrates these aspects of the correlation coefficient. As you can see, the correlation coefficient can have either a positive or negative direction and a magnitude between –1.00 and +1.00.

Negative				Positive		
−1.0	−0.7	−0.3	0	0.3	0.7	1.0
Perfect	Moderately high	Low	Zero	Low	Moderately high	Perfect

Figure 4.1 Attributes of *r*.

© Human Kinetics

Correlation coefficients are used to look at the relationship between variables, such as performance on the pneumatic leg press test and muscular power.

The terms *high* and *low* are subjective and are affected by how the correlation was obtained, by the participant measured, by the variability in the data, and by how the correlation will be used. The sign of *r* simply indicates how the two variables covary (i.e., go together). A positive *r* indicates that participants scoring above the mean on one variable, *X*, will usually be above the mean on the second variable, *Y*. A negative *r* indicates that those scoring above the mean on *X* will generally be below the mean on *Y*. Thus, a correlation of −.5 is not lower than one of +.5. In fact, they are equal in strength but opposite in direction.

Mastery Item 4.1

If you were calculating a correlation coefficient and computed it to be +1.5, what must have happened?

Let's examine the factors of direction and magnitude of the correlation coefficient. Table 4.1 provides the scores for 10 students on three measures: body weight, chin-ups, and pull-ups. As you can see by examining the data, low values for chin-ups are generally paired with low values for pull-ups. If you were to calculate the correlation between these values (described later), you would find *r* to be indicating a **direct relationship** (positive relationship). In contrast, if you examine body weight versus pull-ups, you can see that high body weights are generally paired with low pull-up scores. These measures have an **inverse relationship** (also called a negative relationship). Figures 4.2 and 4.3 provide illustrations of these relationships in the form of scatterplots. A **scatterplot** is a graphic representation of the correlation between two variables. You can create a scatterplot by labeling the axes with the names and units of measurement for each variable, then plotting each pair of scores for each participant.

Table 4.1 Sample Correlation Data

Participant	Body weight	Chin-ups	Pull-ups
1	130	10	8
2	130	9	7
3	140	15	12
4	150	9	10
5	150	7	6
6	160	5	3
7	160	3	4
8	160	8	7
9	170	4	5
10	170	6	3

Note: During a pull-up, palms face away from the person; during a chin-up, palms face toward the person.

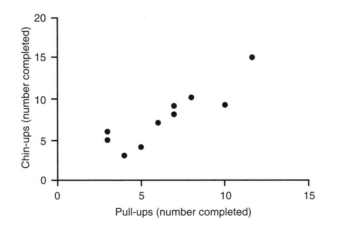

Figure 4.2 Scatterplot of correlation between pull-ups and chin-ups.

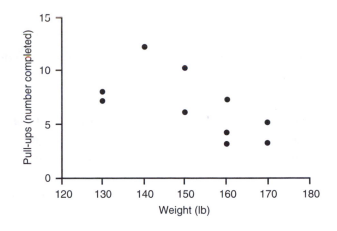

Figure 4.3 Scatterplot of correlation between body weight and pull-ups.

Keep in mind that the correlation coefficient is an index of linear relationship. All paired points must be on a straight line for a correlation to be perfect (–1 or +1). If two variables have a correlation coefficient of 0, or **zero correlation**, their scatterplot would demonstrate nothing even resembling a straight line; in fact, it would look more like a circle of dots. Figure 4.4 illustrates a sample scatterplot for a zero correlation.

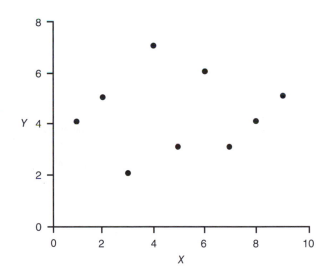

Figure 4.4 Scatterplot of zero correlation.

WWW! **Go to HK*Propel* to complete Student Activity 4.1.**

CALCULATING r

Having developed a fundamental understanding of the correlation coefficient and the ability to illustrate correlations with simple scatterplots, let us turn to the calculation of r.

Steps in Calculating r

1. Arrange your data in paired columns (here, X is the number of chin-ups, and Y is the number of pull-ups).
2. Square each X value and each Y value and place the results in two additional columns, X^2 and Y^2.
3. Multiply each X by its corresponding Y and place the results in a new column (called cross product, or XY).
4. Determine the sum for each column (X, Y, X^2, Y^2, XY).
5. Use the following formula to calculate r:

$$r = \frac{n(\Sigma XY) - (\Sigma X)(\Sigma Y)}{\sqrt{[n(\Sigma X^2) - (\Sigma X)^2][n(\Sigma Y^2) - (\Sigma Y)^2]}} \tag{4.1}$$

Note that everyone must have two scores. If a student completes only one test, his or her score cannot be used in the calculation of r.

Using the data in table 4.2, follow the steps here to calculate the correlation coefficient between chin-ups and pull-ups.

$$r = \frac{10(576) - (76)(65)}{\sqrt{[(10 \times 686 - (76)^2)(10 \times 501 - (65)^2)]}} = .89$$

Table 4.2 Calculating the Correlation Coefficient

Participant	CHIN-UPS		PULL-UPS		
	X	X²	Y	Y²	XY
1	10	100	8	64	80
2	9	81	7	49	63
3	15	225	12	144	180
4	9	81	10	100	90
5	7	49	6	36	42
6	5	25	3	9	15
7	3	9	4	16	12
8	8	64	7	49	56
9	4	16	5	25	20
10	6	36	3	9	18
Σ	76	686	65	501	576

Mastery Item 4.2

How would you describe the magnitude of the correlation between chin-ups and pull-ups? (Hint: Examine figures 4.1 and 4.2.)

The following is another example of calculating *r*. Because obesity is a major health risk factor, researchers were interested in determining a simple way to measure body fat. They first thought to use body weight as a measure, but soon realized that body weight is biased by height. Therefore, they used a new measure called body mass index (BMI), which adjusts the height effect:

$$BMI = Weight\ in\ kg\ /\ (Height\ in\ m)^2$$

To examine if BMI can be used in prediction, they recruited 15 participants for a pilot study. They measured the participants' height (in cm), weight (in kg), and body fat percent (using underwater weighing, a classic and accurate method). Table 4.3 summarizes the data they collected.

Table 4.3 Pilot Study Data for Computing *r*

ID	Height (cm)	Weight (kg)	Fat (%)
1	176	64	17.55
2	167	68	35.77
3	170	67	29.55
4	166	57	16.84
5	166	65	20.08
6	161	53	12.66
7	154	54	24.31
8	159	46	12.78
9	167	60	26.19
10	165	56	26.28
11	162	56	23.68
12	164	52	16.1
13	166	66	34.81
14	175	67	20.53
15	166	53	26.41

Let us use SPSS to compute *r* this time:

1. Start SPSS.
2. Open table 4.3 data.
3. Compute BMI using the SPSS Compute Variable function you learned in chapter 2.
   a. Click Transform → Compute Variable
   b. Type "BMI" in the Target Variable
   c. Insert "weight / (height/100)²" [SPSS expression: weight/(height/100)**2] in the Numeric Expression box
4. Click Analyze → Correlate → Bivariate.
5. Put both "BMI" and "FAT" in the Variables box.
6. Make sure "Pearson Correlation Coefficients," "Two-tailed," and "Flag significant correlations" were selected.
7. Click OK.

The correlation between BMI and body fat percent is .70, as illustrated in figure 4.5.

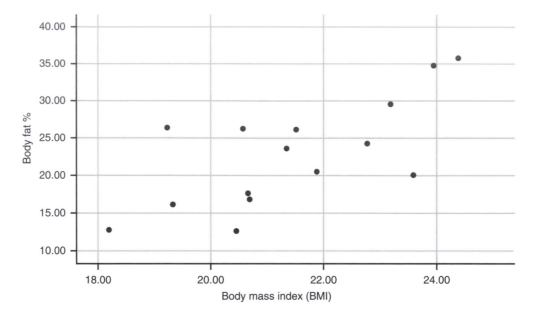

Figure 4.5 Scatterplot illustrating the relationship between BMI and body fat percent.

 Go to HK*Propel* to complete Student Activities 4.2, 4.3, and 4.4.

INTERPRETING *r*

When examining a correlation coefficient, we usually check the direction first and then the absolute value of the coefficient. Recall that *r* can range from –1 to +1. In practice, we express this theoretical range as ±1 and evaluate it as two separate parts: the sign (i.e., positive or negative) and the value (0 to 1), respectively. A positive sign, which is usually omitted in expressions, means that when one variable increases, the values on the other variable generally increase; a negative sign, in contrast, means that when one variable increases, the values on the other variable generally decrease. The *r* = .7 in our example indicates that the relationship between BMI and fat% is a positive relationship.

A negative correlation coefficient can occur for two reasons. First, a negative correlation coefficient can result from two measures having opposite scoring scales to measure the same thing. For example, when measuring aerobic capacity, the distance covered in a 12-minute run (measure 1) and the time required to run 1.5 miles (2.4 km, measure 2) would be negatively correlated. Runners with better endurance will cover more distance (i.e., *greater* distance values) on the 12-minute run; the same runners will run 1.5 miles (2.4 km) faster (i.e., *lower* time values). Runners with poorer endurance will have opposite results.

A second reason for a negative correlation coefficient is that two measures, which measure different things, can have a true negative relationship. A good example is provided by the measures of body weight and pull-ups in table 4.1. In general, heavier people have a more difficult time lifting or moving their body weight than do lighter people.

It is important to remember that a positive or negative sign itself tells nothing about the degree of the correlation. It only indicates how the two variables are related. The sign also has little to do with the directions of the traits measured by the variables. For example, when conducting a running field test to determine $\dot{V}O_2$max, one can obtain a positive correlation if the measurement used is the distance run within a fixed time, or a **negative correlation** if the measurement used is the time to run a fixed distance.

The second part of the interpretation of a correlation coefficient is its value, or degree of the correlation, which could range from 0 to 1 in absolute value. There are three ways we can evaluate the value of a correlation coefficient: (1) a general rule of thumb, (2) a coefficient of determination, or (3) a significance test.

The General Rule of Thumb

The general rule of thumb, summarized below, was developed based on many years of research. Remember, it is the absolute value of the correlation that you are going to interpret.

.80-1.00	High relationship
.60-.79	Moderately high relationship
.40-.59	Moderate relationship
.20-.39	Low relationship
.00-.19	No relationship

Applying the general rule is very straightforward: We simply calculate the correlation coefficient and determine which category it falls into. For example, the coefficient .70 we obtained in our previous example falls into the category of "moderately high relationship." We can, therefore, conclude that BMI has a moderately high positive correlation with body fat percent and, as a result, can be used to predict body fat percent. It should be pointed out that, as is common with any general rule, it is subject to change according to the nature of the relationship being studied, and the interpretation may be misleading under certain circumstances. This will be elaborated on later in this chapter.

Coefficient of Determination

An additional statistic that provides further information about the relationship between two measures is r^2, called the **coefficient of determination**. This value represents the percentage of shared variance between the two measures in question. To understand shared variance, let us examine a specific example. If the correlation between a distance-run test and $\dot{V}O_2$max is $r = .90$, then r^2 would be .81; the percentage of shared variance would therefore be 81% (.81 × 100). This means that performance in the distance run accounts for 81% of the variation in $\dot{V}O_2$max values. Therefore, the remaining 19% of the variance in $\dot{V}O_2$max values is the nonpredicted variance that is unexplained by performance in the distance run. This is error or residual variance, or variance remaining after you have used a predictor (X) to account for variation in the criterion (Y) variable. The coefficient of determination is important in statistics and measurement because it reflects the amount of variation found in one variable that can be predicted from another variable. Figure 4.6 illustrates the variation in $\dot{V}O_2$max that can be predicted from distance run. Note the similarity here to variance, presented in chapter 3. In the BMI and fat% example, the coefficient of determination is $(.7)^2 = .49$, indicating that about 50% of fat% can be predicted by BMI.

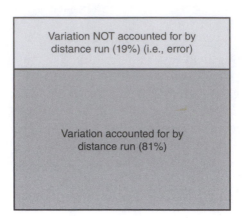

Figure 4.6 Variation in $\dot{V}O_2$max accounted for by distance run when $r = .90$.

Significance Test

Finally, it is possible to test whether the value of a correlation coefficient is statistically significant. This can be done using SPSS. Recall that this information was included in the SPSS report when we previously computed r. For the correlation coefficient of .70 between BMI and body fat percent, the corresponding Sig. value, known as p-value, is .004. Since this is less than .05, we can conclude that .70 is statistically significant. We will learn much more about statistical hypothesis testing in chapter 5. Although the significance test method is widely used in practice, it could be biased by the sample size employed in the computation; that is, a very low correlation can still become statistically "significant" when a large sample size is used. For example, when the sample size is 92, a low correlation coefficient of .21 (the correlation of determination = $.21^2 \approx 4\%$) will be "significant (p <.05)." The interpretation of the value of a correlation coefficient based on this method, therefore, must be used with great caution.

WWW! **Go to HK*Propel* to complete Student Activity 4.5.**

Mastery Item 4.3

What would be the reason for a negative correlation between marathon run times and $\dot{V}O_2$max?

Limitations of *r*

The correlation coefficient is an index of the linear relationship between two variables. If two variables happen to have a **curvilinear relationship**, such as the one depicted in figure 4.7, the PPM correlation coefficient would be near 0, indicating no *linear* relationship between the two variables. However, to say that there is no relationship at all between the variables would obviously be incorrect. Both low and high arousal scores are related to lower performance, and midrange arousal scores are associated with higher performance. This limitation of r is one reason for graphing the relationships between variables using the scatterplot technique.

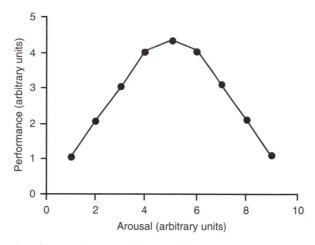

Figure 4.7 Graph of curvilinear relationship between arousal and performance.

A second (and very important) limitation of r is that correlation is not necessarily an indication of a cause-and-effect relationship. Even if the correlation coefficient for two variables is +1 or –1, a conclusion, on the basis of r alone, that one variable is a cause of a measurable effect in another is incorrect. Some third variable may be the cause of the relationship detected by a high r value. For example, if we found the correlation coefficient between athletes' body weight and **power** (the amount of work performed in a fixed amount of time) to be close to +1.00 and assumed a cause-and-effect relationship, we might be tempted to use the following logic: (a) Higher body weights cause higher power; therefore (b) anyone who needs power should gain weight. This logic is not completely right since much of the athletes' weight consists of muscle, which is the third variable here. If people are simply to add lots of fat to their weight, they will not increase in power but instead become so heavy that they would suffer decreases in power. Consider a second example; the PPM correlation coefficient between shoe size and grade level for elementary school children would be high. However, having big feet does not cause a child to be in a higher grade, nor does being in fifth grade cause a child to have big feet. Obviously, both variables are highly related to age, and this is the variable responsible for the high r between shoe size and grade level. Although the lack of a cause-and-effect relationship is relatively easy to explain in this example, the third variable may not be so apparent in most situations.

WWW **Go to HK*Propel* to complete Student Activity 4.6.**

Another limitation of r is the effect of the variance or range of the data on the magnitude of r. Variables with larger variances or ranges tend to have higher r values than variables with restricted variances or ranges. Figure 4.8 demonstrates this phenomenon. The value of r in either the solid or open dots in the figure is much smaller than the value of r for the combined set of data. This is because the variance is greater in the complete set of data than it is in either subset. The take-home message is that the interpretation of the value of r goes beyond just noting its magnitude. Consideration must also be given to other issues, such as linearity, other related variables, and the variability of the data involved.

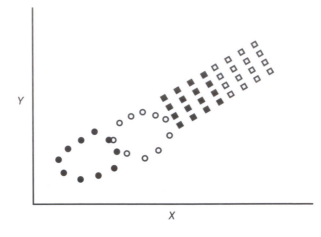

Figure 4.8 Correlation example of range restriction.

WWW. **Go to HK*Propel* to complete Student Activity 4.7.**

Other Correlation Coefficients

Fortunately, other correlations are available when *r* is not appropriate (e.g., when the sample is not random, one of the variables does not represent continuous data, or the observations were not independent). Some examples of such alternative correlation techniques include Spearman's rank correlation coefficient, point-biserial correlation, biserial correlation, phi coefficient, tetrachoric correlation, and Kendall's tau coefficient. We will introduce Spearman's rank correlation coefficient in this chapter; information about other correlation coefficients can be found in the suggested readings (Chen and Popovich, 2002; Mood et al., 2019).

Often the data we collect are ordinal rather than based on a continuous scale (e.g., class or sports team rankings). Furthermore, we sometimes purposely change continuous data into rank order when some required conditions for *r* are not met (e.g., two variables should be linearly related). When the data are expressed in the form of ranks, Spearman's rank correlation coefficient, known also as Spearman's ρ (read as "rho"), should be employed:

$$\rho = 1 - \frac{6(\sum D^2)}{N(N^2 - 1)} \qquad (4.2)$$

where D = the difference between the two ranks for each participant,

D² = the square of the difference between ranks for a participant,

ΣD² = the sum of D² for all participants, and

N = the number of pairs.

Let us use the data in table 4.4 to illustrate how Spearman's rank correlation coefficient is computed. A researcher was interested in determining the correlation between two commonly used upper-body strength measures, push-ups and pull-ups. Eight participants performed both measures, which were measured on a continuous scale (i.e., by the number of push-ups or pull-ups completed). Because the distribution of pull-ups data is often positively skewed, the researcher decided to compute Spearman's rank correlation coefficient for the data.

Table 4.4 Pull-Ups and Push-Ups Data and Related Computational Results

Participant	Pull-ups (X)	Push-ups (Y)	Rank (X)	Rank (Y)	D	D²
A	10	15	1	2	−1	1
B	2	5	7	7	0	0
C	4	11	5.5	4	1.5	2.25
D	6	10	4	5	−1	1
E	7	14	3	3	0	0
F	1	3	8	8	0	0
G	9	16	2	1	1	1
H	4	8	5.5	6	−0.5	0.25
Σ	43	82			0	5.5

First, we rank the scores in each dataset respectively (see the third column of table 4.4). The highest score in each dataset would be given a rank of 1, the second highest 2, and so forth. In the pull-ups dataset, Participants C and H received the same score of 4. The two scores represent two rank positions, but because they are the same, the ranks for the two scores should be identical. These ranks are calculated by averaging the two rank positions—in this case, 5 and 6—and assigning the average rank of 5.5 to each score of 4. Without performing any type of calculation on the ranks, it is possible to obtain a general feeling about the degree of relationship between two variables from a visual examination of the ranks. For example, Participant A would be assigned a rank of 1 on pull-ups and a rank of 2 on push-ups; Participant F would be assigned ranks of 8 and 8, respectively. Although the ranks are not identical for each person, they are similar, which indicates there may be a substantial relationship between the two variables. Note that the sum of the D column also equals 0. This is always true, regardless of the dataset; this is a convenient way to check the numbers in the D column (see the fourth column of table 4.4). Using the derived rankings, equation 4.2 can be applied:

$$\rho = 1 - \frac{6(5.50)}{8(8^2 - 1)} = .935$$

The interpretation of Spearman's ρ is similar to r. We can therefore conclude that pull-up and push-up tests are highly correlated and that either can be used to measure upper-body strength.

PREDICTION

In science, one of the most meaningful research results is the successful forecast. A valuable use of correlations is in **prediction**—that is, estimating the value of one variable from that of another variable. From a mathematical standpoint, if there is a relationship between X and Y, then X can predict Y and vice versa (to some degree). Recall, this still does not mean that X and Y are causally related. To establish a cause-and-effect relationship between X and Y, another type of study and analysis would be necessary (i.e., establishing and testing a hypothesis with an experimental study).

Straight Line

As you may recall from high school geometry, any point on a plane marked with an x-axis and y-axis can be identified by its coordinates (X,Y) on the plane, and a straight line, which describes the relationship between X and Y, can be defined by the equation $\hat{Y} = bX + c$, where $\hat{Y}$ is the predicted value of Y, b is the slope of the line, and c is where the line intercepts the y-axis. The slope indicates how much Y changes for a unit change in X. The Y-intercept represents the value of Y when $X = 0$. In figure 4.9, we have marked six coordinate points: (-1,0), (0,1), (1,2), (2,3), (3,4), and (4,5); these points fall on a straight line. In figure 4.9, the Y-intercept is 1 and the slope of the line is 1. When all of the paired points do not fall in a straight line, a line of best fit can be plotted through the set of points—this is called the *regression line* (also called the *line of best fit* or *prediction line*).

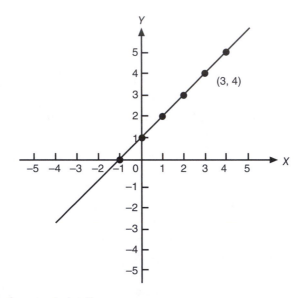

Figure 4.9 Plot of a straight line.

WWW **Go to HK*Propel* to complete Student Activity 4.8.**

Simple Linear Prediction

Simple linear prediction, also called *regression*, is a statistical method used to predict the criterion, outcome, or **dependent variable**, Y, from a single predictor or **independent variable**, X. If the two variables are correlated, which indicates that they have some amount of linear relationship, we can compute a prediction equation. The prediction equation has the same form as the equation for a straight line in plane geometry:

$$\hat{Y} = bX + c \tag{4.3}$$

We must think in terms of $\hat{Y}$ because unless the correlation between X and Y is –1.00 or +1.00, $\hat{Y}$ is only an estimate of Y; generally, $\hat{Y}$ will not equal Y. The following formulas are used for calculating b and c:

$$b = \frac{n(\Sigma XY) - (\Sigma X)(\Sigma Y)}{n(\Sigma X^2) - (\Sigma X)^2} \tag{4.4}$$

$$c = \bar{Y} - b\bar{X} \tag{4.5}$$

For example, using the data in table 4.1, we can calculate *b* and *c* and present the prediction equation for predicting pull-ups from body weight.

Vema, Sajwan, and Debnath (2009) provide an example of the use of simple linear prediction in human performance. They tested a sample of athletes attending the Lakshmibai National Institute of Physical Education and found the correlation coefficient between $\dot{V}O_2max$ (estimated from a 1-mile [1.6 km] run or walk test) and heart rate after a 3-minute step test to be –.70. They then calculated a linear prediction equation to predict $\dot{V}O_2max$ from heart rate for the athletes. Their equation is

$$\text{Predicted } \dot{V}O_2max = -0.24(X) + 114.38$$

where –0.24 is the slope, 114.38 is the *Y*-intercept, and *X* is the athlete's heart rate after the 3-minute step test. $\dot{V}O_2max$ is estimated in $ml \cdot kg^{-1} \cdot min^{-1}$. The 1-minute heart rate is estimated from a 15-second pulse immediately after completing the step test.

Mastery Item 4.4

Using the equation provided in figure 4.10, find the value of *Y* if *X* is 160.

Mastery Item 4.5

Use the data in table 4.1 to hand calculate *b* and *c* and provide the equation for predicting the number of chin-ups from body weight.

WWW **Go to HK*Propel* to complete Student Activity 4.9.**

Errors in Prediction

As stated earlier, unless the correlation coefficient is –1.00 or +1.00, $\hat{Y}$ will not necessarily equal *Y*. The following equation summarizes this:

$$E = Y - \hat{Y} \tag{4.6}$$

E (error) represents the inaccuracy of our predictions of *Y* based on the prediction equation. Figure 4.10 provides a demonstration of the error, or residual score. Participant 3 actually did 12 pull-ups, but the regression equation predicts 7.97 pull-ups for a participant weighing 140 pounds (63.6 kg). Thus, for this participant the residual score (or error) is 4.03.

Residual scores are important for several reasons. First, they represent pure error of estimation or prediction. If these errors can be minimized, prediction can be improved. Second, residual scores can represent lack of fit, which means that the independent variable does not predict a portion of the criterion (*Y*). The predictor should then be examined to reduce this problem. Perhaps more predictors could be added to reduce the error. Finally, residual scores could represent a pure measure of a trait with the predictor statistically removed. In our earlier example of the correlation between pull-ups and body weight, pull-ups could be predicted from body weight. The resulting residual score is interpreted as a person's ability to do pull-ups with body weight statistically controlled—in other words, if everyone was the same weight, how many pull-ups could you do? Positive residual scores would indicate that you did more pull-ups per unit weight than predicted. An important

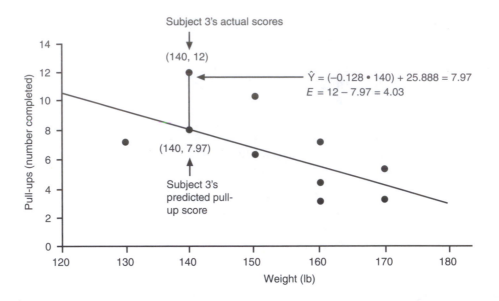

Figure 4.10 Error (residual) scores.

point to remember is that E has a zero correlation with X. This means that the prediction equation is equally accurate (or inaccurate) at any place along the X score scale.

The **standard error of estimate (SEE)**, also called the *standard error of prediction (SEP)* or simply the *standard error (Se)*, is a statistic that reflects the average amount of error in the process of predicting Y from X. Technically, it is the standard deviation of the error, or residual scores. The following formula is used for its calculation:

$$s_e = s_y \sqrt{1 - r^2} \qquad (4.7)$$

s_y is the standard deviation of the Y scores. If we use the data in table 4.1, the SEE for predicting pull-ups (Y) from body weight (X) would be as follows:

$$s_e = 2.953 \sqrt{1 - (.637)^2}$$
$$s_e = 2.953 \sqrt{1 - (.4058)}$$
$$s_e = 2.276$$

Because this is a standard deviation of the error or residual scores, it could be used as follows: If we predicted 7.97 pull-ups for someone weighing 140 pounds (63.6 kg), about 95% of people who weigh 140 pounds will have pull-up scores between approximately 12.5 and 3.4. Remember from chapter 3 that approximately 95% of scores are located ±2 standard deviations from the mean of a normal distribution and error scores are assumed to be normally distributed. Thus,

$$7.97 \pm 2 \times 2.276 = 12.5 \text{ to } 3.4$$

The standard deviation here is actually the SEE because SEE is the standard deviation of the errors of prediction. Note that 7.97 is also the mean value of pull-ups for those who weigh 140 pounds (63.6 kg).

Figure 4.11 illustrates all of these concepts, the line of best fit, the predicted value, and the SEE when estimating DXA-determined percentage of body fat from skinfolds.

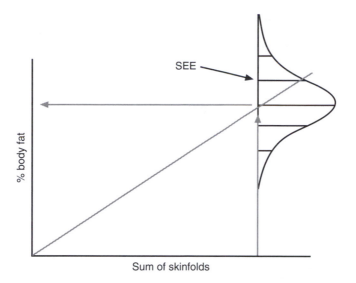

Figure 4.11 Predicting DXA-determined percentage of body fat from skinfolds.

WWW Go to HK*Propel* to complete Student Activity 4.10.

MULTIPLE REGRESSION

Correlation and prediction/regression are two interrelated topics that are based on the assumption that two variables have a linear relationship. We have examined the notion of simple linear prediction that has one predictor, X, of the criterion, Y. A more complex prediction of Y can be developed based on more than one predictor: X_1, X_2, and so on. This is called **multiple correlation**, multiple prediction, or multiple regression. The mathematics of this approach are much more complicated. If X and Y have a curvilinear relationship, then nonlinear regression can be used to predict Y from X.

The general form of the multiple correlation equation is presented below:

$$\hat{Y} = b_1 X_1 + b_2 X_2 + b_3 X_3 + \ldots b_p X_p + c \qquad (4.8)$$

where $\hat{Y}$ is the predicted value, b_1 represents the regression coefficient for variable X_1, b_2 represents the regression coefficient for variable X_2, and so on, and c represents the constant.

Although the mathematics involved in these techniques are beyond the scope of this book and will not be discussed here, we provide two examples of multiple regression equations. Jackson and Pollock (1978) published prediction equations that combined both multiple and nonlinear prediction. Their equation predicted hydrostatically measured body density (BD) of men from age (A), sum of skinfolds (SK), and sum of skinfolds squared (SK^2). The three predictors are the multiple predictors, and the sum of skinfolds squared is the nonlinear component in the prediction.

$$BD = 1.10938 - 0.0008267(\Sigma SK) + 0.0000016(\Sigma SK^2) - 0.0002574(A)$$

Using BD, one can then estimate percentage of body fat.

The second example is based on Kline et al. (1987). Their intent was to estimate $\dot{V}O_2max$ from weight, age, sex, time to complete a 1-mile (1.6 km) walk, and heart rate at the end of the walk. The correlation between measured and estimated $\dot{V}O_2max$ was .88 and the SEE was 5.0 ml · kg^{-1} · min^{-1}.

Their equation was as follows:

$$\text{Predicted } \dot{V}O_2\text{max} = (-0.0769)(\text{weight in lb}) + (-0.3877)(\text{age in yr}) + (6.3150)(\text{sex}) + (-3.2649)(\text{time in min}) + (-0.1565)(\text{heart rate})$$

Males are coded 1 and females are coded 0.

Can you see how the regression coefficients reflect the correlations between the predictors and the predicted value? That is, for increases in weight, age, time, and heart rate, there is a decrease in the predicted $\dot{V}O_2$max. Males, on average, have a $\dot{V}O_2$max 6.3150 units higher than females.

Dataset Application

Return to figure 4.6. This figure illustrates the relationship between $\dot{V}O_2$max and distance run. Consider what would happen to the variation accounted for by the run (and the resultant error variation) if additional variables were added to the regression equation. Assume that you added gender and age. What would happen to the amount of error illustrated in the figure? Why would this happen? What other variables might you include in the model to reduce the amount of error?

Use the chapter 4 large dataset in HK*Propel* to illustrate correlations. Calculate the PPM (go to Analyze → Correlate → Bivariate and move all of the variables to the right) to determine which variables are most related to weekly step counts. Interpret the resulting correlation matrix. Create scatterplots to illustrate the relationships (go to Graphs → Legacy Dialogs → Scatter/Dot and click on Simple Scatter and then Define. Select your x-axis and y-axis variables, then click OK). Create a multiple correlation (go to Analyze → Regression → Linear) and predict weekly step counts from gender and body mass index (BMI). Put "Weekly Steps" in the Dependent cell and "Gender" and "BMI" in the Independent(s) cell, then click OK. What happens to the correlation (and the ability to predict) and the SEE when you use additional predictors? Explain why. Finally, add weight to the prediction equation. Explain what happens to the ability to predict beyond that with gender and BMI only. Would you need or want to add weight to the prediction equation? Why?

MEASUREMENT AND EVALUATION CHALLENGE

Maaz now understands the statistical methods used to estimate his $\dot{V}O_2$max from treadmill time. The type of treadmill test that Maaz completed is called a Balke protocol. Research has shown the correlation between treadmill time with the Balke protocol and measured $\dot{V}O_2$max to be >.90. The prediction equation is $\dot{V}O_2$max = 14.99 + 1.444X (Pollock et al., 1976), where X equals treadmill minutes in decimal form. Note that $\dot{V}O_2$max is actually $\hat{Y}$ (i.e., the predicted value of Y based on X). Thus, Maaz, who ran for 24 minutes and 15 seconds (i.e., 24.25 min), has a predicted $\dot{V}O_2$max of 14.99 + 1.444(24.25) = 50 ml · kg^{-1} · min^{-1}. However, Maaz also realizes that there is some error in the prediction equation because the correlation is not perfect (i.e., ±1.00). The standard error of estimate (SEE) reflects the amount of error in the prediction equation. With this equation, the SEE is about 3 ml · kg^{-1} · min^{-1}. Thus, Maaz can be 68% confident that his actual $\dot{V}O_2$max is between 47 and 53 ml · kg^{-1} · min^{-1} (his predicted score ±1.00 SEE).

SUMMARY

The correlation and prediction statistics presented in this chapter lay the foundation for many necessary skills for the measurement and evaluation process. As you will see, these skills are necessary for generating measurement theory, as well as applying reliability and validity concepts to practical problems in exercise and human performance.

At this point you should be able to accomplish the following tasks:

- Calculate and interpret measures of correlation
- Calculate and interpret a prediction equation
- Calculate the SEE
- Use SPSS or Excel to enter data and to generate and interpret the following:
  - Correlation coefficients
  - Scatterplots of variables
  - Simple linear prediction equations

WWW Go to HK*Propel* for videos, homework assignments, and quizzes that will help you master this chapter's content.

CHAPTER 5

Inferential Statistics

OUTLINE

OBJECTIVES

After studying this chapter, you will be able to

- understand the scientific method and the hypotheses associated with it;

- perform an inferential statistical analysis to test a hypothesis; and

- use selected programs from the SPSS or Excel computer software in data analysis.

www The lecture outline in HK*Propel* will help you identify the major concepts of the chapter.

The descriptive statistical techniques that we have presented thus far are those that you will most commonly use for measurement problems; however, there are a number of other statistics that you may need to use in various measurement situations. The most common of these examines group differences. When these techniques are used to relate the characteristics of a small group (sample) to those of a larger group (population), they are referred to as **inferential statistics**, an extension of the correlational examples presented in chapter 4. Much of human performance research is conducted using inferential statistics.

For example, the researcher Logan read about is interested in determining if there is a relationship between the type of drink used and endurance performance: Does the drink produce a difference in endurance performance? Is endurance performance a function of what the athlete drinks? Draw a figure illustrating this relationship, with drink type on the horizontal axis and cycling performance on the vertical axis. Draw distributions of cycling performance for the two possible results: (1) there is no effect (no difference) and (2) there is an effect (a difference). Can you see how this is similar to examining the relation between variables?

HYPOTHESIS TESTING

The scientific method uses inferential statistics for obtaining knowledge. The **scientific method** requires both the development of a scientific hypothesis and an inferential statistical test of that hypothesis versus another competing hypothesis. A **hypothesis** is a statement of a presumed relation between at least two variables in a population. A **population** is the entire group of people or observations in question (e.g., college seniors). A measure of interest in the population is called a **parameter**. Inevitably, because entire populations are so large and unwieldy (imagine surveying all U.S. college seniors on a certain matter), you study hypotheses about a population by using a subgroup of the population, called a **sample**. The measure of the variable of interest in the sample is called a **statistic**. Using various techniques, you can make an inference—but not an absolute statement—about the whole population from your work with a sample. For example, surveys taken before presidential elections use samples to estimate the percentage of people preferring a particular candidate. Table 5.1 contains the symbols that are commonly used to distinguish sample statistics from population parameters.

Table 5.1   Statistical Symbols

Measures	Population parameters	Sample statistics
Mean	μ	*M*
Standard deviation	σ	*s*
Correlation	ρ	*r*

Consider the following example: A teacher was interested in the minutes of physical activity conducted in a typical physical education class (parameter) of fifth graders in the school. There were 200 fifth graders (population). The teacher randomly selected 50 students (sample) and had them wear pedometers that indicated minutes of moderate-to-vigorous physical activity (MVPA). The MVPA minutes were analyzed, and the sample values (statistic, e.g., the mean) were considered to be representative of the population parameter. Note, however, that there is *always* some error in these techniques.

Mastery Item 5.1

Create a research problem related to something of interest to you. Identify the following: (a) population, (b) sample, (c) parameter, and (d) statistic.

Hypotheses are the tools that allow research questions to be explored. A hypothesis may be one of several types:

• *Research hypothesis.* What the researcher actually believes will occur. For example, assume you believe that training method is related to oxygen uptake. Your research hypothesis is this: *There will be differences in oxygen uptake based on the type of aerobic training one uses.* Or perhaps you are a physical therapist who wants to determine if treatment modality A is more or less effective than treatment modality B. Your research hypothesis is this: *The two modalities are not equivalent.* You can investigate these hypotheses with a t-test or analysis of variance (ANOVA).

• *Null hypothesis (H_0).* A statement that there is no relation (association, relationship, or difference) between variables ($\mu_1 = \mu_2$). In the earlier example, your null hypothesis is that the mean oxygen uptake is not different for training groups that use different training methods. The null hypothesis is the one that you will actually test (and hope to discredit) using the techniques of inferential statistics.

• *Alternative hypothesis (H_1).* A statement that there is a relation (association, relationship, or difference) between variables, typically the converse of H_0. Here, your alternative hypothesis is $\mu_1 \neq \mu_2$ where μ_1 is the population mean for group 1 and μ_2 is the population mean for group 2. Remember that you actually obtain data from samples only, and then infer your results to the population. In this example, the research hypothesis is H_1.

Before you perform the appropriate statistical test, select a probability (*p*) level beyond which the results are considered to be statistically significant. This probability value is called the **significance**, or **alpha level** (α), and allows you to test the probability of the actual occurrence of your result. The alpha level is conventionally set at .05 or .01 (i.e., 5% or 1%). For example, if the researcher sets the alpha level at .05, he or she is saying that the probability of obtaining the statistic just by chance must be less than 5 times out of 100 before he or she will decide the null hypothesis is not tenable. You may recall from chapter 3 that 5% is in the extreme tails (2.5% on each side) of the normal curve. Review

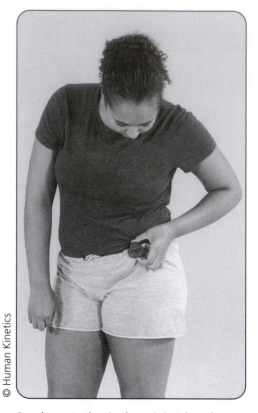

Students' physical activity levels can be measured with inexpensive pedometers. Data can then be used to make hypotheses about a larger group's level of activity.

the normal distribution presented in figure 3.6—about 2.5% of the distribution is outside about ±2 standard deviations. In effect, you assume no relation between variables until you have evidence to the contrary. The statistical data may provide the contrary evidence.

However, the researcher might reach an incorrect conclusion (i.e., say there is a relationship or difference when in fact there is not). The probability of making such an error is the alpha level. This error is referred to as a **type I error**. The alpha level is set at .05 or .01 to make the probability of a type I error extremely small. You can also make a second type of error, a **type II error**, by concluding that there is no relation between the variables in the population when in fact there truly is.

The SPSS computer program (or other statistical software) will calculate the actual alpha level for you. If the probability is less than the preset alpha level of .05 or .01, you conclude that there is a significant relation between the variables. Thus, H_0 is rejected and H_1 is accepted. Essentially, you assume that there is no relationship between the variables of interest (i.e., the null hypothesis is true), then you gather data. If the data you obtain is unlikely to have occurred if the null hypothesis is true (e.g., $p < .05$), then you reject the null hypothesis and conclude that the alternative hypothesis is the true state of the relationship in the population. Figure 5.1 illustrates the types of decisions and errors you might make. You can never know the true state of the null hypothesis in the population, so you always run the risk of making a type I or type II error. You cannot make both a type I and type II error in the same research study. Can you look at figure 5.1 and see why this is the case?

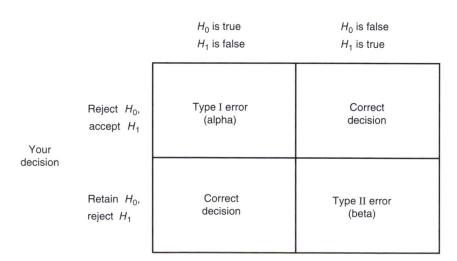

True state in population

	H_0 is true / H_1 is false	H_0 is false / H_1 is true
Reject H_0, accept H_1	Type I error (alpha)	Correct decision
Retain H_0, reject H_1	Correct decision	Type II error (beta)

Your decision

Figure 5.1 Type I and type II errors.

Mastery Item 5.2

A conditioning coach wants to study the best way to develop strength in older adults. He randomly divides a sample into three groups based on the following methods: free-weight strength training, machine strength training, and elastic band strength training. Write the appropriate null and alternative hypotheses for this problem.

Selection of the appropriate statistical technique is based on your research question and the level of measurement of the variables. The number of groups and the level of measurement of the data determine the appropriate statistical procedure to use. Some of the most common are these:

- *chi-square test* (χ^2). Used to examine associations in nominal data.
- *t-test.* Used to examine a difference in a continuous (interval or ratio) dependent variable between two (and only two) groups.
- *ANOVA (analysis of variance).* Used to examine differences in a continuous (interval or ratio) dependent variable among two or more groups (i.e., levels of the independent variable).

INDEPENDENT AND DEPENDENT VARIABLES

The differences between independent and dependent variables are important. The dependent variable is the criterion variable; its existence is the reason you are conducting the research study. The independent variable exists solely to determine if it is related to (or influences) the dependent variable. Independent and dependent variables can be characterized in a number of ways; these are presented in table 5.2.

Table 5.2 Variable Classification

Independent	Dependent
Presumed cause	Presumed effect
Antecedent	Consequence
Manipulated or measured by researcher	Outcome (measured)
Predicted from	Predicted to
Predictor	Criterion
X	Y

If the dependent variable is nominally scaled, the differences between groups (or cells) are measured by frequencies or proportions. If you are dealing with continuous (interval or ratio) data, then differences in mean values are often examined. For example, consider the strength-training example described in Mastery Item 5.2. The variable selected to measure the training effect is strength. This is the dependent variable. The independent variable is method of training and has three levels: free weight, machines, and elastic bands. Can you see why ANOVA would be used in this situation?

WWW **Go to HK*Propel* to complete Student Activity 5.1.**

OVERVIEW OF HYPOTHESES TESTING AND INFERENTIAL STATISTICS

All inferential statistical tests follow the same reasoning and processes:

1. Develop a research hypothesis about the relation between variables. For example, that there is a relation between the type of exercise in which you engage (e.g., moderate or vigorous) and $\dot{V}O_2$max. Alternatively, this could be stated that there is a difference in $\dot{V}O_2$max that depends on whether you engage in moderate or vigorous physical activity.

2. State a null (H_0) hypothesis, reflecting no relation (or difference): There is *no* relation (difference) between the type of exercise in which one engages and $\dot{V}O_2$max.

3. State an alternative hypothesis (H_1), which is the opposite of the null hypothesis. It is a direct reflection of the research hypothesis in step 1. Note: Hypotheses come in pairs and must cover all possible outcomes!

4. Gather data and analyze them based on the research question and the types of variables.

5. Make a decision based on the probability of the null hypothesis being correct given the data you have collected.

6. Compute effect size to further examine the results obtained with the hypothesis testing.

Note that if the null hypothesis is true, then the mean $\dot{V}O_2$max for moderate and vigorous groups would be the same; that is, there is *no* relation between type of exercise and $\dot{V}O_2$max. Further, if the null is true, the difference between the two means would be zero! Recall the normal curve where zero is in the center. However, if the null hypothesis is *not* true, then the difference between the moderate and vigorous $\dot{V}O_2$max means will be nonzero. Consider if the nonzero value was in an extreme tail of the normal distribution. This would suggest that this nonzero finding would be extremely rare *if the null hypothesis is true*. Thus, you conclude that the null hypothesis is *not* true and believe that the alternative hypothesis is true. By doing so, you reject the null hypothesis and conclude that the alternative hypothesis is a more accurate reflection of the data.

This same logic is used regardless of the statistical test that you conduct (i.e., chi-square, t-test, or ANOVA). The chi-square (χ^2), t-test (t), and ANOVA (F) use different distributions, but they are akin and related to the normal distribution. Thus, you can think of the χ^2, t, and F as if they are z-scores—they are not, but they are closely related. When you have a large (positive or negative) z-score, way out in the tail of the distribution, this is a rare occurrence. Thus, in hypothesis testing, if you obtain a χ^2, t, or F that would be rare (e.g., <5 times in 100), you conclude that the null hypothesis is not true. Computer programs typically report the probability associated with the χ^2, t, or F. This is interpreted as the likelihood of obtaining a value this extreme *if the null hypothesis is true*. The researcher would reject the null and conclude that there is a significant relation (or difference) between the levels of the independent variable and the dependent variable if this probability is low (often <.05 or <.01).

This logic can be extended to the most sophisticated statistical inference. In fact, many research studies in human performance use this logic. You will typically see probabilities reported in research reports. Effectively, the researcher sets up a falsifiable hypothesis (the null) and then collects and analyzes data and makes a decision about the truth of the null (or its alternative) based on the sample data.

EFFECT SIZE

Because probability, as briefly mentioned in chapter 4, can be affected by the sample size (e.g., the probability of a small *r* could be less than .05, and therefore be statistically significant when a large sample size was employed in the study), effect sizes should be computed and examined simultaneously when interpreting the findings of a statistical test. The effect size is a quantitative measure of the magnitude of the experimental effect or relationship being examined and can be computed by specific methods associated with a particular statistical test.

SELECTED STATISTICAL TESTS

The following are statistical tests that examine associations or differences between groups. The techniques selected represent common basic inferential tests.

Chi-square (χ^2)

Purpose: To determine if there is an association between levels (cells) of two nominally scaled variables.

Example: An aerobics instructor is teaching two classes, aerobic dance and circuit weight training. The instructor wants to know if the proportion of males and females is the same in each class. The null hypothesis is that there is no association (relation) between gender and type of class in which students are enrolled. The alternative hypothesis is that there is an association. One can only reject the null hypothesis and believe that the alternative hypothesis is the true state of circumstances in the population when the probability of the null hypothesis being true is small (i.e., $p < .05$) based on the sample data. The data are found in table 5.3. Use them in conjunction with the following SPSS commands to calculate the chi-square test of association and verify the results with those presented in figure 5.2. (An Excel template can be found in HK*Propel* in chapters 5 and 7.)

1. Start SPSS.
2. Open table 5.3 data.
3. Click on the Analyze menu.
4. Scroll down to Descriptive Statistics and across to Crosstabs and click.
5. Put "Class" in the rows and "Gender" in the columns by clicking the arrow keys.
6. Click Statistics.
7. Check the Chi-Square box.
8. Check the Phi and Cramer's V box.
9. Click Continue.
10. Click OK.

The resulting SPSS printout is presented as figure 5.2. Although a number of statistics are calculated for us, the test statistic in which we are interested is the Pearson chi-square. The observed chi-square value is 22.5. Think of this chi-square value as if it is a *z*-score (again, it is not, but it is related). Where is a *z*-score of 22.5? It is way out on the extreme end of the normal distribution. This is an unlikely occurrence, particularly if the null hypothesis is true and there is no relation between gender and class. The probability associated with this test is reported to be .000 (labeled as Asymptotic Significance [2-sided] on the SPSS crosstabs output). This is the probability of the cell distribution occurring as it did *if the* null hypothesis is true. In actuality, however, you can never have a probability of zero. It is simply the case that the SPSS program calculates the probability (i.e., significance) to three decimal places. In any case, you should interpret this as .001. Thus, it is exceedingly rare to find the cell frequencies being distributed as they are when, in fact, there is no association between type of class and gender (the H_0). Because of this exceedingly small probability, the aerobics instructor can conclude that there is an association between gender and type of class. The null hypothesis (H_0) of no association is rejected, and the instructor concludes that there is an association between gender and type of class in which a person enrolls. Figure 5.2 shows that 10 of the 12 males registered for circuit weight training,

Table 5.3 Data Entry for χ^2 Example

ID	Gender	Class
1	1	1
2	1	1
3	1	1
4	1	1
5	1	1
6	1	1
7	1	1
8	1	1
9	1	1
10	1	1
11	1	2
12	1	2
13	2	2
14	2	2
15	2	2
16	2	2
17	2	2
18	2	2
19	2	2
20	2	2
21	2	2
22	2	2
23	2	2
24	2	2
25	2	2
26	2	2
27	2	2
28	2	2
29	2	2
30	2	2

Note: Gender code: 1 = male, 2 = female. Class code: 1 = circuit, 2 = dance.

Case processing summary

	Cases					
	Valid		Missing		Total	
	N	Percent	N	Percent	N	Percent
Class enrollment * Gender of the subject	30	100.0%	0	0.0%	30	100.0%

Class enrollment * Gender of the subject crosstabulation

Count

		Gender of the subject		Total
		Male	Female	
Class enrollment	Circuit weight training	10	0	10
	Aerobic dance	2	18	20
Total		12	18	30

Chi-square tests

	Value	df	Asymp. sig. (2-sided)	Exact sig. (2-sided)	Exact sig. (1-sided)
Pearson chi-square	22.500[a]	1	.000		
Continuity correction[b]	18.906	1	.000		
Likelihood ratio	27.377	1	.000		
Fisher's exact test				.000	.000
Linear-by-linear association	21.750	1	.000		
N of valid cases	30				

a. 1 cells (25.0%) have expected count less than 5. The minimum expected count is 4.00.
b. Computed only for a 2x2 table

Symmetric measures

		Value	Approximate significance
Nominal by nominal	Phi	.866	.000
	Cramer's V	.866	.000
N of valid cases		30	

Figure 5.2 SPSS crosstabs output.

whereas all 18 of the 18 females registered for aerobic dance. This association might help the instructor in planning the types of activities for the classes.

The effect size of a chi-square test can be determined by computing Cramer's V (V):

$$V = \sqrt{\frac{x^2}{n \cdot df}} \tag{5.1}$$

where

χ^2 = chi-square value,

n = the number of observations,

df = the minimum value of r – 1 (number of rows – 1) or c – 1 (number of columns – 1), and

r = number of rows and c = number of columns in the contingency table.

Therefore, df of the above chi-square test is min(2 – 1, 2 – 1) = 1 and the effect size is:

$$V = \sqrt{\frac{22.50}{30 \cdot 1}} = \sqrt{0.75} = 0.866$$

V can be evaluated using table 5.4.

We can conclude that the effect size is large, which further supports the finding of the association between gender and types of classes. This Cramer's *V* can be found in figure 5.2.

Table 5.4 Effect Size for Cramer's *V* Based on Degrees of Freedom

df*	Small	Medium	Large
1	.10	.30	.50
2	.07	.21	.35
3	.06	.17	.29
4	.05	.15	.25
5	.04	.13	.22

Reprinted from www.real-statistics.com/chi-square-and-f-distributions/effect-size-chi-square

t-test for Two Independent Groups

Purpose: To examine the difference in one continuous dependent variable between two (and only two) independent groups. Independent groups are groups that are not in any way related (e.g., boys and girls; experimental and control groups).

Example: A high school volleyball coach is selecting players for the varsity team using serving accuracy as a selection factor. After the team is selected, the coach wants to quantify the differences in serving accuracy between varsity and subvarsity players. The serving scores are presented in table 5.5.

Table 5.5 Serving Scores

Varsity	20, 18, 17, 19, 20, 16, 18, 19
Subvarsity	16, 15, 17, 14, 15, 13, 14, 12

The research hypothesis for this study is that there will be a difference in serving accuracy between varsity (*v*) and subvarsity (*sv*) players. The null hypothesis to be tested is that the mean volleyball serving score for the varsity players will be equal to the mean for the subvarsity players (i.e., there is no difference between the two):

$$H_0: \mu_v = \mu_{sv} \tag{5.2}$$

The alternative hypothesis is that the mean for the varsity players will not be equal to the mean of the subvarsity players:

$$H_1: \mu_v \neq \mu_{sv} \tag{5.3}$$

For the coach's purpose, the alpha level is set at .05. The SPSS procedure for the t-test may be used to analyze the data. Use the data in table 5.5 to calculate an independent group's t-test and confirm your results with those presented in figure 5.3.

1. Start SPSS.
2. Open table 5.5 data.
3. Click on the Analyze menu.
4. Scroll down to Compare Means and across to Independent-Samples T Test and click.
5. Put "Score" in the Test Variable(s) box by clicking the arrow key.

6. Put "Group" in the Grouping Variable box by clicking the arrow key.
7. Click the Define Groups button.
8. Put "1" in the Group 1 box and "2" in the Group 2 box.
9. Click Continue.
10. Click OK.

Group statistics

	Team level	N	Mean	Std. deviation	Std. error mean
Serving score	Varsity	8	18.38	1.408	.498
	Subvarsity	8	14.50	1.604	.567

Independent samples test

		Levene's Test for Equality of Variances		t-test for Equality of Means					95% confidence interval of the difference	
		F	Sig.	t	df	Sig. (2–tailed)	Mean difference	Std. error difference	Lower	Upper
Serving score	Equal variances assumed	.095	.763	5.136	14	.000	3.88	.754	2.257	5.493
	Equal variances not assumed			5.136	13.769	.000	3.88	.754	2.254	5.496

Figure 5.3 Sample t-test output: group statistics and independent samples test.

SPSS output is displayed in figure 5.3. Inspection of the means indicates that the varsity players (group 1; mean = 18.38) served significantly (Sig. [2-tailed] = .000) better than the subvarsity players (group 2; mean = 14.50).

A number of statistics in the output from the t-test are beyond the scope of this text. For our purposes, you can ignore the results under Levene's Test for Equality of Variances. Focus your attention on the areas below the heading t-test for Equality of Means. Notice the *t* presented with a value of 5.136 (actually presented twice). Think of this *t* as if it is a *z*-score that you learned about in chapter 3 (again, it is not a *z*-score, but similar). If the *z*-score was large (e.g., greater than 3 in absolute value), you know that the probability of finding a value this large is small. The same can be said of the calculated t-value. Thus, you can see that the *t* is relatively far into the tail of the distribution—generally a rare occurrence. Of most importance to you is the box titled Sig. (2-tailed). This is the probability that the null hypothesis is true given the data from the sample. Because the probability is less than .05, the coach rejects the null hypothesis and retains the alternative hypothesis. This is an example of using a t-test with independent groups.

Excel commands for the independent t-test are found in the appendix.

The effect size of the t-test for two independent groups can be computed using Cohen's D formula:

$$D = \frac{M_1 - M_2}{S_p} \tag{5.4}$$

where M_1 and M_2 are the sample means of Group 1 and 2, respectively, and S_p is the pooled estimated standard deviation. S_p can be estimated using the formula below:

$$S_p = \sqrt{\frac{(s_1^2 + s_2^2)}{2}}$$

(5.5)

So, for our teams' serving accuracy example,

$$S_p = \sqrt{\frac{(1.408^2 + 1.604^2)}{2}} = 1.509$$

$$D = \frac{18.38 - 14.50}{1.509} = 2.571$$

According to Cohen (1988), $D = 0.20$ indicates a small effect, 0.50 a medium effect, and 0.80 a large effect. Therefore, $D = 2.571$ represents a large effect, which supports the conclusion that there is a large difference in serving accuracy between two teams.

Dependent t-test for Paired Groups

Purpose: To compare two related (paired) groups on one dependent variable. Groups can be paired by matching them on some external characteristic (e.g., siblings) or by measuring the same group twice (i.e., pre- and postperformance).

Example: Let's extend the previous independent t-test example. Suppose the coach wanted to examine the serving accuracy of the varsity team during the preseason compared to the end of the season. The research hypothesis is that there will be a difference in preseason and postseason serving accuracy. The null hypothesis is that there will be no difference in the players' serving accuracy over the course of the season. To test the null hypothesis, the SPSS t-test procedure is used again. However, the data are entered differently from the previous example because each person is tested twice (compare tables 5.5 and 5.6). This allows SPSS to properly pair the data so that the correct result is calculated.

Use the data in table 5.6 to calculate a paired (dependent) t-test, and confirm your results with those presented in figure 5.4.

Table 5.6 Data Format for Paired t-test

Preseason	Postseason
18	20
20	24
17	20
16	19
15	20
18	22
19	21
17	21

Paired samples statistics

		Mean	N	Std. deviation	Std. error mean
Pair 1	Postseason performance	20.88	8	1.553	.549
	Preseason performance	17.50	8	1.604	.567

Paired samples correlations

		N	Correlation	Sig.
Pair 1	Postseason performance and preseason performance	8	.775	.024

Paired samples test

		Paired differences							
					95% confidence interval of the difference				
		Mean	Std. deviation	Std. error mean	Lower	Upper	t	df	Sig. (2–tailed)
Pair 1	Postseason performance and preseason performance	3.38	1.061	.375	2.49	4.26	9.000	7	.000

Figure 5.4 Sample t-test output: paired samples statistics, paired samples correlations, and paired samples test.

1. Start SPSS.
2. Open table 5.6 data.
3. Click on the Analyze menu.
4. Scroll down to Compare Means and across to Paired Samples T Test and click.
5. Put "Preseason" and "Postseason" in the Paired Variables box by clicking the arrow key.
6. Click OK.

The mean difference between the posttest and the pretest was found to be 3.38. The observed t-value was found to be 9.000, with an associated probability (alpha level) that approaches zero (Sig. [2-tailed]). Again, think of this t-value as if it were a z-score. A z-score of 9 is way out in the tail, an unlikely occurrence. Thus, the null hypothesis is rejected and the alternative hypothesis is adopted as the true state of differences. The coach can conclude that the differences in serving accuracy from the beginning of the season to the end can be attributed to something other than chance.

Excel commands for the paired t-test are found in the appendix.

The formula for effect size of the paired t-test is:

$$D = \frac{(M_1 - M_2)}{\sqrt{s_1^2 + s_2^2 - (2rs_1s_2)}} \tag{5.6}$$

where r is the correlation between preseason and postseason scores (.775 from figure 5.4). Using the means and SDs of pre- and postseason information and the correlation, D can be determined:

$$D = \frac{(M_1 - M_2)}{\sqrt{s_1^2 + s_2^2 - (2rs_1s_2)}} \frac{(20.88 - 17.50)}{\sqrt{1.553^2 + 1.604^2 - (2 \cdot 0.775 \cdot 1.553 \cdot 1.604)}} = 3.189$$

According to Cohen's standard, $D = 3.189$ represents a large effect size, which confirms the coach's conclusion on the differences in serving accuracy.

Note that the differences in serving accuracy could have been caused by a number of factors. Because a time lag occurred between the first and second measurements, the differences could have been attributable to growth, maturation, or some other factor not under the control of the researcher. In an actual experiment, these factors would have to be controlled.

WWW! **Go to HK*Propel* to complete Student Activity 5.2.**

One-Way ANOVA

Purpose: To examine group differences between one continuous (interval-scaled or ratio-scaled) dependent variable and one categorical (nominal-scaled or ordinal-scaled) independent variable. Unlike the t-test, ANOVA can handle independent variables with more than two levels (groups) of data.

Example: The data for this example are based on the strength-training example previously presented in Mastery Item 5.2. The study participants engage in one (and only one) of the three methods of strength training—free weights, machines, or elastic bands—for 8 weeks. In this example, the independent variable is the strength-training method (coded 1, 2, and 3), and the dependent variable is 1-repetition maximum (1RM) bench press (we realize that the researcher might also assess other strength measures). The problem to be examined is whether there are differences in bench press strength across the three training groups.

The null hypothesis is that means for the 1RM bench press for the three training method groups will be equivalent:

$$H_0: \mu_1 = \mu_2 = \mu_3 \tag{5.7}$$

The alternative hypothesis is that the means will not be equivalent (for at least one of the means):

$$H_1: \mu_1 \neq \mu_2 \neq \mu_3 \tag{5.8}$$

The alpha level was set at .01 rather than .05, indicating that the researcher wanted to reduce the probability of making a type I error and increase confidence that any differences found among the means were not attributable to chance. The data for this problem are presented in table 5.7.

Use the following SPSS procedures to conduct one-way ANOVA on these data:

1. Start SPSS.
2. Open table 5.7 data from HK*Propel*.
3. Click on the Analyze menu.
4. Scroll down to Compare Means and across to One-Way ANOVA and click.
5. Put "1RM Bench Press" in the Dependent List by clicking the arrow key.
6. Put "Group" in the Factor box.

Table 5.7 Input Format for One-Way ANOVA

ID	Group	1RM bench press
1	1	193
2	1	190
3	1	195
4	1	175
5	1	188
6	2	148
7	2	170
8	2	172
9	2	168
10	2	165
11	3	170
12	3	157
13	3	140
14	3	148
15	3	150

7. Click Post_Hoc.
8. Check Tukey.
9. Click Continue.
10. Click Options.
11. Check Descriptive.
12. Click Continue.
13. Click OK.

WWW **Go to HK*Propel* to complete Student Activity 5.3.**

The key information presented in figure 5.5 is the Sig. (significance or probability). The other information is used to obtain the significance. For ANOVA, the significance test is an F or F-ratio (17.145). Again, think of this F-ratio like the z-score you learned about in chapter 3 (not the same, but similar). High values are rare, and the probability of obtaining a high value is reduced when the groups do not differ much from each other. Because the probability level for the observed event is less than .01 (the computer reports it as .000), the null hypothesis is rejected and the alternative hypothesis is adopted. What type of statistical error is possible with this conclusion: type I or type II?

We have run an ANOVA and found that an overall difference exists among the means of three groups. But where does that difference lie? We compare each mean with every other mean and see where the difference occurs with a so-called post hoc test. There are many post hoc tests available using SPSS (e.g., Bonferroni, LSD, Tukey). We have chosen the most common method, the Tukey post hoc test.

Review figure 5.5. Note that the participants in the free-weight group had significantly higher 1RM bench press scores (M = 188.2 lb [85.5 kg]) than the participants from the machine-training group (M = 164.6 lb [74.8 kg]) and the participants from the elastic band group (M = 153 lb [69.5 kg]). Under the heading Multiple Comparisons, you can see this

Oneway Descriptives

1-RM bench press

	N	Mean	Std. deviation	Std. error mean	95% confidence interval for mean		Minimum	Maximum
					Lower bound	Upper bound		
Free-weight	5	188.20	7.855	3.513	178.45	197.95	175	195
Machine-training	5	164.60	9.633	4.308	152.64	176.56	148	172
Elastic band	5	153.00	11.269	5.040	139.01	166.99	140	170
Total	15	168.60	17.614	4.548	158.85	178.35	140	195

ANOVA

1-RM bench press

	Sum of squares	df	Mean square	F	Sig.
Between groups	3217.600	2	1608.800	17.145	.000
Within groups	1126.000	12	93.833		
Total	4343.600	14			

Post Hoc Tests

Multiple Comparisons

Dependent variable: 1-RM bench press

Tukey HSD		Mean difference (I-J)	Std. error	Sig.	95% Confidence interval	
(I) group	(J) group				Lower bound	Upper bound
Free-weight	machine-training	23.600*	6.126	.006	7.26	39.94
	elastic band	35.200*	6.126	.000	18.86	51.54
Machine-training	free-weight	-23.600*	6.126	.006	-39.94	7.26
	elastic band	11.600	6.126	.183	-4.74	27.94
Elastic band	free-weight	-35.200*	6.126	.000	-51.54	-18.86
	machine-training	-11.600*	6.126	.183	-27.94	4.74

* The mean difference is significant at the 0.05 level.

Figure 5.5 One-way ANOVA results.

represents significant pairwise differences (see under "Sig." column) between the free-weight group and machine-training group and between the free-weight group and elastic band group. There is no significant pairwise difference between the machine-training group and the elastic band group. This multiple comparison is important because it allows you to understand the source of the difference among more than two groups.

The effect size of one-way ANOVA can be determined by computing η^2 (eta squared):

$$\eta^2 = \frac{SS_{between}}{SS_{total}} \tag{5.9}$$

For our example,

$$\eta^2 = \frac{SS_{between}}{SS_{total}} = \frac{3217.6}{4343.6} = 0.741$$

There are two ways to interpret η^2:

1. It is the percentage of the total variance accounted for by the treatment effect. The larger the percentage, the stronger the treatment; for our example, this means 74% of the total variance was from the training method, which further supports the researcher's conclusion about the training method difference.

2. Using an evaluation criterion similar to Cohen's D, $\eta^2 = 0.01$ indicates a small effect, 0.06 a medium effect, and 0.14 a large effect. $\eta^2 = 0.74$ therefore indicates a very large effect.

Excel commands for one-way ANOVA can be found in the appendix.

Two-Way and N-Way ANOVA

The one-way ANOVA can be expanded by adding additional independent variables. For example, adding one additional independent variable makes it a two-way ANOVA. In our current example, the researcher might add gender to the variable list, for a total of two independent variables (training method with 3 levels and gender with 2 levels). A researcher might add yet another independent variable, days per week (2 or 3 days per week). This illustrates how one-way ANOVA can be expanded. The ANOVA can be expanded to an N-way ANOVA where N can be literally any number of independent variables. The actual statistical analyses are beyond the scope of this text, but the development of hypotheses is similar to those developed with the one-way ANOVA. If you are more interested in factorial ANOVAs (all independent variables are between-subjects variables) and repeated-measures ANOVAs (one or more independent variables is within-subjects variables), see Kang and Jin (2016a; 2016b). Figure 5.6 illustrates the design of the two-way and three-way ANOVAs presented here.

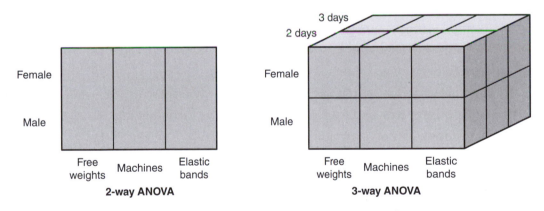

Figure 5.6 Example of two-way and three-way ANOVA designs.

MEASUREMENT AND EVALUATION CHALLENGE

After reading the research article, Logan learned that a t-test was conducted because there was a single independent variable (drink type); the control group drank water only and the experimental group drank the carbohydrate solution. He also learned that the researcher hypothesized that cycling duration (the dependent variable) was a function of which drink was consumed. Logan now knows that $p < .05$ means that a null hypothesis was rejected and an alternative hypothesis was accepted. Based on this sample, the researchers concluded that carbohydrate drinks will improve cycling endurance for most cyclists similar to those from whom this sample of cyclists was drawn. Logan realizes this conclusion is not a certainty because a type I error could have been made, but the probability of such an error is less than 5 in 100. Because the null hypothesis was rejected, it is impossible for the researcher to have made a type II error. Because inferential tests provide evidence only to support or not support the hypotheses, Logan has learned that the researcher's hypothesis about the influence of carbohydrate drinks can never be fully proven with hypothesis testing.

SUMMARY

This chapter provided a brief overview of the tests used in inferential statistics; however, many assumptions regarding these techniques were not discussed. Statistical tests of significance often obscure practical differences. There is no substitute for blending statistical findings with intuitive logic. If you are interested in more on statistical methods, see Mood, Morrow, and McQueen (2020). Thomas, Nelson, and Silverman (2015) also provide excellent examples of research in human performance.

At this point you should be able to accomplish the following tasks:

- Understand and interpret the scientific method
- Write and interpret research, null, and alternative hypotheses
- Use SPSS to enter data and to generate and interpret the following:
  - Chi-square tests of association
  - Independent and dependent t-tests
  - One-way ANOVA

WWW Go to HK*Propel* for videos, homework assignments, and quizzes that will help you master this chapter's content.

Part III

Reliability and Validity Theory

In this part, the information you have learned about basic statistics and computer applications will be extended and applied to issues related to valid decision making. Decisions you make in the field of human performance might be about a person's aerobic capacity, muscular strength, or amount of daily physical activity. You may also need to evaluate a program that you direct or report grades or achievement scores to students, clients, or program participants. Good decisions are based on sound data collected through some form of testing or assessment, which in turn reflects the reliability, objectivity, and validity of the test score or measurement tool employed. Each chapter in part III provides the opportunity to use the SPSS skills you gained in parts I and II to accomplish specific tasks related to these characteristics.

To make accurate decisions about individuals or groups, you must use data that are sufficiently reliable, objective, and valid, because invalid data can result in inappropriate decisions. Chapter 6 provides a general introduction to the concepts of reliability and validity and how to collect data to evaluate them. After introducing the concept of reliability and how to determine it, chapter 6 provides a historical introduction to the evolution of validity and then explains the latest evidence-centered view on validity. This is followed by a detailed description of validity evidence and its applications to human performance.

Chapter 7 addresses these issues from a classification perspective, which is known as criterion-referenced testing. How to determine these categories of measurement or test scores, a procedure known as setting cut scores or standards, is described in detail, including application examples in human performance. With the knowledge and skills learned in part III, your ability to make valid decisions will be enhanced.

CHAPTER 6

Reliability and Validity

OBJECTIVES

After studying this chapter, you will be able to

- discuss the concepts of reliability and validity;
- differentiate among the types of reliability and how to calculate them;
- identify the types of validity evidence that can be used to provide information about a measure's truthfulness and calculate the appropriate statistics;
- describe the relationship between reliability and validity and comment on why these concepts are important to measurement;
- evaluate the evidence for reliability and validity typically presented in the measurement of human performance; and
- use SPSS and Excel to calculate reliability and validity statistics.

WWW. **The lecture outline in HK*Propel* will help you identify the major concepts of the chapter.**

MEASUREMENT AND EVALUATION CHALLENGE

Eva, a YMCA fitness director, wants to assess the cardiorespiratory fitness of young adult members by determining $\dot{V}O_2$max. However, she has heard that the best measure of cardiorespiratory fitness is to have a person run on a treadmill until he or she is exhausted. This test requires collecting the air that participants expire while on the treadmill and therefore requires considerable and expensive equipment. Because of these difficulties, Eva is interested in using a surrogate (field test) measure such as the YMCA 3-Minute Step Test (see chapter 10) or a distance run (e.g., 1 mile or 1.6 kilometers) in place of treadmill performance. She has even heard that nonexercise models can be used to estimate one's $\dot{V}O_2$max. Eva wonders whether she should use that method to save time, money, and equipment, and reduce the risk to participants. However, Eva is concerned that a field test might lack the accuracy that a treadmill test provides. Eva's concern is a very real one. How can she determine if the score that is being obtained with the field test is reliable (i.e., consistent) and valid (i.e., truthful)?

Regardless of the area of human performance in which you work, you will need to make decisions based on the data you collect. Often, these decisions require you to make comparisons among people or report test results to a specific person. For example, Eva may need to report the results of her testing to the programming director or the Y's board of directors to keep a particular fitness program funded. Your decisions and reports therefore need to be accurate.

The accuracy of your decisions depends on whether the test can consistently measure what it is supposed to measure. As you learned in chapter 1, the most important measurement characteristics are objectivity, reliability, and validity. In fact, reliability and validity are the most important concepts presented in this textbook; the accuracy of the many computational, theoretical, and practical examples presented throughout the textbook can be traced back to these concepts. It should be noted that, in the past, reliability and validity were considered to be a feature of a test, but now they are considered to be characteristics of the resulting test scores. This is an important distinction. You will learn about reliability and validity in detail in this chapter; a special application of these concepts when a test is focusing on classification will be presented in chapter 7.

Reliability refers to the repeatability of scores resulting from the application of a testing procedure; it is the degree to which repeated measurements of the same trait are reproducible under the same conditions. Reliability can also be described as consistency, dependability, stability, and precision. A measure is said to be reliable if one obtains the same (or nearly the same) score each time the instrument is administered to the same person under the same conditions. Test reliability will be extremely important as Eva determines which field test to administer.

Validity is the degree of truthfulness of a test score—that is, does the test accurately measure what it is intended to measure? Validity is dependent on two characteristics: reliability and **relevance**, or the degree to which a test pertains to its objectives. Thus, for an instrument or test to be valid, it must measure the particular trait, characteristic, or ability consistently (i.e., reliably), and it must be related to the characteristic reportedly being measured.

Thus, you can see that reliability and validity are important concerns for Eva. She must be certain that the field test produces results that are consistent from trial to trial and also accurately estimates what the $\dot{V}O_2$max would be if it were measured on the treadmill. Can you see how a measure might be reliable yet not valid—that is, return consistent results that are unrelated to what you want to assess? However, if a measure is valid, it must certainly be reliable, because it cannot be valid if it does not measure the characteristic consistently.

Mastery Item 6.1

What variables might Eva obtain if she considered a nonexercise model to estimate $\dot{V}O_2$max? Be sure to consider the reliability and validity of the variables.

A test may be valid in one set of circumstances but not in another. There are many tests that have sufficient reliability but poor validity. For example, assessment of total body weight is typically a reliable measure. It changes little from day to day, and two evaluators would likely report the same or nearly the same value when measuring it. However, total body weight is not a valid measure of body fat because total body weight is made up of fat, bone, and lean tissue. Thus, one's weight depends on the relative proportions of these body components.

Objectivity is a special kind of reliability referring to interrater or intrarater reliability. You have probably taken objective tests (such as those with multiple-choice items) and subjective tests (such as those with essay items). These tests are classified as such because of the type of scoring system used when grading them. Multiple-choice, true–false, and matching test items are objective because they have a high amount of interrater reliability; that is, the scores on these types of items are consistent from one grader to another because there is a well-defined scoring system for the correct (or best) answer. However, a test can be objective in nature yet not accurate or reliable. If the questions are poorly written, a multiple-choice examination may be an unreliable and invalid measure of student knowledge.

The scoring of essay tests tends to be subjective—different readers will give different scores to an answer—but there are ways to increase the objectivity of scoring for essay test items (see chapter 9). **Interrater reliability** refers to consistency between two or more independent judgments of the same performance. This is important in scoring such events as gymnastics, diving, and figure skating, in which several judges rate the same performance. **Intrarater reliability** refers to consistency in scoring when a single rater scores the same test or performance two or more times. When testing motor skills, estimates of intrajudge objectivity are difficult to obtain, because the same performance must be viewed and scored twice. This is usually facilitated by recording the performance for multiple viewings.

WWW Go to HK*Propel* to complete Student Activity 6.1.

RELIABILITY

Many of the basic statistical concepts presented in chapters 3, 4, and 5 help us determine whether a test is reliable and valid. Teachers and researchers generally need specific evidence about a measure's reliability and validity—not general statements suggesting that it is reliable and valid. Numerous statistics are used to provide evidence of reliability and validity. The variance (presented in chapter 3) and the Pearson product-moment (PPM) correlation coefficient (presented in chapter 4) are used to provide evidence of a test's reliability and validity and thus need to be understood fully. Before getting into the number crunching associated with reliability and validity, however, we need to consider these concepts from theoretical perspectives so you are clear about exactly what these constructs are. With a deeper understanding, you will be better able to determine which statistical procedure you should use and how to interpret the results.

Observed, Error, and True Scores

Consider the scores obtained at a recent blood pressure screening (table 6.1). Each of the 10 participants has an observed blood pressure; however, it is possible that errors of measurement may have entered into the recording system so that the observed score is not the person's true blood pressure value. For example, the observed score may be in error because of the amount of experience the tester has, how the actual measurement was taken, when it was taken, where it was taken, the type of instrument used, the time of day, what events took place before testing, and so forth.

Table 6.1 Systolic Blood Pressure Recordings for 10 Participants

Participant	Observed blood pressure	= True blood pressure	+ Error
1	103	105	−2
2	117	115	+2
3	116	120	−4
4	123	125	−2
5	127	125	+2
6	125	125	0
7	135	125	+10
8	126	130	−4
9	133	135	−2
10	145	145	0
Sum (Σ)	1250	1250	0
Mean (M)	125.0	125.0	0
Standard deviation (s)	11.6	10.8	4.1
Variance (s²)	133.6 =	116.7 +	16.9

Note: Units are mmHg.

Data based on an example from Sax (1980).

Although it is unlikely that we can ever know exactly (without any error) what a person's true blood pressure is, imagine that we can develop a method by which to measure it more accurately than is typically done in a laboratory or clinical setting. For example, we might place a pressure-sensitive apparatus directly in an artery to determine the pressure exerted during systole. (Obviously, we would have to ignore the fact that even thinking about such a procedure would undoubtedly alter a person's blood pressure reading.) Assume that we have done this for the participants whose scores are reported in table 6.1. You will note that only 2 of the 10 have observed blood pressures that are equal to their true blood pressures. The other blood pressure readings have various amounts of error associated with them.

Some of the errors result in overestimating the true blood pressure, whereas others result in underestimating the true blood pressure.

A few key points can be seen in table 6.1:

• Each person's observed score is the sum of the true score and the error score. Your true score is without error; it theoretically exists but is impossible to measure. You can think of it as the average of an infinite number of administrations of the test in which you don't get any better because of practice, nor do you get any worse because of fatigue—it represents what you actually know or how well you actually perform. In a sense, your true score never changes for a specific point in time, and it is perfectly reliable. Your error score results from anything that causes your observed score to be different from your true score; it is also a value that theoretically exists but is impossible to measure. Sources of error include individual variability, instrument inaccuracy, cheating, testing conditions, and so forth.

• There is variation in the observed, true, and error scores (the standard deviations and variances are calculated for you).

• Error can be positive (thus increases the observed score) or negative (thus decreases the observed score).

• Error scores contribute relatively little to the observed variation.

• The error score mean is zero.

• The observed score variance (133.6) is equal to the sum of the true score variance (116.7) plus the error score variance (16.9).

Using observed (total), true, and error score variances, the reliability ($r_{xx'}$) is defined as that proportion of observed score variance that is true score variance (i.e., true score variance divided by observed [total] score variance):

$$r_{xx'} = \frac{s_t^2}{s_o^2} = \frac{(s_o^2 - s_e^2)}{s_o^2} \qquad (6.1)$$

where s_t^2 is the true score variance, s_o^2 is the observed (total) score variance, and s_e^2 is the error score variance. In table 6.1, the reliability is 116.7 / 133.6 = .87.

In theory, the true score is perfectly reliable, with $r_{xx'}$ = 1.00. (Certainly, the true score changes if there is a change in the phenomenon being measured, but at any time, the true score is viewed as perfectly reliable and thus contains no error.) Thus, a test is said to be reliable to the extent that the observed score variation is made up of true score variation.

Importantly, you can see that the observed score variance is equal to the sum of the true and error score variances; knowing any two of these variances results in your ability to calculate (or estimate) the third. Using equation 6.1, we see that the limits on reliability are 0 and 1.00. If the observed score is made up of no true score variation, the reliability is 0.00, and if the observed score variation (s_o^2) is made up of only true score variation (s_t^2), the reliability is 1.00. Neither of these two cases generally arises; however, for a test to be valid, it must be reliable, so a test's reliability must be reported. Depending on the nature of the decisions made from the test results, you generally want reliability to be .80 or higher. Although .80 is a goal, other results might be acceptable with lower reliabilities, and others would require reliabilities higher than .80. Consider a police officer's radar gun. Is it reliable? More importantly, is it a valid reading of your speed?

Return now to our original Measurement and Evaluation Challenge. Eva is interested in learning the reliability of the scores from the field test she will use because that measurement will tell her whether the results she obtains are consistent from one testing period to another. If the score is to be of any value to Eva, the results should vary little from one testing session to another. Additionally, the observed differences between people assessed for $\dot{V}O_2max$ obtained from the field test should reflect true differences in $\dot{V}O_2max$ and not be simply a function of measurement errors.

The following practical implications derive from this presentation:

• There should be observed score variance. (If there is none, the reliability is undefined—because of division by zero.)

• Error score variance should be relatively small in relation to the total variance.

• Generally, longer tests are more reliable than shorter tests. This is true because as tests increase in length, there is an increase in observed score variance that is more likely a function of an increase in true score variance than in error score variance.

www! **Go to HK*Propel* to complete Student Activity 6.2.**

You may be saying to yourself, *This is all fine, but how does one ever know a person's true score?* Simply, one can never know the person's true score. However, the observed score is readily available, and there are ways of estimating the error score variation for a set of scores. Therefore, as noted on the right side of equation 6.1, we can estimate the reliability using the observed score variance and error score variance.

www! **Go to HK*Propel* to complete Student Activity 6.3.**

Calculating the Reliability Coefficient

Let's turn to the actual calculation of a reliability coefficient. Reliability coefficients are classified into one of two broad types: interclass reliability coefficients (based on the PPM correlation coefficient, presented in chapter 4) and intraclass reliability coefficients (based on analysis of variance, or ANOVA models, presented in chapter 5).

Interclass Reliability

Let us first look at three **interclass reliability** methods: test–retest reliability, equivalence reliability, and split-halves reliability.

Test–Retest Reliability

Consider the simplest way of determining if a measure is reliable (consistent). We could simply administer the test to participants on two occasions (e.g., on the same day) and then correlate the two sets of observations using the PPM correlation coefficient to see if the correlation is high. This is exactly what is done with the test–retest reliability coefficient. Look at the two trials of sit-up performance presented in table 6.2. The PPM correlation coefficient is calculated to be .927, a correlation between scores from the two test administrations that is certainly high enough to call the measure reliable. The coefficient suggests that 92.7% of the observed score variance is true score variance. If the time between testing occasions is longer (e.g., days or weeks), the test–retest reliability coefficient may be called **stability reliability**, meaning the measure is consistent or stable across time. Importantly, the interclass reliability method should only be used when you are quite certain that there is no change in the mean from one trial to the second trial. You can test for mean differences in two trials with the dependent t-test presented in chapter 5. If there are multiple trials, an extension of ANOVA, called repeated-measures ANOVA, is used to test for mean differences. See Kang and Jin (2016b) for more information about repeated-measures ANOVA.

Table 6.2 Sit-Up Performance for 10 Participants

Participant	Trial 1	Trial 2
1	45	49
2	38	36
3	54	50
4	38	38
5	47	49
6	39	38
7	39	43
8	42	43
9	29	30
10	42	42
Sum (Σ)	413	418
Mean (M)	41.3	41.8
Standard deviation (s)	6.6	6.5
Variance (s²)	43.6	41.7

$r_{xx'} = .927$

Mastery Item 6.2

Use SPSS to confirm the reliability of .927 reported in table 6.2. (Hint: Use SPSS to calculate a correlation coefficient, as illustrated in chapter 4.) For a graphic representation of the data, create a scatterplot of the two trials. Can you see the high relationship demonstrated and why this can be interpreted as a reliability coefficient?

Equivalence Reliability

A second way to determine interclass reliability is by using an equivalence reliability coefficient. Consider a teacher who is concerned about students cheating during a written test. The teacher develops two parallel or equivalent forms of an exam and distributes the exams so that no two adjacent students receive the same examination. However, how should this instructor now determine grades? Should there be two grading procedures for the same class? Does student performance depend on which exam was completed? This teacher should first determine the equivalence of the two examinations. To do so, a test group would be asked to take each of the examinations (both forms) under nearly identical conditions. Half of the participants should take form A first and half should take form B first, so that order does not cause scores to be affected. An assumption must be made that the tests are parallel and that taking the first test neither hinders nor helps a student on the second test. The results from the two administrations are then correlated to determine if there is reliability, or consistency, between the two test forms. Note again that this is simply a PPM calculation in which the two variables correlated are the test scores from the respective forms. This is an equivalence reliability coefficient. As with test–retest reliability, you could use SPSS to display a scatterplot illustrating the relation between scores on form A and those on form B.

WWW **Go to HK***Propel* **to complete Student Activity 6.4.**

Split-Halves Reliability

You may be thinking that the previous two examples are a bit far-fetched, because it is unlikely that teachers will administer tests on more than one occasion (which is a requisite for determining the reliability of a test). You are correct! Teachers will typically administer a test only once because of time constraints and because examinee fatigue might otherwise come into play and negatively affect scores on subsequent trials. Additionally, practice can also affect the subsequent scores and thus the calculated reliability. However, there are ways to make minor adjustments in the equivalence methods and still reach a conclusion regarding a test's reliability. Consider how a teacher might create two equivalent forms of a single test. The equivalent forms can be created after the test is administered by assigning each person a score on two halves of the test (e.g., a score for the odd numbered items and a score for the even numbered items). Thus, the odd and even portions of the test can be perceived of as equivalent forms.

The PPM correlation coefficient can be calculated between the scores on the halves of the test and used as an estimate of the reliability of the measure. Table 6.3 has sample data for calculating split-halves reliability. The split-halves reliability using the PPM correlation coefficient is .639. Because it was suggested earlier that a reliability of .80 or higher is desirable, you might be tempted to dismiss this result as unreliable. However, an additional aspect of the data presented in the table needs to be considered. The reliability coefficient calculated from the data in table 6.3 is the correlation between two halves of the test. Let's assume that each half consists of 13 items, giving a total test length of 26 items. We indicated that longer tests are generally more reliable than shorter ones. Because the .639 calculated in table 6.3 was based on a test of 13 items in length, we must now estimate the reliability of the original 26-item test. You might think that you just multiply the obtained reliability by 2 since the actual test is twice as long. Notice, however, that multiplying .639 × 2 results in 1.278. By definition (see equation 6.1), reliability cannot exceed 1.0, so multiplying by 2 is incorrect. Rather, the Spearman–Brown prophecy formula (equation 6.2) is used to estimate the reliability of a measure when the test length is changed:

$$r_{kk} = \frac{k \times r_{11}}{1 + r_{11}(k-1)} \tag{6.2}$$

where r_{kk} is the predicted (prophesied) reliability coefficient when the test length is changed k times, k is defined as

$$\frac{\text{the number of items for which an estimate of reliability is desired}}{\text{the number of items for which the reliability has been calculated}}$$

and r_{11} is the reliability that has been previously calculated. Thus, to estimate the reliability for the 26-item test, we substitute into the Spearman–Brown prophecy formula and obtain the predicted (estimated) reliability of the measure across all 26 items. Note that $k = 2$ (i.e., 26 / 13) and r_{11} = .639 (our original reliability estimate).

$$r_{kk} = \frac{(26 \div 13) \times .639}{1 + .639([26 \div 13] - 1)}$$

$$r_{kk} = \frac{1.278}{1.639}$$

$$r_{kk} = .78$$

Table 6.3 Odd–Even Scores for 10 Participants

Participant	Odd score	Even score
1	12	13
2	9	11
3	10	8
4	9	6
5	11	8
6	7	10
7	9	9
8	12	10
9	5	4
10	8	7
Sum (Σ)	92	86
Mean (M)	9.2	8.6
Standard deviation (s)	2.2	2.6
Variance (s^2)	4.8	6.7
		$r_{xx'} = .639$

Thus, the estimated reliability for the original 26-item measure is .78. Note that when the reliability is adjusted with the Spearman–Brown prophecy formula, it makes no difference how many items you started with. When the number of items or trials is doubled (i.e., $k = 2$), the predicted reliability will be the same. Had the .639 reliability been obtained for 50 items and you doubled it to 100 items, the predicted r_{kk} would still be .78.

This number can also be estimated from table 6.4, which shows values of r_{kk} calculated from equation 6.2 using numerous values of r_{11} (left column) and k (column heading). The number of times you want to change the test length (k) is listed across the top of table 6.4 (0.25-5.0). You determine the predicted reliability (r_{kk}) by finding where the appropriate row and column intersect. For example, if your calculated reliability (r_{11}) is .40 and you increase the test length by a factor of five, the estimated reliability is .77.

You will note that there are values of k listed in table 6.4 that are less than 1. This indicates that the instructor can estimate the reliability for a test of shortened length. For example, assume that an instructor has a written test of 100 items with a reliability of .92. Randomly splitting the test into equal parts with 50 items each would result in two tests with predicted reliabilities of .85. This would reduce the time to administer and score the examination (and make students happier!). The Spearman–Brown prophecy formula can be used with the interclass reliability estimate as well as with intraclass reliability, which is discussed next.

www. **Go to HK*Propel* to complete Student Activity 6.5.**

Table 6.4 Values of r_{kk} from Spearman–Brown Prophecy Formula

r_{11}	.25	.33	.50	1.50	2.00	3.00	4.00	5.00
				k (change in test length)				
.10	.03	.04	.05	.14	.18	.25	.31	.36
.12	.03	.04	.06	.17	.21	.29	.35	.41
.14	.04	.05	.08	.20	.25	.33	.39	.45
.16	.05	.06	.09	.22	.28	.36	.43	.49
.18	.05	.07	.10	.25	.31	.40	.47	.52
.20	.06	.08	.11	.27	.33	.43	.50	.56
.22	.07	.09	.12	.30	.36	.46	.53	.59
.24	.07	.09	.14	.32	.39	.49	.56	.61
.26	.08	.10	.15	.35	.41	.51	.58	.64
.28	.09	.11	.16	.37	.44	.54	.61	.66
.30	.10	.12	.18	.39	.46	.56	.63	.68
.32	.11	.13	.19	.41	.48	.59	.65	.70
.34	.11	.15	.20	.44	.51	.61	.67	.72
.36	.12	.16	.22	.46	.53	.63	.69	.74
.38	.13	.17	.23	.48	.55	.65	.71	.75
.40	.14	.18	.25	.50	.57	.67	.73	.77
.42	.15	.19	.27	.52	.59	.68	.74	.78
.44	.16	.21	.28	.54	.61	.70	.76	.80
.46	.18	.22	.30	.56	.63	.72	.77	.81
.48	.19	.23	.32	.58	.65	.73	.79	.82
.50	.20	.25	.33	.60	.67	.75	.80	.83
.52	.21	.26	.35	.62	.68	.76	.81	.84
.54	.23	.28	.37	.64	.70	.78	.82	.85
.56	.24	.30	.39	.66	.72	.79	.84	.86
.58	.26	.31	.41	.67	.73	.81	.85	.87
.60	.27	.33	.43	.69	.75	.82	.86	.88
.62	.29	.35	.45	.71	.77	.83	.87	.89
.64	.31	.37	.47	.73	.78	.84	.88	.90
.66	.33	.39	.49	.74	.80	.85	.89	.91
.68	.35	.41	.52	.76	.81	.86	.89	.91
.70	.37	.44	.54	.78	.82	.88	.90	.92
.72	.39	.46	.56	.79	.84	.89	.91	.93
.74	.42	.48	.59	.81	.85	.90	.92	.93
.76	.44	.51	.61	.83	.86	.90	.93	.94
.78	.47	.54	.64	.84	.88	.91	.93	.95
.80	.50	.57	.67	.86	.89	.92	.94	.95
.82	.53	.60	.69	.87	.90	.93	.95	.96
.84	.57	.63	.72	.89	.91	.94	.95	.96
.86	.61	.67	.75	.90	.92	.95	.96	.97
.88	.65	.71	.79	.92	.94	.96	.97	.97
.90	.69	.75	.82	.93	.95	.96	.97	.98
.92	.74	.79	.85	.95	.96	.97	.98	.98
.94	.80	.84	.89	.96	.97	.98	.98	.99
.96	.86	.89	.92	.97	.98	.99	.99	.99

Intraclass Reliability

Interclass reliability, based on the correlation between two measures, is different from **intraclass reliability**, based on ANOVA. The interclass model permits you to correlate only two trials because the PPM is used to correlate only two things at a time. However, you might want to estimate the reliability for more than two trials, given that reliability generally increases as the number of trials increases. In this case, you would use the intraclass model to estimate reliability.

Additionally, if there is a constant difference between two trials (i.e., each person's score goes up or down by the same amount), the interclass reliability would be 1.00, but from a theoretical perspective the results would not be consistent (something does seem wrong). For example, when testing skinfold fat, the measures could become smaller with each measurement if the subcutaneous fat is still compressed from the previous measure. Another example of constant change is demonstrated in table 6.5: The PPM correlation is perfect ($r_{xx'} = 1.00$), yet the reliability (i.e., consistency of measurement) is lacking because each person's score increased by 10 on the second trial. The intraclass reliability model can address this issue. Significant mean differences across trials necessitate a deeper look into the changes across trials. It may be that participant learning or fatigue is affecting the reliability.

Table 6.5 Effect of a Constant Change in Measures

Participant	Trial 1	Trial 2
1	15	25
2	17	27
3	10	20
4	20	30
5	23	33
6	26	36
7	27	37
8	30	40
9	32	42
10	33	43
Sum (Σ)	233	333
Mean (M)	23.3	33.3
Standard deviation (s)	7.7	7.7
Variance (s^2)	59.1	59.1
$r_{xx'} = 1.00$		

The most common intraclass reliability models are Cronbach's alpha coefficient, Kuder–Richardson formula 20 (KR_{20}), and ANOVA reliabilities. Each of these is calculated in a similar manner. The total variance in scores is partitioned into three sources of variation: people, trials, and people-by-trials. People variance is the observed score (total) variance between the participants (s_o^2). Trial variance is based on the variance across the trials. People-by-trials variance is based on the fact that not all participants perform equally differently across the trials. All variance not attributable to people (i.e., trials and people-by-trials variances) can be perceived as error (s_e^2). Reliability is estimated by subtracting error variance from total (observed) variance and dividing the result by the total (observed) variance.

Consider equation 6.1, in which reliability can be estimated from observed score and error score variance. Having estimates of observed and error variance permits you to use equation 6.1 to estimate the reliability of the scores.

The alpha coefficient is calculated as follows:

$$r_{xx'} = \text{alpha coefficient} = \left(\frac{k}{k-1}\right)\left(1 - \frac{\Sigma s^2_{trials}}{s^2_{total}}\right) \tag{6.3}$$

where k is the number of trials, Σs^2_{trials} is the sum of the variance of each trial, and s^2_{total} is the variance for the sum across all trials.

Table 6.6 presents an example of a calculation of the alpha coefficient. The variance calculations are identical to those you learned in chapter 3. Note that the **alpha reliability** (i.e., intraclass reliability) estimates the reliability for the total score (i.e., the sum across all trials). You can then use this result in the Spearman–Brown prophecy formula (equation 6.2) to estimate the change in the reliability coefficient if the number of trials is increased or reduced.

Table 6.6 Calculating the Alpha Coefficient

Participant	Trial 1	Trial 2	Trial 3	Total
1	3	5	3	11
2	2	2	2	6
3	6	5	3	14
4	5	3	5	13
5	3	4	4	11
ΣX	19	19	17	55
ΣX^2	83	79	63	643
s^2	2.70	1.70	1.30	9.50

$$k / (k - 1) \times (1 - (\Sigma s^2_{trials} / s^2_{total}))$$

$$3 / (3 - 1) \times (1 - (2.70 + 1.70 + 1.30) / 9.50)$$

$$3 / 2 \times (1 - 5.7 / 9.50)$$

$$1.5 \times (1 - .60)$$

$$1.5 \times .40 = .60 = \text{alpha coefficient}$$

WWW Go to HK*Propel* to complete Student Activity 6.6.

Mastery Item 6.3

Use the Spearman–Brown prophecy formula to estimate the reliability of six trials from the data in table 6.6. Note that k is $= 2$ (6 / 3) and r_{11} is that obtained with the alpha coefficient (.60).

Mastery Item 6.4

Use SPSS to confirm the reliability estimate reported for the data in table 6.6. Let's calculate the alpha coefficient in two ways with SPSS. The first way uses the variances and equation 6.3. The second way takes advantage of an SPSS program to calculate alpha directly.

1. Start SPSS.
2. Open table 6.6 data from HK*Propel*.
3. Go to Analyze → Descriptive Statistics → Descriptives.
4. Put all three trials and the total in the Variable(s) box.
5. Click on Options.
6. Click *only* the box for Variance in the Dispersion box and click Continue.
7. Click OK.
8. Your output has the four variances necessary for substitution into equation 6.3.

The second way takes advantage of SPSS commands to calculate the alpha coefficient.

1. Start SPSS.
2. Open table 6.6 data.
3. Click on the Analyze menu.
4. Scroll down to Scale and over to Reliability Analysis and click.
5. Highlight "Trial1," "Trial2," and "Trial3" and use the arrow to place them in the Item box. Note: *Do not* include the total in this listing. SPSS will calculate the total for you.
6. Click on OK.

The alpha coefficient can also be used when data are scored as correct (1) or incorrect (0). In this case, the alpha coefficient is referred to as the Kuder–Richardson formula 20 (KR_{20}). The alpha coefficient and KR_{20} are mathematically equivalent. You will learn more about this in chapter 9. Jackson, Jackson, and Bell (1980) also provide an excellent discussion of the alpha coefficient.

www **Go to HK*Propel* to complete Student Activity 6.7.**

Mastery Item 6.5

The following activity illustrates interclass reliability, intraclass reliability, and the Spearman–Brown prophecy formula. Go to HK*Propel* and download the chapter 6 large dataset on reliability. These data represent four consecutive weeks of pedometer counts from a large sample. Do the following:

1. Go to Analyze → Correlate → Bivariate and put all of the variables in the box at the right. Note these will be interclass correlations illustrating consistency of steps across paired weeks. Review the correlations and see that they range from .553 to .762, with a median of about .70. Notice also that week 1 correlates lower with the other weeks.
2. Go to Analyze → Scale → Reliability Analysis and put all 4 weeks in the Items box. Notice the alpha coefficient for all 4 weeks is .885. This is the reliability of the entire 4-week period.

3. How would you estimate the reliability for a single week from this value? (Hint: Use table 6.4.)

4. Use the estimated value .70 obtained for a single week (from step 1) and substitute into the Spearman–Brown prophecy formula (or use table 6.4). Notice that the prophesized value (.90) is similar to that obtained with the alpha coefficient (.885).

Index of Reliability

Another statistic important to the interpretation of the reliability coefficient is the **index of reliability**. The index of reliability is the theoretical correlation between observed scores and true scores and is calculated as the square root of the reliability coefficient (equation 6.4):

$$\text{index of reliability} = \sqrt{r_{xx'}} \tag{6.4}$$

Thus, if the reliability of a test is .81, the theoretical correlation between observed and true scores is .90. Note that if reliability is 1.0, there is perfect correlation between observed and true scores. If the reliability is 0.00, then the correlation between observed and true scores is 0.0!

Standard Error of Measurement

For any given test administration, your best estimate of a person's true score is the obtained score. Although true score can never be actually determined, as suggested earlier in this chapter, it can be thought of as the average of an infinite number of administrations of the test (where neither fatigue nor practice affects participants' scores). If the test is administered twice, you would average the test scores to get a better estimate of the true score, because random positive errors and negative errors will theoretically balance out in the long run.

Regardless of a person's score, it is unlikely in a real-life setting to have a score that is totally without error. Thus, a person's score is expected to change from test administration to test administration. **The standard error of measurement (SEM) reflects the degree to which a person's observed score fluctuates as a result of errors of measurement.** You should not confuse the standard error of measurement (SEM) with the standard error of estimate (SEE) presented in chapter 4. Although the two have similar interpretations (and look quite similar), they are different: The SEM relates to the reliability of measurement, whereas the SEE concerns the validity of an estimate.

The SEM is calculated as follows:

$$\text{SEM} = s\sqrt{1 - r_{xx'}} \tag{6.5}$$

where s is the standard deviation of the test results and $r_{xx'}$ is the estimated reliability.

Assume a test has a standard deviation of 100 and a reliability of .84. The SEM is calculated to be

$$\text{SEM} = 100\sqrt{1 - .84}$$
$$\text{SEM} = 100(.4)$$
$$\text{SEM} = 40$$

If a person scored 500 on a test whose SEM was 40, one can place confidence limits around that person's observed score in an attempt to estimate what the true score is. The SEM, just like the standard error of estimate, is interpreted as a standard deviation: The SEM is the standard deviation of the errors of measurement around an observed score. It reflects how much a person's observed score would be expected to change from test administration to test administration as a result of measurement error. Because error scores are expected to be normally distributed, 68% of the scores would be expected to fall within ±1 SEM of the observed score. In our example, therefore, there is a 68% chance that the person's true score is between 460 and 540 (i.e., 500 ± 40). Note that you could use table 3.4 to place confidence intervals around a particular observed score. You should be able to see that if you take the observed score and add and subtract 2 SEMs to the observed score, you have placed an approximately 95% confidence interval around the observed score. This is because, as you learned in chapter 3, the mean score plus and minus 2 standard deviations results in capturing approximately 95% of the scores in a normal distribution. Remember that the SEM is a standard deviation. As such, it can be used along with table 3.4 to provide evidence of the accuracy of a measure obtained with a test that has specific variability (the test standard deviation) and reliability.

Mastery Item 6.6

You should verify that approximately 95% of true scores are within the range of 420 and 580 when the observed score is 500 and the SEM is 40 points (i.e., ±2 SEM).

www **Go to HK*Propel* to complete Student Activity 6.8.**

A test or measure does not necessarily have reliability in all settings. Said another way, the reliability of a measure is situation specific. Test scores are reliable under particular circumstances, administered in a particular way, and with a specific group of people. It is not appropriate to assume that simply because measures are reliable for one group of participants (e.g., females) they are automatically reliable for another group (e.g., males). The following factors can affect a test's reliability.

- *Fatigue.* Fatigue generally decreases reliability.
- *Practice.* Practice generally increases reliability. Thus, practice trials during teaching and training should be encouraged.
- *Participant variability.* Greater variability in participants being tested results in greater reliability.
- *Time between testing.* Reliability generally decreases as the time between test administrations increases.
- *Circumstances surrounding the testing periods.* Reliability generally increases with a greater similarity between the testing periods.
- *Appropriate level of difficulty for testing participants.* The test should be neither too difficult nor too easy.
- *Precision of measurement.* Adequate accuracy must be assured with the measuring instrument. For example, one should measure the 50-yard (45.7 m) dash not to the nearest second but to the nearest hundredth of a second.
- *Environmental conditions.* Such factors as noise, excessive heat, and poor lighting can negatively affect the measurement process.

Test administrators need to be sensitive to factors that could affect the reliability of any test chosen.

VALIDITY

As previously discussed, the concept of validity is associated with the relevance of a test. If a test is not a relevant measure of the attribute of interest, it will not be useful even if it is a reliable and objective test. If you use a test someone else has developed, it is in your best interest to investigate the validity of the test for your proposed usage. If the validity is inadequate, another test should be considered. If you develop your own test, there are straightforward ways you can examine and describe the validity of your test.

Assume you are a student who has just completed a fitness unit. Mr. Suarez, your teacher, announces that he will give a written test on three main ideas taught during the unit: short-term effects of physical activity, long-term effects of physical activity, and the components of an exercise prescription (frequency, intensity, duration, and type). You carefully study all three areas and feel well prepared to take the test. When the test is administered, you skim through it quickly and find that all the items deal with the long-term effects of physical activity. No questions on short-term effects or exercise prescriptions are included in the test. After completing the test, you might comment that this was not a good test because it did not measure what it was supposed to measure. In other words, the test lacked validity—or, more accurately, the score to be used to make an inference about your knowledge of the short-term effects and exercise prescription of physical activity lacked validity.

Historical View of Validity

To fully understand validity, a quick review of some highlights of the historical evolution of validity will be helpful. Early on, validity focused on how well scores on a test matched the scores from another measurement believed to validly assess the desired characteristics. "The validity of the new test is determined by examining the closeness of agreement between the new test scores and the other objective measure. This other measure is called the criterion" (Bingham, 1937, p. 214). This "criterion-centered" practice still influences many of the testing validation practices in human performance today.

In 1949, Cronbach defined validity as a property of a test, or "the extent to which a test measures what it purports to measure" (p. 48). Cronbach's definition influenced the view of validity for many years. In 1954, the American Psychological Association (APA) published "Technical Recommendations for Psychological Tests and Diagnostic Techniques," the first set of specifications for psychological tests, in which four types of validity were identified: content, predictive, concurrent, and construct. Each of these types of validity was proposed to answer different questions concerning how a test performs.

Cronbach and Meehl (1955) raised the question of whether test validation should focus on whether the test is valid, since the goal is not to validate a test, but rather to make inferences from test scores. The recommendation was revised by the APA in the *Standards for Educational and Psychological Tests and Manuals* (1966), reducing the four types of validity to three: content, criterion, and construct, with the criterion validity encompassing both predictive and concurrent validity.

When the *Standards* were revised in 1974, "types" of validity were changed to "aspects" of validity to better emphasize that the types of validity are related and part of a larger entity. The view of validation, which is a process to collect evidence to support the validity, also changed. No longer was validation an attempt to validate a test, but rather an ongoing process of evaluation of the test scores and the interpretation that can be made from them. Further, validity was not considered a property that can be directly measured, but was inferred from the three related aspects of validity, mentioned above.

The concept of a unified construct validity was further developed in the 1985 *Standards*, which were jointly developed by the APA, the American Educational Research Association

(AERA), and the National Council of Measurement and Evaluation (NCME). Although three types of validity remained, the term "aspect" was now replaced by "evidence." The construct validation of a test is considered to be an effort to collect and accumulate content-, criterion-, and construct-related validities. The description of the validity in most human performance textbooks is based on these three types of validity. To address the uniqueness of skill testing, the concept of logical validity, which can be considered a special case of content validity, is also employed in human performance literature. A summary of these validities is presented in table 6.7. This view of validity lasted for almost 15 years, until the *Standards* were revised again in 1999.

Table 6.7 Types and Definition of Validity Based on 1985 Standards

Type	Subtype	Definition
Content-related	Content	Evidence of truthfulness based on logical decision making and interpretation; or simply whether the content tested is representative of the entire domain a test seeks to measure.
	Logical	The extent to which a test measures the most important components of skill necessary to perform a motor task adequately.
Criterion-related	Concurrent	The relationship between a surrogate measure and a criterion when the two measures are taken relatively close together in time.
	Predictive	The relationship between a surrogate measure and a criterion when the criterion is measured much later.
Construct-related		The highest form of validity; it combines both logical and statistical evidence of validity through the gathering of a variety of statistical information that, when viewed collectively, adds evidence for the existence of the theoretical construct being measured.

Two major changes were made to the consideration of validity when the *Standards* were revised in 1999. The first and perhaps the most important change was that validity was no longer considered to be a property of a test, but rather of the test scores. The change was made mainly based on Messick's arguments (1989, 1996a, and 1996b) that the traditional view of validity is fragmented and incomplete. Because it is the test scores, rather than the test itself, that are interpreted and applied by test users, the validity should not be the property of the test, but rather of the test scores. When test scores are used or interpreted in more than one way, each intended interpretation must be validated.

The second major change was that the validity was no longer considered as a few distinct types of validity (e.g., factorial validity and content validity). Rather, there are multiple sources of validity evidence and validity is a unitary concept, which reflects the degree to which all the accumulated evidence supports the intended interpretation of test scores. The evidences of validity are generally classified into five categories (AERA et al., 1999) based on test content, relations to other variables, internal structure, response processes, and consequence of testing. Note that the evidences are different from those proposed in the 1985 *Standards* in that they are not limited to three types and they are not used to validate the test, but rather the test scores.

The *Standards* were modified again in 2014, defining validity as "the degree to which evidence and theory support the interpretations of test scores for proposed uses of tests" (AERA et al., 2014, p. 11). All five evidences, described next, were maintained in the 2014 *Standards*. Although all the evidences can be collected in practice, we will focus our discus-

Table 6.8 Relationship Between 1985 Validity Types and 2014 Validity Evidences

				Content	Concurrent	Predictive	Construct
				VALIDITY TYPES BASED ON 1985 *STANDARDS*			
					Criterion-related		
Validity evidences based on 2014 *Standards*		**CONTENT**		X			
	Relations to other variables	Test-criterion	Concurrent		X		
			Predictive			X	
		Convergent					X
		Discriminant					X
	Response processes						
	Internal structure						
	Consequences of testing						

sion on the two most often used in human performance research and practice: test content and relations to other variables. Table 6.8 summarizes the relationship between the validity "types" of the 1985 *Standards* and the validity "evidences" of the 2014 *Standards*.

Evidence Based on Test Content

"Test content refers to the themes, wording, and format of the items, tasks, or questions on a test" (AERA et al., 2014, p. 14). Important content-related evidence can be obtained by analyzing the relationship between a test's content and the construct(s) it is intended to measure. In theory, endless content can be created to measure a construct; in practice, however, we can only use a small sample to represent major content domains or categories. Collecting content-related evidence, therefore, consists of determining if the sample of items, tasks, or questions on a test are representative of the defined universe, or domain of content. Therefore, the content-related evidence is the same as the content validity defined in the 1985 *Standards*.

Collecting Content-Related Evidence

How is the content-related evidence of validity of a test collected and evaluated? Either logical or empirical analyses can be employed to determine the adequacy of how well the test content represents the content domain and the relevance of the domain to the proposed interpretation of test scores. Very often, a panel of experts is employed to judge the relevance of test content to its construct. Let us use an example to demonstrate how to collect and evaluate the content-related evidence by analyzing a fitness knowledge test called FitSmart (Zhu et al., 1999).

The content of this test was developed by using a **table of specifications**, sometimes referred to as a *test blueprint*. Figure 6.1 shows the fitness knowledge specifications for high school students. The content areas are listed on the left side of the table and experts' weighting, which is determined by the educational importance attached to each category of content. The largest weighting for a content area in this test is .20 (i.e., 20% of the items in this test are used to measure this content area) and four content areas received this weighting. The area of "Effects of exercise on chronic disease risk factors" received the smallest weighting of .05 (5%), suggesting that, for high school students, the experts viewed this area as less important than the other content areas.

Content experts' weighting

1. Concepts of fitness   .20

    a. Definitions

    b. Relationship to physical activity

    c. Relationship to health

2. Scientific principles of exercise   .20

    a. Physiological (acute)

    b. Physiological (chronic)

    c. Physiological (other)

    d. Psychological

3. Components of physical fitness   .20

    a. CR function

    b. Muscular strength and endurance

    c. Flexibility

    d. Body fatness/leanness

4. Effects of exercise on chronic disease risk factors   .05

5. Exercise prescription   .20

    a. Frequency

    b. Intensity

    c. Duration

    d. Mode

    e. Self-evaluation

    f. Adherence to exercise

6. Nutrition, injury prevention, and consumer issues   .15

Figure 6.1 FitSmart content specifications.

To analyze the content of a test, copies of the test and the table of specifications are needed. First, arbitrarily select several items from the test and determine what section of the table of specifications they represent. Answer the following question about each item: Does the item properly fit the cell designated by the test publisher? For example, suppose you choose to analyze item number 11 from Form 1 of FitSmart:

11. Which of the following physical activities will most likely lead to an increase in muscle size?

    A. Weightlifting*

    B. Walking

    C. Running

    D. Bowling

According to its placement, item 11 is a measure of knowledge about "Concepts of fitness: Relationship to physical activity." Does it measure what it is supposed to measure? If so, the item contributes to the content validity evidence of the test. If not, it is not a valid item. Obviously, this is a demanding job, but a test user is obligated to at least spot-check the test developer's decisions. The following are the major steps to collect and evaluate the content-related evidence for a published test:

1. Answer the test items and score your responses. This step should be a standard procedure for any teacher with any test. If this step is taken seriously, the ability to detect weaknesses in a test, including one's own test, is improved.
2. Review the publisher's statement on content validity.
3. Examine the table of specifications.
4. Assess the appropriateness of the placement of items in the content and behavior categories.
5. Examine the content and behavior categories and respond to the following four questions:
   a. Are important elements of a content area omitted?
   b. Are unimportant elements of the content area erroneously included?
   c. Are any categories in the content and behavior areas weighted improperly? (Weighting is reflected in the number of items included in a category.)
   d. Are all elements of the content and behavior areas educationally important?

More information on how to prepare content-related evidence can be found in chapter 9 if you are going to develop your own tests.

Content-Related Evidence for Physical Performance Tests

Although written tests are used frequently in physical education and the exercise sciences, the most predominantly used tests are those measuring motor skills, performance, and abilities. Instead of using clusters of items to measure various content areas, the test may consist of a single item or task or a repetition of the same item or task, usually referred to as a *trial*. In the past, the term *logical validity* was used in the field of human performance to describe the representation and importance of tasks employed in a physical performance test. Because the test content is now broadly defined and the tasks have been included in this test content (defined earlier), we decided to no longer use the term *logical validity* to distinguish it from content validity. However, some similarities and differences in determining the content-related evidence in physical performance tests are described below.

In general, the initial validation of a physical performance test follows a set of rules similar in concept to content-related evidence described above. The main difference is that, rather than the items and content domain in a written test, the major components of the physical performance become the area of interest. Collecting content-related evidence for a motor skills test therefore focuses on whether the components of the skill being measured correspond with those required to perform the skill adequately. The identification of the important components of a skill is thus important. Many tests in physical education only measure accuracy in performing a skill. Although accuracy is undeniably important in the execution of many skills, it is not the only important factor. Other important aspects of a physical skill performance were likely excluded because these physical tests were not developed according to a logical plan, the way a good written test would be. The following are the major steps to collect content-related evidence for a physical performance test:

1. Review the test developer's statement on the purpose of the test and the components of skill the test is designed to measure. These components should be clearly stated in the test description. Make a list of these components.

2. Examine the test. List the components actually measured in the test.

3. Compare the two lists. Are the components identified and defined by the test developer actually measured by the test?

4. Examine the educational importance of the test:

   a. Are unimportant components of skill measured by the test?

   b. Are important components of skill omitted from the test?

   c. Do certain components of the skill receive inappropriate emphasis in the text?

Consider, for example, a test of the tennis forehand drive. Ms. Spencer, a physical education teacher, describes a good forehand drive as a ball hit so that it travels low over the net and deep into the backcourt. She decides to test her students on this skill using a test in which a sectioned target is positioned in the backcourt so that the areas with the higher scores are closer to the tennis court's baseline. Is this a valid test? The test measures whether the ball is hit deep into the backcourt by giving more points to the ball landing closer to the baseline, but it does not measure the height of the ball's flight over the net. Thus, the content evidence is insufficient, and the test lacks validity. Test validity could be increased by stretching a rope across the middle of the court above the net, so that students must attempt to hit the ball so that it passes between the rope and the top of the net.

This section deals primarily with the content-related evidence for a single skills test. Although it is appropriate to measure specific skills when students are learning a sport, skills tests should not mistakenly be thought to measure playing ability. To measure playing ability, a battery of tests is often used. One of the first steps in validating a test battery of this type involves identifying the important skills that constitute playing ability (the major "components" in this context are skills rather the components of a skill). Tests are then selected to measure each of the identified skills. Of course, each individual test must possess content-related evidence as well. Published tests of sports skills and playing ability frequently lack a statement identifying the important components of a skill. In these cases, the test must be examined and the components being measured must be listed, then the educational importance of the test must be analyzed. If developing your own test is necessary, start with a list of the important components of a skill, which will provide the basis for the design of the test.

Evidence Based on Relations to Other Variables

Validity evidence for a test can also be examined by analyzing the relationship between test scores and variables external to the test. An external variable is a measurement or test that assesses the same or different constructs that the test is supposed to measure. Although the focus is on the relationship between the test and the external variable, research design and statistical methods used to collect this kind of evidence do not have to be cross-sectional and correlational. For example, if a test is designed to measure tennis playing ability, the average test scores of a group of elite athletes should be higher than a group of amateur players. Similarly, test scores of a group of students should improve after effective instruction. Thus, this kind of evidence reflects "the degree to which these relationships are consistent with the construct underlying the proposed test interpretations." Two of the most commonly used methods in collecting evidence related to other variables are test–criterion relationship and convergent/discriminant evidences described below.

Test–Criterion Relationship

Evidence of the relation of test scores to a relevant criterion is demonstrated by comparing test scores with one or more external variables that are considered direct measures of the characteristic or behavior in question (known as the *criterion* or a *criterion test*). Evidence of the relationship of test scores to a relevant criterion is based on having a true criterion measure available. Validity is based on determining the systematic relationship between the criterion and other measures used to estimate the criterion. In practice, the test–criterion relationship is confirmed by establishing a statistical relationship between the test and the trait being measured. In the past, this relationship was often called *criterion-related validity, statistical validity,* or *correlational validity.* These terms are used because criterion-related evidence is based on the PPM correlation between a particular test and a criterion.

For example, refer back to Eva's situation in the Measurement and Evaluation Challenge; she wants to measure the maximal oxygen uptake ($\dot{V}O_2$max) for young adult participants. Eva knows that the best way to measure $\dot{V}O_2$max is to have each person complete a maximal exercise test on a treadmill, cycle ergometer, or the like—however, Eva does not have the equipment and resources for conducting such testing. Therefore, Eva is seeking alternative measures that can be used to estimate $\dot{V}O_2$max—submaximal tests, distance runs, and nonexercise models. These alternative measures must first be validated with the criterion measure. To do this, at some point participants must complete both the criterion test (treadmill-determined $\dot{V}O_2$max) and the alternative test (often called a *field test*) to be used to estimate the criterion. If a strong relationship is found between the criterion and the alternative test, future participants need not complete the criterion measure but can have their value on the criterion estimated from the alternative, or **surrogate measure**. The taking of skinfold measures to estimate body fatness is an excellent way to establish test–criterion relationship. In this case the criterion measure might consist of an underwater weighing procedure to determine body composition, and in particular, percentage of body fat.

Two designs, **concurrent** and **predictive**, are often used in practice to establish the test–criterion relationship. Both are based on the PPM correlation coefficient and the main difference is the time at which the criterion is measured. For the concurrent design, the criterion is measured at approximately the same time as the alternative (surrogate) measure. Using a distance run to estimate $\dot{V}O_2$max is an example of concurrent validity. For the predictive design, the criterion is measured in the future; it might be assessed many weeks, months, or even years after the original test is conducted. For example, the prediction of heart disease development in later life is based on a predictive design. It has been shown that lack of exercise or physical activity, high body fat, smoking, high cholesterol, and hypertension are all predictive of future heart disease. However, the criterion—development of heart disease—is not measured until many years later. (Of course, these same variables can be used to predict if one currently has heart disease. Thus, the time at which the criterion is measured and the interpretation of the correlation help identify whether the criterion-related validity is concurrent or predictive in nature.) The following list provides some examples of concurrent and predictive designs to establish the test–criterion relationship in human performance, exercise science, kinesiology, and education. Each criterion is followed by a list of possible predictors.

Concurrent Design

- $\dot{V}O_2$max (criterion: oxygen consumption)
  - Distance runs, which could be measured by the distance achieved in a fixed-time run or by time to complete fixed-distance run (e.g., 9 min, 12 min, 1.0 mi [1.6

km], 1.5 mi [2.4 km], 2 km [1.2 mi], 20-m [21.9 yd] shuttle)
- Submaximal tests (e.g., cycle, treadmill, swimming)
- Nonexercise models (e.g., self-reported physical activity)
- Body fat (criterion: dual-energy X-ray absorptiometry, DXA; hydrostatically determined body fat)
  - Skinfolds
  - BMI
  - Anthropometric measures (e.g., girths, widths, and lengths)
- Sport skills (criterion: game performance, expert ratings)
  - Sport skills tests (e.g., wall volley tests, accuracy tests, and total body movement tests)
  - Expert judges' evaluation at skill performance

Predictive Design
- Heart disease (criterion: heart disease developed in later life)
  - Present diet, exercise behaviors, blood pressure
  - Family history of heart disease or related health issues
- Success in graduate school (criterion: grade point average or graduation status)
  - Graduate Record Examination scores
  - Undergraduate grade point average
- Job capabilities (criterion: successful job performance)
  - Physical abilities
  - Cognitive abilities

Sport skills tests are also good examples of test-criterion validation procedures. Green, East, and Hensley (1987); Hensley, East, and Stillwell (1979); Hensley (1989); and Hopkins, Schick, and Plack (1984) provide excellent examples of the procedures used to validate sport skills tests. Although these tests are all over 30 years old, they retain their value as robust procedural models. A criterion measure must first be developed, then a variety of skills tests (i.e., a test battery) must be correlated with the criterion measure to determine which are the most valid and most helpful in estimating the criterion. If a series of tests is used to estimate the criterion, multiple correlational procedures (see chapter 4) are used rather than the simple PPM correlation coefficient. However, the logic is the same: An attempt is made to account for variation (i.e., increase the coefficient of determination) in the criterion measure from more than one measure. Consider a golf test. The criterion could be the average score from several rounds of golf. Thus, a study could be conducted in which everyone completes several rounds of golf to obtain the criterion measure. Each participant then completes a variety of skills tests (e.g., driving, long irons, short irons, chipping, and putting), which are then correlated with the criterion measure to determine which measure or combination of measures best estimates the criterion measure. Note that there will always be some error in all of the measures (criterion and estimators).

Interpretation of the coefficient of the test–criterion relationship depends on its absolute value. Because the coefficient is simply a PPM correlation coefficient, it must range from −1.00 to +1.00. However, the closer the absolute value of the validity is to 1.00, the greater the validity. For example, Green at al. (1987) reported validities for items on a golf skill test battery ranging from .54 for a pitch shot to .66 for a middle-distance shot.

Let's return to the standard error of estimate (SEE) presented in chapter 4. The SEE is often reported in the concurrent design. For example, assume that you have a submaximal test that estimates $\dot{V}O_2max$ from a timed distance run of 1 mile (1.6 km) and the SEE is 4 ml · kg^{-1} · min^{-1}. If someone has a predicted $\dot{V}O_2max$ of 50 ml · kg^{-1} · min^{-1}, you can place confidence limits around the predicted score: You can be 68% confident that the actual $\dot{V}O_2max$ is between 46 and 54 (i.e., 50 ± 4) ml · kg^{-1} · min^{-1}. Kline et al. (1987) provide an excellent example of concurrent design for estimating $\dot{V}O_2max$. Their reported correlation coefficient is .88 and the SEE is 5 ml · kg^{-1} · min^{-1}. Note that the SEE reflects the accuracy of estimating your score on the criterion measure—in other words, it is a validity statistic.

Development of the criterion measure is extremely important in establishing the test-criterion validity evidence. Examples of how criterion measures can be obtained include the following:

• *Actual participation.* One can actually complete the criterion task (e.g., play golf, shoot archery, conduct job-related activities).

• *Known valid criterion.* One can use a criterion (e.g., running on a treadmill, obtaining underwater weight) that has been previously shown to be valid.

• *Expert judges.* Experts judge the quality of the criterion performance. This is often used with team activities (e.g., volleyball) in which it is difficult or impossible to obtain a number that reflects performance on the task being measured.

• *Tournament participation.* Rankings of abilities can be determined when everyone participates with everyone else (best used when the skilled event is an individual sport).

• *Known valid test.* Participants can complete a test that has been previously shown to be valid.

Convergent and Discriminant Evidence

The attribute or trait one is interested in studying can sometimes be complex and impossible to directly measure. Consider an attribute like athletic ability: Certain aspects of athletic ability can be measured, but as a whole it defies precise measurement. Although such a construct cannot be measured precisely, indicators of the construct behavior are often measured. One such example is anxiety, often an attribute of concern in athletics. Anxiety cannot really be measured directly, but one can measure indicators of anxiety, such as measuring the sweatiness of palms, administering an anxiety inventory, and checking heart rate or blood pressure. How is this type of test validated? Since no criterion test is available, indirect logical deriving is necessary. Convergent and discriminant evidence are often collected to indicate validity.

Relationships between test scores and other measures intended to assess similar constructs provide **convergent evidence**, whereas relationships between test scores and measures intended to assess different constructs provide **discriminant evidence**. Collecting convergent and discriminant evidence is needed to determine validity of a test by comparing it with external variables that have established validity and reliability. Let's assume we wish to measure physical activity, and we would like to use a questionnaire rather than a more direct measure such as an accelerometer or heart rate monitor. We would expect our questionnaire to correlate fairly well with other questionnaires measuring physical activity. This is an example of convergent evidence. On the other hand, we would expect the questionnaire to have a low correlation with a questionnaire measuring another construct, such as self-efficacy. This is an example of discriminant evidence.

Finally, another method to collect convergent or discriminant evidence is the **group differences method**, in which two or three groups are compared to verify the construct being measured. The idea behind this method is that if a group (e.g., trained) has a known better ability or trait than another group (e.g., not trained), the test that measures that ability or trait should be able to detect the group difference. The method is, in fact, often used when collecting convergent or discriminant evidence experimentally. Let's use Kenyon's Attitude Toward Physical Activity Inventory (Kenyon, 1968) as an example. One of the six scales in the Kenyon Inventory is labeled "Vertigo." Essentially this scale is a measure of the risk-taking capacity of an individual in situations involving vigorous physical activity. How should this scale be validated? Logically, persons who participate in forms of vigorous physical activity involving risk would be more likely to obtain high scores on the scale than sedentary individuals. For example, a group of mountain climbers should score significantly higher on the scale than a group of sedentary office workers. If not, the validity of the scale is questionable. Comparing two groups' scores, which are expected to differ, is one way of providing evidence of construct validity.

Evidence Based on Internal Structure

Evidence based on the internal structure of a test refers to the degree to which the relationship among test items and the components within a test conform to the construct on which the proposed test score interpretations are based (AERA et al., 2014). The conceptual framework or current understanding of the construct that a test is designed to measure often determines the design of the internal construct of the test. If, in return, the evidence collected can confirm the internal structure, the test is likely measuring the construct it is designed to measure. For example, if a test is designed to measure a single ability or trait, the score of each item in the test should highly correlate with the total score of the test. If, on the other hand, the test is designed to measure several factors or components related to an ability or trait, an item in the test should correlate with a factor or component that it is measuring, but moderately correlate with the total score of the test. Collecting evidence based on internal structure is extremely important to psychological and affective measures since there is often no criterion measure of this kind of construct. In practice, the internal structure of the construct is often verified using more sophisticated statistical techniques (e.g., factor analysis and structure equating modeling). The description of these methods is beyond the level of this text.

Evidence Based on Response Processes

Evidence based on response processes refers to the degree to which the processes of test takers or scorers are consistent with the intended interpretation of scores. It often focuses on the fit between the construct and the detailed nature of performance or response actually engaged in by examinees. Usually, the evidence can be collected by analyzing individual responses (e.g., systematic observations of the person's response behavior). For example, we can assess students' problem-solving strategies by asking them to report their decision-making process using think-aloud protocols. If the test is designed to measure students' knowledge and strategies in a specific area, the recorded decision-making processes should match the construct being measured. Technology also provides us with many objective measures to assess response processes. Using a heart rate monitor or accelerometer, for example, we can now reliably record the amount of low-, moderate-, and vigorous-intensity physical activities a person engages in during a typical day. By adding a GPS device, we can even record where a specific activity was engaged in and the link between one activity (driving) and another (e.g., walking in a shopping mall).

Studies of response processes are not limited to test takers only. When raters or observers are involved in testing, their scoring process can also be studied to determine if their judgment is consistent with intended interpretation or construct definition. To help students learn and improve their performance, many teachers have switched their assessment focuses from the product (e.g., accuracy, distance, speed) to process (e.g., how a task was completed) by rating student performance using developed scoring rubrics (see chapter 15). To make sure their ratings are not influenced by factors that are irrelevant to the intended interpretation, teachers' judging process can also be studied. For example, teachers can record their rationale for giving a specific score to a student using the think-aloud method. The teacher recordings, student performances, and scoring rubrics then can be reviewed in a group discussion format so that irrelevant factors that may bias teachers' judgment (e.g., distance and angle of the camera, poor reliability of the task, omitted description of the trait in scoring rubrics) can be identified.

Evidence Based on Consequence of Testing

The results of testing or assessment, as well as decisions associated with testing and assessment, will have consequences for the test takers and related parties (e.g., parents, teachers, and schools). For example, a student may be upset by the grade of a nonvalidated test; a teacher might change his or her instruction methods because of students' performance on the test; the body composition of a special population could be incorrectly predicted using an equation developed for another population; or a test taker might not get a job by failing an employment test. The evidence based on test consequence thus focuses on scoring meaning and the intended and unintended consequences of assessment use. A numerical qualitative (e.g., questionnaires, observation, and case study) and quantitative (e.g., group difference by t-test) can be used to collect and evaluate the evidence of test consequence.

One of the challenges in evaluating this kind of consequence is that it could be easily confused with a social policy decision. Gender equality, for example, is a required social policy for all employment tests. If a large proportion of female test takers failed a job-related physical fitness test, one may criticize the test having a poor test consequence. This may not be correct; the higher female failure rate may not be caused by the inappropriate test design, but rather by true difference between male and female test takers. Thus, the key issue here is whether the consequence (failing more female test takers) is relevant to the validity of the test. If the test is designed based on a comprehensive job analysis and the test tasks sampled represent the physical fitness, skills, and abilities needed for the job, the consequence of the test reflects biological differences and should be accepted. The correct reaction to the consequence should therefore be to provide appropriate accommodations for the job tasks to make it easier or help prepare female test takers. For example, if the failure is mainly due to relatively lower strength of females, the profession may consider providing more lightweight equipment or offering a training program so that female applicants can achieve the required strength within a short period.

If, on the other hand, the test design is not based on job analysis, or if selected tasks are male-favored (e.g., daily activities they are familiar with or testing equipment more suited to their physical characteristics), the test consequence becomes test related and therefore not acceptable. The right action for this kind of consequence should focus on test modification so that factors that cause the biased results can be eliminated. Other test-related information (e.g., the benefits of a required physical fitness test for certain jobs) should also be collected to serve as the evidence of test consequence.

Integrating Validity Evidence

Finally, after learning about different types of validity evidence, a question may naturally be raised: Should we collect all the evidences when validating a test? In the past, the most common practice was for test developers to conduct a validation study, in which a single type of validity was usually collected from a single sample (occasionally split into two for cross-validation purposes). Based on the results of the study, a conclusion was made: the test is valid or the test is not valid.

Based on the newer validity-argument framework (Kane, 1992), however, this kind of practice is no longer accepted—a test will not be always valid for all purposes or in all situations, especially when the conclusion is based on a single piece of information. Instead, the new practice believes that "a sound validity argument integrates various strands of evidence into a coherent account of the degree to which existing evidence and theory support the intended interpretation of test scores for specific uses" (AERA et al., 2014, p. 21). This means that, instead of using a "one-shot" study design to collect validity evidence, long-term efforts should be made to continuously collect multiple types of validity evidence, especially when applied to new subgroups and under different contexts.

Lack of continuous and accumulated effort to validate tests and assessment is unfortunately a common practice in the field of human performance. On the other hand, it is almost impossible to collect all the validity evidence at one time, and certain evidences may be more important than others. Content-based evidence, for example, is important for all types of tests and should be treated as the most important evidence to collect. When the purpose of a developed field test is to substitute for a criterion measure, test–criterion relationship is the critical evidence to collect. If there is no criterion test available for a construct or trait being measured (e.g., a psychological trait), multiple evidences, such as those based on test–criterion relationship, internal structure, and response processes, may have to be collected simultaneously.

Figure 6.2 illustrates the key measurement characteristics of a test or measure we have introduced in this chapter, including reliability (shown on the left side) and validity (shown on the right side). Reliability coefficients can be classified as subjective and objective measures, which can further be classified as interclass or intraclass. Three interclass reliability methods are test–retest reliability, equivalence reliability, and split-halves reliability. Intraclass reliability is based on ANOVA and results in alpha and KR_{20}. Validity (relevance) evidences are based on test content, response procedure, internal structure, relationship to other variables, and consequence of testing. The evidence of relationship to other variables can further break down as test–criterion relationship (including concurrent and predictive designs), convergent, and discriminant.

Dataset Application

To learn how to apply the SEE to establishing the test–criterion relationship validity evidence, go to the chapter 6 large dataset on validity in HK*Propel*. The dataset contains a number of variables that *might* be related to underwater weighing. Use the information that you learned in chapter 4 to determine which variables are most and least related to hydrostatically determined body fatness. Can you also calculate the SEE for some of the variables and interpret the accuracy of the body fatness estimation?

[www] **Go to HK*Propel* to complete Student Activity 6.9.**

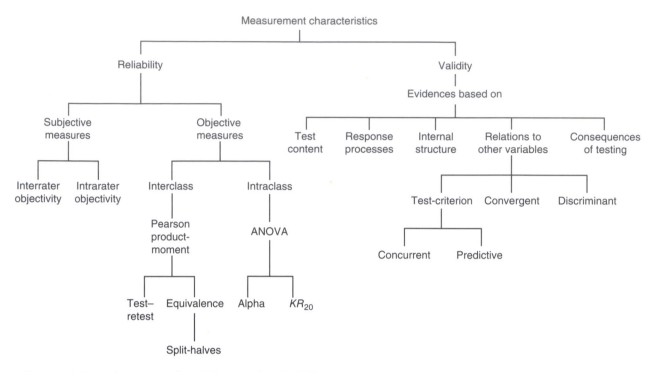

Figure 6.2 Diagram of validity and reliability terms.

Much of the information presented in this chapter has related to the Pearson product-moment (PPM) correlation coefficient, which is also introduced in chapter 4. In some cases, the PPM is interpreted as a reliability coefficient; in other cases it could be an objectivity coefficient or a criterion-related (either concurrent or predictive) validity coefficient. In all cases the PPM correlation coefficient is calculated as you learned in chapter 4. The difference in the interpretation depends on what two things are correlated. This is described in figure 6.3. Essentially, if two trials of the same thing are correlated but measured at different times, then this PPM is interpreted as a reliability (stability) coefficient. If two raters are correlated when they scored the same test, this PPM is interpreted as an objectivity coefficient—either interrater (between raters) if more than one rater is involved or intrarater (within raters) for the same rater on more than one occasion. If two forms of the same test are being correlated, this is an estimate of equivalence. If one of the measures being correlated is a criterion, then you are working with validity. Whether the PPM calculated is a concurrent or predictive validity coefficient depends on when the criterion was measured. This illustrates the generalized use of the PPM for estimating reliability, objectivity, and validity. Distinguishing among these correlations is important. See Odom and Morrow (2006) for further illustrations of this concept and how to interpret the correlation coefficient.

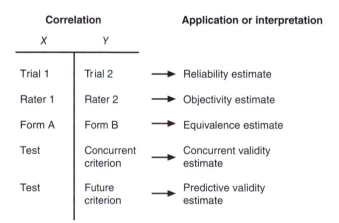

Figure 6.3 Applications of the Pearson product-moment (PPM) correlation in reliability and validity.

[www] **Go to HK*Propel* to complete Student Activity 6.10.**

APPLIED RELIABILITY AND VALIDITY MEASURES

Let's look at some examples of reliability and validity in human performance. Green and colleagues (1987), for example, used correlation to develop a golf skills test that provided validity information, but their validity matrix contained no reliability results. Remember, to estimate reliability, you must administer the same thing on *at least* two occasions. Green and colleagues further investigated validity through the use of the multiple skills test to predict golf performance and identified the middle-distance shot, pitch shot, and long putt combined to have a concurrent validity of .76. Adding a fourth skills test, the chip shot, only increased the concurrent validity to .77 (Green et al., 1987). Their golf test batteries were determined with multiple regression techniques (see chapter 4) to determine which test items best combine to account for variation in the criterion measure (golfing ability). The test administrator would therefore need to determine if it is worth the time and effort to measure four skills (validity = .77) rather than three (validity = .76). Table 6.9 provides illustrations of reliability estimates from physical performance and a variety of sport skills tests.

Table 6.10 provides concurrent validity coefficients for estimating $\dot{V}O_2$ max from a variety of measures. Some of the authors used a single measure to estimate $\dot{V}O_2$ max, whereas others used multiple regression. Look at the results from Murray and colleagues (1993) presented in table 6.10. Can you explain why the correlation increases with additional items? Can you also explain why the SEE decreases with increasing number of items?

Table 6.9 Reliability Measures from Sport Skills and Fitness Tests

Author	Test item	Reliability (r_{xx})
Engelman and Morrow (1991)	Traditional pull-up (boys) Traditional pull-up (girls) Modified pull-up (boys) Modified pull-up (girls)	.83 to .92 .91 to .92 .68 to .83 .77 to .83
Green, East, and Hensley (1987)	Golf—chip shot (males) Golf—chip shot (females) Golf—long putt (males) Golf—long putt (females) Golf—short putt (males) Golf—short putt (females)	.85 .86 .87 .93 .54 .46
Hensley, East, and Stillwell (1979)	Racquetball—short volley (males) Racquetball—short volley (females) Racquetball—long volley (males) Racquetball—long volley (females)	.77 .86 .85 .82
Hensley (1989)	Tennis—serve (males) Tennis—serve (females) Tennis—volley (males) Tennis—volley (females)	.86 & .95 .79 & .88 .70 & .72 .69 & .79
Hopkins, Schick, and Plack (1984)	Basketball—shot (males) Basketball—shot (females) Basketball—pass (males) Basketball—pass (females)	.84 to .95 .87 to .95 .88 to .96 .82 to .91
Nelson, Yoon, and Nelson (1991)	Modified push-up (boys) Modified push-up (girls)	.78 to .89 .77 to .91
Rikli, Petray, and Baumgartner (1992)	1/2 mile (0.8 km) (boys) 1/2 mile (0.8 km) (girls) 3/4 mile (1.2 km) (boys) 3/4 mile (1.2 km) (girls) 1 mile (1.6 km) (boys) 1 mile (1.6 km) (girls)	.65 to .82 .32 to .77 .48 to .94 .58 to .83 .44 to .87 .34 to .90
Schick and Berg (1983)	Golf—5 iron	.90

Note: All reliabilities are intraclass.

Table 6.10 Concurrent Validity Measures for $\dot{V}O_2max$

Author	Criterion	Predictor(s)	Validity (r)	Standard error of estimate (ml · kg⁻¹ · min⁻¹)
Getchell, Kirkendall, and Robbins (1977)	$\dot{V}O_2max$	1.5-mi (2.4 km) run	.92	2.38
Kline et al. (1987)	$\dot{V}O_2max$	1-mi (1.6 km) walk Sex Age Body weight	.88	5.00
Murray et al. (1993)	$\dot{V}O_2peak$ $\dot{V}O_2peak$ $\dot{V}O_2peak$	20-min steady-state run 20-min steady-state run Sex 20-min steady-state run Sex Weight	.68 .73 .79	5.32 4.96 4.45
Jurca et al. (2005)	Maximal cardiorespiratory fitness	Sex Age BMI Resting heart rate Self-reported physical activity	.76-.81	6.90-5.08
Wier et al. (2006)	$\dot{V}O_2max$	Sex Activity code Age BMI	.80	4.90

Note: BMI = body mass index.

WWW **Go to HK*Propel* to complete Student Activities 6.11 and 6.12.**

ESTIMATING AGREEMENT BETWEEN MEASURES USING THE BLAND–ALTMAN METHOD

Bland and Altman (1986) argue that using correlation and regression statistical methods has serious drawbacks in demonstrating the degree of agreement between two measures of the same attribute—specifically, that two measures might have a high correlation but still have clinically important differences in terms of the magnitude and the distribution of those differences. (Note the earlier discussion of this concern in the intraclass reliability section, illustrated in table 6.5.) The Bland–Altman approach is therefore based on calculating the absolute differences between the two measures and plotting those differences against the averages of the two measures.

In order to understand their approach, examine the data in table 6.11. The data represent two clinical measures of systolic blood pressure. One was taken by a trained clinician and the other was taken by an automated device. The differences are calculated by consistently subtracting the clinician's measure from the automated value. The averages represent the average systolic blood pressure for each participant across both measures. To represent the standard error of the agreement between measures, the standard deviation of the differences is calculated. For this example, the standard deviation equals 4.11.

Table 6.11 Example of the Bland–Altman Method

Participant	Clinician	Automated	Differences	Averages
1	103	105	2	104
2	117	115	–2	116
3	116	120	4	118
4	123	125	2	124
5	127	125	–2	126
6	125	125	0	125
7	135	125	–10	130
8	126	130	4	128
9	133	135	2	134
10	145	145	0	145
Mean (*M*)	125	125	0	125
Standard deviation (*s*)	11.56	10.80	4.11	10.99

The next step is to develop a scatterplot with the average values on the x-axis and the differences on the y-axis (figure 6.4). Three horizontal reference lines are added to the scatterplot:

• The first line indicates zero or no differences between the measures. Notice from table 6.11 that there are two participants with zero difference scores; those points are located on the line, indicating no difference between the measures.

• The second line is slightly above a difference of +8. This is 0 + 2 × 4.11 (the standard deviation of the differences).

• The third line is slightly below a difference of –8. This is 0 – 2 × 4.11.

Examine the scatterplot in figure 6.4 to see the number of large errors in agreement. These would be points above the top reference lines or below the lower reference line. Then examine the distribution of the errors across the range of the average blood pressures. Example questions to be answered include these:

• Are the errors consistently above, below, or equally distributed in relation to the zero error reference line? (Errors are somewhat equally distributed.)

• Are the magnitudes of the errors consistent across the range of blood pressure measures? Are errors larger or smaller at high blood pressures? Are errors larger or smaller at low blood pressures? (There are larger positive differences at mid-range blood pressures.)

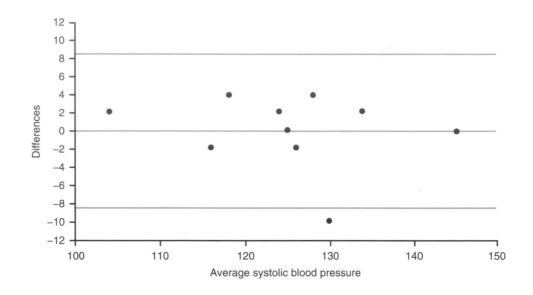

Figure 6.4 Bland–Altman plot of systolic blood pressure (clinician and automated).

Another consideration that Bland and Altman (1986) emphasized was the clinical relevance of the differences. For instance, were the differences (i.e., lack of agreement) between the measures large enough to cause issues in clinical interpretation? In our example, only one of the participants had a systolic blood pressure error below the bottom reference line. Thus, overall, clinical interpretations would be similar between the two measures. However, examine the modified data in table 6.12. The standard deviation of the differences is now 12.86, which is much larger due to the larger magnitude of differences in the revised data. A systolic blood pressure of 140 indicates a person with hypertension. Only participant 10 is consistently indicated as hypertensive by both methods. The second Bland–Altman plot in figure 6.5 illustrates that the larger errors are at the higher blood pressure measures.

Table 6.12 Modified Systolic Blood Pressure Data

Participant	Clinician	Automated	Differences	Averages
1	103	105	2	104
2	117	115	−2	116
3	116	120	4	118
4	123	125	2	124
5	147	125	−22	136
6	125	125	0	125
7	151	125	−26	138
8	126	130	4	128
9	133	151	18	142
10	145	145	0	145
Mean (*M*)	128.6	126.6	-2	127.6
Standard deviation (*s*)	15.38	13.34	12.86	12.88

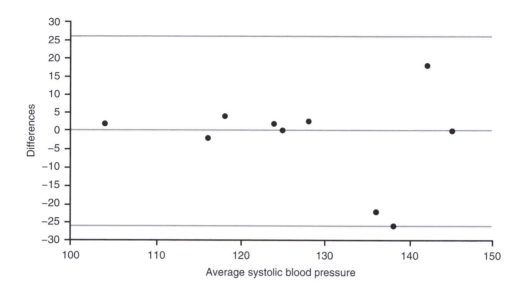

Figure 6.5 Bland–Altman plot of systolic blood pressure (modified).

MEASUREMENT AND EVALUATION CHALLENGE

You should now be able to determine the steps that Eva must take to select and administer a reliable and valid field test of aerobic capacity to her young adult members. She must first determine if the test she has selected is reliable—that is, are the results consistent from one administration to another, assuming that the participants have not actually changed training or activity levels? She must also be sensitive to the standard error of measurement (SEM).

Next, she must determine the concurrent validity between the proposed field test and the actual treadmill performance of the participants. Such information might be available in the research literature, or she may need to actually work with a researcher to obtain this vital information. She should also concern herself with the types of participants she is testing in comparison to those used in the validation process. If they are similar, she should feel confident that the field test results will provide a fairly accurate estimate of the participants' aerobic capacities. However, the field test will not provide an exact measure of the participants' aerobic capacities; she must be concerned about the standard error of estimate (SEE) in estimating actual $\dot{V}O_2$max from the surrogate (i.e., field) measure.

SUMMARY

The issues of reliability, objectivity, and validity are the most important ones that you will encounter when you test and evaluate human performance, whether the performance is in the cognitive, affective, or psychomotor domain. Reliability coefficients represent the consistency of response and range from 0 (totally unreliable) to 1.00 (perfectly reliable). Likewise, objectivity (interrater reliability) values range from 0 to 1.00. The standard error of measurement (SEM), a reliability statistic, reflects the degree to which a person's score will change as a result of errors of measurement. A validity coefficient represents the degree to which a measure correlates with a criterion. Statistical validity coefficients range from –1.00 to +1.00. The absolute value of the validity coefficient is important: A value of 0 indicates no validity; 1.00 represents perfect correlation with the criterion. The standard error of estimate (SEE), a validity statistic, indicates the degree to which a person's predicted or estimated score will vary from the criterion score.

Finally, you should realize that reliability and validity results are not typically generalizable. The reliability or validity obtained is specific to the group tested, the environment of testing, and the testing procedures. You must study whether the reliability and validity results that you obtain can be inferred to another population or setting before making such an inference.

Now that you are familiar with concepts related to reliable and valid assessment, you should be able to better evaluate the instruments that you might use in human performance testing.

WWW. **Go to HK*Propel* for videos, homework assignments, and quizzes that will help you master this chapter's content.**

Criterion-Referenced Tests: Cut Scores, Reliability, and Validity

OUTLINE

OBJECTIVES

After studying this chapter, you will be able to

- define a criterion-referenced test;
- explain the approaches for developing criterion-referenced standards;
- explain the advantages and limitations of criterion-referenced measurement;
- select appropriate statistical tests for the analysis of criterion-referenced tests;
- interpret statistics associated with criterion-referenced measurement;
- discuss and interpret epidemiologic statistics; and
- use SPSS and Excel to calculate criterion-referenced statistics.

WWW The lecture outline in HK*Propel* will help you identify the major concepts of the chapter.

MEASUREMENT AND EVALUATION CHALLENGE

Christina is an elementary school teacher whose school district has adopted the FitnessGram tests to determine student fitness levels. Christina has completed FitnessGram test administration training and has chosen to use the FitnessGram PACER test to assess aerobic capacity in her students. However, she is concerned with the reliability of the testing procedure. She is familiar with determining reliability for a norm-referenced procedure, but less so for the criterion-referenced testing procedures that FitnessGram uses. She would like to understand the process of estimating the reliability of FitnessGram's criterion-referenced tests.

As discussed in chapter 1, there are two common ways to interpret and evaluate the meaning of a test score in human performance: through norm-referenced (NR) evaluation or through criterion-referenced (CR) evaluation.

With NR evaluation, a person's score is compared with that of peers (e.g., same age and sex). In practice, the score is often compared with a previously developed norm, which results in either a specific percentile rank or a specific ranking category related to others (e.g., average, above average, or below average). Because the classification is based on how one's performance compares to others, NR evaluation is a relative classification (see figure 7.1). If the measurement interest is to rank or determine a person's relative position in the population, the use of NR evaluation is appropriate. In fact, NR evaluation has been the main way to interpret test scores in human performance for many years.

The advantage of NR evaluation is that the norms can be easily established as long as the data from the targeted reference population are available and the information is regularly updated. In practice, however, NR evaluation often has several challenges and limitations. It is difficult to update norms due to cost, time, and personnel constraints; in fact, many norms now used with human performance data (e.g., Hoffman, 2006) are outdated. Furthermore, the referenced population has to be "normal" or "healthy"; otherwise, derived classifications could become misleading (e.g., a classification of the average as "healthy" is misleading if the entire population is unhealthy; see Zhu, 2012a; 2013 for additional details). Finally, NR evaluation (e.g., set the 85th percentiles as the passing scores) tends to reward the top group of test takers and discourage the lower group (Zhu et al., 2011). Fortunately, these limitations can be overcome by employing criterion-referenced (CR) evaluation.

The concepts of CR evaluation and a related measurement practice, known as *mastery testing*, were introduced to education by Glaser in 1963, but real development and applications were not implemented until the late 1970s (see Popham, 1978) and 1980s (see Berk, 1980a,b; Livingston and Zieky, 1982). In contrast to the NR framework, in which the evaluation of a test taker's competency is judged relative to other test takers' performance, CR evaluation compares the test taker's performance with a predetermined absolute criterion or criterion behavior. As a result, CR tests are used to make categorical decisions, such as passing or failing, or classifying the participant as a master or a nonmaster of the trait measured (see figure 7.1 for an illustration of the CR evaluation). In assessment practice, the absolute criterion behavior could be whether a student has mastered the information taught in a specific subject, grade, or course (e.g., is a test taker skilled enough to be certified as a personal trainer?). Thus, the nature of the CR evaluation is absolute and is not affected by the performance or status of the test taker's peers. In addition, the standard developed for a CR evaluation often focuses on the minimally required competency

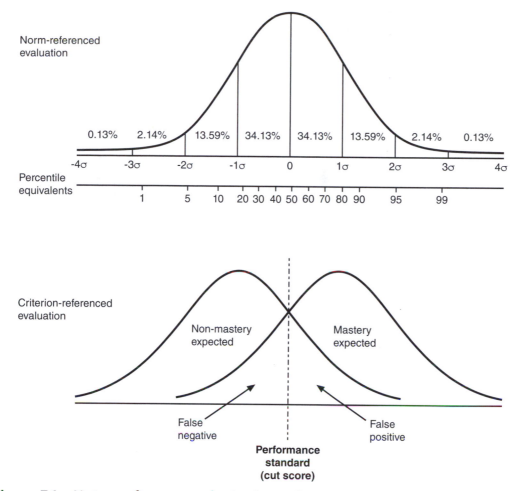

Figure 7.1 Nature of norm- and criterion-referenced evaluations.

of the trait measured. A test based on CR evaluation is known as a **criterion-referenced test (CRT)**. A CRT is one that is deliberately constructed to yield measurements that are directly interpretable in terms of specific performance standards. Performance standards are generally specified by defining a class or domain of tasks that should be performed by each participant (Nitko, 1984, p. 12).

Although CR evaluation has a number of measurement advantages over the NR evaluation, it has its own issues and challenges. Setting a valid "absolute" cut score—also known as the standard—is perhaps the most challenging one. This is because the consequence of the decision based on such a standard could be very serious (e.g., if a test taker is qualified for a job). The standard developed has to be connected with the construct measured and support the classification made. In addition, the standard should be able to stand alone and be valid for any individual from the targeted population. No matter how carefully a standard is developed, two kinds of errors are possible: a false positive (or false acceptance), in which nonmasters are classified as masters (e.g., an unqualified person is mistakenly passed on the test), and a false negative (or false rejection), in which masters are classified as nonmasters (e.g., a qualified person is mistakenly classified as not qualified) (see figure 7.1). As a result, it is important to provide reliability and validity evidence for a CRT.

[www] **Go to HK*Propel* to complete Student Activity 7.1.**

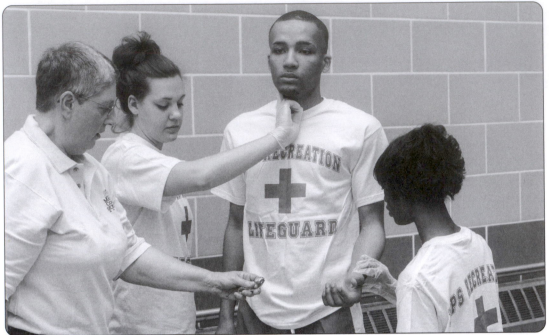

© Human Kinetics

Before he could become a lifeguard, this candidate had to pass many tests by demonstrating specific performance standards. These types of tests are known as CRTs.

When specific, well-defined goals are identifiable, CRTs may best measure the reliability and validity of that particular item. CRTs are not limited to nominal measurement. Often, continuous variables can be used with CR testing methods. For example, push-up or sit-up performance can be evaluated using criterion **cut scores** rather than NR methods. Historically, programmed instruction that centers on **behavioral objectives**—specifically written goals with instructions on how they can be obtained—is well suited for this type of measurement approach. Mastery instruments based on behavioral objectives are best exemplified by tests involving licensing, such as the test you took to get a driver's license and, in the human performance area, Red Cross standards for CPR, lifesaving, and swimming certification. It is easy to see in these examples that a minimum standard must be obtained before competency is proclaimed and a license is granted.

Another excellent example is the physical ability test used by many fire departments as part of the selection process for candidates to become firefighters. For example, a fire department might choose seven physical performance tests requiring balance, coordination, strength, endurance, and cardiorespiratory fitness. In these cases, cut scores are referenced to a standard based on a theoretical minimal performance level associated with being a successful firefighter. Achieving the standard (or minimal cut score) provides evidence that a test taker is qualified, passes, or is sufficient in some manner.

SETTING CRITERION-REFERENCED STANDARDS

Four basic approaches are used to set CR standards for tests of human performance (Safrit, Baumgartner, Jackson, and Stamm, 1980):

1. The *judgmental approach* is based on the experience or knowledge of experts and reflects what they believe to be an appropriate standard based on their background and

experience. For example, many high school volleyball coaches require players to be able to serve overhand to play on the varsity team. The coach might therefore set a cut score to make the varsity team, such as placing 8 out of 10 overhand serves in the court.

2. The *normative approach* uses NR data to set standards. For example, research from the Cooper Institute in Dallas, Texas, indicates that those who are in the top 80% of fitness levels for their age and sex are less likely to incur negative health outcomes. This criterion is based not only on experts' opinions—and research—but also on the available norms.

3. The *empirical approach* relies on the availability of an external measure of the criterion attribute, with cut scores directly established based on the data available for this measure. A concrete example of a pass/fail item using the empirical approach is a firefighter having to scale a 5-foot (1.5 m) wall to perform his or her duties. Another excellent example of this approach is the work of Cureton and Warren (1990) on cardiorespiratory fitness, presented later in this chapter. The FitnessGram Healthy Fitness Zone (see chapter 10) standards were established in this manner. The empirical approach is the least arbitrary of the four; however, it is seldom used because of the lack of a directly measurable external criterion.

4. The *combination method* involves using all available sources: experts, prior experience, empirical data, and norms. Usually, experts' opinions and norms are the basis for making CR decisions in human performance. The combination method is perhaps the best because it employs multiple methods to determine the most appropriate cut score.

An example of setting CR standards is cholesterol levels determined by professional associations. The American Heart Association and the National Heart, Lung, and Blood Institute have established cut scores for blood cholesterol levels related to the risk of coronary heart disease. They are as follows:

- Low risk: <200 mg/dl
- Moderate risk: Between 200 mg/dl and 240 mg/dl
- High risk: >240 mg/dl

A physician who is counseling a patient about the risk of coronary heart disease would use the patient's blood test results and compare them with these standards. The physician might advise the following:

- No need for concern (patient's level = 180 mg/dl).
- Increase physical activity levels and eat a low-fat diet (patient's level = 215 mg/dl).
- Increase physical activity levels, eat a low-fat diet, and take prescription medication (patient's level = 300 mg/dl).

In another example, the *Physical Activity Guidelines for Americans* (USDHHS, 2018) calls for all adults to engage in 150 minutes of MVPA each week for health benefits. If someone does not meet this criterion, specific suggestions and prescriptions can be provided to help achieve these guidelines. On the other hand, if a person does 150 minutes, he or she might simply be unmotivated to do additional physical activity because he or she sees the minimum as the goal. The physical activity guidelines specifically state that additional health benefits are achieved with increased amounts of physical activity (essentially, a dose response). Thus, the criterion serves a good purpose (a goal) but can also be problematic (unmotivated to do beyond the minimum amount).

WWW **Go to HK*Propel* to complete Student Activities 7.2 and 7.3.**

DEVELOPMENT OF CRITERION-REFERENCED TESTING

As mentioned above, the primary goal of an NR test (NRT) is to establish a range of behavior to discriminate among levels of knowledge, ability, or performance. If a certain level of performance is necessary, then NRTs do not provide this information in the most efficient way. For this reason, NRTs are not well suited to the assessment of specific objectives. For example, if an NR approach is used to determine who receives a driver's license, then your ability to pass the test would be based on the driving ability of the group tested and not on your absolute ability to drive a car. CRTs, on the other hand, are usually structured to assess far fewer objectives than a traditional NRT and therefore can be set up to identify specifically enumerated goals for behavioral items. For example, how many sit-ups should a 10-year-old boy be able to complete to be considered physically fit?

The primary difference between NRTs and CRTs is that CRTs are evaluated categorically. The traditional statistical techniques used to establish the reliability and validity of NRTs (presented in chapter 6) cannot be used with CRTs. Therefore, you must choose specific techniques that best estimate the reliability and validity of CR measures. Indexes of reliability associated with CR tests are called *indexes of dependability*. The indexes allow you to determine not only **proportion of agreement (P)** (which refers to the consistency with which performances are categorized across methods or trials) but also the consistency with which decisions are made. Specific examples of indexes of dependability are presented later in this chapter.

Cureton and Warren (1990) summarized the advantages and limitations of CR measurement as follows:

Advantages

• CR standards represent specific, desired performance levels that are explicitly linked to a criterion. In human performance, this criterion is often a health outcome.

• Because CR standards are absolute standards, they are independent of the proportion of the population that meets the standard. Regardless of the prevalence of people meeting the standard, the standard is still valid. For example, regardless of the percentage of the population that smokes, a goal is to reduce the number.

• If standards are not met, then specific diagnostic evaluations can be made to improve performance to the criterion level. For example, if a person does not meet the standard of 150 minutes of moderate-to-vigorous physical activity (MVPA) per week, you can provide specific feedback regarding what he or she should do.

• Because the degree of performance is not important, competition is based on reaching the standard, not on besting someone else's performance level. For example, has a person stopped smoking or achieved the 150 minutes of MVPA per week?

The following are other key advantages:

• Performance is linked to specific outcomes.

• People know exactly what is expected of them.

Limitations

• Cut scores always involve some subjective judgment. Because few criteria are clear-cut, philosophical guidelines can drastically affect the selection of the performance criterion. Authorities often disagree, so cut scores are sometimes arbitrarily determined.

• The consequences of misclassifications can be severe. For example, misclassification of patients could have severe health consequences if doctors prescribe medication based on a CR standard.

- Because cut scores must be set at some concrete level, those who attain the cut scores may not be motivated to continue to improve. Conversely, those who never attain the cut scores could become discouraged and lose interest.

To examine some of these limitations, Cureton and Warren (1990) studied CR standards for the 1-mile (1.6 km) run/walk test, for which the FitnessGram (Cooper Institute for Aerobics Research, 1987) and Physical Best (AAHPERD, 1988) both provide CR standards. To examine the validity of these standards, these authors developed an external criterion:

The criterion was defined as the lowest level of $\dot{V}O_2$ max in children and adolescents consistent with good health, minimized disease risk, and adequate functional capacity for daily living. Because no empirical data specifically identifies the minimum level [for children and adolescents], the criterion $\dot{V}O_2$ max was based primarily on indirect evidence relating aerobic capacity to health disease/risk. (p. 10)

Essentially, Cureton and Warren determined 1-mile (1.6 km) run/walk speeds that corresponded to criterion levels of $\dot{V}O_2$ max and converted these speeds to run times. The authors evaluated data on 581 boys and girls aged 7 to 14 against the FitnessGram criterion and the Physical Best criterion. Their results are presented in table 7.1. The table indicates that 496 of the 581 cases (85%) were properly classified by the FitnessGram standards, whereas only 357 (61%) were properly classified by the Physical Best standards. Fifteen percent were misclassified by the FitnessGram standards, and 39% were misclassified by the Physical Best standards. This analysis highlights the importance of setting standards correctly.

Table 7.1 Comparison of FitnessGram and Physical Best Standards for 1-Mile (1.6 km) Run/Walk Times

FITNESSGRAM		
Run/walk test result	Above the criterion $\dot{V}O_2$	Below the criterion $\dot{V}O_2$
Did achieve the standard	472 (81%)	61 (11%)
Did not achieve the standard	21 (4%)	24 (4%)
PHYSICAL BEST		
Run/walk test result	Above the criterion $\dot{V}O_2$	Below the criterion $\dot{V}O_2$
Did achieve the standard	227 (39%)	201 (35%)
Did not achieve the standard	23 (4%)	130 (22%)

STATISTICAL ANALYSIS OF CRITERION-REFERENCED TESTS

Not only is the procedure for setting cut scores critical, but so is the selection of statistical tests for examining the appropriateness of the cut scores. Selection of the statistical tests to be used to analyze CRTs is based on the same principles as those used for selecting NRTs. The first factor to consider is the level of measurement of the variables involved. With CRTs, data are categorized into nominal variables; therefore, you must select statistical tests appropriate for this level of measurement. Remember that nominal variables are categorical in nature. In order for tests that are measured on a continuous scale to be evaluated with CR instruments, the scores must first be categorized above and below the cut score criterion.

Contingency Table

For CR testing, the primary tool for analysis is a statistical technique using a **contingency table** (a 2 × 2 chi-square; see figure 7.2). Since reliability is a necessary condition for validity, let us look first at how to use the contingency table to examine a CRT's reliability. To determine the reliability, we administer a CRT twice at different times to determine if the person achieves the standard or does not achieve the standard each time. The greater the number of times the same classification is obtained on each test administration, the higher the CR reliability. Figure 7.2 illustrates the stability (dependability) of the CRT over 2 days. People classified as meeting the standard (n_1) on both days or not meeting the standard (n_4) on both days are consistently classified. Those classified as meeting the standard on one day and not meeting the standard the next (n_2) or vice versa (n_3) are misclassified. **Marginals** are the sum of observations for a specific row ($n_1 + n_2$ or $n_3 + n_4$) or column ($n_1 + n_3$ or $n_2 + n_4$) of a contingency table (see figure 7.2).

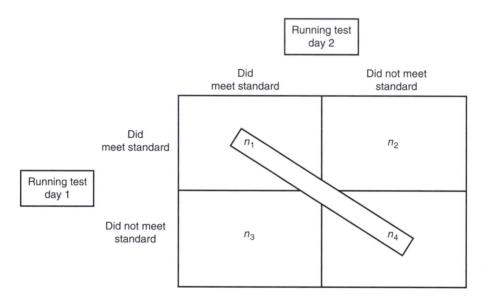

Figure 7.2 A 2 × 2 contingency table for a CRT taken over 2 days (reliability).

The next factor to consider in analysis is the specific characteristics or quality of a test. These characteristics are the same as those associated with NR testing. To establish the reliability of a CRT, you first need to determine whether you're concerned with the equivalence or the stability of the test. To measure the validity, you must have a criterion measure. The criterion measure reflects the true state of circumstances regarding the test being investigated. Recall the Measurement and Evaluation Challenge from the beginning of this chapter: Christina now knows she must administer the PACER test twice to estimate the CR reliability.

The same contingency table can also be used to determine the validity of a CRT. Figure 7.3 illustrates the accuracy (validity) of a running test, which serves as a predictor, to classify the true states of aerobic capacity measured by $\dot{V}O_2$max, which is the criterion measure. People classified as meeting the standard (n_1) on both tests or not meeting the standard (n_4) on both tests are consistently classified. Those classified as meeting the standard on the running test, but not meeting the standard of $\dot{V}O_2max$ (n_2) or vice versa (n_3) are misclassified. Since $\dot{V}O_2max$ represents the true status of one's aerobic capacity, n_2 represents a "false positive" misclassification and n_3 a "false negative" misclassification.

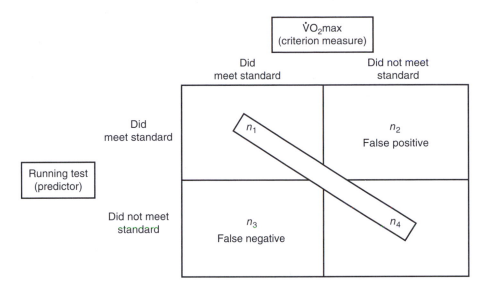

Figure 7.3 A 2 × 2 contingency table for a CRT versus its criterion measure (validity).

Commonly Used Statistics for Reliability and Validity of Criterion-Referenced Tests

Several statistics are used to estimate the reliability and validity of CRTs. In this text we present the techniques of chi-square (chapter 5), proportion of agreement (P), the phi coefficient (actually a Pearson product-moment correlation between two dichotomous variables), and Kappa (K). These are techniques that reflect association and agreement and can be used with data measured on a nominal scale.

As illustrated in chapter 5, the chi-square test is a test of association between nominally scaled variables. A positive association between how a person does on the first attempt of a CRT and on the second attempt demonstrates CRT stability reliability. Likewise, an association between how a person does on a field test of a measure and how he or she would do on a truthful measure (i.e., the criterion) of the characteristic being measured demonstrates CRT criterion-related validity. Recall from chapter 5 that the null hypothesis in both of these tests is that there is *no* association (or relation), but rejection of the null hypothesis results in deciding there *is* an association between the variables. Obviously, you would like there to be a positive relationship between how one is scored multiple times on the same CRT.

Note that the variables are scored 0 or 1 for both of the measures. You can calculate the Pearson product-moment (PPM) correlation coefficient (chapter 4) between the dichotomously scored variables. This special case of the PPM correlation coefficient is called the **phi coefficient**. The phi coefficient has limits of –1.00 and +1.00, with a value closer to 1.00 in absolute value indicating increased association, and a value close to zero indicating no association.

SPSS produces the chi-square and phi coefficient as statistics options within the Cross-tabs routine; this will be illustrated later in the chapter.

The proportion of agreement (P) value is established by adding the proportions in the cells that are consistently classified; thus P is equal to the number of agreements $(n_1 + n_4)$ divided by the total number $(n_1 + n_2 + n_3 + n_4)$:

$$P = \frac{(n_1 + n_4)}{(n_1 + n_2 + n_3 + n_4)} \tag{7.1}$$

The value of P ranges from 0 to 1.00, and the higher the value, the more closely the data are consistently (correctly) assigned to cells. However, the major problem with this statistic is that it does not consider the fact that some of these agreements (values up to .50) could be expected purely because of chance.

The **Kappa (K)** value is a widely used technique that allows for the correction for these chance agreements. Whereas P is a rough estimate of agreement or association between two nominal variables, K takes chance agreement into account and therefore gives a more conservative estimate of the association between two nominal variables. It is closely associated with the phi (ϕ) coefficient, which is the PPM correlation calculated on nominal data. K is most appropriately used to assess interobserver agreement but can be used in test–retest situations or to examine the agreement between a predictor and a criterion (i.e., test validity) that are both nominally scaled. The formula for K is

$$K = \frac{(P - P_c)}{(1 - P_c)} \tag{7.2}$$

where P is the proportion of observed agreement and P_c is the proportion of agreement due to chance. Consider the following reliability example: Four hundred elementary students performed the 1-mile (1.6 km) run on each of 2 days. Their instructor wanted to know if the test could consistently measure the students' abilities to achieve the cut scores established in FitnessGram. Table 7.2 presents these data. Note that this is a reliability example because it involves administration of the same test on more than one occasion.

Table 7.2 CRT Test–Retest Reliability Example

		DAY 2		
		Did achieve the standard	Did not achieve the standard	Total
DAY 1	Did achieve the standard	250	50	300
	Did not achieve the standard	20	80	100
	Total	270	130	400

$\chi^2 = 137.13$, $df = 1$, $p < .001$; phi coefficient = .586

For this example, P is calculated to be

$$(250 + 80) / 400 = 330 / 400 = .825$$

K is then calculated to correct for chance. The proportion of chance values (P_c) is determined by multiplying the marginals and dividing by n^2, as shown here:

$$(270 \times 300) / (400 \times 400) = .506$$

and

$$(130 \times 100) / (400 \times 400) = .081$$

The sum of these properties is .587. Therefore, K = (.825 – .587) / (1 – .587) = .238 / .413 = .576. This value is substantially lower than the P value of .825. This illustrates why it is a good idea to calculate chi-square, phi coefficient, proportion of agreement, and Kappa values to give the most information about the association involved.

Thus, given a 2 × 2 table, the proportion of observed agreement (P) is determined by summing the number of agreements that appear on the diagonal of the table and dividing by the total number of paired observations. These proportions are summed across all the cells to obtain a total proportion of chance agreement (P_c). Then the proportion of agreement and the proportion of chance agreement are substituted into the Kappa formula.

The values of K can theoretically range from –1.00 to +1.00; however, a negative value of K implies that the proportions of agreement resulting from chance are greater than those attributable to observed agreements. For that reason, K practically ranges from 0.00 to 1.00. The magnitude of the K is interpreted just as any other reliability or validity coefficient, with the higher the values, the better. However, because of the adjustment for chance agreement, values seldom exceed .75. Kappas of .20 or below indicate slight or poor agreement. Values ranging from .41 to .60 are often considered to be moderate, whereas values in the .61 to .80 range are usually considered to be substantial (Viera and Garrett, 2005).

A serious disadvantage of the K coefficient is that it is highly sensitive to low values in the marginals and small contingency tables because chance values are so high. It is also limited to square contingency tables. Again, SPSS can provide the Kappa coefficient as one of the statistics options within the Crosstabs routine.

Criterion-Referenced Reliability

For the most part, the same types of measurement characteristics (i.e., reliability and validity) exist for CRTs as with NRTs. Equivalence reliability as well as stability reliability can be estimated (see chapter 6).

Equivalence Reliability

Mahar and colleagues (1997) examined the CR and NR reliability of the 1-mile (1.6 km) run/walk and the PACER (Progressive Aerobic Cardiovascular Endurance Run) test, which are both used on the FitnessGram. The sample consisted of 266 fourth- and fifth-grade students who were administered two trials of the PACER test and one trial of the 1-mile (1.6 km) run/walk. Equivalence reliability was examined between the 1-mile (1.6 km) run/walk and each trial of the PACER test for the total sample and also by sex. Both P and K values were calculated for all cases. The results are presented in table 7.3.

Inspection of the results indicates that fairly high P values (.65 ≤ P ≤ .83) are associ-

Table 7.3 Criterion-Referenced Equivalence Reliability Between the 1-Mile (1.6 km) Run/Walk and PACER

TESTS	TOTAL SAMPLE	BOYS	GIRLS
Trial 1			
P	.76	.83	.66
K	.51	.65	.33
Trial 2			
P	.71	.76	.65
K	.43	.52	.30

Note: For trial 1, n = 126 boys, n = 95 girls, and total (both) N = 221; for trial 2, n = 122 boys, n = 91 girls, and total (both) N = 213.

ated with varying levels of K (.30 ≤ K ≤ .65). Remember that you expect the K values to be more conservative than the P values. Whereas the equivalence reliability looks to be at least acceptable for the total sample and with boys alone, the values for the girls are much lower (P values of .66 and .65 and K values of .33 and .30). This result points out not only the nature of CRT reliability estimates, but also the importance of examining specific reliability situations.

Stability Reliability

Rikli, Petray, and Baumgartner (1992) examined the reliability of distance-run tests for students in kindergarten through grade 4. Test–retest reliability estimates using both NR (intraclass reliability) and CR techniques were calculated. Data on the 1-mile (1.6 km) and half-mile (0.8 km) run/walk tests were gathered in the fall (for 1229 students—621 boys, 608 girls) and the following spring (1050 students—543 boys, 507 girls). The P values for these data were calculated using the Physical Best and FitnessGram cut scores. The results are presented in table 7.4.

Inspection of the results indicates that all reliability estimates fall in the acceptable

Table 7.4 Criterion-Referenced Reliability Estimates

		AGE									
		5		6		7		8		9	
		F	S	F	S	F	S	F	S	F	S
PHYSICAL BEST											
1/2 mi (.8 km)	M	.79	.86	.98	.95	.92	.86	.97	.83	.89	.90
	F	.88	.74	.98	.90	.89	.91	.96	.91	.92	.75
1 mi (1.6 km)	M	.70	.70	.94	.89	.95	.92	.90	.94	.95	.93
	F	.75	.88	.88	.73	.81	.87	.95	.94	.92	.90
FITNESSGRAM											
1 mi (1.6 km)	M	.75	.70	.76	.66	.85	.77	.91	.85	.86	.83
	F	.69	.51	.71	.45	.81	.85	.90	.84	.83	.94

Note: F is fall semester; S is spring semester.

Reprinted by permission from R.E. Rikli, C. Petray, and T.A. Baumgartner, "The Reliability of Distance Run Tests for Children in Grades K-4," *Research Quarterly for Exercise and Sport* 63 (1992): 270-276. Reprinted by permission of Taylor & Francis (Taylor & Francis Ltd, www.tandfonline.com).

range (P >.70) except the FitnessGram standards for 5-year-old girls (fall = .69, spring = .51) and for 6-year-olds in the spring (boys = .66, girls = .45). These CR values are consistently higher than the associated NR values. This is understandable because P values are not corrected for chance. Rikli and colleagues (1992) also explained this as follows: "The higher values for Physical Best are not surprising because *P* is always larger when there is a large percentage of scores that either meet or do not meet the standard on both the test and retest" (p. 274).

Morrow, Martin, and Jackson (2010) used the just-described CR reliability procedures with school-based teachers to see if repeated administrations of the FitnessGram test by the same teachers resulted in students achieving or not achieving the FitnessGram Healthy Fitness Zone consistently on two occasions. Using aerobic fitness, body composition, and muscle-strengthening items from the FitnessGram resulted in percentage of agreements between .74 and .97 (median = .84); modified Kappas between .48 and .94 (median = .68);

and phi coefficients between .40 and .92 (median = .65). All chi-square results were significant (P <.001). In total, these results suggested that there was good teacher reliability with FitnessGram assessments.

Criterion-Referenced Validity

Two types of validity evidence are often examined for CRT. The first one is called *criterion-related validity*, which is usually established by confirming some type of test–criterion relationship, either concurrent or predictive. The second one is called *construct-related validity*, which can be demonstrated by examining the overlap of two divergent groups measured on a continuum.

Criterion-Related Validity

An example of the criterion-related validity approach, in this case concurrent validity, can be seen in the work of Cureton and Warren (1990). Remember, Cureton and Warren studied criterion-referenced standards for the 1-mile (1.6 km) run/walk test. The FitnessGram (Cooper Institute for Aerobics Research, 1987) and Physical Best (AAHPERD, 1988) tests were used. Both tests provided criterion-referenced standards. The data were presented in table 7.1.

The results from these two CRT examples are presented in table 7.5. These results illustrate some of the problems of interpreting CRT results. Both tests have significant chi-square results, the phi coefficient is higher for the Physical Best standards, and the proportion of agreement and Kappa coefficient are higher for the FitnessGram analysis.

Look again at table 7.1, which shows that 85% of participants were correctly classified by

Table 7.5 Comparison of Two CRT Validities

	FitnessGram	Physical Best
Chi-square result	$\chi^2 = 55.35$, $df = 1$, $p < .001$	$\chi^2 = 66.41$, $df = 1$, $p < .001$
Phi coefficient	.309	.338
Percent agreement (P)	.85	.61
Kappa	.288	.277

the FitnessGram test. Of the remaining 15%, 11% achieved the standard on the run/walk test but were below the criterion $\dot{V}O_2$. These results are *false positives*—that is, a participant is said to meet the standard on the run/walk (i.e., the field test), but in actuality (i.e., the criterion) he or she is below the standard. Notice too that 4% of participants did not meet the standard on the run/walk but were above the criterion $\dot{V}O_2$. These participants are referred to as *false negatives* because the field test results indicate they do not meet the standard but their performance on the criterion is above the standard.

The impact of false negatives and false positives can be important in determining which field test you might use. For example, consider a field test of cholesterol that involves a simple finger stick to obtain a drop of blood. The criterion method for estimating cholesterol would be from drawing venous blood. The results of your finger stick (i.e., the field test) can be accurate (you have been correctly identified as having a healthy or unhealthy cholesterol level) or inaccurate. If the field test reports that your cholesterol level is healthy when in fact it is not, the results are a false positive. If the field test results indicate that your cholesterol level is too high when it is actually in the healthy range, the result is a false negative.

Construct-Related Validity

Setting cut scores is a difficult undertaking. The divergent group method, in which one uses two groups that are clearly different from each other, can be used as a construct validation procedure. To establish a cutoff using this technique, we plot the distributions of scores for the divergent groups, using the point in the curves where the scores overlap as the criterion cut score (see figure 7.4). A theoretical application of this approach would be to select two groups, one made up of participants who are physically active enough to obtain a health benefit, and the other who are not active enough for a health benefit. Obtaining data on the amount of physical activity for each of these groups and then graphing it should help set a cut score for a minimum amount of physical activity needed for a health benefit. For more about this method, see Plowman (1992).

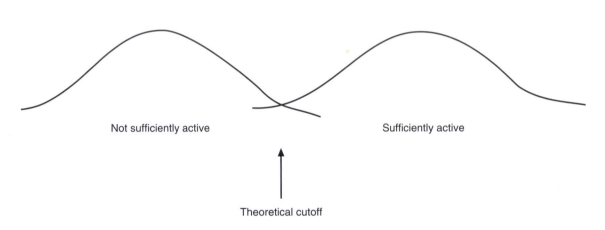

Figure 7.4 A theoretical example of the divergent group method.

Tarter and colleagues (2009) provide another example of construct validity using a construct-related CRT approach called *receiver operating curve analysis* (ROC). The authors were examining the use of an aggregate performance measure called the composite physical fitness index (CPFI) to predict potential for playing in the National Hockey League (NHL). Playing in the NHL was defined as playing at least five games within a 4-year period after a player was drafted. The ROC technique maximizes the number of discriminations between those who achieved the criterion and those who did not. It was found that setting the cutoff at the 80th percentile of the CPFI for defensemen yielded a 70% probability of success, whereas setting it at the 80th percentile for forwards yielded only a 50% probability of making the NHL. When the CPFI scores were adjusted to the 90th percentile, the probabilities for defensemen and forwards shifted to 72% and 61%, respectively. This study again points to the importance of setting cut scores appropriately.

Morrow, Going, and Welk (2011) present 11 manuscripts in a supplement issue of the *American Journal of Preventive Medicine* in which similar ROC procedures were used to determine criterion-referenced standards for FitnessGram aerobic capacity and body composition measures. Effectively using the procedures described above, they were able to determine cut scores for the FitnessGram Healthy Fitness Zone based on whether or not the children and youth had metabolic syndrome. Thus, metabolic syndrome served as the criterion; cut scores were determined for aerobic capacity and body composition that differentiated those with and without metabolic syndrome. This example also illustrates a criterion-referenced validity study.

CRITERION-REFERENCED TESTING EXAMPLES

The use and interpretation of reliability and validity procedures with CRTs are similar to those presented in chapter 6. If two trials of the same measure are administered, then stability reliability is being assessed. With CRTs, this is the reliability or dependability of classification. If two tests that are thought to measure the same thing are being compared, equivalence is being assessed. With CRTs, the equivalence is whether the two tests result in equivalent classifications for those being assessed. There is no analogy to internal consistency reliability (i.e., alpha reliability) with CR testing. If one of the measures is a criterion, the matter being investigated is validity. As we have pointed out several times in this chapter, determination of the criterion is the most difficult aspect of CR testing to establish. However, when the analysis is to determine if the measure is significantly associated with a criterion, test validity is being investigated.

The following examples present specific applications of selected techniques for the assessment of reliability and validity using CR testing. Try to calculate P and K for the following mastery items.

Mastery Item 7.1

Assume that two CR physical fitness tests have been developed to establish the cut scores for physical performance on sit-ups. For test 1, participants perform sit-ups with their hands clasped on their chests ("handches"). For test 2, they perform sit-ups with their hands behind their heads ("handhead"). Are the tests equivalent? A sample of participants is administered test 1 and test 2, and a 2 × 2 contingency table is developed to determine if there is equivalence in the classification on the two tests. The data are presented in table 7.6. Use the following steps to obtain the chi-square, phi coefficient, and K statistics. SPSS does not calculate P, so you will have to do that by hand from the output you receive. (An Excel template is provided in HK*Propel* in chapters 5 and 7.)

1. Start SPSS.
2. Open the table 7.6 data from HK*Propel*.
3. Go to Analyze → Descriptive Statistics → Crosstabs.
4. Put "handches" in the rows and "handhead" in the columns by clicking the arrow keys.
5. Click Statistics.
6. Check the Chi-square, Phi and Cramer's V, and Kappa boxes.
7. Click Continue.
8. Click OK.

Table 7.6 Example of Equivalence Reliability

Participant	handches	handhead	Participant	handches	handhead
1	1	1	21	0	0
2	1	1	22	0	0
3	1	1	23	0	0
4	1	1	24	0	0
5	1	1	25	0	0
6	1	1	26	0	0
7	1	1	27	0	0
8	1	1	28	0	0
9	1	1	29	0	0
10	1	1	30	0	0
11	1	1	31	0	0
12	1	1	32	0	0
13	1	1	33	0	0
14	1	0	34	0	0
15	1	0	35	0	1
16	1	0	36	0	1
17	1	0	37	0	1
18	1	0	38	0	1
19	1	0	39	0	1
20	1	0	40	0	1

Note: 0 = fail; 1 = pass.

Mastery Item 7.2

Human performance specialists use test–retest reliability—the stability of a test over successive administrations—more frequently than equivalence to determine the reliability of CRTs. Let's assume that we select one test and administer it on Friday (day 1) and then administer it to the same group of students on the following Monday (day 2). The data are presented in table 7.7. We are concerned with the consistency of classification across the two testing periods. Use the SPSS commands from Mastery Item 7.1 to calculate the statistics.

Table 7.7 Example of Stability Reliability

Participant	Friday	Monday	Participant	Friday	Monday
1	1	1	21	0	0
2	1	1	22	0	0
3	1	1	23	0	0
4	1	1	24	0	0
5	1	1	25	0	0
6	1	0	26	0	0
7	1	0	27	0	0
8	1	0	28	0	0
9	1	1	29	0	0
10	1	1	30	0	0
11	1	1	31	0	1
12	1	1	32	0	1
13	1	1	33	0	0
14	1	1	34	0	0
15	1	1	35	0	0
16	1	1	36	0	0
17	1	1	37	0	0
18	1	1	38	0	1
19	1	0	39	0	1
20	1	0	40	0	1

Note: 0 = failed to meet criterion; 1 = met the criterion.

Mastery Item 7.3

From a validity standpoint (whether it be predictive validity, concurrent validity, or construct validity), the application of the 2 × 2 contingency table is appropriate. For example, let's assume that we have a standard for body composition and we suspect that if people achieve the FitnessGram Healthy Fitness Zone for body composition, they are less likely to have metabolic syndrome. Alternatively, if they are in the Needs Improvement Zone, they are more likely to have metabolic syndrome. Therefore, we want to determine if the FitnessGram body composition cut score can properly classify those who have metabolic syndrome as opposed to those who do not. The data are presented in table 7.8. Use the same SPSS commands presented in Mastery Item 7.1 to interpret the validity of the cut score for body composition to predict metabolic syndrome. Hint: Put the Needs Improvement Zone in the rows and the metabolic syndrome in the columns. Note these data are for illustrative purposes only.

Table 7.8 Example of Statistical Validity

Participant	Metabolic syndrome	Needs Improvement Zone	Participant	Metabolic syndrome	Needs Improvement Zone
1	0	1	31	1	1
2	0	1	32	1	1
3	0	1	33	1	1
4	0	1	34	1	1
5	0	1	35	1	1
6	0	1	36	1	1
7	0	1	37	1	1
8	0	1	38	0	0
9	0	1	39	0	0
10	1	0	40	0	0
11	1	0	41	0	0
12	1	0	42	0	0
13	1	0	43	0	0
14	1	0	44	0	0
15	1	0	45	0	0
16	1	0	46	0	0
17	1	0	47	0	0
18	1	0	48	0	0
19	1	0	49	0	0
20	1	0	50	0	0
21	1	1	51	0	0
22	1	1	52	0	0
23	1	1	53	0	0
24	1	1	54	0	0
25	1	1	55	0	0
26	1	1	56	0	0
27	1	1	57	0	0
28	1	1	58	0	0
29	1	1	59	0	0
30	1	1	60	0	0

Note: 0 = yes; 1 = no.

APPLYING CRITERION-REFERENCED STANDARDS TO EPIDEMIOLOGY

Epidemiology is the study of the distribution and determinants of health-related states and events in populations and the applications of this study to the control of health problems (Last, 1992). This fundamental science of public health uses hypothesis testing, statistics, and research methods to develop an understanding of the frequency and distribution of mortality (death) and morbidity (disease or injury), and more importantly, the risk factors that are causally related to mortality and morbidity (Stone, Armstrong, Macrina, and Pankau, 1996). Descriptive epidemiology seeks to describe the frequency and distribution of mortality and morbidity according to time, place, and person—for instance, what was the rate of breast cancer in adult women in the United States during the 1990s? Analytical epidemiology pursues the causes and prevention of mortality and morbidity—for example, does obesity increase the risk of breast cancer in women? In women who are obese, does moving into a healthy weight range lower the risk of breast cancer? Epidemiology uses research approaches that are both prospective (tracking a study group into the future) and retrospective (looking back at a database of previously collected data). Epidemiologists use a variety of research designs, some of which are depicted in table 7.9.

Table 7.9 Research Designs in Epidemiology

Type	Description
EXPERIMENTAL	
Randomized clinical trial	Randomly assigns participants to treatments or exposures
Community trial	Randomly assigns whole communities to treatments or exposures
OBSERVATIONAL	
Case series	Notes cases at a particular time or place
Cross-sectional	Takes a snapshot of identifiable groups at one time
Proportionate mortality or morbidity study	Compares results of a study group to the population
Case-control	Compares known cases of mortality or morbidity with matched noncases
Cohort	Longitudinal; generally tracks populations long term

In the field of human performance, modern epidemiologic research has clearly discovered the increased risk for a variety of chronic diseases related to a sedentary or physically inactive lifestyle (Ainsworth and Matthews, 2001; Caspersen, 1989; USDHHS 1996; 2008). This type of research is becoming increasingly popular in human performance measurement. It is closely related to CRT because the variables are often nominal in nature and some of the statistics used are those calculated from a 2 × 2 contingency table. The criterion measure is categorical (e.g., alive or dead; has a disease or does not have a disease). The predictor variables can be nominal (e.g., gets sufficient physical activity or does not get sufficient physical activity) or continuous (e.g., weight). It is when the predictor and criterion variables are both nominal that epidemiologic statistics are most like those of CRT (and can even be calculated with SPSS Crosstabs or Excel).

Epidemiology requires the use of advanced statistics and complicated multivariate models to understand the relationships between risk factors and mortality and morbidity while controlling for confounding factors or extraneous variables. Those types of analyses are beyond the scope of this text and are not necessary for us to know at present. However, we do need to know some basic procedures and statistics to understand how CR standards play a role in epidemiology. Two basic statistics are the calculations of incidence and prevalence.

• *Incidence.* The number, proportion, rate, or percentage of *new* cases of mortality and morbidity. Incidence could be calculated in a randomized clinical trial or a prospective, longitudinal cohort study.

• *Prevalence.* The number, proportion, rate, or percentage of total cases of mortality and morbidity. Prevalence is calculated in a cross-sectional study.

Values of incidence and prevalence are often expressed as a rate, which is the number of cases per unit of the population. An example would be 10 cases per 1000 in the population or 100 deaths per 100,000 in the population. The value of expressing incidence and prevalence as a rate is that two populations of different sizes can be compared—for example, the mortality rate in Dallas, Texas, can be compared with that in New York City.

In analytical epidemiology, we convert measures of incidence or prevalence into estimates of risk:

• *Absolute risk.* The risk (proportion, percentage, rate) of mortality or morbidity in a population that is exposed or not exposed to a risk factor.

• *Relative risk.* The ratio of risks between the exposed or unexposed populations. This statistic is calculated with incidence measures.

• *Odds ratio.* An estimate of relative risk used in prevalence studies.

• *Attributable risk.* The risk of mortality and morbidity directly related to a risk factor. This risk can be thought of as the reduction in risk related to removing a risk factor.

Let's combine CR standards with an example of a simple analysis in epidemiology by examining the results of a theoretical epidemiologic study about the relationship of cholesterol and mortality attributable to heart attack. High cholesterol is defined by the American Heart Association and the National Heart, Lung, and Blood Institute as a value of 240 mg/dl or above (e.g., the CR standard for total cholesterol). Examine table 7.10, which is a 2 × 2 contingency table. We have labeled the cells A, B, C, and D to make all descriptive and analytical calculations quite simple. We also conduct our analyses on incidence and prevalence bases. In this study, 56 participants with high cholesterol and 44 without high cholesterol are compared. All had a genetic history of early coronary heart disease. Note that both variables are categorical in this example.

Table 7.10 Results of a Hypothetical Study Relating Cholesterol and Heart Attack Mortality

Exposure	OUTCOME	
	Heart attack deaths	No heart attack deaths
High cholesterol	A 25	B 31
No high cholesterol	C 7	D 37

If you examine all the results in figure 7.5 you can observe the following:

• All calculations can be made from the easy-to-follow formulas using the A, B, C, and D cell identifiers.

• The absolute risk for heart attack death was 32% for all participants, 45% for those with high cholesterol, and 16% for those without high cholesterol.

• If a participant had high cholesterol, the relative risk of 2.81 indicated that high cholesterol elevated the risk of heart attack mortality by a multiplier of 2.81.

• If a participant had high cholesterol, the odds ratio indicated elevated odds of heart attack mortality by a multiplier of 4.26.

• The attributable risk indicated that high cholesterol contributed to 64% of the heart attack mortality. Thus, heart attack mortality could be reduced by 64% if high cholesterol were no longer present in people of this population.

The example used in table 7.10 and figure 7.5 serves as a simple demonstration of some basic concepts and analyses in epidemiology. However, research studies using epidemiologic methods have demonstrated strong relationships between levels of physical activity and fitness and a variety of mortality and morbidity outcomes from chronic diseases. Chapter 10 will discuss some of those specific findings in more detail.

$$Total = \frac{A+C}{A+B+C+D} = \frac{25+7}{25+31+7+37} = \frac{32}{100} = .32 \text{ or } 32\%$$

$$High = \frac{A}{A+B} = \frac{25}{25+31} = \frac{25}{56} = .45 \text{ or } 45\%$$ Absolute risk

$$Not\ high = \frac{C}{C+D} = \frac{7}{7+37} = \frac{7}{44} = .16 \text{ or } 16\%$$

$$RR = \frac{A \div (A+B)}{C \div (C+D)} = \frac{.45}{.16} = 2.81$$ Relative risk

$$OR = \frac{AD}{BC} = \frac{25 * 37}{7 * 31} = \frac{925}{217} = 4.26$$ Odds ratio

$$AR = \frac{[A \div (A+B)] - [C \div (C+D)]}{A \div (A+B)} = \frac{.45 - .16}{.45} = .64 \text{ or } .64\%$$ Attributable risk

Figure 7.5   Statistical analysis of epidemiological data in table 7.10.

Mastery Item 7.4

Confirm that you can calculate the odds ratio and relative risk by using the following Cross-tabs routine.

1. Start SPSS.
2. Open the table 7.10 data from HK*Propel.*
3. Go to Analyze → Descriptive Statistics → Crosstabs.
4. Put "cholesterol" in the rows and "heart attack" in the columns by clicking the arrow keys.
5. Click Statistics.
6. Check the Risk box.
7. Click Continue.
8. Click OK.

When you review the SPSS results, you should see that the odds ratio and relative risk values are presented in the SPSS output.

Note that in table 7.10, a positive exposure (i.e., classified as having high cholesterol) is listed in the first row, followed by negative exposure in the second row. Likewise, the positive outcome (record showed heart attack death) is listed in the first column and the negative outcome is listed in the second column. We recommend that you construct the contingency table in this fashion. You can reorder either or both of the variables and arrive at similar conclusions, but setting it up as we suggest will generally make your interpretation more understandable.

Mastery Item 7.5

Table 7.11 shows a 2×2 contingency table containing results from a study conducted by Bungum, Peaslee, Jackson, and Perez (2000). The study examined the relationship of physical activity during pregnancy and the risk of Cesarean birth compared with normal vaginal birth. Perform an analysis like the one presented in figure 7.5 with these data.

Table 7.11 Results of a Study Relating Physical Activity During Pregnancy and Type of Birth

Exposure	OUTCOME	
	Cesarean section birth	Vaginal birth
Sedentary	A 26	B 67
Active	C 7	D 37

Mastery Item 7.6

Let's now apply these epidemiologic statistics to the data in table 7.8. Remember that you want to set up the contingency table as suggested previously for the negative and positive exposures and negative and positive outcomes. Run SPSS Crosstabs on the data. Click on Statistics and then check the Risk box. Confirm that the risk of having metabolic syndrome goes up by 294.9% if one is in the Needs Improvement Zone (notice that the Odds ratio = 3.949).

Dataset Application

The chapter 7 large dataset in HK*Propel* consists of data for school-age children on two Fit-nessGram tests: body mass index (BMI) and skinfolds (to estimate percentage of fat). This is an example of equivalence reliability. Theoretically, it should not make any difference if either of these tests is used; a person who is overweight or at risk should be identified as such with each test. Use SPSS to calculate chi-square, the phi coefficient, and Kappa. You will need to calculate proportion of agreement from the 2 × 2 table that you produce with SPSS. What is your interpretation of the results of these two tests? Are they equivalent? Do you get similar results if you conduct the analyses separately for boys and girls?

MEASUREMENT AND EVALUATION CHALLENGE

Christina now sees the similarities between reliability testing in NR and CR settings. Additionally, she understands why she needs to administer the FitnessGram PACER test to her students on two occasions. She will then classify them as meeting the FitnessGram Healthy Fitness Zone or not on each test administration. She can then use the procedures illustrated in this chapter (i.e., chi-square, phi coefficient, proportion of agreement, and Kappa) to confirm that the obtained results have evidence to support the reliability of the testing process and results.

SUMMARY

Several specific measurement objectives in human performance are well suited for CR measurement. CR testing is the method of choice when variables are categorized and where an obvious level of proficiency must be achieved before proceeding to the next level (e.g., flotation and treading water skills need to be mastered before one enters the deep end of the pool). The primary problem associated with CR testing, however, is establishing a criterion or cut score. Because few measurement problems in human performance have concrete criterion scores associated with them, cut scores have to be established from experts' opinions or normative data. Cut scores can often be arbitrary, thus affecting empirical validity. The establishment of these scores also affects the reliability and the statistical validity of the test. Therefore, criterion scores must be set with a high degree of caution.

In the area of youth fitness testing, CR standards have been established by test developers (e.g., FitnessGram). In other areas of human performance, such as sport skills testing, such standards have typically not been set. In epidemiological research and practice, many cut scores have been established that are directly related to health risks. See Morrow, Zhu, and Mahar (2013) for a summary of CR testing use with the FitnessGram and Zhu (2013) for a more detailed description on setting standards in human performance.

CR statistical techniques are used to analyze data. The typical statistics used with CR testing reliability and validity are chi-square, phi coefficient, proportion of agreement (P), and Kappa (K), which adjusts the proportion of agreement for chance.

Finally, you learned how epidemiologic statistics are closely related to CR testing procedures. Epidemiology is a powerful method for identifying risk factors for various disease outcomes.

[www] **Go to HK*Propel* for videos, homework assignments, and quizzes that will help you master this chapter's content.**

Part IV

Human Performance Applications

Part IV of this book is about what you will do after graduation. Some of you will work in school-based instructional settings, others in athletics, and still others in a wide variety of health professional and human performance job settings, such as physical therapy, health clubs, corporate fitness or wellness programs, hospitals, and graduate schools. The particular measurement and evaluation tasks that you will encounter will vary from job to job. However, knowledge of reliability and validity and the skills to analyze data using SPSS will serve you well, whatever your career.

We introduce part IV with a brief description of the cognitive, psychomotor, and affective learning that you will be evaluating after graduation. As you learned in chapter 1, each of these domains is reflected in a taxonomy, and each level of the taxonomy is built on the levels below it. Measurement protocols for each domain must therefore be carefully considered so that the measurement tasks that you conduct reflect the appropriate level of learning or performance expected of the people with whom you are working. A key skill in measurement is the ability to design and use measurement protocols that discriminate among people who are actually at different levels of achievement.

In part I, you were introduced to tests and measurement and the use of computers to help you make evaluation decisions. In part II, you learned basic statistical concepts, including descriptive statistics, correlation and prediction, and inferential statistics. In part III, you learned about reliability and validity theory. Now you have the background, theory, tools, and information with which to make valid decisions. In part IV, we turn to the various domains within which you will make these decisions.

Chapter 8 contains important information on grading and evaluation, including how to make decisions leading to valid student assessments and grade reports, as well as how to properly obtain composite scores. The contents learned in this chapter can be applied to all domains. Chapter 9 provides information about developing valid written tests and surveys; this chapter relates to the cognitive domain.

Because the medical and scientific literature on the relationships among physical fitness, physical activity, and the prevention of disease increases almost daily, exercise scientists must fully understand the issues of reliability and validity as they relate to human performance testing. Chapter 10 addresses reliability and validity in the measurement and evaluation of health-related physical fitness, whereas chapter 11 deals with measurement and evaluation of performance-related physical fitness. Chapter 12 presents techniques for reliable and valid measurement of ability and skills assessment for sport and human performance, and chapter 13 focuses on the measurement of physical activity and energy expenditure. All of these chapters can be applied to the psychomotor domain. Chapter 14, on the affective domain, provides guidelines for making valid decisions when using psychological measurements in sport and exercise psychology. Finally, chapter 15 presents examples of alternative strategies that can be used to assess student achievement; the strategies learned in this chapter can also be applied in all three domains.

Evaluation: Theory and Practice

OUTLINE

OBJECTIVES

After studying this chapter, you will be able to

- list the proper criteria for grade assignment;
- illustrate the methods for assigning grades to one test or performance;
- use the methods for assigning final grades;
- apply evaluation principles and practices to a variety of nonschool settings; and
- understand that there are many uses for grades, including motivating and guiding the learning, educational, and vocational plans of students, as well as their personal development; communicating to students and parents about student progress; condensing information for use in determining college admissions, graduate school admissions, and scholarship recipients; communicating a student's strengths and limitations to possible employers; and helping the school tailor education to student needs and interests and evaluate the effectiveness of teaching methods.

WWW The lecture outline in HK*Propel* will help you identify the major concepts of the chapter.

> ## MEASUREMENT AND EVALUATION CHALLENGE
>
> Tomas is getting ready to teach a unit on tennis to his high school physical education class. He wants to weight the cognitive and the psychomotor aspects of the unit equally. Tomas plans to give two written tests during the unit to measure cognitive objectives; the first examination will count for 20% of the final grade, and the final examination will count for 30% of the final grade. He plans to use two skills tests to measure the psychomotor objectives of the unit; the serving test will be weighted as 35% of the grade, and the ground stroke test will be 15% of the final grade. What steps should Tomas follow to ensure that the final composite score for each student will reflect these weightings and provide fair and appropriate feedback to his students on their tennis competency?

EVALUATIONS AND STANDARDS

Tomas' task, in fact, is involved in another part of decision making in physical education practice: evaluation. In chapter 1 we defined several terms important to the measurement and evaluation process, including formative and summative evaluation. As discussed in that chapter, evaluations result from a decision-making process that places a judgment of quality on a measurement. In physical activity instruction, teachers make formative evaluations at the beginning of and during the instructional process and use them to detect weaknesses in student achievement and to direct (or redirect) future learning activities. A formative evaluation can be a formal measurement activity, such as a pretest before instruction, or an informal or subjective evaluation given by an instructor (e.g., verbal feedback during tennis practice). This chapter focuses on formal summative evaluations, including the proper steps for conducting such evaluations, which result in a final grade being determined for a student's overall performance or achievement.

Corresponding to formative and summative evaluations, the evaluations in physical education can be also classified as *process evaluation* and *product evaluation*. During process evaluation, the teacher observes how the student completes a task or skill in order to provide specific feedback. Students are allowed to make mistakes, since they are an important part of the evaluation process. Therefore, it is often a step-by-step interaction between teacher and student, with ongoing input and output. Because of its interactive nature, the process evaluation is appropriate for use in the formative stage. The decision making is often subjective and thus some training may be required for evaluators. The Test of Gross Motor Development (Ulrich, 2016) is a good example of process evaluation in physical education. For example, to evaluate children's running skill development, the evaluator is asked to rate children's performance while running using the following performance criteria:

1. Arms move in opposition to legs with elbows bent
2. Brief period where both feet are off the surface
3. Narrow foot placement landing on heel or toes (not flat-footed)
4. Nonsupport leg bent about 90 degrees so foot is close to buttocks

During the product evaluation, in contrast, the focus is only on the results of a performance (e.g., how fast a student can run, how high a student can jump, or how far a student can throw). In fact, most of the tests in physical education are product-centered. With a standardization, the advantages of product evaluation are that it is easy to administer, therefore saving time, and is more objective than process evaluation. As a result, product evaluation is usually used during summative evaluation.

For a judgment to be made about the quality of performance, the performance must be compared with a standard. In chapters 1, 6, and 7, we discussed norm-referenced and criterion-referenced standards for evaluation. (Additional information on these topics also appears in chapters 10 and 11.) To review briefly, a norm-referenced standard is established by comparing a person's performance to the performances of others of the same sex and age or another well-defined group. The establishment of this standard usually requires some type of data analysis. In a norm-referenced evaluation, participants are ranked from excellent to poor based on their position among the comparison participants' scores. A criterion-referenced standard, on the other hand, is a specific predetermined level of performance that has been established from past databases or expert opinion. In a criterion-referenced evaluation, a participant either achieves the standard (passes) or does not achieve the standard (fails).

[WWW] **Go to HK*Propel* to complete Student Activity 8.1.**

EVALUATION IN SCHOOL SETTINGS

Grading is certainly the most commonly used evaluation practice in the school setting. The grading process can be a challenging one that many beginning (and even veteran) teachers do not look forward to completing. However, prospective teachers must realize that assigning grades is a professional obligation and an important component of physical education teaching.

Over the years there have been attempts to devise standard curricula for many public school subjects. Most of these attempts have been at the local or state level, but occasionally some have been at the national level. Although the majority of these efforts have been in classroom-based subjects such as math, English, the sciences, and social studies, a few have attempted to define a national physical education curriculum. The fewer attempts in physical education probably have to do with the large variability in philosophies of what should be taught and required, availability of equipment and facilities, and local customs.

One comprehensive example is PE Metrics, produced by the National Association for Sport and Physical Education (NASPE, 2008; SHAPE America, 2019), an association of the American Alliance for Health, Physical Education, Recreation and Dance (now named SHAPE America). This program is a standards-based, cognitive, and motor skills assessment package that provides reliable and valid measuring instruments for physical educators to use in the assessment and grading process.

Along with the need for accurate measuring instruments, there are many other nuances to consider in the process of assigning a final grade to each student. Grades, which are assessments of one person by another, are based on subjective evidence. Sometimes the amount of subjectivity is large, such as grades based on an assessment of a student's answers to an essay examination and participation in class discussions. Even grades determined through seemingly precise and objective methods (e.g., how many sit-ups completed in a set time, time required to finish a 1-mile (1.6 km) run, or how many baskets made in a 1-minute shooting test) require subjective decisions, such as determining which skills to include and which response is most correct for each specific test. Often instructors devise a precise formula for combining certain student achievements, but this formula is generally based on a subjective weighting of the importance of the course objectives, similar to those Tomas selected in the measurement and evaluation challenge described at the beginning of this chapter. Thus, although it is true that some grades involve a greater amount of subjectivity than others, it is also true that all grades involve some degree of subjectivity.

Generally, the greater the degree of subjectivity involved in the grading system, the greater the unreliability. In other words, a subjective system, if repeated, would probably not assign the same grades to the same students. This lack of objectivity and consequent lowering of reliability lead to a second, more serious, concern about grades: the lack of a common understanding of what a particular grade represents. Grades are affected by the type of marking system used, the particular instructor who assigns the grade, the makeup of the class, the institution where the grades are earned, and many other related factors. For example, an E might stand for excellent in one grading system but for failing in another; a grade of B from an instructor who rarely assigns a B has a meaning different from that of a B given by an instructor who seldom gives anything lower; an A obtained in a class of superior students may represent a more significant achievement than an A received in a class with less competition; and the same level of performance might be given an A at one institution but a lower grade at another.

Whereas the use of insufficient objective evidence reduces the *reliability* of grades, the lack of a clear and generally accepted definition of what a particular grade represents affects the *validity* of grades. Grades, which should reflect the degree to which a person has achieved the course objectives or goals, are often contaminated by many other factors. Failure of grades to validly reflect course achievement is an important concern.

No specific rules for grading can be firmly established because differences always exist in areas such as teaching techniques, course objectives, equipment and facilities available, and type of students in the class. However, an effective, consistent grading process allows an instructor and the enrolled students to have confidence in the validity of the assigned grades. If a student understands and accepts the teacher's grading system, he or she may not be happy with a poor grade but can accept it as fair. Were you ever unhappy when you received a grade different from the one you had expected? Was it easier to accept a poor grade if you felt a teacher's grading method was fair to all students? The challenge of this chapter is to provide you with the skills and knowledge required for developing and using good grading practices in your instructional programs, whether you will use these practices in an academic setting or any similar situation that requires a final assessment of the degree to which objectives have been met.

In addition, there are many ways to use grading in practice, such as motivating and guiding students' learning, educational, and vocational plans and their personal development; communicating to students

© Human Kinetics

Teachers must work hard to ensure reliability of assessments and relevance of topics so that grades truthfully reflect achievement.

and parents about students' progress; condensing information for use in determining college admissions, graduate school admissions, and scholarship recipients; communicating a student's strengths and limitations to possible employers; and helping the school tailor education to student needs and interests and evaluate the effectiveness of teaching methods.

WWW **Go to HK*Propel* to complete Student Activity 8.2.**

PROCESS OF GRADING

On what should students or subjects be graded? As explained in chapter 1, there are three domains of potential objectives:

1. *Psychomotor domain.* Concerns physical performance or achievement.
2. *Cognitive domain.* Concerns mental performance or achievement.
3. *Affective domain.* Concerns attitudes and psychological traits (also called the psychological domain).

Teachers of subjects such as mathematics, science, English, and history have instructional objectives limited mostly to the cognitive domain. Thus, in one sense their evaluation process is simpler than that of a physical education instructor. To evaluate students effectively, an instructor must have a clear understanding of the instructional objectives of the unit he or she is planning to teach. The instructor then must select and administer tests and measurements that are relevant to these objectives, compare the resulting test scores to appropriate standards, and finally, determine grades. Figure 8.1 illustrates this continuum.

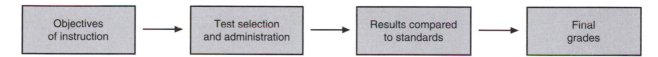

Figure 8.1 The grading process.

An effective and successful grading process requires that students understand the course objectives, the tests and measurements used for grading, and the method with which test scores will be combined to determine final grades. Inform your students of these factors at the beginning of instruction, and use formative evaluation techniques to update them on their progress throughout the course. Students should always be aware of the nature of the evaluations being conducted in the class and should ultimately not be surprised by the grades assigned.

WWW **Go to HK*Propel* to complete Student Activity 8.3.**

DETERMINING INSTRUCTIONAL OBJECTIVES

There are three questions to consider when determining whether a potential objective should be a part of your instructional unit and thus also a part of your grading process:

1. Is the objective defensible as an important educational outcome?
2. Does every student have an equal chance to demonstrate his or her ability on the objective?
3. Can the objective be measured relatively objectively, reliably, relevantly, and validly?

Grading in Physical Education

Often students in physical education are graded on such objectives as attendance, correct uniform (dressing out), shower taking, leadership, attitude, fair play, participation, team rank, or improvement. Figure 8.2 summarizes a study finding (Hensley and East, 1989) on the most commonly reported attributes used for grading. As you ask the preceding three questions of each attribute, you will see that some are largely inappropriate as bases for students' grades and thus should be eliminated, wheras others are appropriate.

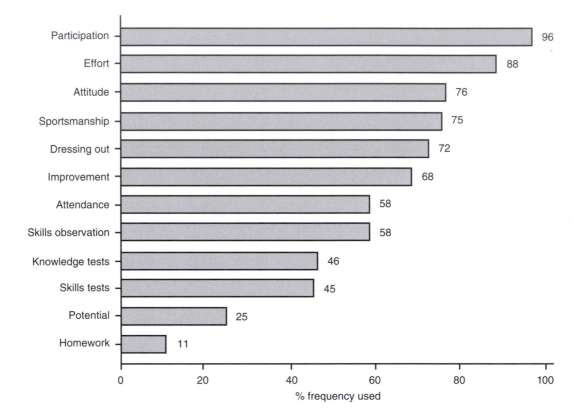

Figure 8.2 Attributes used for grading physical education.

What Not to Grade On

Research indicates that many of the parameters in figure 8.2 have been widely overused as objectives in physical education. Let's briefly analyze why they are inappropriate.

Although attendance, correct uniform, and shower taking are obviously worthwhile and necessary for appropriate physical education instruction, they fail to meet the first criterion: having an important educational outcome. Students failing to meet these requirements should face consequences, but it should not be by a direct lowering of their grades. Students in mathematics may have to attend class and bring their textbooks, but they are seldom graded on these required factors, nor should they be.

Leadership, attitude, fair play, and participation are worthwhile objectives of any physical education or athletic program. However, to grade these factors reliably and validly, an instructor would need to implement a formal and systematic program of measurement and evaluation. Implementing such a program takes time and expertise, which the physical education instructor likely does not have. When these factors are used for grading, they are usually based on random teacher observations, which tend to be subjective and possibly biased (unless some of the techniques explained in chapter 15 are used). The result is grades that lack objectivity, reliability, and validity and are thus unfair to students.

Grading on team rank in team sport classes fails to meet the second criterion—provision of an equal opportunity to demonstrate ability. A poor performer might be placed on a good team and receive an A, whereas a good performer could be placed on a poor team and receive a low grade. This strategy is not fair to all students because their grades depend on the performance of other students, over which they have little or no control. From a measurement and evaluation prospective, grading students in team sport classes may be one of the most problematic tasks a teacher faces. Sport skills tests are often used to address this problem (see chapter 12). However, an individual sport skills test that is isolated from the team game may lack relevance and validity. Evaluating players as they participate in game conditions, or performance-based assessment (see chapter 15), may also be used but requires well-developed rating scales and rubrics.

Improvement is one of the most attractive objectives to use in grading in physical education. As an instructor you want your students to improve, but grading on improvement presents some difficult problems. One such problem is illustrated in figure 8.3: A beginner (James) can improve at a much greater rate than a more advanced student (Robert). This is a natural phenomenon, because the beginner has more room for improvement. A new marathon runner will typically show a much more improved time between his or her first and second marathon than the experienced runner will between his or her 19th and 20th marathon. Another problem is that improvement scores tend to be less reliable than either the pretest or posttest from which they are computed. A third problem is that some students, knowing that they are to be graded on improvement, may provide false initial performance scores that will inflate their improvement scores. Some teachers attempt to support improvement grading by suggesting that a student's improvement on a final performance be compared with his or her potential. However, the means for validly determining potential are often unreliable or simply don't exist.

Why do you suppose educators, and especially physical educators, like to grade on

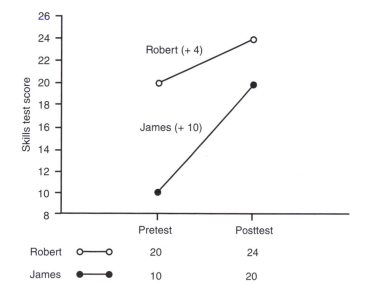

Figure 8.3 The problem with grading on improvement.

improvement? It is probably because this practice gives many students a chance to earn high grades, regardless of their initial ability level. A great number of educators are reluctant to give low grades because they believe low grades discourage effort, which increases the probability of low grades, resulting in a cycle that continues until students dislike the subject. The concern for physical educators is that when students become turned off by the subject, they may be less likely to continue physical activity during their lifetimes.

However, little, if any, evidence exists to support this contention. It is doubtful that students feel rewarded when given high grades for improvement when their actual performance level is lower than most of their peers. For example, poor swimmers know they are poor swimmers without being told so (and so do their peers); an honest student knows that in the long run it is the actual level of achievement that is important, not the rate of improvement. Suppose it is true that low grades cause a student to dislike school—or, more specifically, that low physical education grades destroy incentive and reduce the probability of a student engaging in physical activity. The solution of not giving low grades is like a physician treating the symptoms rather than the cause of the illness. Instead, instructors need to determine why a student is performing poorly and provide opportunities for success. This might apply to other fitness situations in clinical or community settings as well; you do not want clients to be discouraged from attending fitness classes or programs.

WWW **Go to HK*Propel* to complete Student Activity 8.4.**

First, an instructor may want to increase the number of opportunities for achieving success by broadening the curriculum to a larger variety of activities. Because the skills required to succeed in physical education are relatively heterogeneous compared with those required in other subjects, varying the physical education curriculum will probably result in a higher percentage of students achieving success than in the other academic areas. For example, learning to derive a square root is dependent on knowing how to divide, but learning how to swim a particular stroke is not dependent on knowing how to do a forward roll.

A second means of encouraging success lies in ability grouping and enrolling students in courses suitable to their skill levels. If differences in initial ability are small, grading on improvement becomes unimportant (because it becomes almost the same as grading on achievement). Other subjects, such as math, English, and the sciences, often make use

© Human Kinetics

Introducing students to atypical physical education activities is one way to help them discover new activities and encourage success.

of ability grouping; to adopt this approach, physical educators need to develop valid and reliable measuring devices to classify students by ability.

A third approach involves educating students that a grade is simply an expression of the level of achievement of the course objectives attained by a student; it is not a reward or a punishment. Furthermore, not every student can excel in every course. One way to help ensure that grades remain a clear indication of a student's performance is to think of grades as measurements rather than as evaluations. In grading, a measurement is a quantitative description of a student's achievement, whereas an evaluation is a qualitative judgment of a student's achievement. In other words, a grade, as a measurement, represents the degree to which a student has achieved the course objectives. If a grade is considered an evaluation, it should indicate to some degree how adequate a particular student's level of achievement is.

There are several advantages to treating grades as measurements rather than evaluations. Judgments of the adequacy of a student's achievement depend not only on how much he or she achieved but also on the opportunity for achievement and effort. This makes it difficult to report evaluations precisely in a standard grading system. Additionally, it is probably of more value to a future teacher or employer to know that a student was outstanding, average, or poor in a particular course of study than to know that this student did as well as could be expected or that he or she failed to live up to the expectations of a teacher. Given valid and reliable measurement of a student's achievement in various areas, a future teacher or employer can make his or her own evaluations in light of current circumstances, which would probably be more valid than having evaluations made by others under totally different conditions. The worth of physical education must be conveyed to students through the teaching procedures, personal example, and the curriculum, not through the awarding of high grades regardless of the level of proficiency.

Remember from the Measurement and Evaluation Challenge described at the beginning of this chapter that Tomas has decided to award one-half of the total grade based on the cognitive aspects of the tennis unit. Thus, he is broadening the opportunities for students to succeed, and his grading method does not reflect the less defensible objectives of attendance, improvement, and so forth.

What to Grade On

Grades should be based on reliable and valid tests and measurements that are representative of important instructional objectives. In individual or team sport classes, sport skills tests, expert observations, rating scales, and round-robin tournament performance can be effective criteria for grading. In individual sports, such as bowling, golf, and archery, a student's achievement of objectives can be evaluated through performance. In fitness classes, field tests of physical fitness provide reliable and valid data for evaluation. Because the cognitive domain is important in physical education, knowledge testing should be part of the overall grade. Chapters 9 through 15 provide you with a selection of reliable and valid tests and testing procedures for use in human performance. In summary, use the following procedures to make grading fair, reliable, and valid:

- Carefully determine defensible objectives for each course before it begins.
- Group students according to ability in the physical skills necessary for the course, if possible.
- Inform students about the grading policies, procedures, and expectations.
- Construct tests and measurements as objectively as possible, realizing that all tests are subjective to some degree.
- Remember that no matter how well constructed a test is, no test is perfectly reliable.

- Realize that the distribution of grades for any one class does not necessarily fit any particular curve, but that over the long run physical skills are probably fairly normally distributed.

- Determine grades that reflect only the level of achievement in meeting the course objectives and not other factors.

- Establish grades on the basis of achievement, not improvement.

- Avoid using grades to reward the good effort of a low achiever or to punish the poor effort of a high achiever.

- Consider grades as measurements, not evaluations.

CONSISTENCY IN GRADING

A goal of any grading system should be to achieve consistency in grade determination. Theoretically, a student's grade should not depend on any of the following:

- *A particular section of a course.* If students are taking PE 100–Beginning Tennis, their grades should not be affected by whether they are in the 9:00 a.m. or the 2:00 p.m. section.

- *A particular semester of the class.* A student's level of performance should warrant the same grade whether the course is taken in the fall semester or the spring semester. A student's grade depends on his or her performance only and should not be influenced by the performance of other students in the class. This objective suggests that grading on the curve is inappropriate.

- *A particular instructor.* Two different instructors should give the same grade to the same level of performance.

- *The method of course delivery.* It should not make a difference if the course is delivered online, face-to-face, or blended (assuming each method is an equally effective teaching style).

The goal of consistency is extremely hard to achieve because of student and teacher differences, which are natural phenomena of any instructional setting. As you will discover, lack of consistency in grading is one of the weaknesses of several grading schemes.

WWW **Go to HK*Propel* to complete Student Activity 8.5.**

GRADING MECHANICS

Four steps are involved in the process of arriving at a grade to represent each student's level of achievement of course objectives:

1. Determine the course objectives and their relative weights.
2. Measure each student's achievement of the course objectives.
3. Combine measurement results to obtain a composite score for each student.
4. Convert the composite scores to grades.

Each step may be accomplished in a variety of ways, depending on different situations and philosophies.

Step 1: Determine Course Objectives and Their Relative Weights

Determining course objectives and their significance is the most important of the four steps and requires a considerable amount of thought. In fact, doing so is basic to every aspect of teaching a course, not just grading. Determining course objectives precedes planning the sequence of presentation of material, necessary equipment, teaching procedures, assignments, and grading procedures. It should be based on as much knowledge of the abilities of the prospective students as you can obtain; the general objectives of physical education; such practical considerations as the number of students, the facility limitations, and the duration of the course or unit; and the number, frequency, and duration of meeting times.

The objectives that you finally decide on for a physical education course can be classified into the psychomotor, cognitive, and affective domains (see chapter 1). Stating the objectives in behavioral terms will help make the second step (measuring achievement) easier. In other words, list the actual performance levels, the specific knowledge, and the social conduct expected to be achieved by students. Sample objectives for a badminton unit are as follows:

Sample Badminton Objectives

- Cognitive objective: Know the setting rules when the game becomes tied at certain scores.
- Psychomotor objective: Be able to place at least 4 out of 5 low, short serves into the proper portion of the court.
- Affective objective: Be aware of proper etiquette when giving the shuttlecock to an opponent at the end of a rally.

How you weight the course objectives will vary greatly from one situation to another, depending on your philosophy, the age and ability of your students, and so on. For example, you might place less emphasis on the affective domain objectives for a physical fitness unit for 10th-grade boys or girls (single gender) than for a coeducational volleyball unit at the same grade level. The actual weight you give to each course objective should result in a balanced and defensible list of goals to be achieved by each class member.

Informing students of what is expected of them at the outset facilitates the planning of student experiences, teaching methods, and grading procedures and should reduce student anxiety.

WWW Go to HK*Propel* to complete Student Activity 8.6.

Step 2: Measure the Degree of Attainment of Course Objectives

Recall that measurement is the procedure of assigning a number to each member of a group based on some characteristic. In this case, the characteristic involved is the degree of achievement of each course objective. Unlike step 1, steps 2 through 4 occur after the teaching and, presumably, the learning have taken place. This is not to imply that all measuring should be of a summative or product nature. On the contrary, there is merit in obtaining measures throughout a unit (formative or process evaluation) because doing so leads to increased awareness on the part of both the students and teacher of progress toward the course objectives. However, the testing or measuring used for grading students' degree of achievement should obviously not occur before the completion of instruction and learning. Chapters 9 and 15 provide ideas for constructing, evaluating, selecting, and administering tests and other devices to arrive at the numerical value to assign to each student that most accurately represents his or her level of achievement of the course objective being measured.

Step 3: Obtain a Composite Score

Seldom does a course have a single objective. Often, more than one measurement is made to determine the amount of achievement for each objective. For these reasons, it is usually necessary to combine several scores to arrive at a single value representing a student's overall level of achievement of the course objectives. This **composite score** is then normally converted into whatever grade format is being used (e.g., A-B-C-D-F; pass/fail). The correct method to obtain the composite score depends on several factors, the most important of which is the precision of the scores that are to be combined.

In the case of performance scores, it is apparent why a composite score cannot be obtained by simply totaling the various raw scores. Units of measurement often differ: Feet cannot be added to seconds; the number of exercise repetitions cannot be added to a distance recorded in inches. As mentioned in chapter 3 under the discussion of standard scores, scores from distributions bearing differing amounts of variability, even if they have the same unit of measure, cannot simply be added together because the variability affects the weight each score contributes to the composite score. The following example illustrates why adding raw scores of tests that actually do have the same units (such as written tests) may not result in the desired composite score: Imagine that the scores from a class of 25 students on three tests worth 9 points, 9 points, and 27 points, respectively, were distributed as shown in table 8.1. This rather extreme example was chosen to make a particular point: that a score from a set of scores with a large variability will have more weight in the composite score than one from a set with little variability, regardless of the absolute values of the scores. The variability of the scores on the first two tests is greater than the variability for test 3. It would seem that, because the total number of points possible on test 3 is triple that of tests 1 or 2, the score achieved on test 3 would have the most influence on a student's composite score. Notice, however, some possibilities:

Student A was one of the seven students achieving the highest score made on test 3; scores on the first two tests were average. Student A received a composite raw score of 34. The composite raw score for student B—who scored above average on the first two tests but was among the lowest scorers on test 3—is 37, higher than that for student A. Student C scored at the average on two tests and above average on one test, as did student A. However, because student C's above-average performance came on test 1 (on which the scores were more variable than the test on which student A achieved above average, test 3), student C's composite raw score, 37, is also higher than student A's. Finally, even though student D scores as high as anyone in the class on test 3, the low scores made on the first two tests lowered the composite raw score, 30, below the others.

Unless two (or more) sets of scores are similar in variability, summing a student's raw scores to arrive at a composite score may lead to some incorrect conclusions.

Assume that, instead of representing the distributions of scores on three written tests, the values in table 8.1 describe the distributions of the measures of how well cognitive, affective, and psychomotor objectives were met by the students. Further assume that you had decided that achievement of the cognitive objectives, the affective objectives, and the psychomotor objectives would represent 20%, 20%, and 60% of the final grade, respectively. To achieve this weighting, you simply made the number of points possible for each of the

Table 8.1 Test Score Distributions from a Class of 25 Students

TEST 1 (9 POINTS)		TEST 2 (9 POINTS)		TEST 3 (27 POINTS)	
Score	Frequency	Score	Frequency	Score	Frequency
9	1	9	1	27	
8	2	8	2	26	
7	3	7	3	25	
6	4	6	4	24	7
5	5	5	5	23	11
4	4	4	4	22	7
3	3	3	3	21	
2	2	2	2	20	
1	1	1	1	19	
Student	**Score on test 1**	**Score on test 2**	**Score on test 3**	**Total raw score**	
A	5	5	24	34	
B	8	7	22	37	
C	9	5	23	37	
D	3	3	24	30	

three objectives the desired percentage of the total number of possible points (i.e., 9, 9, and 27). As illustrated previously, unless the variability of the three sets of scores is quite similar, the actual weighting will be different from that originally planned. The solution is to do the weighting after a score for each objective has been obtained rather than attempt to build the weighting factor into the point system (unless you can assume that equal or nearly equal variability among the sets of scores will occur).

If calculating composite scores by combining raw scores is not feasible because of different units of measurement or a lack of equal variability among the sets of scores, how can you form composite scores? As described in chapter 3, convert each set of scores into the same standard distribution so that a common basis is established to make comparing, contrasting, weighting, and summing of scores from several sets of scores possible. There are several methods of doing this (three of which are discussed shortly); selecting the best method depends on the precision of the measurement involved and the assumption of normality.

In the case of the precision of scores, determine whether the scores are on an ordinal scale or an interval or ratio scale. As outlined in chapter 3, when you use an ordinal scale of measurement (such as the rankings from a round-robin tournament), it is only possible to say that A is greater than B. However, with interval and ratio scales (such as the number of free throws made), it is possible to state how much greater A is than B, because these scales have equal-sized units.

The second consideration involves determining whether the distribution of scores approximates a normal distribution. If it does not, is it because the trait being measured is itself not normally distributed, or is it because, even though the trait is normally distributed, the sample at hand for some reason does not reflect this?

There are five possible situations involving these two considerations (table 8.2). The reason there are not six possible situations is that if the scores are ordinal measures, the

Table 8.2 Methods of Obtaining a Composite Score Based on Shape of Distribution and Scale of Measurement

Shape of distribution of sample	SCALE OF MEASUREMENT	
	Ordinal	Interval or ratio
Nonnormal	Rank method	Rank method
Nonnormal but trait is normally distributed	Normalizing method	Normalizing method
Approximately normal	(Not possible)	Standard score method

distribution of scores—a simple ranking of the students—cannot approximate a normal distribution. Each situation is associated with one of the three methods of converting scores to a standard distribution: rank, normalizing, and standard score. Although statistical tests are available to determine whether a distribution of interval or ratio scores is significantly different from a normal distribution, such tests are beyond the scope of this book. However, a visual inspection of the frequency distribution of the sample is usually sufficient for revealing its closeness to a normal distribution. If you are still uncertain whether a distribution of scores approximates the normal distribution closely enough, you can use the normalizing method.

The end result of the rank method is simply a ranking of the students, whereas the end result of the other two methods is a set of standard scores. Note: It is not possible to arrive at a composite score if some of the sets of scores are converted to ranks and some are converted to standard scores. Therefore, if one of the several scores being summed to obtain a composite score must be expressed as a rank, then all the scores must be expressed as ranks. For this reason, you should plan in advance the type of measuring that you will use during a unit of instruction.

Rank Method

The simplest method, requiring the least precise measurement, is numerical ranking of the performance of the students on each test taken. However, this lack of precision also makes this method less reliable than others, and thus this method should be avoided, if possible. The only time this method should be used is when, due to a lack of adequate measuring devices, it is the only way that students' performances can be evaluated. In the rank method, the best performance is given a 1, the second-best performance a 2, and so on until the worst performance is given a rank equal to the number of students being measured. To obtain a composite score for each student in the numerical ranking system, simply sum the ranks for each student. The lowest total represents the overall best achievement.

In the event that one or more ranks are missing for a student, you may use a mean rank for that student, which is obtained by dividing the sum of the ranks by the number of values contributing to that sum. If you wish to weight the rankings, add in those considered most important more than once to arrive at the total. For example, assume that three rankings were obtained, the first to count 10%, the second 40%, and the third 50% of the final grade. For each student, a composite score would be obtained by summing the first rank, four times the second rank, and five times the third rank. As before, the lowest total would represent the best overall achievement. This may seem to contradict what you studied in chapter 3; however, the robustness of ranks in this situation permits performing mathematical operations on them even though they represent ordinal data.

A variation of the numerical ranking system is to rate each student as belonging to one of a set number of categories. For example, the best five achievements might be rated as 1, the next five as 2, and so on. Another variation goes one step further by giving the categories letter grades rather than numerals. This latter procedure does not require a particular number of students to be classified into each category. It has some merit in that it is more informative to the students than pure numerical rankings, but the principles used to arrive at a composite score are the same. In fact, a composite score is obtained in this system by changing the letter grades to numerals and proceeding in a fashion similar to that described for the numerical ranking system. For example, the categories A+, A, A–, B+, B, B–, C+, C, C–, D+, D, D–, and F are given the values 12, 11, 10, 9, 8, 7, 6, 5, 4, 3, 2, 1, and 0, respectively. (Note that in this method, a higher number is better than a lower one.) A student received an A+ on the affective objectives (10%), a B– on the cognitive objectives (40%), and a C– on the psychomotor objectives (50%). To arrive at a composite score for a student, the letter grades are converted to their numerical equivalents, multiplied by the weight of the corresponding objective, and summed:

$$(12 \times 1) + (7 \times 4) + (4 \times 5) = 12 + 28 + 20 = 60$$

Dividing this sum by 10 (the sum of the weights: $1 + 4 + 5 = 10$) and comparing the resulting value to the categories converts this student's performance to a C+ ($60 / 10 = 6 = \text{C+}$).

Mastery Item 8.1

Using the system and weighting scheme just shown, what grade would you assign a student receiving grades of B+, C, and A–, respectively?

Normalizing Method

Use the normalizing method when the scores obtained in measuring a trait that is known or believed to be normally distributed do not appear to result in an approximation of the normal curve. For example, you would expect a distribution of the number of basketball free throws made in a given time period for males in the 10th grade to approach normality as the number of data points increases. If the measurement takes the form of ranks, the normal curve will not be approximated. Other reasons, such as not obtaining a representative sample, may cause a distribution to be nonnormal even though the trait being measured should be normally distributed. The normalizing method is much like converting a set of raw scores into a distribution of some standard score. The difference, however, is that in the present case the raw scores are first converted to percentiles, which are converted into a standard score scale (usually T-scores). A description of the procedures for converting a set of raw scores into percentiles appears in chapter 3, and the resulting percentiles can be converted into T-scores using table 8.3. To illustrate how this procedure converts a nonnormal distribution into a normal distribution, notice that a score that falls 34.13% above the mean (1 standard deviation above the mean in a normal curve)—regardless of its raw score standard deviation distance from the mean—is equivalent to a T-score of 60, which is 1 standard deviation above the mean in the T-score scale. An example of the normalizing method resulting in a T-score corresponding to each raw score is shown in table 8.4. Once each set of scores is converted to T-scores, the T-scores can be weighted as desired and then combined to arrive at a composite score for each student. These composite scores can then be converted to the appropriate grade of the particular format being used.

Table 8.3 Conversion of Percentiles to T-Scores

Percentile	T-score	Percentile	T-score	Percentile	T-score
0.02	15	13.57	39	90.32	63
0.03	16	15.87	40	91.92	64
0.05	17	18.41	41	93.32	65
0.07	18	21.19	42	94.52	66
0.10	19	24.20	43	95.54	67
0.13	20	27.43	44	96.41	68
0.19	21	30.85	45	97.13	69
0.26	22	34.46	46	97.72	70
0.35	23	38.21	47	98.21	71
0.47	24	42.07	48	98.61	72
0.60	25	46.02	49	98.93	73
0.82	26	50.00	50	99.18	74
1.07	27	53.98	51	99.38	75
1.39	28	57.93	52	99.53	76
1.79	29	61.79	53	99.65	77
2.28	30	65.54	54	99.74	78
2.87	31	69.15	55	99.81	79
3.59	32	72.57	56	99.87	80
4.46	33	75.80	57	99.90	81
5.48	34	78.81	58	99.93	82
6.68	35	81.59	59	99.95	83
8.08	36	84.13	60	99.97	84
9.68	37	86.43	61	99.98	85
11.51	38	88.49	62		

Note: Although the T-score scale theoretically extends from 0 to 100, in practice, T-scores lower than 15 or higher than 85 are rare and thus are not included in the table.

Table 8.4 Example of Normalizing Method

Raw score	f	cf	cfm	Percentile	T-score from table 8.3
85	1	6	5.5	91.7	64
74	1	5	4.5	75.0	57
63	1	4	3.5	58.4	52
59	1	3	2.5	41.7	48
53	1	2	1.5	25.0	43
47	1	1	0.5	8.4	36
	6				

Note: In this table, the f column is simply the frequency of occurrence of each score. The cf column is the cumulative frequency associated with each score (beginning with the lowest score). To obtain the values in the cfm (cumulative frequency of the midpoint) column, add one half of the f value for each interval to the cf value from the next lowest score. For example, the value of 3.5 in the cfm column for the interval called 63 is obtained by adding 0.5 (1/2 of the frequency of 1 for this interval) to 3.0 (the cf of the interval below the interval of 63). Finally, to express the values as percentiles, divide 100 by n (in this case, 6) and multiply each cfm value by the resulting quotient.

Standard Score Method

If you judge it safe to assume that the raw score distribution closely approximates a normal distribution, convert the raw scores to a standard score scale such as the T-score scale as described in chapter 3. As in the normalizing method, the net result of this procedure is a T-score corresponding to each raw score. If you convert all sets of scores to T-scores, a common basis exists for comparing scores made on two tests—even though the two raw scores were expressed in different units or had differing variability—because they all have the same standard deviation (recall that the standard deviation of T-scores is always 10). Furthermore, you can now add T-scores representing various achievements to obtain meaningful composite scores that can be used to determine final grades. As with the rank method, you can also weight the various tests (i.e., the respective objectives on which they are based) by multiplying the T-score by the appropriate weights you determined previously.

Once you obtain a composite score for each student, the final step is to change each score to the appropriate grade of the particular format being used. As in the other three steps, various factors will affect how you do this. The decision of what procedures to follow is based on the form of the composite score, policies of the school system or department, and your individual philosophy.

Step 4: Convert Composite Scores to a Grade

At the end of step 3, the composite scores will be in one of two forms: Each student will have a total or average ranking or a total or average standard score. In effect, both forms are an ordering of the students, although in the rank method the lowest total (or lowest average) usually represents the best achievement, whereas in the normalizing and standard score methods the highest total (or highest average) represents the best achievement. The procedures for converting a set of composite scores to grades are the same regardless of the form of the composite scores and involve answering two related questions:

1. Is the class to be graded below, at, or above average in achievement of the course objectives in comparison to similar classes?

2. What percentage of students should receive each grade?

If tests and measuring devices were absolutely reliable and valid and if course objectives remained constant over time, teachers would not need to answer this question before converting composite scores to grades. However, measurements are not perfect, objectives change, unexpected or unplanned events occur, facilities and equipment change over time, and several other factors make it impossible to compare the achievement of the current class to previous classes on a strictly objective basis. Grading is especially difficult for new teachers because they lack experience on which to base their answers to these questions. After arriving at some subjective answers, you can use several methods to convert the composite scores to grades.

Observation

Observation is one of the simplest methods for determining grades. List the composite scores from best to worst. Examine the list for natural gaps or breaks in the scores. Table 8.5 lists the scores for 15 high school boys. As you see, two gaps appear in the data. Those observed gaps are used for cutoffs for the letter grades A, B, and C.

This method usually works well for a small number of scores, because gaps in the data often appear. However, observation is not useful for a large number of scores, because observable gaps may not be present. Furthermore, this method does not ensure consistency. The natural gaps could be quite different in two classes: A grade of A in the fall semester class might fall in the B category in the spring.

Table 8.5 Scores Graded by the Observation Method

Scores	Frequency	Grade
150	1	A
140	2	
110	3	B
100	2	
90	2	
80	1	
70	1	
40	1	C
30	1	
20	1	

Predetermined Percentages

The predetermined-percentages method may be used with composite scores in the form of rankings or standard scores because it is not the value of the score that matters but its position in the distribution. Once you decide on the percentage of students to whom each grade will be assigned, you need only multiply the percentage values by the number of students in the class, as follows:

$$k = N \times (P / 100) \qquad (8.1)$$

where k is the number of students to receive a particular grade, N is the total number of students, and P is the percentage of a particular grade. The resulting product is the number of each letter grade to be assigned. Table 8.6 shows this procedure for a class of 45 students, which, on the basis of test scores and other evidence, has achieved substantially above the average of other similar classes. Notice that this decision is, in effect, answering the first question under step 4. The teacher decides to allot 15% As, 25% Bs, 45% Cs, 10% Ds, and 5% Fs. If the composite scores are in the form of standard scores or the sum (or average) of several rankings (but not if the composite scores are a single rank in which the best student has a rank of 1, the second-best student has a rank of 2, and so on), you can modify the predetermined-percentages method slightly by using the natural breaks in the distribution of the composite scores. As the last column of the table indicates, the actual number of scores for a grade may be slightly different from the calculated k, but the cut points are selected to result in numbers of students close to the predetermined number.

As with the observation method, the predetermined-percentages method of grading does not ensure consistency in grade assignment from class to class or semester to semester. Even if the predetermined percentages remain consistent, because scores vary from class to class and semester to semester, the actual cut point (in terms of scores) will vary. Thus, a score that translates into an A in the fall semester might be a B in the spring. You can adjust for this, however, by changing the percentages of each grade to be given.

Grading on the Curve

Although *grading on the curve* is a phrase often heard in educational settings, the process is not well understood. Like the rank and normalizing methods, grading on the normal curve also belongs to the relative method (i.e., comparing a score with a reference group or population). Although these methods should not be used in classroom grading practice, they can sometimes help in selection and classification (e.g., obesity status based on the percentile categories). Actually, grading on the normal curve is a variation of the predetermined-percentages method in which the assumption is made that the differences

Table 8.6 Grades Determined by Preset Percentages

Grade to be given	Percentage, preset by teacher	Number of students to receive grade (k)	Scores	Frequency	Total actual number of students receiving grade
A	15%	45 × .15 = 6.75 → 7	120 110 100	2 2 2	6
B	25%	45 × .25 = 11.25 → 11	90 80 70	3 4 5	12
C	45%	45 × .45 = 20.25 → 20	60 50 40 30	8 5 5 3	21
D	10%	45 × .10 = 4.5 → 5	20	4	4
F	5%	45 × .05 = 2.25 → 2	10	2	2

in student abilities in a class are normally or at least approximately normally distributed and therefore that the percentages of each grade assigned can be determined through use of the normal curve. However, differences in student abilities are rarely normally distributed, so this method of determining final grades implies a precision that is seldom true.

Some confusion about grading on the curve stems from the fact that it is possible to use the normal curve in two ways. The practical limits of the normal curve (±3 standard deviations) can be equally divided into the number of categories of grades to be assigned, or certain standard deviation distances from the mean may simply be selected as the limits for each symbol assigned. Refer to table 8.7 for data used to illustrate each approach. In this example, the composite scores of 65 students have been compiled, and you wish to use the normal curve to determine the cut points for assigning the grades into the five categories of A, B, C, D, and F. The mean and standard deviation of the composite scores are 63.4 and 15.09, respectively.

Table 8.7 Composite Scores for 65 Students

98	78	71	64	60	52	40
93	78	70	64	59	51	38
91	77	69	63	57	50	37
88	76	68	63	57	48	36
86	75	67	63	56	47	26
85	74	67	63	56	47	
83	73	66	62	55	46	
81	73	65	62	55	45	
81	72	65	61	54	44	
79	71	65	61	53	41	

First Approach

For practical purposes, the normal curve may be considered to extend ±3 standard deviation units above and below the mean. (Recall that 99.74% of the area under the normal curve is found between these two points.) This total width of 6 standard deviation units is divided equally into the same number of parts as there are categories of grades to be assigned—in this case, five (A, B, C, D, F). Each grade thus encompasses a width of 1.2 standard deviation units (6 / 5 = 1.2) (figure 8.4). Because the grading format used in this illustration has an uneven number of possible grades, one half of the middle grade (C) falls on either side of the mean. A grade of C will be assigned to those students whose composite scores lie between 0.6 standard deviation units above and 0.6 standard deviation units below the mean. A grade of B will be assigned to students whose composite scores are between 0.6 and 1.8 standard deviation units above the mean; a grade of D will be assigned to those whose composite scores lie between −0.6 and −1.8 standard deviation units below the mean. If this process is continued, it will result in the limits of 1.8 and 3.0 standard deviation units above the mean for grades of A and −1.8 and −3.0 standard deviation units below the mean for grades of F. To account for extreme scores, you can specify that composite scores more than 3.0 standard deviations above the mean also correspond to As and that composite scores more than −3.0 standard deviations below the mean also correspond to Fs.

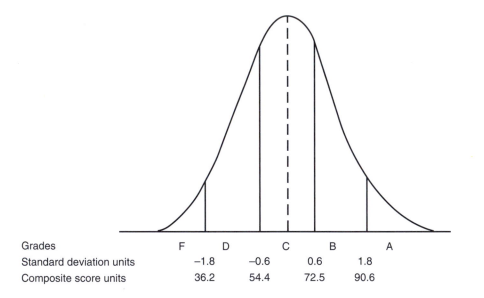

Grades	F	D		C		B	A
Standard deviation units		−1.8		−0.6		0.6	1.8
Composite score units		36.2		54.4		72.5	90.6

Figure 8.4 Relationship among grades, standard deviation units, and composite score units.

The final step involves expressing the standard deviation units in terms of the composite score values. Because in this example, 1 standard deviation is equivalent to 15.09 composite score units, the cut point between C and B and between C and D must lie 9.05 composite score units above and below the mean, respectively. (The value of 9.05 results from multiplying 15.09 by 0.6.) The two cut points are thus 72.5 and 54.4 (63.5 ± 9.05). The cut points between B and A and between D and F are obtained by adding to and subtracting from the mean the product of 15.09 and 1.8. The result is 27.16, and thus the cut points are 90.56 and 36.24 (see figure 8.4). To simplify, you would probably round the cut points; you could also establish a conversion chart for assigning grades to the composite scores (table 8.8).

Table 8.8 Conversion of Composite Scores to Grades: Example of the First Approach

Composite score	Grade
91 and over	A
73-90	B
54-72	C
36-53	D
35 and below	F

If a grading format involving five grades is used in conjunction with the normal curve, the teacher is, in effect, deciding to assign approximately 3.5% of the class As, 24% Bs, 45% Cs, 24% Ds, and 3.5% Fs. (To verify this, review how to use table 3.4.) Applying the conversion chart in table 8.8 to the 65 composite scores results in the assignment of 3 As (4.6% of the class), 15 Bs (23.1%), 30 Cs (46.1%), 14 Ds (21.6%), and 3 Fs (4.6%). Slight differences occur between the normal-curve percentages and the actual percentages because of the rounding in establishing the conversion chart and the fact that the set of 65 scores is not exactly normally distributed.

Second Approach

The second approach to grading on the curve involves selecting certain standard deviation distances from the mean as the limits for each grade. In this approach, instead of dividing the 6 standard deviations (the practical limit of the normal curve) into equal units for each grade category, select distances that conform to your notion of the percentage of each grade to be assigned. One possibility, based on the data from table 8.7, is depicted in table 8.9.

The selection of these particular standard deviation distances would result in assigning approximately 7% As, 25% Bs, 36% Cs, 25% Ds, and 7% Fs to a set of normally distributed scores. As in the first approach, these standard deviations must be converted to the units of the composite scores by multiplying the selected constant by the standard deviation of the composite scores. For the data found in table 8.7 (mean = 63.4, standard deviation = 15.09), the resulting cut points are displayed in table 8.9. (As before, slight differences between the normal-curve percentages and the actual percentages occur because of rounding and the fact that the 65 scores are not exactly normally distributed.)

Although any set of standard deviation distances may be chosen, if the assumption regarding normality is generally true, the selected values will necessarily result in symmetrical assignment of grades. If a nonsymmetrical distribution of grades is desired, the

Table 8.9 Curve Grading: Example of the Second Approach

Grade	Standard deviation distance for the grade (predetermined by instructor)	Percentage of students to receive grade (use table 3.4)	Number of students to receive the grade (N = 65)	Composite scores corresponding to the grade
A	Above +1.48	7%	.07 × 65 = 5	≥86
B	+0.47 to +1.47	25%	.25 × 65 = 16	71-85
C	−0.47 to +0.46	36%	.36 × 65 = 23	57-70
D	−1.47 to −0.46	25%	.25 × 65 = 16	41-56
F	Below −1.48	7%	.07 × 65 = 5	≤40

Note: The first and last columns would constitute a simple conversion table.

second approach to curve grading actually becomes the predetermined-percentages method described previously. Note that the normal-curve method does not ensure consistency in grading from class to class because the mean and standard deviation may change from class to class.

Mastery Item 8.2

The mean and standard deviation of a set of composite scores are 38.5 and 6.6, respectively. Calculate the normal-curve grading scale if you are to have a grade distribution of 10% of the students receiving As, 20% Bs, 40% Cs, 20% Ds, and 10% Fs.

WWW! **Go to HK*Propel* to complete Student Activity 8.7.**

Definitional Grading

As mentioned previously, determining relevant course objectives and their weighting should be one of the first tasks in preparing to teach any course. Special care must be taken in assigning weights to various objectives so that a superior performance on the assessment of one objective does not totally offset a poor performance on another objective. As an extreme example, a student could achieve a grade of C by earning an A on the written test on rules and strategy and nearly failing an assessment of the physical performance if each of the two objectives has an equal weighting of 50% of the grade.

One method to ensure this does not occur is through the use of *definitional grading*. Briefly, in this grading system, an instructor sets a minimum performance required for each objective. The level of achievement on each objective determines the grade for each objective, and the final grade is equal to the lowest grade achieved on any objective.

Table 8.10 illustrates a simplified possible set of minimums for grades of objectives in the three domains of a tennis unit. In this class the instructor plans to measure the cognitive objective with a 50-item written test on tennis rules and strategies, the psychomotor objective by counting the number of good serves made out of 10 attempts, and the affective objective with a 5-point rating scale applied while observing students' etiquette during tennis matches.

To achieve an A in this class, a student must answer at least 45 questions correctly on the written test, make 8 or more good serves on the serving test, and be given a rating of 5 on the observation of etiquette. Although tables can be constructed to allow various weightings of the objectives to determine the final grade, one option is to have the final grade for a student be determined by the lowest level of performance on any of the three objectives. For example, a student receiving a 37 on the written test, a 9 on the serving test, and a 4 on the etiquette observation would receive a grade of C. For more on this grading system and how these tables can be constructed, see Looney (2003).

Table 8.10 Illustration of Definitional Grading System

	Cognitive	Psychomotor	Affective
A	45+↑	8+↑	5
B	40-44	7	4
C	35-39	6	3
D	30-34	5	2
F	25-29	4	1

Norms

As discussed in previous chapters, norms are derived from a large number of scores and measurements from a specifically defined population. The scores are analyzed to produce a descriptive statistical analysis, which allows the production of percentile norms or standard score norms. Norms are available for many physical fitness and sport skills tests; for example, the National Children and Youth Fitness Studies I and II produced percentile norms on various physical fitness tests for American youth (Pate, Ross, Dotson, and Gilbert, 1985; Ross et al., 1987). Although it is not recommended that you grade on these characteristics, such data can help you interpret student performance, assign performance grades, and provide students with information for goal setting. For example, if you were teaching a volleyball unit, it would be helpful if you had normative data for various volleyball skills.

[www] **Go to HK*Propel* to complete Student Activity 8.8.**

Norms can be used both for assigning grades on individual test items and for making your own norms for composite scores to arrive at final grades. For example, table 8.11 provides the 75th and 25th percentile norms for 1-mile (1.6 km) walk test scores for 400 males and 426 females aged 18 to 30. You might assign an A to the top quartile (above P_{75}), a B to the two middle quartiles (P_{25}-P_{75}), and a C to the last quartile (below P_{25}). National and other published norms can be used to establish grading scales. However, the fairest norms for grade determination are local norms. To be used in grading, norms must be representative of the students' sex, age, training, and instructional situation.

Table 8.11 Percentile Norms for the 1-Mile (1.6 km) Walk Test (min:sec)

Percentile	Males	Females
75	11:42	12:49
25	13:38	14:12

Data from Jackson, Solomon, and Stusek (1992).

You can develop local norms for composite scores at your facility by following these steps:

1. Determine the objectives of your instructional or training program.

2. Select the most accurate possible tests and measurements to assess these objectives.

3. Administer the same tests with standardized procedures for several years.

4. Collect enough data to have at least 200 scores for each sex and age you intend to evaluate.

5. Conduct a statistical analysis of the data and establish percentile or standard score norms.

If you follow these steps, you will be able to establish norms that are representative of your students and their learning situation. The grading standards will provide consistency in grade assignment between classes, semesters, and instructors if all instructors are involved in the development and use of the norms. However, until you have sufficient data to establish norms, you will have to use some other grading technique. Ultimately, you should work toward establishing local norms, because this method will reduce most of the inconsistencies in grading.

Dataset Application

Open the chapter 8 large dataset in HK*Propel*. Assume that the 500 scores there represent total points earned in a class across several semesters and you are interested in determining what the distribution of points looks like. Do the following:

1. Create a histogram of the "TotalPoints" to see if it is normally distributed (chapters 2 and 3).

2. Compare the distributions for the females and males in your classes by creating two histograms (chapter 3) using Graphs → Legacy Dialogs → Histogram and entering "Total-Points" in the Variable box and "Gender" in the Panel by Rows area. Do the distributions of scores appear similar for females and males?

3. Compare the gender means to see if they are significantly different (refer to the procedures you used for independent t-tests in chapter 5). Confirm that the mean for the females is 496.83 and that the mean for males is 503.46. Are they significantly different?

4. Calculate the mean and standard deviation with SPSS. Use the z-table in chapter 3 (table 3.4) to determine cut points for various grades that you might assign based on the percentage of letter grades.

Arbitrary Standards

The previous techniques for converting scores to grades have involved data analysis of the observed measurements. Using arbitrary standards for grade assignment does not require data analysis of test scores. With this process, criterion-referenced standards are established for each grade. Table 8.12 provides an example of such standards on a 100-point knowledge test.

The biggest advantage of this type of grading system is that it provides consistency in grading. A 90 on the knowledge test will be an A in whatever class or semester a student takes the test. The system is simple and easy to understand. However, setting the standards in physical performance tests without knowledge of expected student performance levels results in a best guess process that may result in an undesirable grade distribution. If the standards you use are accurate reflections of student achievement levels, the arbitrary system can be used to good effect, but, as with the norms method, it relies on a relatively long period of data accumulation.

A specific version of using arbitrary standards in evaluation is the pass/fail assignment. Student performance is compared with a specific criterion-referenced standard that represents a minimum level of acceptable performance or ability. However, the setting of these minimum pass/fail standards is not an easy task and often takes years of experimentation and data collection.

Table 8.12 Arbitrary Standards for Grading a 100-Point Knowledge Test

Point range	Grade
100-90	A
89-80	B
79-70	C
69-60	D
59-0	F

Dataset Application

Return to the chapter 8 large dataset in HK*Propel* and see if you can determine grades for the 500 students there based on some of the methods that have been presented. For example, can you set cut points based on the distribution for grades of A, B, C, D, and F?

www. **Go to HK*Propel* to complete Student Activity 8.9.**

Mastery Item 8.3

Consider a course you might teach. Develop grading criteria and standards.

EVALUATION IN NONSCHOOL SETTINGS

Although this chapter is primarily directed to students who plan to become physical education teachers, many of the concepts described above are relevant to many nonschool settings, including sports programs, fitness centers and clubs, worksite health promotion programs, community-based organizations, and hospitals and medical clinics.

In addition to the typical formative and summative evaluation often used in school settings, assessment practice in nonschool settings can be generally classified as follows (Park and Kang, 2014):

1. *Placement or grouping.* An individual is placed into a specific category or group based on his or her test result. For example, a coach could assign a player to a team of similar-level players so that appropriate and targeted training can be arranged. Norm-based evaluation could serve well for the placement purpose.

2. *Achievement confirmation.* An individual is tested to determine if a preset goal was met. Both norm- and criterion-based evaluation can serve well here.

3. *Prediction.* An individual's future performance or health status is estimated based on a predeveloped equation, for example, prediction of a fitness club member's health risk based on current aerobic capacity. For the equation itself, the regression-like methods learned in chapter 4 will be needed. For the outcome of the prediction, the criterion-based evaluation can be employed.

4. *Selection.* An individual is selected based on the order of his or her performance in a group or a relative rank corresponding to a group or population (norm-based evaluation). The methods of placement, achievement confirmation, and prediction above can also be used for the selection.

5. *Diagnosis.* An individual's current weakness, problem, or injury is determined through assessment. For example, a stress test may be used to determine if a person has any potential risk of a cardiac event. The outcome is often a criterion-based evaluation, but the assessment itself and related diagnosis criterion are often quite sophisticated. (For more details on diagnosis assessment, see ACSM et al., 2018.)

MEASUREMENT AND EVALUATION CHALLENGE

After thinking through what objectives he wants his students to achieve, Tomas decides to weight two cognitive tests and two psychomotor skills tests to make up certain percentages of the final tennis unit grade for each student. To accomplish this, he has to convert the scores on all the tests to a standard score format (e.g., T-scores) and adjust these converted scores in accordance with the percentages he has chosen. The four adjusted standard scores are then added together to arrive at a total composite score representing each student's class achievement. Tomas then can select one of the methods described in the last section of this chapter to convert the composite scores to final grades.

SUMMARY

Determining and issuing grades to students is difficult. However, because grades are vital to a great number of important decisions in evaluation in a school setting, you must take them seriously and determine them fairly and in a meaningful way. This chapter identifies many of the issues involved in the process of grading and suggests methods for you to consider when faced with this task. Determine your own answers to some of the philosophical issues of assessing the performance of others, and use the mechanical tools provided here for measuring, weighting, and combining various scores, as well as determining the final grade. Methods for applying evaluation principles in nonschool settings were also introduced.

www. **Go to HK*Propel* for videos, homework assignments, and quizzes that will help you master this chapter's content.**

Developing Written Tests and Surveys

OBJECTIVES

After studying this chapter, you will be able to

- plan high-quality written tests;
- construct high-quality written tests;
- score written tests efficiently;
- administer written tests properly;
- analyze written tests; and
- understand concerns associated with planning, constructing, and enhancing the return of questionnaires.

[www] The lecture outline in *HKPropel* will help you identify the major concepts of the chapter.

MEASUREMENT AND EVALUATION CHALLENGE

Kate, an educational researcher, is conducting an experiment on the effectiveness of using computers to teach basic statistical concepts. She randomly assigns students taking the basic statistics course to one of three sections of the class. One group will be taught by the traditional lecture method. A second group will receive all class lectures and activities online using a newly developed multimedia approach. This group will not attend the general lectures. The third group will complete the course in a blended fashion: They will receive 50% of their coursework in the traditional method and 50% through online instruction and activities. What decisions must Kate consider and what steps does she need to follow to measure how well students in each group learn basic statistical concepts?

One common primary objective of a research project or a physical education curriculum is to increase participant knowledge and understanding of various aspects of physical activity. To determine if this objective is being met, making measurements in the cognitive domain is necessary, generally through the use of a written test. Another common objective of research is to assess people's attitudes, opinions, or thoughts about a particular topic, which is most often accomplished through the use of a questionnaire. Although the focus of this chapter is on the process for developing paper-and-pencil tests and questionnaires, all of the concepts apply equally well to online tests, which are now widely used in testing.

There are many sources for written tests. In some disciplines, nationally normed standardized tests are available. Textbook publishers often provide prewritten tests or banks of test questions from which you can build your own tests. Some state agencies provide written tests for statewide testing programs, making comparisons possible among schools or districts. In human performance, however, outside sources of written tests are uncommon. In physical education, the lack of standardized tests is partly due to the great variety of activities embedded in physical education curricula and the fact that there are fewer textbooks available in physical education than in such classroom subjects as English and math. In our discipline, the most common source of written tests and questionnaires is undoubtedly the researcher or teacher interested in measuring cognitive objectives. This is not all bad, because the person making the assessment should be able to construct the most valid measuring instrument (one that measures what it is intended to measure). However, knowing what to measure and knowing how to measure it are different things. There are five requirements for constructing effective written tests:

1. You must be knowledgeable in the proper techniques for constructing written tests. Various types of questions have differing efficiencies and uses in certain situations.

2. You must have a thorough knowledge of the subject area to be tested. Without this knowledge, it is difficult to construct meaningful test questions.

3. You must be skilled at written expression. Test questions devised by people lacking good writing skills are often ambiguous. This ambiguity reduces the validity and reliability of a written test because there is no way of distinguishing whether an incorrect response is chosen due to the participant's lack of knowledge or to an error in interpretation of the question.

4. You must be aware of the level and range of understanding in the group to be tested so that you can construct questions of appropriate difficulty. As will be explained later, not having such awareness can affect the efficiency of the test.

5. As the prospective test constructor, you must be willing to spend considerable time and effort on the task. Effective written tests are not put together overnight. Santiago and Morrow (2020) provide an illustration of how the five requirements listed previously are used to develop a reliable and valid test of preservice teachers' content knowledge of health-related fitness.

If you examine these five requirements carefully, you will notice that the last four are also qualities of a careful researcher or a dedicated teacher. However, we will limit ourselves in this chapter (and book) to presenting information about the first requirement, proper techniques for constructing a written test. Properly constructed tests can result in reliable and valid decisions about the cognitive ability being assessed. For Kate to determine whether the manner in which statistical concepts are taught results in different learning outcomes, she must first be able to accurately measure students' levels of understanding of these concepts. Kate can use the information in the following sections as she considers how to construct such a written test.

PLANNING THE TEST

First, consider the differences between mastery tests (criterion-referenced tests, or CRTs) and achievement tests (norm-referenced tests, or NRTs). A **mastery test** is used to determine whether a student has achieved enough knowledge to meet some minimum requirement set by the tester. It is not used to determine the relative ranking of students' cognitive abilities but rather to determine each student's compliance, or lack of compliance, with some previously determined standard or criterion. A familiar example of a mastery test is the type of spelling test in which the expected score is perfect or nearly perfect (all words spelled correctly). Other examples are the written portion of the test for obtaining a driver's license or other health professional licensures, on which you must get a certain minimum number correct to pass.

© Human Kinetics

In the psychomotor domain, not every student will be able to master every tennis skill; in the cognitive domain, not every student will be able to master every cognitive objective. In both cases, achievement tests are designed to ascertain each student's level of accomplishment.

The purpose of an **achievement test**, on the other hand, is to discriminate among levels of cognitive accomplishment. Because it is usually not reasonable to expect every student to achieve 100% of every cognitive objective put forth, identifying each student's progress toward meeting the objectives is of great importance.

In human performance, both types of tests have key uses. For example, in a potentially dangerous activity such as gymnastics or swimming, using a mastery test of safety rules is prudent. For the most part, however, this chapter deals with the various phases of constructing and using achievement tests, which are more commonly used for human performance assessments than mastery tests.

WWW **Go to HK*Propel* to complete Student Activity 9.1.**

There are two important decisions to make when planning your written test. The first and more significant of these involves determining what is to be measured. One technique for ensuring that a written test measures the desired objectives and that the correct emphasis is given to each objective is to develop a table of specifications. The second fundamental decision in planning a written test involves answering several mechanical questions concerning how the objectives are to be measured, the frequency and timing of testing, the number and type of questions, and the format and scoring procedures that will be used.

What to Measure

The question of *what* the test will measure should be answered before instruction begins. The objectives of a course of study, the experiences used to meet these objectives, and the implementation and sequence of these experiences must all be determined in advance if an instructional unit is to be effective. You can alter these elements as instruction progresses, but radical changes should not be necessary. In any event, testing will allow you to measure the degree to which the objectives of the course are being achieved and to evaluate where problems may exist. When the objectives to be assessed lie in the cognitive domain, the initial step in designing a written test is the development of a table of specifications.

As described in chapter 6, the table of specifications is to the test writer what the blueprint is to the home builder. It provides the plans for construction. The table of specifications identifies the relative importance of each content area on the test by assigning it a percentage value. It is a two-way table, with the **content objectives** of the instructional unit along one axis and **educational objectives** along the other. Content objectives are the specific goals determined by an instructor, and educational objectives are generic topics suggested by various experts. A table of specifications helps ensure a test's content validity (the extent to which the items on a test adequately sample the subject matter and abilities that the test is designed to measure).

Let us look at an example that demonstrates the process of formulating a table of specifications for a 60-item test to be used with an instructional badminton unit. The content objectives of the instructional unit and the instructor's decision about their relative importance might be as follows:

History	5%
Values	5%
Equipment	10%
Etiquette	10%
Safety	10%
Rules	20%

Strategy	15%
Techniques of play	25%
Total	**100%**

The educational objectives (see also chapter 1) and the tester's weighting of each might be as follows:

Knowledge	30%
Comprehension	10%
Application	30%
Analysis	20%
Synthesis	0%
Evaluation	10%
Total	**100%**

Once an instructor has determined the content and educational objectives and their relative importance, the table of specifications can be constructed.

For this example, the result is shown in table 9.1. The content objectives and their relative weights are located in the rows, and the educational objectives and their weights are in the columns. The weight associated with a single cell of the table is found by determining the product of the intersection of the appropriate row and column. For example, the weight for knowledge of history is determined by multiplying 5% (the weight of history) by 30% (the weight of knowledge) to get 0.015, or 1.5%. This product for any cell is an expression of the approximate percentage of the test that should be made up of items combining the two types of objectives intersecting in that cell. The actual number of questions of each combination is found by multiplying the obtained percentage by the proposed length of the test. In this case, we wanted a 60-item test, so for knowledge of history we would multiply 0.015 by 60 to get 0.9. In table 9.1, each cell is divided into two halves; the upper number represents the percentage of the test made up of items combining the appropriate combination of objectives, and the bottom number represents the number of questions of this type based on a total test length of 60 items.

Obviously, it is not possible to include on the test 0.9 of a question dealing with knowledge of the history of badminton; the numbers in the table of specifications are to be used as guides, and usually some rounding and adjusting are required. For example, the 0.9 would logically result in one question of the 60 focusing on knowledge of the history of badminton. If the table of specifications is followed closely, the resulting test will contain questions in proportion to the percentages of weighting for each category.

Mastery Item 9.1

Based on the table of specifications in table 9.1, how many questions involving the analysis of techniques of play would be included on a 100-item test?

Various educators and test-construction experts have identified educational objectives that may be used in tables of specifications. The educational objectives in table 9.1 come from a list published in *Taxonomy of Educational Objectives* (Bloom, 1956). Recall from chapter 1 that the cognitive taxonomy consists of knowledge, comprehension, application, analysis, synthesis, and evaluation. Briefly, knowledge is defined as remembering and being able to recall facts; comprehension is the lowest level of understanding; application is the

Table 9.1 Table of Specifications for a 60-Item Written Test on Badminton

Content objectives		Weight	Knowledge 30%	Comprehension 10%	Application 30%	Analysis 20%	Synthesis 0%	Evaluation 10%	Totals for content objectives 100%
	History	5%	1.5%	0.5%	1.5%	1.0%	0%	0.5%	
			0.9	0.3	0.9	0.6	0	0.3	3
	Values	5%	1.5%	0.5%	1.5%	1.0%	0%	0.5%	
			0.9	0.3	0.9	0.6	0	0.3	3
	Equipment	10%	3.0%	1.0%	3.0%	2.0%	0%	1.0%	
			1.8	0.6	1.8	1.2	0	0.6	6
	Etiquette	10%	3.0%	1.0%	3.0%	2.0%	0%	1.0%	
			1.8	0.6	1.8	1.2	0	0.6	6
	Safety	10%	3.0%	1.0%	3.0%	2.0%	0%	1.0%	
			1.8	0.6	1.8	1.2	0	0.6	6
	Rules	20%	6.0%	2.0%	6.0%	4.0%	0%	2.0%	
			3.6	1.2	3.6	2.4	0	1.2	12
	Strategy	15%	4.5%	1.5%	4.5%	3.0%	0%	1.5%	
			2.7	0.9	2.7	1.8	0	0.9	9
	Techniques of play	25%	7.5%	2.5%	7.5%	5.0%	0%	2.5%	
			4.5	1.5	4.5	3.0	0	1.5	15
	Totals for educational objectives	100%	18	6	18	12	0	6	Test total = 60

Note: The top number in each cell of the table body is the percentage of questions for the combined content and educational objectives for that cell; the bottom number is the actual number of questions (of the 60 in total) that the percentage represents.

use of abstractions in real situations; analysis is the division of material into its parts to make clear the relationship of the parts and the way they are organized; synthesis is putting together elements and parts to form a whole; and evaluation is making judgments about the value of ideas, works, solutions, methods, and materials.

The following list of questions or tasks gives you an idea of how Bloom's taxonomy would apply to a written test for basketball.

• *Knowledge.* What is the height of a regulation basketball hoop?

• *Comprehension.* What area of the court is the forwards' responsibility in a zone defense?

• *Application.* What defense should be used when the opposing team is much faster than your team?

• *Analysis.* Prioritize the following basketball skills for each player position: blocking shots, dribbling, passing, shooting.

• *Synthesis.* Design a practice schedule for the first 3 weeks of the season for a boys' high school team with 35 players, 4 baskets in the gym, 5 days per week to practice, and 90 minutes for each practice.

• *Evaluation.* Present arguments for and against the following statement: Rather than have separate boys' and girls' basketball teams at the junior high school level, the school should have coed teams.

In 2001, Bloom's taxonomy was expanded by Anderson and Krathwohl (2001) to combine the cognitive process with knowledge dimensions. They refer to the highest level as *creating*, and slight modifications were made to the categories. However, the basic concepts for constructing a table of specifications, as already described, remain valid.

Another list of educational objectives includes the categories of terminology, factual information, generalization, explanation, calculation, prediction, and recommended actions (Ebel, 1965). Bloom's and Ebel's examples indicate some of the educational objectives that can be used in constructing the table of specifications. You can also devise your own lists.

WWW **Go to HK*Propel* to complete Student Activity 9.2.**

How to Measure

As mentioned previously, determining how to measure usually involves answering several mechanical questions. The answers are often partially resolved by deadlines or by practical considerations, but frequently the answers require understanding the outcomes of various testing procedures.

When to Test

For testing that occurs in a school system, institutional policies may dictate the times testing is done. The type and frequency of grade reporting, a requirement to set aside certain class periods for testing, and various class-scheduling practices may influence the decision of when to test. Most frequently, tests are administered during a regularly scheduled class period at or near the end of each unit of study, and the lengths of the units are designed to coincide with the school's grading periods. These practices are justifiable for the achievement type of test discussed in this chapter. However, there may be valid reasons for administering tests at other times during instruction.

Deadlines, too, usually determine the appropriate time to administer a written test related to a research project a student is completing. Depending on the hypothesis being tested in the research, an investigator may plan cognitive assessments before, at the conclusion of, throughout, or at both ends of an instructional unit embedded in a research project.

Test frequently enough to ensure that you obtain reliable results, but not so frequently that you needlessly use valuable instructional or student time. For obvious reasons, there is no set amount of time that you should reserve for measurement purposes, but it is likely that more errors are made by providing too little time for testing than by incorporating too much time.

How Many Questions to Include

Generally, the reliability of an achievement test increases as its length increases. Reliability increases with increases in test length because the more often an assessment of achievement is made, the less overall effect chance has on the results. Flipping a coin twice and obtaining two heads is meager evidence to support the contention that the coin has two head sides. However, if the coin is tested 50 times and heads occur 50 times, the contention becomes tenable because the chance occurrence of such an event with a normal coin is extremely remote.

The length of a test is a function of other factors in addition to the desire for reliable results. Three other important factors determining the number of questions on a test are

1. the time available for testing,
2. the type of questions used, and
3. the attention span of the students.

In most school situations, the length of the class period is the limiting factor on the length of an achievement test. Often, only the typical 45 to 60 minutes of class time is available. The number of questions that can be answered in this time largely depends on the type of questions used. For example, only a few essay questions requiring extensive answers can be completed within one class period, but many more essay questions requiring a one- or two-sentence response can be included. A test can realistically include more factual multiple-choice questions that only require recall than multiple-choice items that require examinees to apply knowledge to novel situations, which require additional thinking and reflection. Finally, differences in the attention spans of examinees influence the decision of how many questions to include on a test. Schools often account for differences in attention spans by adjusting the length of class periods according to the grade level of the students. A researcher has more flexibility than a schoolteacher in varying test length, so the factor of attention span typically becomes the most important limiting factor for any researcher.

Another aspect to consider when determining the number of questions to include on a test is that not all students work at the same rate. What percentage of students should be able to complete the test? In most situations, all or nearly all of those being tested should be able to finish the test. With a few exceptions—such as a sport-officiating course or an emergency room diagnosis unit in which an objective is to acquire the ability to make rapid and correct decisions—it is generally true that a measurement of the ability to answer questions correctly is of more value than a measurement of the speed with which correct answers can be produced. Furthermore, to construct a test containing more questions than can be completed by most or all examinees is an inefficient use of your time, because the questions near the end of the test will be seldom used.

The numerous combinations of the factors of available time, question type, attention span, and work rate make it inevitable that a certain amount of trial and error will occur in determining the number of questions on a test. However, we suggest some general guidelines that you may adjust to meet your particular situation. Most students of high school age and older should be able to complete three true–false questions, three matching items, one or two completion questions, two recognition-type multiple-choice items, or one application-type multiple-choice item in 1 minute. For younger students, these estimates should be reduced appropriately. Few guidelines can be given regarding the number of essay questions; however, allow enough time for a student to be able to organize an answer and then put it on paper (or online). Also, in general, including many short essay questions measures achievement more effectively than using a few lengthy ones.

Although technology such as online testing has greatly improved the efficiency of the testing and survey process, these major elements concerning the number of questions to include remain key considerations when constructing a test or survey.

Mastery Item 9.2

Approximately how much time should be allotted for a college-aged student to complete a written test containing a combination of 25 true–false questions, 25 recognition-type multiple-choice questions, and 25 application-type multiple-choice questions?

What Test Format to Use

Although more and more achievement tests are moving online, printed tests are still being used, especially in classroom settings. Convenience, minimizing opportunities for cheating, and concern for vision- or hearing-impaired examinees are the key considerations that affect the decision of what format to use. The format selected should maximize the opportunity for every examinee to understand and complete the required task or tasks.

Oral presentation of test questions is, in general, an unsatisfactory procedure for most types of items, with the possible exception of true–false questions. Although the expense and your preparation time for this format are minimal, all examinees are forced to work at the same set pace, and there is little or no opportunity for examinees to check over answers. Projecting the test using PowerPoint has the same disadvantages as oral presentation. In addition, this format introduces some expense and time-consuming preparation. Probably the most common, efficient, and preferred method of presenting achievement tests is in written form, in which each examinee receives a copy of the test questions. Although this method requires advance preparation (in this case, typing, proofreading, duplicating, and possibly compiling), it maximizes convenience for the examinees. Each examinee can work at his or her own rate, the answers can be checked if time permits, and the questions can be answered in any order. You are free to monitor the test. Online tests are also quickly becoming popular. Although it may slightly increase the time required for test preparation, this format can be extremely efficient, especially in the administration of the test and scoring and analysis of the test results.

Paying attention to the way you lay out your test can help you cut costs and test-preparation time as well as enhance the accuracy of responses. When an examinee actually knows the correct answer to a question but makes an incorrect response because of an illegible test copy, the reliability and validity of the test are reduced. Carefully proofreading your test before administering it also eliminates the need to orally correct errors in test items, which wastes valuable testing time. Here are some additional tips to consider:

- Provide students with advance notice about the number and nature of the test items.
- Provide extensive directions for completing the test (and review these the day before the test, if possible).
- If various types of questions are used in one test, group together questions of the same type to reduce fluctuation among the types of mental processes required of examinees.
- Group together questions of similar content (i.e., subject area) on achievement tests.
- Although ordering test questions from easiest to hardest is not generally recommended, including a relatively simple question or two at the beginning of a test may benefit students by reducing anxiety.

Two interesting variations of the typical written test are the open-book or open-notes test and the take-home test. Each has advantages and disadvantages and under certain conditions can be used effectively. The greatest benefit of both is the reduction of student anxiety. In addition, an open-book test typically requires you to ask fewer trivial questions and more application questions; it forces you to invent novel situations rather than present questions based entirely on circumstances presented in the textbook or lectures. Furthermore, open-book tests reduce the possibilities for cheating because students are allowed to use books, notes, and other materials.

One possible disadvantage of an open-book test is that it may reduce a student's incentive to learn and reduce the time a student spends preparing for the test. Students tend to rely on being able to obtain answers from their notes and books during the test and may thus spend less time studying. Because examinees can look up answers, you will need to set time limits on open-book tests, or some examinees (usually unprepared and in effect studying while taking the test) will take an inordinate amount of time to finish. If an open-book test is well constructed, most examinees will find that the textbook and notes are of little value except for looking up formulas and tables. Examinees should not be able to answer open-book test items simply by turning to a specific page in the textbook and finding the answer.

Take-home tests can be used in situations in which more time is required to complete a test than is available in a controlled setting. The major problem with them lies in the impossibility of ensuring that each person does his or her own work. Thus, take-home tests should not be used to measure achievement but might be used as study guides and home-work assignments. Similarly, online tests should be administered in a supervised setting to avoid the possibility of someone other than your students or subjects completing the test.

WWW. **Go to HK*Propel* to complete Student Activity 9.3.**

What Type of Questions to Use

Questions can be classified into three general categories: semiobjective, objective, and essay. There are three types of semiobjective questions: short-answer, completion, and mathematical questions. For these questions, examinees must compose the correct answer; the answer is so short that little or no organization is necessary. Some subjectivity may be involved in scoring (e.g., awarding partial credit for the incorrect spelling of the correct answer, or for a mathematical problem that gives a wrong answer but uses the correct procedures). Scoring procedures are generally similar to those used for objective questions: The response is checked to see whether it matches the previously determined correct answer.

Types of questions classified as objective include true–false, matching, multiple-choice, and classification items. Characteristically, the task of an examinee responding to an objective question is to select the correct (or best) answer from a list of two or more possibilities provided. This type of question is considered objective because scoring consists of matching an examinee's response to a previously determined correct answer; that is, this type of scoring is relatively free of any subjective or judgmental decision.

When responding to an essay question, an examinee's task is to compose the correct answer. Usually the question or test item provides some direction by including such terms as *compare* or *explain*. The item may also constrain the answer by including such phrases as *Limit your discussion to . . .* or *Restrict your answer to the year. . . .* Essay questions are subjective because scoring usually involves making judgmental decisions.

Several differences among the categories have consequences for either instructors or examinees. For examinees, much of the time available for testing is consumed in writing (essay questions), reading (objective or semiobjective questions), or calculating the answer (mathematical problems). Because reading is less time consuming than writing or calculating, a greater number of objective questions can usually be included on a test than questions from the other two categories. Also, examinees who are weak in one of these areas (writing, reading, or calculating) may be at a disadvantage in taking tests composed mainly of questions requiring the skill in which they are weak. A poor reader, for example, may do worse on an objective test than on an essay test over the same material.

From the test constructor's point of view, essay and semiobjective questions are easier to prepare than objective questions but harder to score. In addition, the quality of an objective test depends almost entirely on your ability as a test constructor rather than as a scorer, whereas the situation is reversed for an essay or semiobjective test. Thus, your decision about what type of test to construct might be influenced, in part, by the time you have available to construct and score it, or whether your abilities lie in constructing or in scoring tests.

It is plausible that students study differently for different types of tests (though evidence for this is inconclusive); for example, some believe that objective tests promote the study of factual and general concepts. However, this belief rests mainly on the mistaken assumption that objective questions cannot measure depth of achievement. Although often more difficult to construct, a test composed of objective questions can usually measure achievement as well as a test made up of essay questions. In short, the type of studying promoted by a test is more a function of the *quality* of the questions than the *type* of questions.

It is true, though, that some types of questions are more efficient than others in a particular situation. It would be difficult, for example, to conceive how the quality of one's handwriting might be measured efficiently with an objective test, or how the ability to solve mathematical problems might be measured any more validly than by a test composed of mathematical problems. However, the fact that it may be more efficient to use objective questions to measure factual knowledge and essay questions to measure the organization and integration of knowledge has stereotyped the way certain questions are used. Also, other testing factors may preclude the use of what appears to be the most efficient type of question. For example, it is often impractical to correct an essay test given to a large number of people. Thus, an objective test may be used even though the measurement involves more than just factual information. Although many nationally standardized tests include some essay questions, their heavy reliance on objective questions is an example of this situation.

Despite the names of the three categories of questions, remember that subjectivity is a part of every test. Subjective decisions are required in the scoring of essay questions and, to a lesser degree, semiobjective questions. Your decisions in determining both what questions to ask and how to phrase them are subjective in nature. To increase the reliability of written tests, reduce the amount of subjectivity involved in their construction and scoring as much as possible. Practices such as formulating a table of specifications (see What to Measure) and consulting with colleagues can ensure that you accomplish this.

WWW Go to HK*Propel* to complete Student Activity 9.4.

Mastery Item 9.3

Review the term *objectivity* discussed in chapter 6. How does the concept of objectivity apply to the administration of a written test? Can you list some of the procedures used in the administration of the ACT or SAT that are meant to increase objectivity?

Regardless of the type or types of questions used on a test, the usefulness of the resulting score depends on its stability (i.e., reliability). A test is intended to measure the achievement of certain objectives, and the scores resulting from the test are supposed to express the degree of achievement. If a different construction, administration, or correction of the test by you or a different person were to result in a different set of scores, the stability and thus the usefulness of the score is reduced. The type of questions included on a test affects the stability of the scores in various ways.

For example, if two people were told to construct a test over the same unit of instruction, it is more likely that the two tests would contain similar questions if the two were told to construct an essay test rather than an objective or semiobjective test. On the other hand, if two people each scored an objective test, a semiobjective test, and an essay test, concurrence is much more probable for the objective test than for the semiobjective or essay test.

Understanding the similarities and differences among the types of questions and being aware of the advantages and disadvantages of each type of question (covered in the next section) are necessary in order to select the most efficient types of questions for a particular situation. This knowledge, in addition to proficiency in the general requirements of test construction, will allow you to develop valid and reliable written achievement tests.

Kate, from the Measurement and Evaluation Challenge, has decided to develop a table of specifications to ensure the proper emphasis and weighting of the concepts her test will assess. She also will probably choose to have as lengthy a test as possible and to use either multiple-choice questions, mathematical problems, or a combination of both.

CONSTRUCTING AND SCORING THE TEST

There are many ways to construct and score the various types of questions to increase their efficiency. As we've discussed, essay questions are relatively easy to construct and time consuming to score, whereas multiple-choice questions are the opposite.

Semiobjective Questions

As stated, the three types of semiobjective questions are *short-answer questions*, *completion questions*, and *mathematical problems*. The short-answer question and the completion question differ only in format: The completion item is presented as an incomplete statement (a fill-in-the-blank), whereas the short-answer item is presented as a question. The task required to answer a mathematical problem can be specified by symbols or by words (as in a story problem). Because of their similarities, we will describe the uses, advantages, and limitations and provide construction and scoring suggestions for all three types of questions in this section.

Uses and Advantages

Semiobjective questions are especially useful for measuring factual material such as vocabulary words, dates, names, identification of concepts, and mathematical principles. They are also suitable for assessing recall rather than recognition because each examinee supplies the answer. The advantages of semiobjective questions include relatively simple construction, simple and rapid scoring, and a low likelihood of examinees guessing the correct answer by chance.

Limitations

Because of the limited amount of information that can be given in one question or incomplete statement, it is often necessary to include additional material to prevent semiobjective questions from being ambiguous. Even when a fair amount of detail is included, the danger of ambiguity is not completely removed, especially for completion items. Occasionally, a blank left in a sentence can be filled by a word or phrase that is technically correct even though it is not precisely what the test constructor desired. For example, consider the following completion item: "Basketball was invented by _____." The name *James Naismith*, the phrase *a male*, and the date *1900* are three possibilities that correctly complete the sentence. When this situation occurs, an instructor must decide whether to award credit. With mathematical questions, instructors may have to decide whether to award no credit, partial credit, or total credit if a student followed correct procedures but gave the wrong answer. Similar decisions are necessary when an examinee provides the correct answer but it is unclear how it was derived. These situations introduce some subjectivity and thus the possibility of inconsistency in the scoring procedure. Specific construction techniques can help reduce (but seldom completely eliminate) this problem.

Recommendations for Construction

Of the three types of semiobjective questions, ambiguity is most likely to occur with completion questions. Rephrasing the incomplete sentence into a question—that is, converting it into a short-answer item—often resolves several problems. However, if you prefer a completion question, these suggestions may reduce some ambiguities.

• Avoid or modify indefinite statements for which several answers may be correct and sensible. Do this, in part, by specifying in the incomplete statement what type of answer is required. For example, "Basketball was invented by _____" can be reworded as "The name of the person who invented basketball is _____." A similar method for eliminating ambiguity is to present the item this way: "Basketball was invented by _____ (person's name)."

• Construct the incomplete sentences, when possible, so that the blank occurs near the end of the statement. This technique better identifies the specific type of answer required than when the blank space occurs early in the statement. For example, in the item "The _____ system of team play in doubles badminton is recommended for beginners," the desired correct answer is *side-by-side*, but the blank could logically be filled in with the phrase *least complex* because it is not clear that the name of the system is desired. Rewording the statement so that the blank occurs near the end solves this problem: "The type of team play recommended for beginners in doubles badminton is called the _____ system."

• Do not leave so many blanks in one statement that the item becomes indefinite. Consider this extreme example: "The name of _____ who invented _____ is _____." As the example demonstrates, the more blanks in the statement, the less information given; answering the question becomes a guessing game. Give additional information by either explaining what is required or making several items from the one.

• Do not give inadvertent clues. Occasionally the phrasing of the statement or the use of a particular article (*a* or *an*) or verb reduces the number of possible words or phrases that might complete a statement. Use the following format for the indefinite article: "Basketball was invented by a(n) _____ (nationality)." If more than one blank occurs in a statement, each blank should be the same length to avoid giving students information about the length of the correct response.

• If a numerical answer is required, indicate the units and degree of accuracy desired (e.g., "Round to the nearest one-tenth."). Specifying this information simplifies the scorer's task and eliminates one source of confusion for examinees.

• Use short-answer questions where possible to reduce ambiguity. For example, using a short-answer question such as "An athlete from what country won the gold medal in the pentathlon in the 2016 Olympics?" rather than the completion item, "The gold medal in the pentathlon in the 2012 Olympics was won by _____ (country)" increases the probability that the country will be identified rather than other possible information. Scoring consistency is enhanced because the examinees' task is typically more clearly identified. You should also phrase short-answer items in such a way that the limits on the length of the response are obvious.

Recommendations for Scoring

If semiobjective questions are well constructed and you encounter no problems (e.g., when two or more answers are plausible for one item), the scoring process is simple, objective, and reliable. The answers can be scored easily by persons other than the test maker.

If the test consists of completion items, you can prepare an answer key by using a copy of the test and cutting out a rectangular area where each blank occurs. Write the correct answer immediately below or adjacent to the rectangular area. When the answer key is superimposed on a completed test, each response can be quickly matched with the keyed answer.

For short-answer items, having students use separate answer sheets speeds the scoring process. Because only one-word or short-phrase answers are expected, you can distribute, along with the test itself, a previously prepared answer sheet with a numbered blank space corresponding to each test item. Usually, you can place two columns of answers on one side of a standard-sized piece of paper. To score short answers efficiently, construct an answer key by recording the correct responses on a copy of the answer sheet and place this alongside each answer sheet. This procedure eliminates the need to search through the pages of each test to locate the answers.

www **Go to HKPropel to complete Student Activity 9.5.**

Objective Questions

Questions requiring the selection of one of two or more given responses can be scored with minimal subjective judgment and are thus categorized as *objective questions*. Although there are many similarities among types of objective questions, we give separate consideration to true–false, matching, and multiple-choice questions because of their differences.

True–False Questions

Perhaps unfortunately, true–false questions are widely used, probably because these questions are relatively easy to construct and score. Although there are advantages to true–false questions and situations in which their use is justifiable, they are the least adequate type of objective question because of several weaknesses.

Uses and Advantages

Like the various semiobjective questions, true–false items are particularly suited for measuring factual material such as names, dates, and vocabulary words. The advantages of using true–false items include the ease of construction, administration, and scoring and the fact that more true–false items can be answered than any other type of question within a given time.

Limitations

Many of the major weaknesses of true–false questions stem from the fact that an unprepared examinee might answer half of the items correctly by chance alone. This makes it difficult to assess a test taker's level of achievement. A correct answer could be an indication of complete understanding of the concept, a correct blind guess, or any shade of understanding between these two extremes. In addition, the inordinately excessive influence of chance lowers the amount of differentiation among good and poor examinees and consequently the reliability of the test.

To be fair and avoid ambiguity, a true–false item should be absolutely true or absolutely false. It is difficult to meet this requirement except when factual knowledge is involved. True–false questions are not well suited for measuring complex mental processes. Because of this, ill-composed true–false tests can include trivial questions and reward sheer memory rather than understanding.

Recommendations for Construction

Generally, writing good true–false questions involves avoiding ambiguity. Examples of good and poor true–false questions are given at the end of the section. Some specific suggestions for constructing good true–false questions include the following:

• Avoid using an item whose truth or falsity hinges on one insignificant word or phrase. To do so results in measuring alertness rather than knowledge.

• Beware of using indefinite words or phrases. A question whose response depends on the interpretation of such words or phrases as *frequently*, *many*, or *in most cases* is usually a poor item.

• Include only one main idea in each true–false question. Combining two or more ideas in one statement often leads to ambiguity. If the combination introduces the slightest amount of falsity in an otherwise true statement, the examinee must decide whether to mark true or false on the basis of the amount of truth rather than absolute truth.

• Avoid taking statements directly out of textbooks or lecture notes. Out of context, the meaning of the resulting item can be confusing. In addition, using textbook sentences as true–false items results in rewarding memorization rather than learning.

• Use negative statements sparingly and avoid double negatives completely. Inserting the word *not* to make a true statement false borders on trickery and may result in a measurement of vigilance rather than knowledge. Statements containing double negatives, especially if false, are often needlessly confusing.

• Beware of giving clues to the correct choice through **specific determiners** or statement length. Specific determiners are words or phrases that inadvertently provide an indication of the truth or falsity of a statement. For example, true–false items containing words such as *absolutely*, *all*, *always*, *entirely*, *every*, *impossible*, *inevitable*, *never*, or *none* are more likely to be false than true because an exception can usually be found to any such sweeping generalization. Conversely, such qualifying words as *generally*, *often*, *sometimes*, or *usually* are more common in true statements than in false statements. Because it often takes several qualifications to make a statement absolutely true, take care to avoid a pattern of long statements being true and false statements being short.

• Include approximately the same number of true statements and false statements on a test. Having too many of one or the other can bias responses. There is some evidence that false statements are slightly more discriminating, perhaps because an unprepared examinee is more inclined to mark true. For this reason it may be advantageous to include a slightly higher percentage of false statements.

• Avoid arranging a particular pattern of correct responses. Regulate the placement of true and false statements by chance to minimize the possibility that examinees will detect a pattern of responses.

• Arrange for a colleague to review the true–false questions before administering them. Doing so may help you remove ambiguity from questions.

Modifications

Test makers have attempted to modify true–false questions with the intent of reducing excessive blind guessing. One method is to require the examinee to identify the portion of a false statement that makes it false. A further modification requires the correction of the inaccurate portion. These modifications are called *corrected true–false*. Although these two modifications partially eliminate the effect of chance on the final score, they simultaneously introduce other problems. Ambiguity may result, as in the following: "James Naismith invented the game of volleyball." The statement is false but may be corrected by replacing the name *James Naismith* with the name *William Morgan*, or by replacing the word *volleyball* with the word *basketball*. These kinds of true–false questions can introduce some subjectivity into the scoring. Furthermore, the advantage of quick scoring is lost.

Another way to modify true–false questions involves changing the answering and scoring procedure to reflect the degree of confidence examinees have in their responses. The intent is to discriminate between those who get an answer wrong because they do not know the correct response and those who know something but not enough to prevent a bad luck choice. Several scoring systems have been devised to accomplish such *confidence weighting* of the response to a true–false item, such as the one shown in table 9.2. For example, if the examinee marks A, and the correct answer is "true," the examinee is awarded two points, but if the correct answer is "false," two points are deducted from the examinee's score. This modification, although increasing the discriminatory power of a true–false test, may introduce some undesirable variables. For example, differences in personality traits among examinees (some more willing to gamble than others), the importance of knowledge of the content being tested, and an awareness of the nature of one's knowledge become factors influencing the final test results. Thus, these modifications may well increase the reliability and discriminatory power of a true–false test but simultaneously reduce its validity.

In this example, the reason half a point is awarded for choosing C is that if a totally unprepared examinee guessed at every question on a normally scored true–false test, his or her score would be 50% correct (in the long run).

Table 9.2 System for Confidence Weighting Answers to True–False Questions: Scoring Procedure

Response	Mark	POINTS AWARDED OR SUBTRACTED	
		Correct	Incorrect
Definitely true	A	2.0	–2.0
Likely true	B	1.0	0.0
Omit or don't know	C	0.5	0.5
Likely false	D	1.0	0.0
Definitely false	E	2.0	–2.0

Recommendations for Scoring

As is true of most semiobjective and objective questions, using a separate answer sheet facilitates the scoring procedure. Because of the similarity between the letters *T* and *F*, it is not a good idea to have the examinees write the letters *T* and *F* on a sheet of paper when completing a true–false test. A previously prepared answer sheet on which examinees block out, circle, or underline the correct response eliminates such scoring problems. Special answer sheets that can be scored by machine are available for most objective questions, including true–false questions. You (or even someone not familiar with the subject matter) can efficiently score the test by hand by matching each response on an answer sheet with a previously prepared correct answer sheet.

WWW Go to HK*Propel* to complete Student Activity 9.6.

TRUE–FALSE QUESTIONS FOR BASKETBALL

Good Questions

1. Kicking the ball is a team foul. (False)

2. It is generally better to dribble than to pass. (False) (Earlier, *generally* was identified as a specific determiner, and its use was discouraged. However, notice that in this question it is used in a false statement rather than the normally expected true statement.)

3. A double violation occurs when a player commits two fouls at the same time. (False)

Poor Questions

1. Basketball was first introduced in 1901. (False) (Very trivial; it was introduced in 1891)

2. The overhead pass should always be used by short players. (False) (Use of the specific determiner *always*)

3. The shovel, underhand, and hook passes are made while holding the ball with both hands. (False) (Part of the statement is true and part is false)

4. In most cases, teams play one-on-one or zone defense. (True) (Use of indefinite phrase *in most cases*)

5. A time-out shouldn't be wasted when the team isn't in trouble. (True) (Use of a double negative)

Matching Questions

Matching questions generally involve two lists of words or phrases. The examinee's task is to match an item in one list with the item most closely associated with it in the second list.

Uses and Advantages

As with true–false questions, matching questions are most efficient for measuring relatively superficial types of knowledge. Measurements of vocabulary, dates, events, and simple relationships, such as authors to books, can be effectively obtained with matching questions. Matching questions are used to measure who, what, where, and when rather than how or why. Among the advantages of matching questions are relative ease of construction and the rapidity, accuracy, and objectivity of scoring. Matching questions require developing a cluster of similar questions and similar answers. The most discriminating matching questions often are those used in conjunction with a graph, chart, map, diagram, or similar device for which labels on the illustration are matched with functions, names, or similar categories of answers.

Limitations

It is difficult, although not impossible, to construct matching questions that require the examinee to use higher-order mental processes. However, the most limiting aspect of matching questions is that they require similarity within each of the two lists that make up the items. As compliance to this requirement lessens, the discriminating power of the matching items also usually diminishes.

Recommendations for Construction

Because it is easier to write questions that measure relatively superficial knowledge than it is to write questions that measure higher-order cognitive processes such as application, analysis, and evaluation, refer often to the table of specifications developed for a test when constructing matching questions. This ensures that you will achieve the desired balance among the areas measured. Unless you exercise caution, a test composed mainly of matching items may concentrate more heavily on factual material than warranted by the table of specifications. Here are some additional suggestions for constructing matching questions:

• Present clear and complete directions. In general, include three details in the instructions:

1. The basis for matching the item in the two lists
2. The method to record the answers
3. Whether a response in the second column may be used more than once

• An instruction such as "Match the statements in column 1 with those in column 2" does not include any of the three points; contrast it with the following, much clearer instruction: "For each type of physical activity listed in column 1, select the physical benefit from column 2 that is most likely to be derived from it. Record your choice on the line preceding the question number. An item in column 2 may be used once, more than once, or not at all."

• Avoid providing clues. Every word or phrase in one column must be a logically and grammatically acceptable answer to every question in the other. Use the same verb tense, either singular or plural words, and the same articles (if possible) in all questions.

• Avoid including too many questions in one matching item. To be effective, the list of questions and the list of answers in a matching item must be somewhat homogeneous. As the length of the list of questions or answers increases, meeting the requirement of homogeneity becomes increasingly difficult. In most cases five or six questions are the practical limit for each matching item.

• Make sure all questions and answers appear on the same page of the test.

• Include a greater number of answers than questions or allow the repeated use of some answers. This removes the possibility of using the process of elimination to obtain the answer to one question of a matching item.

• Keep the parts of the matching questions as short as possible without sacrificing clarity. Because examinees must completely reread the list of possible answers as they respond to each item, having them read needlessly long answer choices consumes valuable testing time.

• Arrange the two lists of questions and answers in a random fashion. There should not be any particular pattern to the sequence of correct responses. An exception is placing dates in chronological order if dates are in the matching column (see the next point).

• Place the answer choices in a logical order (e.g., alphabetical, chronological) if one exists. This allows an examinee who knows the answer to locate it quickly.

Examples of good and poor matching questions are given at the end of this section.

Recommendations for Scoring

Because matching questions are generally answered on the test itself rather than on a separate answer sheet, arrange the items on the test so that a key can be placed next to the margin for quick scoring. Scoring a matching item can be done by someone not familiar with the subject matter.

MATCHING QUESTIONS

Good Question

For each person listed in column 1, select the sport from column 2 for which he or she is most noted. Record your choice on the line preceding the question number. A sport in column 2 may be used once, more than once, or not at all.

_____ 1. Hank Aaron	a. Baseball
_____ 2. Lebron James	b. Basketball
_____ 3. Serena Williams	c. Cycling
_____ 4. Phil Mickelson	d. Football
_____ 5. Bela Karolyi	e. Golf
_____ 6. Babe Ruth	f. Gymnastics
_____ 7. Mia Hamm	g. Soccer
_____ 8. Roger Bannister	h. Swimming
	i. Tennis
	j. Track

Poor Question

Match column 1 and column 2.

_____ 1. Sit-and-reach	c. Muscle fibers
_____ 2. 50-yard dash	h. Golf
_____ 3. Pull-up	f. Tennis
_____ 4. Shuttle run	a. $\dot{V}O_2max$
_____ 5. Balke treadmill	e. Agility
_____ 6. Dyer backboard volley	g. Arm strength
_____ 7. Disch putting	d. Speed
_____ 8. Biopsy	b. Flexibility

This is a poor matching question because

- the instructions do not indicate the basis for matching, how to record answers, or how many times items in column 2 may be used;
- the items in each column are too heterogeneous, making the answers too obvious;
- both columns contain the same number of items, so the last item could be answered by process of elimination; and
- some matches are too obvious (e.g., Dyer backboard volley and tennis, Disch putting and golf)

www Go to HK*Propel* to complete Student Activity 9.7.

Multiple-Choice Questions

A multiple-choice question includes two parts: the *stem*, which may be in the form of a question or an incomplete statement, and at least two *responses*, one of which best answers the item or completes the statement. The task is to select the correct or best response to the item presented in the stem.

Uses and Advantages

Multiple-choice questions make up a large portion of almost all nationally standardized written tests for several reasons:

- Questions can be scored and analyzed efficiently, quickly, and reliably.
- Questions of this type often create less ambiguity than other kinds of questions.
- Questions with more than two possible responses are not as susceptible to chance errors caused by blind guessing.
- Questions can be used to measure the higher-order cognitive processes, such as application, analysis, synthesis, and evaluation.
- Questions can measure almost any educational objective.
- Questions can be analyzed to determine their contribution to the test's reliability and validity.

Because multiple-choice questions are capable of measuring all levels of cognitive behavior, are applicable to nearly any subject or grade level, and can be used to measure virtually any educational objective, they can be used in almost any situation. If you are testing a large group of examinees or are planning to reuse a test, multiple-choice tests are the most efficient to construct, administer, score, and analyze. Because scoring is quick and accurate and can even be done by someone not familiar with the subject area, multiple-choice tests are also useful in the event that rapid feedback is important. Generally, you can include a fairly large number of multiple-choice questions on a test because the time required to answer each item is short. Because of this, and because multiple-choice questions can be constructed to measure most educational objectives, it is easier to construct a test fitting the table of specifications by using multiple-choice questions than any other type of question.

Limitations

Because of their versatility, multiple-choice questions do not have many intrinsic weaknesses. However, the required investment of time makes multiple-choice questions inefficient for small groups or one-time use. A few objectives are not as efficiently measured by multiple-choice questions as by other types of questions. For example, although appropriate multiple-choice questions could probably be devised, writing characteristics such as organization of an answer or grammatical construction would be better measured by essay questions.

Recommendations for Construction

Writing good multiple-choice questions requires paying careful attention to many aspects, such as constructing the stem and the responses and avoiding clues. Examples of good and poor multiple-choice questions are given at the end of this section. General considerations include the following:

- As you write the initial draft, realize that each question will probably require revisions.
- Set up a computer file to allow for revision and the addition of information. Record the educational and course objectives each question measures so you quickly can determine its place in the table of specifications. Also record the location of the source material around

which the question is built, since this information is often lost with the passage of time.

• Base each question on an important, significant, and useful concept. Usually the most successful multiple-choice questions are those based on generalizations and principles rather than on facts and details. For example, a question requiring knowledge of the general organization of Bloom's taxonomy is more valuable than a question requiring the examinee to know that the third category of the taxonomy is called *application*.

• Use novel situations when possible. Generally, effective questions result when you avoid the specific illustrative materials used in the textbook or lectures and require the application of knowledge to new ideas instead.

• Phrase each question such that one response can be defended as being the best of the alternatives. It is not always necessary that the response keyed as correct is the best of all possible answers to the question, but it must be able to be defended as the best of the choices listed. Also in this regard, avoid asking a question that requests an opinion, because this results in a no-one-best-answer situation. For example, consider the following stem: "What do you consider to be the best defense against the fast break in basketball?" Because this asks for the test taker's opinion, any choice must be regarded as correct, whether or not it agrees with the opinions of basketball authorities or the test constructor.

• Phrase each question clearly and concisely. Ideally, the stem should contain enough information that examinees understand what is being asked and yet be brief enough that no testing time is wasted reading unnecessary material. Occasionally, it is necessary to include a sentence or two to clarify a situation and avoid ambiguity. However, avoid the practice of teaching on the test, including inserting unnecessary information (called *window dressing* by some test-construction experts), or using flowery and imaginative language. Flowery language can increase possible interpretations of questions, which then can lead to ambiguity.

• Avoid negatively stated questions as much as possible. When you do use them, capitalize or underline the negative words. The purpose of asking a question is to determine whether the examinee knows the answer, not to see who reads carelessly and who can work through the confusion that sometimes arises with negatively stated questions.

• Do not include a question that all examinees will answer correctly unless you determine that the question must be included to increase the validity of the test. A question that every examinee answers correctly (or incorrectly) is of little value on an achievement test because no discrimination results. In fact, it can be shown mathematically that maximum discrimination can occur only when a question is of medium difficulty—that is, when approximately half the examinees answer the question correctly and half incorrectly. Although it is difficult to estimate the proportion of examinees who will answer a question correctly the first time the question is used, you should attempt to structure multiple-choice questions so they will be of medium difficulty. (Recall that one of the requirements for writing good test questions is to be aware of the level and range of understanding of the group being tested.) The difficulty of a multiple-choice question is most effectively altered by changing the homogeneity of the responses; the more homogeneous the responses, the more difficult the question. A method for obtaining an index describing the difficulty of a multiple-choice question is presented in the Item Analysis section later in this chapter (see equation 9.3 later in the chapter).

• Arrange to have the questions reviewed by someone knowledgeable in the subject. An independent reviewer can often locate ambiguities, grammatical mistakes, idiosyncrasies, and clues—all of which can affect a test negatively. If it is not possible to arrange for another person to review the questions, reread them yourself a few days after you have written them. (You should not write the questions the night before the test is to be administered. One of the requirements for writing good test questions is the willingness to spend a considerable amount of time on them.)

• Consider layout issues when formatting and printing out the test. List each response choice on a new line rather than immediately following one after another. Also, unless each response is long (an unlikely event), print the items in two columns instead of across the page. Use letters instead of numbers to identify the responses (this avoids confusion between questions and answers). Keep all the responses to a question on the same page as the question stem. Separate groups of related questions from other questions by a space or dotted line.

• Review the items and responses once you have constructed them to be certain that any correct letter option does not appear too often in a series of questions and that the correct options are fairly evenly distributed across all questions. Students will begin to second-guess themselves when more than two or three B (or whatever) responses appear as correct answers in a row. Sometimes, students think that the best option if they do not know is to pick C. For all these reasons, spread the correct responses out among the possible choices.

Writing the Stem

If a multiple-choice item is to be meaningful and important, you must keep in mind a definite concept around which the item is built. The most important part of the multiple-choice item to express this concept is the stem, and it is the first part constructed.

As stated, the stem can take two forms: a direct question or an incomplete sentence. It is usually wise (especially for novice question writers) to use questions rather than incomplete stems so that the examinee's task is clearly defined. No matter which form is used, by the time an examinee finishes reading the stem, a definite problem needs to have been identified so that the search for the correct response can begin. A stem such as "Badminton experts agree that . . ." does not provide a specific question or task, because badminton experts agree on many things. The examinee is forced to read through all the responses to determine what exactly is being asked. This stem would not be improved greatly by changing it to the question "On what do badminton experts agree?" If the stem is revised to "Badminton experts agree that beginning badminton players should learn the rotation strategy . . ." the examinee can begin reading the responses to locate the correct one rather than determining what is being asked. Using incomplete stems is more likely to result in incomplete specification of tasks than is using direct questions. The suggestions provided previously under Recommendations for Construction are especially germane to writing the stem of a multiple-choice question.

Writing the Responses

After the stem of a multiple-choice question are usually three, four, or five potential responses. One of the responses is predetermined to be the correct response (usually called the **keyed response**). The remaining responses are labeled **foils** or **distractors**. When constructing a multiple-choice question, write the keyed response immediately after you write the stem. Following this procedure helps ensure that the question is based on an important concept. On the test, of course, the position of the keyed response among the responses should be determined by some random procedure.

There is no reason that a multiple-choice question must contain any set number of responses or that all the multiple-choice questions on a test have to have the same number of responses. Three, four, or five responses are commonly used because this represents a compromise between the problem of finding several adequate, plausible possibilities and including enough responses so that, as happens with true–false questions, chance does not become an important factor.

There is some evidence (Rodriguez, 2005) that three responses may be the most efficient number for constructing the most accurate test. This is because the reduction of reading time permits the addition of an increased number of questions (recall that longer tests generally tend to be more reliable than shorter tests). However, having only three responses does increase the possibility of answering a question correctly by chance. The most important consideration regarding the best number of responses is whether all are reasonable and plausible choices to answer the question. A distractor that is chosen by no examinee is worthless and reduces the discriminating power of the test.

The distractors, the last part of a multiple-choice question developed, should not be constructed for the purpose of tricking the knowledgeable examinee into selecting one of them. However, one should make the distractors attractive to the unprepared examinee. All the responses should be plausible answers to the item. Use statements that are true but do not answer the question or those that include words or phrases that are common to whatever content area the test is assessing. The unprepared examinee may be attracted to a distractor containing such words or phrases simply because they sound familiar. As mentioned, using a ridiculous distractor unlikely to be chosen by any examinee is a waste of testing time.

Take care not to word the keyed response more precisely than the distractors. Recall that the keyed response needs only to be the best of the listed choices, not unequivocally correct under any circumstances. Keep all responses as similar as possible in appearance, length, and grammatical structure to avoid the selection of any response for reasons other than its correctness. As with the stem, keep the responses simple, clear, and concise to avoid ambiguity and to keep reading time to a minimum. If a natural order exists among the responses (such as with dates), list them in that order to remove one possible source of confusion.

Distractors need to appear to be equally correct to an examinee who is not familiar with the content of the item. However, examinees who fully understand the concept being tested should be able to ascertain the correct response. You want an item to appear ambiguous to the ill-prepared student (i.e., to have **extrinsic ambiguity**). If an item appears ambiguous to well-prepared examinees, it has **intrinsic ambiguity**. Extrinsic ambiguity is desirable; intrinsic ambiguity is not. Figure 9.1 depicts the difference between these types of ambiguity.

When sufficient plausible distractors are difficult to invent, it is tempting to use "None of these" as the final response. To avoid confusion, however, do not use this unless the keyed response is absolutely correct (as in a mathematical problem) and not merely the best response. When some or all the responses are partially correct (even if one of them is more correct than the others), the response "None of these" might be defended as being correct because none of the partially correct answers is absolutely correct. Without the "None of these" response in this situation, the most correct response is defensible as the best and thus the correct answer. A similar problem exists with the response "All of these." When there is not an absolutely correct answer and all responses contain some element of correctness, "All of these" might be the keyed response, but the examinee is put in a difficult position if one of the responses is a little more correct than the others. If you use these types of responses, make sure they are the keyed response occasionally (especially at the beginning of the test), so that examinees realize that they are to be considered seriously as possible correct answers.

Clues

Ideally, an examinee will answer a multiple-choice question correctly only if he or she knows the answer and incorrectly if he or she does not. Two factors, however, can adversely affect this situation. An examinee may blindly guess the correct answer to a question—there is no way to determine whether a correct response indicates a high degree of knowledge or luck. However, in the long run, everyone has an equal chance to be lucky, and the effects of chance can be mathematically accounted for. The second and more serious factor is that

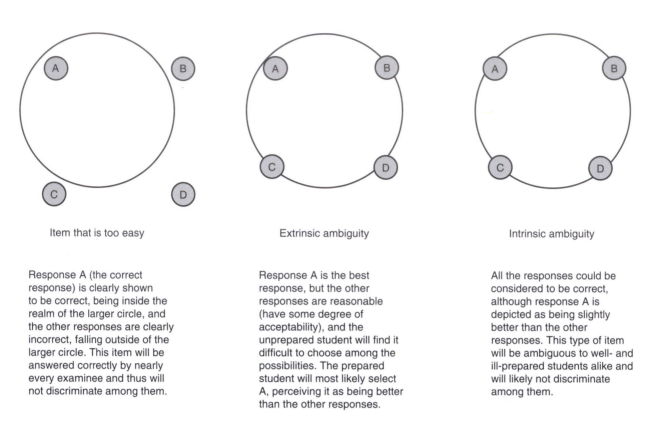

Item that is too easy   Extrinsic ambiguity   Intrinsic ambiguity

Response A (the correct response) is clearly shown to be correct, being inside the realm of the larger circle, and the other responses are clearly incorrect, falling outside of the larger circle. This item will be answered correctly by nearly every examinee and thus will not discriminate among them.

Response A is the best response, but the other responses are reasonable (have some degree of acceptability), and the unprepared student will find it difficult to choose among the possibilities. The prepared student will most likely select A, perceiving it as being better than the other responses.

All the responses could be considered to be correct, although response A is depicted as being slightly better than the other responses. This type of item will be ambiguous to well- and ill-prepared students alike and will likely not discriminate among them.

Figure 9.1 The difference between extrinsic and intrinsic ambiguity. The keyed correct response in each example is A.

of clues included within multiple-choice questions or tests. Because all examinees are not equally adept at spotting clues, the effects are not as predictable as those resulting from chance. The only way to eliminate the problem is to eliminate the clues.

Some clues are rather obvious; others are subtle. For example, it is usually easy to spot the use of a key word in both the stem and the correct response, or a keyed response that is the only one that grammatically agrees with the stem (e.g., the stem calls for a plural answer and all but one of the responses are singular). Words that belong together (e.g., bats and balls, shoes and socks, up and down) are often relatively difficult for the test constructor to spot but provide immediate clues to test takers. We suggest using these types of words or phrases as a method of securing attractive distractors. However, do not use these in the correct response, because an unprepared student may select the keyed response because it sounds good rather than because he or she knows it to be the correct answer.

In the process of asking one question, test constructors may inadvertently give information that answers another item on the test. Such interlocking questions provide clues for a test-wise examinee. This is especially likely to happen if you construct a test by selecting several questions from a file of possible questions, or if you add new questions or revise old questions on an existing test. To prevent interlocking items, read the test in its entirety once you have assembled it.

Variations

Several variations of multiple-choice questions have been devised to meet the needs of particular situations. For example, the classification item is an efficient form of the multiple-choice format if the same set of responses is applicable to many items. An example of a classification item follows:

For questions 89 through 92, determine the type of test best described by each statement or phrase. For each item, select answer

   a. if an essay test is described.
   b. if a true–false test is described.
   c. if a matching test is described.
   d. if a classification test is described.
   e. if a multiple-choice test is described

89. Test limited by difficulty in securing sufficiently similar stimulus words or phrases. (c)
90. Responses generally cover all possible categories of the concept being tested. (d)
91. Quality of the test is determined by skill of reader of the answers. (a)
92. Student can answer most items per minute. (b)

Another variation of the multiple-choice question involves using pictures or diagrams. This is illustrated in figure 9.2. If the shaded circle represents a top view of a tennis player making a crosscourt forehand stroke, in what location should the ball be when contacted by the racket: A, B, C, or D? (b)

You can create other variations to serve particular functions as long as examinees are able to understand their task in answering. Most of the suggestions presented previously will apply to these variations.

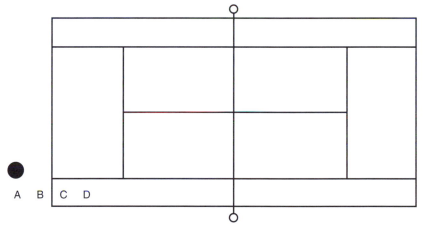

Figure 9.2 Sample of a diagram used on a written test.

Recommendations for Scoring

Typically, examinees record the answers to multiple-choice questions on the test itself or on a separate answer sheet. Having students mark directly on the test slightly reduces the chances of mismarking an answer and is convenient when discussing a test after it has been administered. If this procedure is used, you can facilitate the scoring process by arranging the questions so that the answers are recorded along the margins of the test and by using an answer key overlay spaced to match each page.

Although not as convenient for an examinee, recording answers on a separate answer sheet has many advantages for a scorer. You can score tests quickly and accurately by constructing a key from one of the answer sheets. Punch holes corresponding to the positions of the keyed responses on the answer sheet. When the key is superimposed on an examinee's answer sheet, you can count the number of correct responses. You can also use machine-scorable answer sheets that allow data to be scored and analyzed by the machine and a computer program.

MULTIPLE-CHOICE QUESTIONS FOR RACQUETBALL AND HANDBALL

Asterisk denotes the correct answer.

Good Questions

Notice that these questions are printed in two columns with each item completed in the column or on the page on which it starts. Notice also that responses are stacked and identified by letters.

1. What is called if a served ball hits the server's partner who is standing in the correct area?
    a. short
    b. fault
    c. handout
    d. *dead ball

2. How many outs are there in the first inning of a doubles game?
    a. one
    b. two
    c. *three
    d. four

3. Which of the following shots is used to get your opponent to move into the backcourt?
    a. kill shot
    b. passing shot
    c. *clear shot
    d. front-wall angle shot

4. What is called if the server stops a served ball that has hit the front wall and has bounced twice in front of the service line?
    a. handout
    b. hinder
    c. fault
    d. *short

5. What is called if a served ball hits the server on the rebound from the front wall?
    a. short
    b. fault
    c. *handout
    d. dead ball

Poor Questions

1. In handball
    a. One short and one hinder put the server out. *b. The fist may be used to hit the ball. c. Receivers can make points. d. The game may be played by only two or four persons.

The stem does not ask a question and thus the examinee must read everything to determine what is being asked. Moreover, the responses have been run together and printed across the page.

2. What is called if a served ball that hits the front wall, side wall, floor, back wall, and the other side wall is not returned by the receiver?
    1. Lucky *2. Point 3. 911 4. Strikeout

All distractors are not plausible answers. Also the question has been printed across the page, numerals have been used to identify responses, and the responses have been run together and printed across the page.

3. What is called if a player gets in the way of his or her opponent?
   a. Point
   b. Short
   c. Kill
   d. *Hinder

"Gets in the way of" is an obvious clue for "Hinder." Most examinees would answer correctly with little or no knowledge of the game.

4. What is called if, in receiving a serve, one receiver causes the ball to hit his or her partner before it touches the front wall or floor?
   a. Hinder
   b. Fault
   c. Handout
   d. *Point, because hitting your partner is your own fault

Wording the keyed response more precisely than the distractors to ensure its correctness might cause examinees to select it even if they are not sure of the answer.

5. What is the best shot to use in racquetball?
   a. Passing shot
   b. *Ceiling shot
   c. Kill shot
   d. None of these

"None of these" could be defended as correct because there is no absolute "best" shot in every situation.

Mastery Item 9.4

Using the suggestions just presented, write five multiple-choice questions on a topic of your choice. Critique your classmates' questions.

Essay Questions

To complete an essay question, an examinee must read the question, conceive a response, and write the response. The essay question has many uses, such as requiring the examinee to give definitions, provide interpretations, make evaluations, contrast and compare concepts or other topics, and demonstrate knowledge of relationships. For the response to an essay question to be scored accurately, the evaluator must be knowledgeable in the topic being assessed.

Uses and Advantages

Although almost any type of question can effectively measure the ability to organize, analyze, synthesize, and evaluate information, essay questions are easier to construct for this purpose than any other type. The contention that essay questions promote the study of generalizations rather than facts seems reasonable but has not been and probably never will be conclusively substantiated. Essay questions can also effectively measure opinions and attitudes. Although we are seldom interested in measuring these attributes in an instructional unit, it may be desirable to assess them in a research setting. In some situations, using essay questions is more efficient or convenient, regardless of the mental processes or subject matter involved. For example, the total time required to construct and correct an essay test is often less than for other types of questions.

You should also consider your personal preferences. If you are confident in your ability to construct and score essay questions but lack confidence in using other types of questions, you should probably use essay tests. However, be aware of the limitations of essay questions and the ways you can eliminate or minimize those limitations. Finally, when scheduling circumstances dictate little time for test construction but ample time for test correction, use essay tests.

Limitations

Even with careful preparation and scoring methods, at least three problems can arise when essay questions are used to measure achievement:

1. *Inability to obtain a wide sample of achievement.* Because of the time required to organize and write answers, it is not always possible to include enough essay questions on a test to measure the achievement of each content and educational objective. Consequently, there is some lack of content validity. You can alleviate this problem by constructing a table of specifications, using several essay questions requiring relatively short answers rather than a few questions requiring extended answers, and testing frequently to reduce the amount of material measured by each test.

2. *Inconsistencies in scoring procedures.* The most serious problem associated with essay questions is the unreliability of the scoring procedures. Not only does it take a substantial amount of time to correct an essay question properly, but several factors cause inconsistencies in the scores obtained. Because of the freedom an examinee has in constructing the essay answer, it is often necessary for you to decide, sometimes subjectively, whether that examinee has achieved an objective. You can reduce (although not completely eliminate) the subjectivity if you are knowledgeable in the subject matter tested and if you make it clear what task is required of the examinee for each question. Also, prior to the correcting of an essay question, construction of a rubric (see chapter 15) as a scoring guide can help eliminate inconsistent scoring.

Another problem is the halo effect—the part of an examinee's score that reflects your overall opinion of him or her. Giving the benefit of the doubt on one question to an examinee who has done well on most of the other questions on a test or to an examinee who has impressed you favorably in the past is an example of this phenomenon. Handwriting, spelling, and grammar may also positively or negatively affect the scoring of an essay answer. Unless these are specific objectives of the test, the score should not reflect these elements. By devising a coding system so that examinees' names do not appear on the answer sheets and by correcting the test question by question rather than paper by paper, you can diminish the consequences of this problem.

3. *Difficulties in analyzing test effectiveness.* After you have constructed, administered, and corrected a test, you will want to analyze how well the test measured what it was intended to measure, especially if you will use the test again. Analyzing a test generally includes obtaining indications of the overall reliability, validity, and objectivity of the test and the strengths and weaknesses of the test's individual items. Although some of these characteristics can and should be investigated for essay tests, essay questions do not lend themselves to this scrutiny as conveniently as do objective questions.

www **Go to HK*Propel* to complete Student Activity 9.8.**

Recommendations for Construction

The following suggestions for constructing essay questions will help you overcome some of the weaknesses and problems associated with their scoring.

• Phrase the question such that the mental processes required to respond to it are clearly evident. The objective of a question might be to determine whether mastery of factual material has occurred (e.g., "What are the outside dimensions of a regulation tennis court?"), to ascertain the degree to which a student can apply learned material to novel situations (e.g., "If the rules were changed to allow the shot-put circle to be raised by 2 feet [61 cm], would this increase or decrease the distance the shot would travel if all other factors were kept equal? Why?"), or to evaluate the ability to organize an answer in a logical manner (e.g., "Trace the development of public school physical fitness tests from the Kraus and Weber Low Back Test to the FitnessGram"). The examinee should be able to recognize the type of answer required by the manner in which a question is stated.

• Use several essay questions requiring relatively short answers rather than a few questions requiring extended answers. This practice usually leads to two positive results: a wider sampling of knowledge and a test made up of more specific questions, the answers to which can normally be scored with increased reliability.

• Phrase the question so that the task is specifically identified. Avoid asking for opinions when measuring educational achievement. Begin essay questions with such words or phrases as *Explain how*, *Compare*, *Contrast*, and *Present arguments for and against*. Do not start essay questions with such words or phrases as *Discuss*, *What do you think about*, or *Write all you know about*. Also, unless the purpose of a question is to measure the mastery of relatively factual material, do not begin an essay question with such words as *List*, *Who*, *Where*, or *When*.

• Set guidelines to indicate the scope of the answer required. Build limiting factors into the question: "Illustrate, through words and figures, how health-related physical fitness is associated with academic achievement in school . . ." or "Limiting your answer to team games only, compare. . . ." You may also indicate the approximate number of words examinees should write (e.g., 50 words; 150 words; 1 paragraph) or the amount of time they should spend on each item (e.g., 5 minutes; 10 minutes).

• Indicate how many points each essay question is worth. Students will then spend more time on those items that are worth more points.

• Prepare an ideal answer for each question. Because this requires that you identify exactly what the question is intended to measure, ambiguities often become apparent. As mentioned previously, this practice also increases the reliability of the scoring process.

• Avoid giving a choice of essay questions to be answered. If an essay test is designed to measure achievement of the objectives for a group of students who received the same instruction, each student should be required to answer the same questions. Optional questions add another variable and increase the possibility of inaccurate assessment.

• Avoid asking students for their opinions. Although the intent is to grade each response based on the substantive support provided for the answer, it is difficult to separate opinions from truth—and who is to say which opinion is best? (As you might imagine, it is typically an instructor's opinion that is perceived as best.)

Recommendations for Scoring

Certain practices reduce some of the unreliability inherent to the process of scoring an essay answer. Several of these procedures are related to or follow from the previous suggestions for construction.

- Decide in advance what the essay question is intended to measure. If an essay question is designed to measure application of facts, evaluate the answers to the question on that basis and not on the basis of organization, spelling, grammar, neatness, or some other standard. Ignore elements not dealing with the question's objective.

- Use the ideal answer you previously prepared as a frame of reference for scoring. This is especially important if you secure an independent rating of the answers (see the last bullet in this list for additional information).

- Determine the method of scoring. Use one of three systems:

    1. **Analytic scoring** involves identifying specific facts, points, or ideas in the answer and awarding credit for each one. An answer receiving a perfect score would include all the specific items in your ideal answer. This type of scoring is especially effective when the question's objective is to measure whether students have acquired the factual material.

    2. **Global scoring** consists of reading the answer and converting the general impression obtained into a score. In theory, the general impression is a function of the completeness of the answer in comparison to the ideal answer. Of the three grading methods, this is the most subjective and the one most likely to be affected by extraneous factors.

    3. **Relative scoring** consists of reading all students' answers to one question and arranging the papers in order according to their adequacy. You may accomplish this by setting up several categories (such as good, fair, and poor or excellent, above average, average, and below average) and assigning each answer to one of the categories. Second, third, and possibly more readings may be necessary to arrange the papers within each category, and you may occasionally shift one paper to another category. The end result is an ordering of all the papers with respect to the correctness of the answers to the one question evaluated. After sorting, scores may be assigned to each answer. There is no reason the top paper must be assigned an A or the bottom paper an F; your evaluations should also be influenced by the comparison of each answer to the ideal answer. This ordering of the answers enhances consistency in the scoring procedure and is especially effective when the objective of a question is to measure relatively complex mental processes. Repeat the procedure for each of the remaining questions.

- Develop a system so that you don't know whose paper is being scored. Examinees could sign their names on a piece of paper next to a number corresponding to the number on their test booklet, or they could mark their test copy with unique designs or patterns that only they will recognize. Having each answer recorded on a separate sheet of paper also eliminates the bias caused by noticing the scores given to a previous answer. If several answers do occur on one answer sheet (as would be the case if short answers are required), record points awarded for each answer on a separate sheet of paper, thus helping to eliminate the halo effect. This procedure is also useful if tests are rescored to check reliability: The second reader, who may or may not be you, will not be influenced by the score previously awarded.

- Score everyone's answers to one question rather than score one complete paper at a time. This process is required if you use global or relative scoring. Although not required for analytic scoring, the process usually leads to the most consistent scoring because it is easier to compare all the answers to one question at one time.

- Ensuring the reliability and objectivity of essay test scoring requires that each answer be scored twice and the two scores compared. Ideally, to assess the test's reliability these two scores should be awarded by two scorers to ensure that they are independently obtained (refer back to chapter 6 for a discussion of interrater reliability). If it is possible to arrange for another person knowledgeable in the content area to score the responses, provide that scorer with the ideal answers so that the two scores have a common basis. However, if it is not feasible to obtain an independent scorer, score the answers yourself on two occasions, perhaps separated by a week, in an effort to secure some evidence about the consistency of the scoring procedures used.

As should be obvious by this point, the process of constructing and scoring a reliable essay test can be tedious and time consuming. However, to be fair to examinees, the procedures explained here should be followed if an essay test is used to measure cognitive objectives.

WWW **Go to HK*Propel* to complete Student Activity 9.9.**

ADMINISTERING THE TEST

As we have noted, there are potential problems involved in testing. Some examinees can experience anxiety, cheating can occur, and they may experience feelings of humiliation or haughtiness after learning their scores. However, these undesirable circumstances do not *have* to occur. The suggestions presented here should help eliminate or reduce many of the problems that are often associated with test administration. Although the written test and the scoring procedures used have some influence on these occurrences, the administration of the test itself probably has the greatest impact on whether problems will arise before, during, and after the test.

Before the Test

• *Prepare the examinees for the test.* Generally, less anxiety is associated with a test announced well in advance than with surprise tests, and discussing the content of an upcoming test with students can help reduce their apprehension. It is not logical (or ethical) to test students on topics that have not been covered or assigned. Items such as the general areas to be measured, the approximate amount of testing time to be devoted to each area, the types of questions that will be asked (e.g., essay, multiple-choice), and the length of the test represent legitimate concerns of examinees. A well-constructed written test should be a precise expression of the objectives of the instructional unit. It is difficult to imagine a situation in which knowledge of these objectives should be withheld from examinees.

• *Eliminate the advantage for test-wise examinees.* Use proper test-construction techniques outlined previously in this chapter (avoiding grammatical clues, specific determiners, interlocking items, and the like) and provide examinees with test-taking suggestions. For example, the following recommendations might be made to examinees:

- Realize that all the material measured by a good test cannot be learned the night before the test. Spend this time reviewing, not learning.
- Read the instructions to the test before beginning to answer the questions. Know how the test will be scored. Be aware of (a) whether all questions have the same value; (b) whether neatness, grammar, and organization will be accounted for in the score; and (c) whether a correction-for-guessing formula will be applied.
- Pace yourself.
- Plan an essay answer before starting to write it down.
- Check often to see that you are writing answers in the correct place on the answer sheet.
- Check over your answers if time permits.

For more, see the Test-Taking Skills highlight box.

- *Give any unusual or lengthy instructions before test administration time.* Doing so will save time on the day the test is given and, more important, will enable examinees to begin the test as soon as possible. Providing instructions in advance reduces the time available for anxiety to build, especially for those who feel pressured by time.

- *Proofread the test before it is reproduced.* Proofreading helps to ensure that each examinee will receive a legible copy free of typographical, spelling, and other errors. It also eliminates or reduces the time spent during the test clearing up these errors.

- *Give a practice test to reduce examinee anxiety.*

During the Test

- *Organize an efficient method for distributing and collecting the tests.* With a small group, this is seldom a concern. However, with 60 or so examinees spread out in a large room, an efficient distribution procedure is necessary so that all examinees are given approximately the same amount of time to complete the test, and an efficient collection procedure is vital to keep the test secure.

- *Help examinees pace themselves.* This can be accomplished by quietly marking on a blackboard the time remaining as well as a rough estimate of the portion of the test on which the examinees should be working.

- *Answer individual questions carefully and privately.* To avoid disturbing others, answer individual questions at the examinee's or the proctor's desk. However, take care that your response does not give any examinee an advantage over others.

- *Control cheating.* Obviously, cheating negates the validity of test scores. Of more serious concern, though, are the negative attitudes generated toward those who cheat, toward the proctor who does not control cheating, and toward testing in general.

- *Control the environment.* Any factor that prevents an examinee from doing his or her best on a written test lowers the reliability, validity, and usability of the resulting set of scores. Some of these factors, such as examinee motivation and reading habits, are not under the direct control of the tester. You can, however, provide adequate lighting, eliminate noise distractions, maintain a comfortable temperature, and provide adequate space in which to work.

- *Describe the rules and features of online testing.* If the tests are administered online, the students should be informed of the rules and features of the test, such as which browsers they can use and whether the test or individual items will be timed.

After the Test

- *Correct the tests and report the scores as quickly as possible.* The rapidity of this operation depends, of course, on the type and length of the test administered. However, examinees generally appreciate prompt results.

- *Report test scores anonymously.* You can use a confidential identification number system if you post scores publicly or write a student's score on the second page of the exam if you pass them out in the classroom.

- *Avoid misusing and misinterpreting test scores.* By following the suggestions in this section, you will be able to improve the reliability and validity of your written tests. However, remember that no test is perfectly reliable and valid and you should not base crucial decisions on the results of one written test. For example, do not interpret a 1-point variation between two examinees' scores on a written test as showing a significant difference between the examinees. (Refer back to chapter 6 for more about the standard error of measurement.) Such an interpretation is a misuse of test scores. Along with other forms of measurement, consider written test results when evaluating people, but allow these results to influence the evaluations only to the degree their accuracy permits.

TEST-TAKING SKILLS

Preparing for the Test

- Schedule your time—plan ahead when good study times will be available.
- Know when, where, and how you will be tested. Ask your instructor.
- Go to your instructor when you encounter a difficult or problematic area while studying.
- Be in the best physical and mental shape possible. Get a good night's rest, do not use stimulants or tranquilizers, and do not drink a lot of fluids or eat a big meal just before the test.
- Be motivated and positive in your attitude toward the test.
- Be prepared with adequate study space, pencils, text, and notes.
- Practice, practice, practice by using practice tests. Generally people who are more familiar with taking tests perform better. Practice efforts are generally better on timed tests. The less the interval between practice and testing, the better the practice effect.
- Carefully read the summaries of each chapter. Look at highlighted text, figures, and tables.
- Study with classmates.
- Avoid cramming.
- Be test wise; it will help on poorly constructed tests.
- If studying for an essay exam, make up and answer practice questions in advance.
- If preparing for an open-book test, mark particularly important pages or sections with tags to make them easy to find during the test.
- If studying for a take-home test, find out what sources you will be permitted to use for help.
- Get to the testing area early and familiarize yourself with the area.
- Avoid last-minute panic questions. Panic is contagious. Do not talk with friends immediately before the examination.
- Relax.

Getting Started and Taking the Test

- Sit where you feel comfortable—whether that means near a window, near an outlet, or where you generally sit in class. Don't sit near distracting people.
- Read and listen to instructions carefully, including oral directions or corrections.
- Find out how the test will be scored, whether some items are worth more than others, whether guessing is penalized, and whether neatness or spelling counts.
- Check to see that you have all the pages and items before beginning.
- Look quickly through the test before starting so that you can plan ahead and gauge your time.
- Know how much time you have to complete the test, and be aware of time remaining. Pace yourself and budget your time. Leave yourself time at the end to review your answers.
- Concentrate on the test; do not pay attention to others in the room.
- Stay calm if you don't know the answer; make an educated guess.
- Ask your instructor if you do not understand something.
- If you are stuck, move on to the next question and come back to a difficult one later. Activity reduces anxiety.
- Do not worry about other students (e.g., if they leave earlier than you or ask questions).
- If you are using a separate answer sheet, check often that you are in the correct column and row. Check that you answered all the items before you submit the answer sheet—make sure that the number of answers selected equals the number of test items.

After Taking the Test

- Write down what you remember about the test.
- If you think you did poorly, ask to review the test with the instructor.
- Appeal when you feel that you are correct about an answer that was marked as incorrect.

WWW Go to HK*Propel* to complete Student Activity 9.10.

ANALYZING THE TEST

To determine the amount of confidence that can be placed in the set of scores resulting from a test administration, examine the reliability and the validity of the test (i.e., how correctly and consistently it measures what it is intended to measure). Evidence for the reliability and validity of a test is both global (overall test performance) and specific (quality of individual questions).

Reliability

If a test were perfectly reliable, each examinee's observed score would be an exact representation of his or her true level of achievement; each observed score would be a true score, uncontaminated by error. In actuality, of course, an observed score consists of the true score as well as the error score, which may be positive or negative (as discussed in chapter 6). As the error portion for the observed score increases, reliability decreases. Unfortunately, there are several sources of error in written tests:

• *Inadequate sampling.* The questions that appear on a test represent only a sample of the infinite population of possible questions. Error is introduced if the sample does not adequately represent the desired population of possible questions. An examinee's failure to be credited with understanding—or being penalized for not comprehending—a particular objective because no question was included on the test to measure that comprehension is an example of how sampling error might reduce test reliability (and validity).

• *An examinee's mental and physical condition.* Illness, severe anxiety, overconfidence, or fatigue can alter one's score and thus lower the reliability of a test.

• *Environmental conditions.* Poor lighting, poor temperature control, excessive noise, or any other similar variable that negatively affects concentration can cause observed scores to misrepresent true scores.

• *Guessing.* Because each examinee theoretically has the same chance for good or bad luck when blindly guessing during an objective test, the total effect should balance out over the long run and not introduce error. However, one administration of a test does not represent the long run, and test reliability might be reduced because some examinees could, on one administration of a test, be luckier guessers than their peers.

• *Changes in the field.* Sometimes error is introduced not by the measuring instrument but by the fact that the variable being measured is changeable. Lack of consistent definitions (e.g., disagreement by authorities on the definition of physical fitness) and fluctuations in the attribute being measured (e.g., attitudes toward physical activity) can make constructing reliable tests difficult.

Thus, many factors can introduce error and consequently reduce the reliability of a written test. As indicated in chapter 6, there are several methods of calculating a coefficient to express the reliability of a test; each of these methods reflects one or more of the sources of error. If test questions are scored correct (1) or incorrect (0), the alpha coefficient (identical to the Kuder–Richardson formula 20, or KR_{20}) can be used to estimate test reliability. The KR_{20} is actually the average of all possible split-halves reliability coefficients and, as such, is a relatively conservative estimate of a test's reliability. Obtaining a satisfactory reliability coefficient when using a conservative procedure is good because using other less conservative procedures would only result in higher estimates. KR_{20} is defined as

$$KR_{20} = \frac{K}{K-1}\left[1 - \frac{\Sigma pq}{s^2_{total}}\right]$$ (9.1)

where K is the number of test items, s^2_{total} is the variance of the test scores, and Σpq is the sum of the difficulty (p) times q where q is defined as $(1 - p)$. You will learn more about p (Difficulty or Diff.) in the Item Analysis section that follows.

Another method of estimating a written test's reliability when it is reasonable to assume that all items on the test are equally difficult and discriminating is the KR_{21}. The formula follows here:

$$KR_{21} = \frac{K}{K-1}\left[1 - \frac{M(1 - \bar{p})}{s^2_{total}}\right]$$

(9.2)

where K is the number of test items, s^2_{total} is the variance of the test scores, M is the mean test score, and $\bar{p}$ is the average difficulty defined as M/K. Note the similarity between KR_{20}, KR_{21}, and the alpha coefficient (see equation 6.3). The alpha coefficient is actually equivalent to KR_{20}. The KR_{21} reliability estimate is relatively easy to calculate, but the assumption of equally difficult and discriminating items is rarely true. Violation of this assumption results in underestimating the test's reliability. The KR_{21} formula is a more conservative estimate of the test reliability; thus KR_{20} will always be greater than or equal to KR_{21}. Obtaining a satisfactory reliability coefficient when using a conservative procedure is a good idea because using less conservative procedures would only result in higher estimates.

Mastery Item 9.5

Use the KR_{21} formula to estimate the reliability of a 60-item test having a mean of 45 and a standard deviation of 6.

Dataset Application

The chapter 9 large dataset in HKPropel consists of 400 responses to 10 items. Open the chapter 9 large dataset in SPSS and do the following:

1. Use Analysis → Scale → Reliability to estimate KR_{20} for the 10 items.

2. Use Analysis → Descriptive Statistics → Descriptives and obtain the variances (chapter 3) for each of the items and the total score. Substitute these values into the KR_{20} formula just presented (or the alpha coefficient found in chapter 6) and confirm the results that you obtained in step 1.

3. Use the results from the SPSS output to calculate KR_{21} using formula 9.2.

4. Are you pleased with the reliability you have obtained? If not, what would you do? (For a hint, turn to the Spearman–Brown prophecy formula found in chapter 6.)

Validity

Although it may measure something consistently, if a written test does not measure what it is designed to measure, the resulting scores are of little value. As noted in chapter 6, there are various types of validity and several methods of assessing them.

For a written test, one of the most important types of validity is content validity. This is generally determined subjectively by the extent to which the individual test items sufficiently represent the course's educational and content objectives. Following the proper procedures for constructing a written test, especially using a table of specifications, helps ensure that your test will have content validity. Having another content expert review your test items provides additional input on the test's content validity.

The preceding presentations on test reliability and validity refer to the overall test. But the quality of the entire test is determined by the quality of the individual items. We now turn to item analysis, which will help determine the quality of the individual items and how they might contribute to overall test reliability and validity.

ITEM ANALYSIS

Analyzing the responses to test items is important for several reasons, but especially to continually improve the items and consequently the test. The difficulty level and the discriminating power (ability of the question to separate strong from weak examinees) of each item are the keys to item improvement. **Item analysis** can also improve your instruction by identifying weaknesses in the examinees as a group, in your instructional methods, or in the curriculum. It can also improve your skill in constructing written tests. Most of the illustrations and examples presented involve multiple-choice questions because they are efficient to analyze; software programs can even calculate these values for you if you use scannable answer sheets. However, you can modify most of the following steps of item analysis for other types of objective items, and you can apply the principles involved to most types of questions. The procedures for completing an item analysis by hand follow.

- *Step 1.* Score the tests.
- *Step 2.* Arrange the answer sheets in order from high to low score.
- *Step 3.* Separate the answer sheets into three subgroups: (a) the upper group, which consists of approximately the top 27%; (b) the middle group, which consists of approximately the middle 46%; and (c) the lower group, which consists of approximately the lowest 27%. The upper and lower groups should contain the same number of answer sheets. You will use only these two extreme groups—the upper and the lower—in the item analysis. Test authorities suggest that these groups should each include 27% of the tests to include as many responses as possible and maximize the differences between the types of responses; however, as long as there is an equal number in each of these groups, use the most convenient number of answer sheets between 25% and 33%. For example, if 60 answer sheets were available for analysis, the highest and lowest 15 to 20 could be used.
- *Step 4.* For each item, count and record the frequency of selection of each possible response by the upper group.
- *Step 5.* For each item, count and record the frequency of selection of each possible response by the lower group.

Steps 4 and 5 are the most time-consuming portions of the item analysis. Several procedures can reduce the tedium of this task:

- Use previously prepared scorecards for each item.
- Use a computer or an optical scanner to speed the process of reading and recording responses.
- Cooperate with another scorer, with one person reading and the other recording.

An example of a possible organization of the resulting data is shown in figure 9.3. (These data were obtained from a question included on a nationally standardized test of physical fitness knowledge administered to college senior physical education majors.)

At the completion of step 5, the necessary data are available to calculate an **index of difficulty** (the estimated percentage of examinees who answered the item correctly) and an **index of discrimination** (an estimate of how well the item discriminates among exam-

Source: *Handbook of Physical Fitness*	Topic: Physical fitness
First draft: In the opinion of most authorities, three of the following factors have contributed to a lowering of the national level of physical fitness. Which has NOT had this effect? A. An increase in life span B. A decrease in the physical effort required for daily living C. An increase in the number of occupations involving sedentary activity *D. An increase in school consolidation	*Revision:* In the opinion of most authorities, three of the following have contributed to a lowering of the national level of physical fitness. Which has NOT had this effect? A. An increase in the number of senior citizens B. A decrease in the physical effort required for daily living C. An increase in the number of occupations involving sedentary activity *D. An increase in school consolidation

Item 5	Test: Form D trial	Date: 6/98	*n* = 185						Item 25	Test: Final form A	Date: 9/00	*n* = 1112					
Responses	A	B	C	D*	E	Omit	Diff.	Net D	Responses	A	B	C	D*	E	Omit	Diff.	Net D
Upper 27% = 50	28	2	1	19		0			Upper 27% = 300	69	10	5	216		0		
Lower 27% = 50	24	8	1	17		0	36%	4%	Lower 27% = 300	89	52	54	104		1	53%	37%

Figure 9.3   One way to organize data for item analysis.

inees who have been categorized by some criterion) for each item. The data in figure 9.3 illustrate the calculation of these two indexes and how change suggested by the response pattern can improve an item. In this example, the left side of the figure contains the initial draft of the question and the data resulting from its administration to approximately 185 examinees. The right side contains the revised question and the data resulting from its administration to more than 1000 examinees.

• *Step 6.* Calculate and record the index of difficulty for each item using the following formula:

$$\text{Diff} = \frac{U_c + L_c}{U_n + L_n} \times 100 \tag{9.3}$$

where Diff is the index of difficulty, U_c is the number of examinees in the upper group answering the question correctly, L_c is the number of examinees in the lower group answering the question correctly, U_n is the number of examinees in the upper group, and L_n is the number of examinees in the lower group (recall that $U_n = L_n$).

The index of difficulty is the percentage of examinees who answered the question correctly; thus, the higher the index, the easier the question. The following examples illustrate the use of the index of difficulty formula (again using the data in figure 9.3). You can compare item difficulties to course and student learning objectives to help determine which objectives students are meeting and which might need additional reinforcement or would benefit from changes in course delivery strategies.

First-draft results: $n = 185$; therefore $U_n = L_n = 185 \times 0.27 = 50$.

Revision results: $n = 1112$; therefore $U_n = L_n = 1112 \times 0.27 = 300$.

The maximum amount of discrimination can occur only when an item has an index of difficulty of exactly 50%. If this criterion were met by every question on a test, the mean score of the test would be equal to one-half the number of items on the test. For example, the mean score of a test containing 80 items would be 40.

This ideal 40, however, would not occur because another factor (chance guessing of the correct answer) will occur. On an 80-item multiple-choice test on which each item has four possible responses, the random guessing should produce approximately 20 correct responses (i.e., $1/4 \times 80 = 20$). When the consideration of chance is added, the mean score of the test just described should be 50. This value is obtained by determining what score lies halfway between the chance score and the highest possible score (80 items – 20 correct by chance = 60 items; if an examinee answers 50% of these 60 items correctly, he or she would have 30 items correct, plus the 20 by chance, resulting in a score of 50). If the index of difficulty of each of the 80 items was 62.5%, the mean score for the test would be 50 ($80 \times 0.625 = 50$).

It is not possible, especially on the first draft, to produce an item that exactly meets some predetermined difficulty index. The point is, to maximize an item's discrimination power, an attempt should be made to write each item in such a manner that half or slightly more than half of the examinees will be able to answer it correctly, although this is very difficult to do. One further point should be noted. Maximum discrimination can occur only for an item of medium difficulty, but meeting this condition does not necessarily guarantee that it will occur. Figure 9.4 describes the relationship between discrimination and difficulty: As difficulty increases from 0 to 0.50, the potential discrimination increases. However, as difficulty continues to increase from 0.50 to 1.0, potential discrimination decreases.

• *Step 7.* Calculate and record the index of discrimination for each item.

$$\text{Net D} = \frac{U_c - L_c}{U_n} \times 100 \tag{9.4}$$

(Note that either U_n or L_n may be used in the denominator.) The index of discrimination used here, known as the **Net D**, is only one of nearly 100 discrimination indexes that have been devised. The most commonly cited discrimination indexes are correlational techniques for quantifying the relationship between the score on a particular item and a criterion score (usually the total test score). However, the Net D uses the same data that is used to determine the difficulty index and is fairly simple to calculate and interpret. The following examples, again using the data presented in figure 9.3, illustrate the use of the Net D formula.

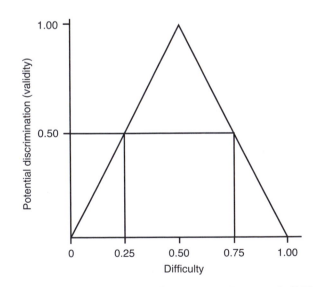

Figure 9.4 The relationship between discrimination and difficulty.

First-draft results: $n = 185$; therefore $U_n = L_n = 50$.

Revision results: $n = 1112$; therefore $U_n = L_n = 300$.

The criteria usually used to examine the discriminating power of an item are the scores on the entire test on which the item appears. Note that the higher the value of Net D, the higher the discriminating power of the item, and note also that the Net D formula could produce a negative number, indicating an item that discriminates negatively. In fact, the value obtained is actually the net percentage of good, or positive, discriminations achieved by an item (thus the name *Net D*). Figure 9.5 illustrates this concept.

In general, if the examinees who performed well on the entire test did well on the item and the examinees who did poorly on the entire test did poorly on the item (a correlation of .4 or above), the item is considered a good discriminator. If approximately the same number of high-performing and low-performing examinees answer an item correctly (a correlation between 0 and .19), it is considered to have little or no discriminatory power. If the item is answered correctly by more of the low-performing examinees than by the high-performing examinees, it is considered a negative discriminator, which indicates some problems—either the item was poorly written or the item was miskeyed in the data analysis; either way, a careful follow-up check is needed.

For example, in figure 9.5, no discrimination occurred between Darius, Rhea, Tyson, Layla, Thanh, and Jesus, because all answered the item correctly. Similarly, no discrimination occurred between Jacob, Minako, Pranav, and Sadie, because all answered the item incorrectly. The discrimination that occurred between Darius (or Rhea, Tyson, or Layla) and Minako (or Pranav or Sadie) is considered a positive discrimination because the groups in which these examinees have been placed are based on their total test scores. Altogether a total of 12 (4 × 3) positive discriminations occurred. Conversely, the discrimination that occurred between Jacob and Thanh (or Jesus) is considered a negative discrimination because Jacob is in the upper group and Thanh and Jesus are in the lower group. A total of two (2 × 1) negative discriminations occurred. The maximum number of discriminations possible with five examinees in each group is 25 (5 × 5). Of these 25, 12 were positive, 2 were negative, and 11 did not occur. Subtracting the 2 negative discriminations from the

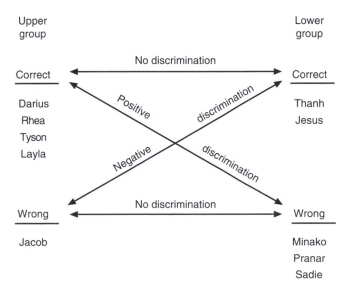

Figure 9.5 Positive and negative discrimination.

12 positive discriminations results in a net of 10 positive discriminations. The ratio of net positive discriminations to the total possible (10 / 25) is 40%. Using equation 9.4 for Net D results in this same value:

$$\text{Net D} = \frac{4-2}{5} \times 100 = 40\%$$

Attempt to keep the index of discrimination of an item on an achievement test as high as possible. Most test-construction authorities agree that an item with a discrimination index of 40% or higher is a good item. Discrimination indexes between 20% and 39% are acceptable but may indicate the need for revision, especially as the value approaches 20%. Items with an index of discrimination below 20%, and especially those with negative discrimination indexes, are poor and should probably be discarded from future tests.

• *Step 8*. Examine the pattern of responses to determine how an item might be improved.

According to the previous suggestions for retaining and discarding questions based on their discrimination index, the initial draft of the question in figure 9.3 probably should have been discarded. However, the response pattern of the examinees revealed a possible solution. Although it is often difficult to understand why certain responses are selected or ignored and even more difficult to determine possible alterations of the responses or stem that will improve an item, examining the response patterns often suggests possibilities. For example, response A to the first draft of the item displayed in figure 9.3 was chosen by more than 50% of the combined upper and lower groups of examinees, even though it is incorrect. Rewording this distractor in the revision resulted in the keyed response becoming more attractive than the first response, especially to the examinees in the upper group. The positive changes in the difficulty and discrimination indexes indicate that the alteration of this one response improved the item considerably. Remember, discrimination is the most important characteristic of an item. A test cannot be reliable or valid unless the individual items discriminate among examinees.

Mastery Item 9.6

How many answer sheets should be used in an item analysis for a written test taken by 250 examinees?

Mastery Item 9.7

Calculate the index of difficulty and the Net D discrimination index for a multiple-choice question answered correctly by 40 of the 60 examinees in the upper group and by 10 of the 60 examinees in the lower group.

Mastery Item 9.8

To demonstrate the relationship between the difficulty and potential discrimination of an item, calculate the indexes of difficulty and discrimination for the following five items:

Item no.	Upper group $n = 10$	Lower group $n = 10$
1	2 correct	0 correct
2	5 correct	5 correct
3	10 correct	5 correct
4	10 correct	0 correct
5	5 correct	10 correct

Table 9.3   Difficulty and Discrimination Indexes

	Difficulty index	Discrimination index
Lowest	.00 (0%)	−1.00 (−100%)
Highest	1.00 (100%)	1.00 (100%)
Desired	.50 (50%)	1.00 (100%)*

*As indicated in the text, values at or above .40 (40%) are considered quite good and discriminating.

Table 9.3 illustrates the lowest, highest, and desired values for difficulty and discrimination.

Positive discrimination values result in increased test reliability, negative discriminations result in decreased test reliability, and zero discriminations result in no change in test reliability. Figure 9.6 illustrates how the correlation between the item score (0 is incorrect and 1 is correct) and the total test score is also an estimate of item discrimination. Note that the three items' total score correlations presented in figure 9.6 are for positive (a), negative (b), and no (c) discrimination. These result in increased reliability (a), decreased reliability (b), and no effect on reliability (c).

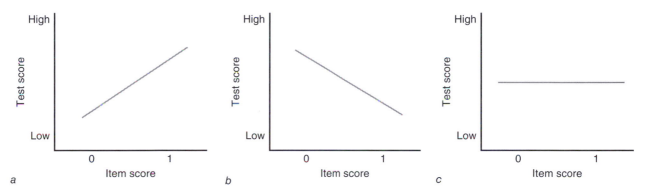

Figure 9.6   Illustration of how correlation can be used to estimate test item discrimination.

Mastery Item 9.9

Using the chapter 9 large dataset in HK*Propel*, calculate the item–total score correlation (i.e., correlation between item score and total test score) for each of the 10 items. (For a refresher on computing correlation using SPSS, see chapter 4.)

SOURCES OF WRITTEN TESTS

Generally, the number of sources for written tests in the field of human performance is limited. When a written test is given, the chances are great that it was constructed locally. With a few exceptions, nationally standardized written tests are not available.

When you are constructing a written test, it is often helpful to examine similar tests to obtain ideas for questions. Some possible sources for similar tests are professionally constructed tests, textbooks, periodicals, theses, and dissertations. Zhu, Safrit, and Cohen (1999) have published a test on physical fitness knowledge that was developed using many of the concepts presented in this chapter.

QUESTIONNAIRES AND SURVEYS

The questionnaire is a close relative of the written test. Both of these data collection instruments require careful construction and thoughtful analysis of the data they produce. The written test is primarily designed to assess a student's knowledge and to discriminate among students on the basis of their cognitive behavior, whereas questionnaires are typically used to measure affective domain concerns such as attitudes, opinions, and behaviors. For example, you might conduct a survey to determine how many minutes of moderate-to-vigorous physical activity (MVPA) people engage in each week to determine if they meet public health recommendations, or you might want to assess attitudes toward physical activity before and after exposure to a particular curriculum.

Questionnaire responses provide the independent and dependent variables for survey research. Cox and Cox (2008) provided an extensive presentation of questionnaire development. Thomas, Nelson, and Silverman (2015) listed eight steps for conducting survey research:

1. Determining the objectives
2. Delimiting the sample
3. Constructing the questionnaire
4. Conducting the pilot study
5. Writing the cover letter
6. Sending the questionnaire
7. Following up
8. Analyzing the results and preparing the report

Using a questionnaire to collect information has both advantages and disadvantages. On the plus side, the questionnaire can be relatively efficient in terms of money and time, and each respondent is exposed to exactly the same instrument. Because a questionnaire can be sent to all respondents simultaneously, data collection can be completed over a period of a few weeks. As an alternative to mailing a questionnaire, it is now increasingly common to use the Internet to gather information. For both written and Internet surveys, respondents can be widely spread geographically, and they can respond at their own convenience. For the best response to a written or electronic questionnaire, keep the survey short and specific.

On the negative side, the value of questionnaire data can be reduced by low response rate, inability to clarify a question that the respondent finds ambiguous, unanswered questions, and lack of assurance regarding who actually completes the questionnaire. Some of these concerns can be addressed through careful planning, but they can never be totally eliminated.

Planning the Questionnaire

There are now numerous online survey companies. Many of them have surveys already constructed for many common concerns such as customer satisfaction, faculty satisfaction, and educational outcomes. These companies also offer help in designing your own questionnaire and collecting and analyzing the resulting data.

Time spent planning a questionnaire is invaluable. Before constructing a questionnaire, clarify the purpose of the instrument by formulating relevant hypotheses so it will be possible to determine specifically which data each item on the questionnaire is designed to obtain. Unfortunately, this direct link between the items on the questionnaire and their exact purpose is not always carefully considered, resulting in either the collection of unnecessary information, the inability to respond to some hypotheses, or both. To prevent this, ask yourself exactly how each item on the questionnaire is to be analyzed. If you can't answer this question for a particular item, it should be omitted.

As with a question on a written test, it is difficult to know how an item on a questionnaire will function the first time it is used. This is why it is necessary to do a few pilot studies before finalizing your questionnaire. Perhaps the best advice is to conduct pilot work with the questionnaire as you are developing it. In the first trial run, ask some colleagues to examine potential items for ambiguities, personal idiosyncrasies, and problems in the directions to the respondents. After these are addressed, get feedback on the next draft of the instrument from a focus group (a small sample from the potential respondent pool). Their task is not only to respond to the questionnaire but also to indicate any problems they encounter. Then address these problems and analyze the resulting data to determine if the correct information to address the hypotheses is being secured and if there are any data entry problems (e.g., multiple responses to items, inappropriate responses).

WWW **Go to HK*Propel* to complete Student Activity 9.11.**

Constructing the Questionnaire

One of the first decisions in constructing a questionnaire is to decide whether to use open-ended or closed-ended questions. Open-ended questions are those for which response categories are not specified. Essentially, the respondent provides an essay-type answer to the question. An example is "What benefits do children gain from participation in an organized sports program?" Closed-ended questions are those requiring the respondent to select one or more of the listed alternatives—for example, "How many days a week should elementary school children participate in physical education classes? 1 2 3 4 5." Both types of questions have advantages and disadvantages.

Open-Ended Questions

The advantages of open-ended questions are that they
- allow for creative answers and allow the respondent freedom of expression,
- allow the respondent to answer in as much detail as desired,
- can be used when it is difficult to determine all possible answer categories, and
- are probably more efficient than closed-ended questions when complex issues are involved.

The disadvantages of open-ended questions are that they
- do not result in standardized information from each respondent, making analysis of the data more difficult;
- require more time of the respondent, which ultimately could reduce the return rate of the questionnaire;
- are sometimes ambiguous because they attempt to solicit a general response, which can result in the respondent not being certain what is being asked; and
- can provide data that are not relevant.

Closed-Ended Questions

The advantages of closed-ended questions are that they

- are easy to code for computer analysis;
- result in standard responses, which are easiest to analyze and to use in making comparisons among respondents;
- are usually less ambiguous for the respondent; and
- require less time of the respondent, which increases the return rate of the questionnaire.

The disadvantages of the closed-ended questions are that they

- may frustrate a respondent if an appropriate category is omitted,
- can result in a respondent selecting a category even if he or she doesn't know the answer or have an opinion,
- may require too many categories to cover all possible responses, and
- are subject to possible recording errors (e.g., the respondent checks B but meant to check C).

Deciding which type of question to use depends on several factors, such as the complexity of the issues involved, the length of the questionnaire, the sensitivity of the information sought, and the time available for construction and analysis of the questionnaire. In general, closed-ended questions work best when the responses are discrete, nominal in nature, and few in number. Descriptive information such as gender, years of education, and marital status is most often measured with closed-ended questions. Closed-ended questions may also be more useful when seeking sensitive information: For example, a respondent may be willing to provide information about annual income if it is done by checking the appropriate range (e.g., between $80,000 and $100,000) rather than by listing the salary specifically. Closed-ended questions should be used if you have more time to construct the questionnaire than to analyze it. Open-ended questions are relatively simple to construct, but the coding and interpreting necessary once the responses are returned are anything but simple.

Other Item Concerns

Questionnaire items need to be simple and avoid containing more than one element. For example, how would you answer yes or no to this question: "Do you think physical education or music should be kept or dropped from the curriculum?" Avoid ambiguous questions. Consider the responses to this item: "Do you think the punishment was appropriate? Yes or No." If a person answers no, you do not know if they thought the punishment was too light or too harsh. To avoid misinterpretation on the part of the respondent, avoid vague terms, slang, colloquial expressions, and lengthy questions. Be certain that the level of wording is appropriate for those who will be responding. Avoid the use of leading questions. For example, how would you likely respond to this item: "Most experts believe that regular moderate exercise produces health benefits. Do you agree?"

WWW! **Go to HK*Propel* to complete Student Activity 9.12.**

Factors Affecting the Questionnaire Response

Besides the instrument itself, many ancillary materials and techniques affect the success of collecting data with a questionnaire. Computers are widely used to obtain survey data. However, these data may still be problematic in that the sample choosing to respond may not be representative of the population about which you want to generalize, and not everyone feels comfortable answering personal questions on the computer because of perceived security and privacy issues.

Sample Representation

The respondents must be representative of the population that was sampled. Having data to provide evidence that the sample reflects the population and is not systematically biased in some way is important. For example, do the ethnicity, age, and sex of your respondents look like the group to whom the original questionnaire was sent?

Cover Letter

Next to the questionnaire itself, the most important item sent to the respondent is the cover letter. This brief document has the important task of describing the nature and objectives of the questionnaire and soliciting the cooperation of the respondent. It is also becoming increasingly common with computer-disseminated questionnaires to alert the possible respondents via e-mail a few days before sending the questionnaire; such an introductory e-mail can take the place of the cover letter in a mailed questionnaire. If possible, personalize cover letters, introductory e-mails, or online survey invitations (address them to the respondent rather than "Dear Sir or Madam"), use some slight form of flattery (e.g., "Because of your extensive background . . ."), and include the endorsement of someone known to the respondent. In addition, be sure that the cover letter or e-mail is short, neat, and attractive. These strategies enhance the probability that the questionnaire will be returned.

Ease of Return

For mailed questionnaires, provide clear instructions as to how and when the questionnaire is to be returned; enclosing a self-addressed postage-paid envelope has been shown to produce increased response rates. If you are sending a large number of questionnaires, it may be advantageous to set up a postage-due arrangement with the post office. Under this arrangement, you pay only for the questionnaires that are returned rather than putting postage on return envelopes that may never be returned. The ease of mailing for the respondent is the same in either case. If information is being collected via the Internet, be sure to clarify whether the respondent can stop and return to complete the questionnaire at a later time.

Neatness and Length

It is logical that if you spend time making the questionnaire and cover letter neat, easy to read, free of grammatical errors, and uncluttered, the respondent may be more inclined to spend time responding. The shorter you can make the questionnaire, the more likely it will be returned. If information is being collected via the Internet, make sure to consider adding ways to indicate how much the respondent has completed and how much yet remains.

Inducements

The inclusion of a pencil or a pen ("so you won't have to locate one"), a penny ("for your thoughts"), a dollar ("for a cup of coffee while you complete the questionnaire"), or a raffle ticket ("winner to be chosen from returned questionnaires") are examples of inducements people have used to encourage respondents to return mailed questionnaires. At the University of Colorado, a $2 bill was enclosed each year with a mailed student satisfaction survey. The hope is to instill a sense of obligation—for some people, it may be difficult to put the money in their wallet and the survey in the wastebasket. As with mailed questionnaires, Internet surveys often include the inducement that returning the survey will result in an entry into a lottery to win a relevant prize.

Timing and Deadlines

It is best not to have the questionnaire arrive just before a major holiday or other significant event, such as the beginning or end of the school year (the return rate of the University of Colorado student satisfaction survey mentioned previously would likely be low if it arrived to students during finals week). The inclusion of reasonable deadlines should enhance return rates. Receiving the questionnaire the day before it is due back gives the respondent an easy excuse for not completing the questionnaire, whereas receiving it too long before it is due to be returned may cause the respondent to put aside the instrument and forget about it.

Follow-Up

It is generally believed that at least one follow-up procedure can enhance the return rate of questionnaires. After one or two reminders, the effectiveness of follow-up procedures generally decreases dramatically. A typical procedure to potentially increase the response rate is to send the original questionnaire and cover letter, wait until the responses trickle down to a few, and then send out a reminder letter or e-mail. If additional follow-up seems necessary, it is common to next send a duplicate questionnaire and cover letter; if still no response is received, a telephone reminder is the next possibility. Follow-up procedures beyond this tend to be unsuccessful.

Analyzing the Questionnaire Responses

How data obtained from returned questionnaires are analyzed will depend on the types of questions (open or closed) that are asked. Open-ended responses require a reader to carefully decipher each response and make subjective judgments about the answers. A reader may need to make a list of ideas that are expressed along with the frequency of the occurrence. Closed-ended responses are most often tabulated and presented in some descriptive way, such as through percentages or graphs. Statistical methods learned in chapters 3, 4, and 5 can also be employed for the analysis:

• A quick frequency analysis can be run to determine the overall response rate, as well as the response rates by major subgroups important to your study (e.g., age, sex, racial and ethnic category). Presenting the results in graphic form can also be quite effective.

• Descriptive statistics can be conducted for each individual question, which is similar to the item analysis for written tests introduced earlier in this chapter. For example, if everyone selects 5 in a rating scale of 1-5 on a specific question, this question will have no discrimination among the respondents and therefore will provide limited information for the survey. Similarly, the mean and SD from the descriptive statistics could also provide useful information about the function of a specific question. If, for example, a question has a mean close to 3 and SD of 0.1, it indicates that almost everyone selected "3" from the "1-5" rating scale, whereas if the mean is close to 3 and the SD is 1.5, it indicates the same overall rating but there is quite a bit of difference in respondents' responses. Finally, inferential statistics (e.g., t-test or chi-square test) can be employed to determine if there are group differences in responding to specific survey questions or to the whole survey.

For mailed questionnaires, creating tables or graphs requires inputting the data to a computer in order to calculate such descriptive statistics. For computerized questionnaires, the data are typically easily converted into relevant spreadsheet formats.

The mechanics of how the data are going to be analyzed should be quite straightforward if the questionnaire constructor has followed the suggestions for planning presented earlier. Checking to see that the questions will lead to obtaining data that can be used to evaluate each of the objectives and hypotheses in the planning stage will result in clear directions for analyzing the responses.

Questionnaire Reliability

The procedures presented in chapter 6 are most often used to validate and estimate the reliability of questionnaire responses. To estimate the reliability of a single item, you must ask the specific item on at least two occasions. However, affective and cognitive domain subscales completed on a questionnaire can have their reliability estimated with the alpha coefficient. An important issue is the stability reliability of the responses. To estimate stability reliability, you must administer the questionnaire to the same people on two or more occasions. The typical time between administrations to determine stability reliability is 2 to 4 weeks; any longer could result in actual changes in respondents' opinions. If there are changes in responses that reflect true changes of opinion, the estimated reliability will be reduced. The specific type of reliability estimate to use depends on the nature of the questions asked. For example, are the items nominally (e.g., "What is your sex?") or intervally scaled (e.g., a series of questions about attitudes)? See chapter 6 for the specific methods to estimate reliability for the obtained responses.

Dataset Application

Use the chapter 9 large dataset, MVPA example, in HK*Propel* to determine the alpha reliability (refer to chapter 6) of the results from a questionnaire that asked people to self-report weekly minutes of MVPA for 3 weeks. How reliable are these data? What would the estimated reliability be if you obtained data for 4 weeks instead of 3 weeks? (Recall the Spearman–Brown prophecy formula from chapter 6.) What steps might you take to estimate the validity of these self-reports?

Questionnaire Validity

The most important issue of a questionnaire, as with any measuring instrument, is the validity of the responses. It is important that the respondents truthfully respond to the items and do not respond based on what they believe the socially acceptable response would be. Developing quality questionnaire items, having items reviewed by experts, conducting pilot testing, and ensuring confidentiality or anonymity are ways to increase the validity of responses. Most questionnaires are validated with content-related procedures (presented in chapter 6). However, there are ways to cross-check the responses with additional data to determine if the respondent is answering truthfully. For example, if a respondent says that he or she voted for a certain candidate in an election, there is no way to determine the actual vote. However, you can verify through public records whether the respondent is actually registered to vote and voted in the specific election.

Finally, whether the respondent sample is representative of the population about which one desires to generalize is an important validation issue to consider. See Santiago and Morrow (2020) for an example on how to determine the content validity of a fitness knowledge test. Booth, Okely, Chey, and Bauman (2002) also provide an example of estimating the reliability and validity of a questionnaire in their examination of the Adolescent Physical Activity Recall Questionnaire.

MEASUREMENT AND EVALUATION CHALLENGE

To measure how well three groups of students have learned basic statistical concepts via different teaching methods, Kate has constructed a 60-item multiple-choice test based on a table of specifications she had developed before her research began. The table of specifications reflects the content and importance of the material. She initially developed a 100-item test and then conducted a pilot test followed by an item analysis to evaluate the individual items and the overall test. Using the item analysis, she was able to select 60 items that each demonstrated acceptable difficulty and discrimination indexes and, as a total test, an acceptable reliability. She also asked two experts who have taught statistical concepts for nearly 20 years to evaluate the items. These experts' suggestions helped her ensure that the items had acceptable content validity.

The test will be administered to students in each of the three teaching-method groups. The mean score on the test will be used as the dependent variable in an analysis of variance (see chapter 5) to see if the groups differed in their knowledge of statistics.

SUMMARY

When a research project or a human performance curriculum requires the assessment of objectives in the cognitive domain, the instrument of choice is usually a written test. When the assessment of aspects of the affective domain, such as attitudes or opinions, is desired, a questionnaire is typically used. The procedures for planning, constructing, scoring, administering, and analyzing the results from application of such instruments have been presented in this chapter. All of the procedures described in this chapter focus on making the written test or the questionnaire as objective, reliable, and valid as possible.

[www] **Go to HK*Propel* for videos, homework assignments, and quizzes that will help you master this chapter's content.**

Assessment of Health-Related Physical Fitness

OBJECTIVES

After studying this chapter, you will be able to

- identify and define the components of health-related physical fitness;

- identify and define the risks of fitness testing;

- use reliable and valid methods of measurement of aerobic capacity, body composition, muscular fitness, flexibility, and bone density; and

- identify and use test items and batteries specifically developed for older adults, adults, children and youth, and persons with disabilities.

www The lecture outline in HK*Propel* will help you identify the major concepts of the chapter.

MEASUREMENT AND EVALUATION CHALLENGE

Jamal, a new graduate with a major in kinesiology, is interviewing with the YMCA for a position as physical fitness director. The executive director of the YMCA is interested in having high-quality fitness programs but also wants to have a good fitness assessment program available to all members. He asks Jamal if he is competent in fitness assessment and if he can help develop an entry assessment program for their branch. They are going to meet again in a week, and at the end of the interview the executive director concludes, "Jamal, if you can present a satisfactory assessment program, I think that we will ask you to join our team." Jamal is excited but also nervous. Because his career may depend on his responses to the executive director, Jamal has to do some research and thinking.

No more important objective exists in the sport and exercise sciences than the attainment of physical fitness, which is a state of being that reflects a person's ability to perform a specific physical task or function and is related to the person's present and future health. Physical fitness is a multifaceted objective, with different meanings to different people—a cardiologist might define physical fitness differently than a gymnastics coach.

A BRIEF HISTORY OF PHYSICAL FITNESS TESTING

Physical fitness—being able to walk, run, balance, jump, crawl, climb, lift, carry, throw, and catch—was essential to human survival for centuries. Gradually, however, the necessity for physical fitness to survive disappeared from modern society, particularly after the Industrial Revolution. After the importance of maintaining physical fitness became widely recognized in the early 1900s, however, it became the focus of physical education. In 1915, U.S. Surgeon General Dr. F.C. Smith stated that "exercise is necessary for all except those actually and acutely physically ill, at all ages, for both sexes, daily, in an amount just short of fatigue" (Smith, 1915, p. 4). This recommendation remains similar today. Testing during that period of time focused on physical abilities such as "motor ability," including jumping, climbing, lifting, vaulting, and running. One prominent test, the Playground and Recreation Association of America's Athletic Badge Test, was introduced for boys in 1913 and girls in 1916 (Playground and Recreation Association of America, 1913; 1916).

Both World War I and World War II became driving forces for physical fitness promotion, in part because a large proportion of young men were found physically unfit for military service. This realization resulted in new test developments and a tremendous extension of physical education requirements for youth in the 1930s. In 1956, President Eisenhower established the President's Council on Youth Fitness in response to a cross-country study (Kraus and Hirschland, 1953; 1954), which found that American children were less fit than their European peers. This was confirmed by the American Alliance for Health, Physical Education and Recreation (AAHPER) Youth Fitness Project in 1958, which resulted in the establishment of the AAHPER Youth Fitness Test. This norm-referenced test (e.g., Morrow et al., 2009) measured cardiovascular endurance, muscular strength and endurance, and athletic performance (e.g., 50-yard dash) as well as "military preparedness" of youth (the softball throw, for example, was considered good preparation for throwing grenades).

Faced with the Cold War threat from the Soviet Union, President John F. Kennedy was an advocate for promoting youth fitness, and his 1960 article "The Soft American," published in *Sports Illustrated*, made a significant impact on the awareness and development of physical fitness in both youth and adult populations. In 1966, President Lyndon B. Johnson introduced the President's Challenge, which was designed to help motivate participants to improve regardless of activity and fitness level. The publication of *Aerobics* by Dr. Kenneth H. Cooper in 1968, as well as his well-known 12-minute running test and aerobic point system to measure and assess aerobic fitness (Cooper, 1968), also significantly influenced the development of the fitness movement in the United States and around the world.

In the mid-1970s, due to an increased concern of chronic diseases, efforts were made to develop a new fitness construct and the concept of *health-related fitness* was developed (Jackson, 2006a). In order to distinguish them from the newly developed health-related fitness tests, the then-existing motor- and sport-centered fitness tests were relabeled as *performance-related fitness tests* (discussed in chapter 11).

In 1980, the American Alliance for Health, Physical Education, Recreation and Dance (AAHPERD) published the initial test for health-related physical fitness, which it defined as "a multifaceted continuum extending from birth to death. Affected by physical activity, it ranges from optimal ability in all aspects of life through high and low levels of different physical fitness, to severely limiting disease and dysfunction" (p. 3). The test components were body composition, cardiorespiratory function, and musculoskeletal function, and the test employed criterion-referenced evaluation, though it was based on norm-referenced percentiles (e.g., those below the 25th percentile "should receive special attention and be strongly encouraged to improve" (p. 11). During this same period, Dr. Charles L. Sterling developed a physical fitness "report card," which was later further developed into a true criterion-referenced health-related fitness test called FitnessGram by the Cooper Institute (Plowman and Meredith, 2013).

With the goal of promoting health and preventing disease in mind, the U.S. Public Health Service conducted two National Children and Youth Fitness Studies (NCYFS I and II) in 1985 and 1987, respectively, and body composition, measured by skinfolds, was included in both studies. In 1989, using data from the Aerobics Center Longitudinal Study (now called the Cooper Center Longitudinal Study), Blair and others (Blair, Kohl, et al., 1989) found that even a modest increase in aerobic fitness can substantially reduce the chance of dying from heart disease, cancer, or other causes (collectively called *all-cause mortality*). In 1996, the U.S. Department of Health and Human Services (USDHHS) released *Physical Activity and Health: A Report of the Surgeon General*, a landmark scientific presentation summarizing the physiological and psychosocial benefits of a physically active lifestyle and being physically fit and providing guidelines for the amount of physical activity individuals should engage in to accrue health benefits. Many other nations around the world subsequently adopted similar guidelines.

Many industrialized countries have adopted the position that the general public should have sufficient levels of physical activity and fitness to accrue health benefits and deal with daily physical challenges. The U.S. government established public health objectives for improved levels of physical activity and fitness, largely through its initiative called *Healthy People 2000*—which was later extended to *Healthy People 2010, 2020*, and now *2030* (https://health.gov/healthypeople)—as well as through community-based fitness assessment and promotion programs (e.g., YMCA Fitness Testing and Assessment).

Although state-based fitness testing programs and studies existed, such as FitnessGram (e.g., Morrow, Martin, Welk, et al., 2010; Thompson et al., 2019), there were no national studies on youth fitness in the United States until more than two decades after the NCYFS reports. In 2012, the Centers for Disease Control and Prevention conducted the NHANES

National Youth Fitness Survey (www.cdc.gov/nchs/nnyfs/index.htm) to evaluate the health and fitness of U.S. children ages 3 to 15. In September 2012, the Institute of Medicine, at the request of the Robert Wood Johnson Foundation, created the Committee on Fitness Measures and Health Outcomes in Youth to assess the relationship between youth fitness items and health outcomes and recommend the best fitness test items for use in national youth fitness surveys as well as fitness testing in schools. The results of this effort, a report called *Fitness Measures and Health Outcomes in Youth* (Committee on Fitness Measures and Health Outcomes in Youth, 2012), is summarized in the accompanying sidebar.

FITNESS MEASURES

For National Youth Fitness Surveys

- To measure body composition, national surveys should include
  - body mass index (BMI) as an estimate of body weight in relation to height,
  - skinfold thickness at the triceps and below the shoulder blade as indicators of underlying fat, and
  - waist circumference as an indicator of abdominal fat.
- To measure cardiorespiratory endurance, national surveys should include a progressive shuttle run, such as the 20-meter (21.9 yd) shuttle run. If physical space is limited, cycle ergometer or treadmill tests are valid and reliable alternatives.
- To measure musculoskeletal fitness, national surveys should include handgrip strength and standing long jump tests.

For Fitness Testing in Schools

- To measure body composition, schools should use BMI measures.
- To measure cardiorespiratory endurance, schools should include progressive shuttle runs.
- To measure musculoskeletal fitness, schools should include handgrip strength and standing long jump tests.

For schools, additional test items that have not yet been shown to be related to health but that are valid, reliable, and feasible may be considered as supplemental educational tools (e.g., distance or timed runs for cardiorespiratory endurance, modified pull-ups and push-ups for measuring upper-body musculoskeletal strength, curl-ups for measuring core strength). A measure of flexibility, such as the sit-and-reach test, may also be included.

From a measurement and evaluation perspective, there are some important things to note about these recommendations:

1. The recommendations are consistent with the importance of measuring the following health-related physical fitness components:
   a. Cardiorespiratory fitness
   b. Body composition
   c. Musculoskeletal fitness

2. Differences exist between the assessment of body composition for national youth fitness surveys and fitness testing in schools.

    a. National surveys should include BMI, skinfold measures of the triceps and shoulder blade, and a waist circumference measure.

    b. Fitness testing in schools should be limited to BMI.

3. Based on these recommendations, handgrip strength and vertical jump (with the standing long jump test as an alternative) were added into the FitnessGram test battery. The Eurofit test battery also includes handgrip strength and the standing long jump.

Finally, a recent health-related fitness highlight occurred in 2016, when the American Heart Association (AHA), after an extensive review of the scientific literature, issued a statement on the importance of assessing cardiorespiratory (CR) fitness in clinical practice and called for treating CR fitness as a "vital clinical sign" (Ross et al., 2016, p. 1). Four years later, the AHA issued another statement declaring that CR fitness is also an important marker for youth health (Raghuveer et al., 2020).

TYPES OF FITNESS

There are many types of fitness. Because physical fitness is multifaceted, an effective definition must be broad and encompassing. Two factors, the purposes of the tests and the defined population, provide a framework for defining physical fitness for any person. The purposes of fitness assessment are related to the specific population to be tested. As can be seen from table 10.1, there might be many objectives for different groups of people. Different levels of capacity or function result in different definitions of physical fitness. For example, one who is engaged in high-level athletic activities will need a greater level of physical (performance) fitness. Thus, we can define physical fitness based on who and what are to be measured.

Table 10.1 Populations and Purposes of Physical Fitness Testing

Population	Health-related	Performance-related	Diagnosis	Military preparation	Functional capacity
Youth	*	*		*	*
Adults	*				*
Older adults	*				*
Special					
Persons with mental and physical disabilities	*		*		*
Athletes		*	*		
Persons who are ill or injured			*		

In this chapter we primarily address health-related and functional aspects of fitness. The relationships among performance-related fitness, health-related fitness, functional fitness, and impairment are illustrated in figure 10.1. In general, fitness is a multifaceted continuum. Depending on the purpose and criterion employed, it becomes performance-related fitness (and can classify the population as "high" and "typical"), health-related fitness (within the "health fitness zone" or "need improvement"), functional fitness ("functioning" or "at risk"), or impairment ("mobile/functioning" or "ill") (see Morrow et al., 2009 for more information). Within this book, we define health-related physical fitness as the attainment

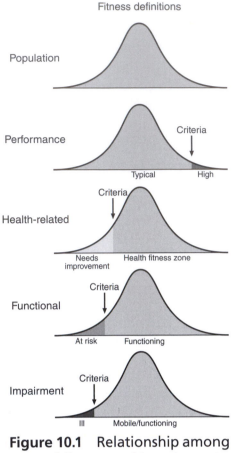

Figure 10.1 Relationship among type of fitness and impairment.

or maintenance of physical capacities that are related to good or improved health and are necessary for performing daily activities and confronting physical challenges. This definition is consistent with the health-related physical fitness definition presented by Pate (1988) and supported by the ACSM (2014a).

Mastery Item 10.1

Consider table 10.1. What would be some reasons for both a 30-year-old mother and a 17-year-old cross-country runner to take a test to determine their cardiorespiratory endurance?

HEALTH-RELATED PHYSICAL FITNESS

In the past, health-related fitness was often considered to have three components: cardiorespiratory endurance (CR), body composition, and muscular fitness (comprising muscular strength, endurance, and flexibility). In 2004, the landmark report "Bone Health and Osteoporosis: A Report of the Surgeon General" (U.S. Department of Health and Human Services, 2004) highlighted the importance of bone health, especially in older adults and women. Bone health has since been included in fitness tests such as FitnessGram and added as a component of health-related fitness (listed in table 10.2 and described in subsequent sections of this chapter).

The evidence to support these components as related to health are derived from the branch of medicine called **epidemiology**, which examines the incidence, prevalence, and distribution of disease. For example, a large majority of epidemiologic studies have indicated that physically fit and active groups have lower relative risk of developing fatal cardiovascular disease (CVD) than less fit and active groups (Ross et al., 2016). **Relative risk** refers to the risk of mortality (death) or morbidity (disease) associated with one group compared to another. Physically fit and active groups have higher levels of **cardiorespiratory endurance**, or ability to extract and use oxygen in a manner that permits continuous exercise, work, or physical activity. Studies have also shown an inverse relationship between death rates and cardiovascular endurance (Blair, Kohl, et al., 1989; USDHHS, 1996; Ekelund et al., 1988). Figure 10.2 demonstrates these findings—the death rate of those in the poorest cardiorespiratory endurance quartile was 8.5 times higher than those in the most fit quartile.

WWW Go to HK*Propel* to complete Student Activity 10.1.

Because those who suffer from obesity have higher rates of CVD, cancer, and diabetes, body composition is included in a health-related **fitness battery** to determine percent body fat and the presence of obesity (ACSM, 2010). Muscular fitness, including muscular strength, muscular endurance, and flexibility, is positively related to good health; main-

Table 10.2 Health-Related Fitness Components and Benefits

Component	Benefits
Cardiorespiratory endurance	Reduction in risk of cardiovascular disease and all-cause mortality Reduction in risk of some types of cancer
Body composition	Reduction in risk of cardiovascular disease, type 2 diabetes, and metabolic syndrome
Muscular fitness, including muscular strength, muscular endurance, and flexibility	Reduction in risk of all-cause mortality Reduction in risk of lower-back pain and injury Reduction in the prevalence and incidence of obesity Reduction in the prevalence and incidence of type 2 diabetes and metabolic syndrome Maintenance of or increase in bone mass Improvement in posture and functional capacity Improvement in glucose tolerance Maintenance of ability to conduct daily activities Increase in free-fat mass and resting metabolic rate
Bone health	Support for body mobility Protection of brain, heart, and other organs from injury Storage of minerals (e.g., calcium and phosphorous) Protection from bone fractures Protection from osteoporosis

Modified from ACSM (2014a).

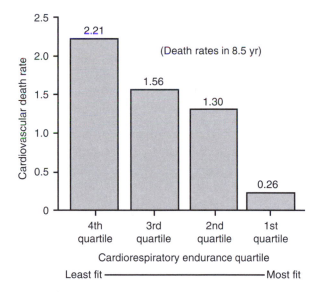

Figure 10.2 Relationship between cardiorespiratory endurance and cardiovascular death rate.

taining a minimum level of muscular fitness is essential for accomplishing daily activities and being prepared to deal with expected or unexpected physical challenges (see table 10.2). Research has indicated inverse relations between muscular strength and obesity, as well as all-cause mortality, after controlling for cardiorespiratory endurance (FitzGerald et al., 2004; Jackson et al., 2010; Ruiz et al., 2008).

NORM-REFERENCED VERSUS CRITERION-REFERENCED STANDARDS

When fitness assessment in the past focused on performance-related fitness, the evaluation standards employed were mainly norm-referenced fitness standards—levels of performance relative to a specifically defined group (typically sex and age). When the interest shifted to health-related fitness, a corresponding change was made to determine fitness achievement based on criterion-referenced fitness standards—specific predetermined levels of performance. Although the PCPFS (1999) used criterion-referenced standards for evaluation of youth fitness scores in the President's Challenge program, FitnessGram was the first nationally recognized fitness test battery that presented health-related criterion-referenced standards (Cooper Institute for Aerobics Research, 1987; 1992; Cooper Institute, 1999; Plowman and Meredith, 2013). Safrit and Looney (1992) and others have taken the position that health-related criterion-referenced standards are useful and appropriate but that normative data on youth fitness scores are also useful for evaluating a program, identifying excellence in achievement, and identifying the current status of people either locally or nationally. Zhu (2013) provided a comprehensive review on issues related to setting standards in kinesiology.

HEALTH SCREENING FOR FITNESS TESTING AND EXERCISE

When the participant is undergoing a fitness assessment or starting an exercise program, one of the critical issues is determining whether medical clearance or physician supervision should be required. The latest view on medical clearance is that exercise is beneficial for most people and they can do it without first visiting a physician. The ACSM (2022) recommends focusing on the following two screening components to determine if medical clearance is needed and what exercise intensity is appropriate:

1. Do the participants currently exercise regularly?
2. Do they have any of the following major signs or symptoms suggestive of cardiovascular, metabolic, and renal disease?

   - Pain, discomfort in the chest, neck, jaw, arms, or other areas that my result from myocardial ischemia or other recent onset pain of unknown origin
   - Shortness of breath at rest or with mild exertion
   - Dizziness or syncope
   - Orthopnea or paroxysmal nocturnal dyspnea
   - Ankle edema
   - Palpitations or tachycardia
   - Intermittent claudication
   - Known heart murmur
   - Unusual fatigue or shortness of breath with usual activities

For example, if the participants do not currently exercise regularly and have signs or symptoms of disease, they need medical clearance. If they do not have disease signs or symptoms, there is no need for medical clearance, but they should start exercise at light to moderate intensity and gradually progress to vigorous exercise. ACSM's health screening algorithm is summarized in table 10.3. For more information, see the preexercise evaluation chapter in ACSM's *Guidelines for Exercise Testing and Prescription* (2022).

Table 10.3 Summary of Preparticipation Screening Algorithm

Exercises regularly	Has suggestive signs or symptoms of diseases	Asymptomatic diseases		Has signs or symptoms suggestive of diseases		Recommended exercise intensity		Medical clearance
		Yes	No	Yes	No	Light to moderate	Vigorous	
No	Yes							X
	No					X		
Yes	Yes	X				X	X	
			X	X				X
			X		X	X	X	
	No					X	X	

Note: The algorithm is modified from ACSM's exercise preparticipation health screening recommendations chart. More detailed information can be found here: www.acsm.org/education-resources/trending-topics-resources/resource-library/detail?id=3c40ff70-9b2a-4ae6-9ee7-24067d7f31a5

Based on ACSM (2022).

In the following sections of this chapter, we examine some of the tests and protocols available for assessing health-related fitness. It is impossible to cover every test, but we emphasize some of the more important test protocols and measurement issues related to them. First, we provide examples of laboratory (criterion) methods and then provide examples of field (surrogate) measures. Although they are often used because they are more feasible to administer than laboratory methods, field tests, by their nature, will have less validity than laboratory methods, and it is important to be aware of this decline. Finally, we will introduce a few commonly used fitness test batteries for various populations (adults, older adults, youth, and persons with disabilities).

MEASURING AEROBIC CAPACITY

As mentioned earlier, physical activity and cardiorespiratory endurance are related to the risk of CVD. The exercise physiologist's concept of cardiorespiratory endurance is a person's aerobic capacity, or **aerobic power**, which is the ability to supply oxygen to the working muscles during physical activity.

Laboratory Methods

Fitness assessment in laboratories and clinical settings involves expensive and sophisticated equipment and exacting test protocols. From a measurement perspective, these tests are often criterion referenced.

Measuring Maximal Oxygen Consumption

The single most reliable ($r_{xx'} > .80$) and valid measure of aerobic capacity is the **maximal oxygen consumption ($\dot{V}O_2max$)** (ACSM, 2010; Safrit et al., 1988). $\dot{V}O_2max$ is a measure of the maximal amount of oxygen that can be used by a person during exhaustive exercise.

In laboratory testing of $\dot{V}O_2max$, participants perform a **maximal exercise test** on an ergometer, such as a treadmill, stationary cycle, step bench, swimming flume, or arm crank device (figure 10.3). Each participant performs the exercise under a specific protocol

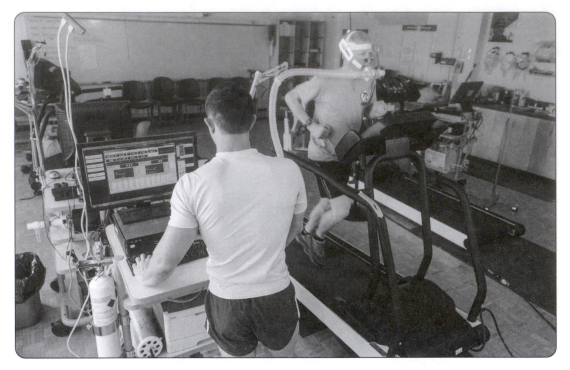

Figure 10.3 Gas exchange analysis during a maximal exercise test.

© Human Kinetics

until he or she reaches exhaustion. While the participant is exercising, expired gases are monitored with a gas analysis system. Most modern exercise physiology laboratories use an automated and computerized system.

A variety of exercise protocols are available in the literature for determining $\dot{V}O_2max$; all focus on increments in work rate until a participant reaches exhaustive levels of physical exertion (ACSM, 2014a). $\dot{V}O_2max$ is achieved when the work rate is increased, but the oxygen consumption ($\dot{V}O_2max$) does not increase or has reached a plateau. Other indications of $\dot{V}O_2max$ are a respiratory exchange ratio (RER) greater than 1.1 and heart rates near age-predicted maximal levels. When these physiological criteria are not clearly achieved, then the maximal oxygen consumption measured during the test is called $\dot{V}O_2peak$. $\dot{V}O_2max$ and $\dot{V}O_2peak$ are highly correlated and represent a valid measure of a participant's aerobic capacity. Blair, Kohl, and colleagues (1989) estimated that $\dot{V}O_2max$ values of 31.5 ml · kg^{-1} · min^{-1} for females and 35 ml · kg^{-1} · min^{-1} for males represent the minimal levels of aerobic capacity associated with a reduced risk of disease and death for a wide range of adult ages. Table 10.4 shows minimal $\dot{V}O_2max$ values associated with lower risks of CVD for specific age groups. Figure 10.4 demonstrates the Balke treadmill protocol for determining $\dot{V}O_2max$. Table 10.5 provides evaluative norms for $\dot{V}O_2max$.

Table 10.4 Minimum Levels of $\dot{V}O_2$max (ml · kg^{-1} · min^{-1}) for Reduced Risks of Morbidity and Mortality

Sex	AGE (YEARS)			
	20-39	40-49	50-59	60+
Men	36.4	34.7	29.8	25.2
Women	28.7	26.6	23.5	20.3

Data based on Sui et al. (2007).

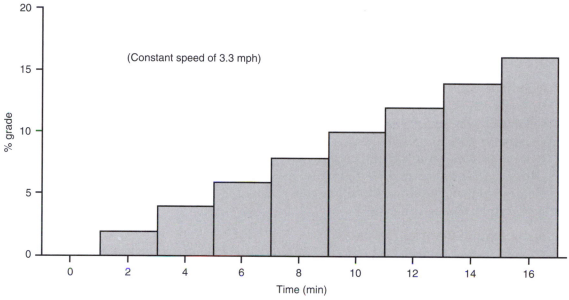

Figure 10.4 The Balke treadmill protocol.

Table 10.5 Male and Female Norms for $\dot{V}O_2$max (ml · kg^{-1} · min^{-1})

Male rating	AGE (YEARS)					
	18-25	26-35	36-45	46-55	56-65	66+
Excellent	100-65	95-60	90-55	83-49	65-43	53-38
Good	60-53	55-50	49-45	45-40	40-37	34-32
Above average	50-48	48-44	43-40	39-36	35-33	31-29
Average	45-43	42-39	38-36	35-32	32-30	28-26
Below average	42-38	38-34	35-31	31-29	28-26	25-23
Poor	36-32	33-30	30-27	27-25	25-22	22-20
Very poor	30-20	27-15	24-14	24-13	21-12	18-10
Female rating						
Excellent	95-59	95-58	75-50	72-45	58-40	55-34
Good	56-50	53-48	46-42	41-36	36-33	31-29
Above average	47-44	45-43	41-37	35-32	32-30	28-26
Average	42-39	41-37	36-33	31-29	28-26	25-23
Below average	38-35	36-34	32-29	28-26	25-23	22-20
Poor	33-30	32-28	28-25	25-22	22-19	19-17
Very poor	27-15	25-14	24-12	20-11	18-10	16-10

Data from Golding (2000).

Mastery Item 10.2

What level of $\dot{V}O_2$max would you like to have based on the norms in table 10.5?

Estimating $\dot{V}O_2$max

Although $\dot{V}O_2$max is the criterion measure of aerobic capacity, it is a difficult measure to determine because it requires expensive metabolic equipment, exhaustive exercise performance, and a substantial amount of time. Consequently, researchers in exercise science have developed valid and reliable techniques for estimating, or predicting, $\dot{V}O_2$max. The estimations are calculated from measurements of maximal or submaximal exercise performance or submaximal heart rate; the same or similar exercise protocols and ergometers discussed previously are used.

Maximal Exercise Performance

$\dot{V}O_2$max can be accurately estimated from the maximum exercise time of a maximal treadmill exercise test (Pollock et al., 1976). Although this procedure requires exhaustive exercise, it does not require the metabolic measurement of expired gases; thus, the test is greatly simplified and expensive equipment is not necessary. Published correlations (concurrent validities) between $\dot{V}O_2$max and maximal exercise time exceed .90. Baumgartner, Jackson, Mahar, and Rowe (2016) provide $\dot{V}O_2$max estimates for maximal treadmill times for several treadmill protocols.

Submaximal Exercise Testing

Submaximal estimates of $\dot{V}O_2$max are based on the linear relationship among heart rate, workload, and $\dot{V}O_2$max. Such estimates are based on **submaximal exercise tests**, which require less than maximal effort. As figure 10.5 indicates, a participant with good aerobic capacity has a higher $\dot{V}O_2$max than one with poor aerobic capacity (both have a maximal heart rate of 200 beats/minute). The slopes of the lines representing the linear relationship between heart rate and $\dot{V}O_2$max are different for each participant. In figure 10.6, we see the difference in workloads each participant can achieve for a fixed submaximal heart rate, 160 beats/minute.

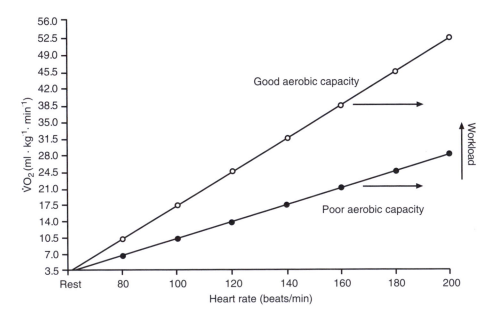

Figure 10.5 Linear relationship among oxygen consumption, heart rate, and workload.

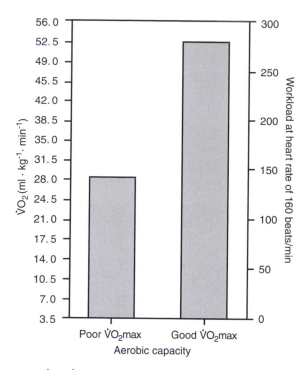

Figure 10.6 How maximal oxygen consumption affects a submaximal workload.

Several exercise test protocols are available for estimations of $\dot{V}O_2$max (ACSM, 2014a). One of the classic procedures is the Åstrand–Rhyming nomogram (Åstrand and Rhyming, 1954), which was originally established as a cycle ergometer test that coordinated workload and heart rate responses into a prediction of $\dot{V}O_2$max. Baumgartner and colleagues (2016) converted the nomogram into an equation that can be used to produce the same predictions of $\dot{V}O_2$max from cycling or treadmill tests (Jackson et al., 1990), allowing computer calculations of predicted aerobic capacity. The ACSM (2010; 2014a) and the YMCA (Golding, 2000) provide descriptions of specific treadmill and cycle test protocols for estimating $\dot{V}O_2$max.

www↘ **Go to HK*Propel* to complete Student Activity 10.2.**

Mastery Item 10.3

In figure 10.5, out of the two participants, who would achieve the higher heart rate and rating of perceived exertion (RPE) for any submaximal workload?

Perceived Exertion During Exercise Testing

Borg (1962) pioneered the measurement of **perceived exertion**, or the mental perception of the intensity of physical work. To assess perceived effort during exercise testing, Borg (1998) presented rating of perceived exertion (RPE) scales. Several versions of the RPE scale with specific purposes and applications are available. RPE scale values correlate with exercise variables such as heart rate, ventilation, lactic acid production, percent $\dot{V}O_2$max, and workload (ACSM, 2014b). The participant simply gives a verbal or visual score from the scale during the exercise test as the workload increases or as time progresses. RPE is typically monitored during exercise tests and is used in exercise prescription to control exercise intensity. The *Physical Activity Guidelines for Americans* presents a simple relative intensity scale (figure 10.7) to aid a participant in setting the intensity of a physical activity or exercise period (USDHHS, 2008). The relative intensity is a person's level of effort relative to his or her fitness level.

- Relatively moderate activity is a level of effort of 5 or 6 on a scale of 0 to 10, where 0 is the level of effort of sitting, and 10 is maximal effort.
- Relatively vigorous activity is a 7 or 8 on this scale.

0	1	2	3	4	5	6	7	8	9	10
Sitting					Moderate intensity		Vigorous intensity			Maximal effort

Figure 10.7 Relative intensity scale.
Data from U.S. Department of Health and Human Services (2008).

Laboratory tests for assessment of aerobic capacity tend to be reliable and valid. However, as in any other testing situation, sources of measurement error are present even in these laboratory situations. Participants, the test and protocol, or test administrators can all be sources of measurement error. The following list provides some important facts you should know concerning laboratory assessment of aerobic capacity.

- Equipment, treadmills, cycles, and gas analysis systems should be calibrated and checked regularly.
- Test administrators should be trained and qualified.
- Practice test administrations should be required for both the participant and administrator to become familiar with test protocols and equipment.
- Standardized test procedures should be established and followed; this creates a focused test environment.
- Treadmill $\dot{V}O_2$max values will be greater than values from cycle ergometer tests for most participants.
- Many Americans seldom ride bicycles, so cycle exercise tests can produce artificially low values of $\dot{V}O_2$max attributable to test cessation from localized fatigue in the legs.
- Typically, submaximal estimates of $\dot{V}O_2$max have a standard error of estimate greater than 5.0 ml $\cdot$ kg^{-1} $\cdot$ min^{-1}.

Field Methods

Field methods include ways to assess aerobic capacity and are feasible for mass testing. Generally, field methods require little equipment and are less expensive in time and costs than laboratory methods.

Distance Runs

Distance runs to achieve the fastest possible time or greatest distance covered in a fixed time are some of the more popular field tests of aerobic capacity. For adults, distances of 1 mile (1.6 km) or greater are used. Safrit and colleagues (1988) indicated that distance runs tend to be reliable ($r_{xx'} > .78$) and have a general concurrent validity coefficient of .741 ± .14. The 12-minute run developed by Cooper (1968) is an example of a distance-run test. AAHPERD has published norms for the 1-mile (1.6 km) run for college students (AAHPERD, 1985; see table 10.6).

Table 10.6 Percentile Norms for the 1-Mile (1.6 km) Run for College Students (min:sec)

Percentile	Males	Females
90	5:44	7:26
75	6:12	8:15
50	6:49	9:22
25	7:32	10:41
10	8:30	12:00

Data based on American Association of Health, Physical Education, Recreation and Dance (1985).

Distance runs are useful for educational situations in which entire classes will be tested in a short amount of time. However, to ensure reliability, validity, and safety (i.e., to correct pacing and provide proper physical conditioning), participants should receive aerobic training and practice trials on the test. It is important that distances, timers, and recording procedures be used in a standardized manner. Adults age 65 or older or people with poor aerobic capacity should undergo one of the other field tests or procedures discussed next.

A multi-stage run is another form of distance run; the 20-meter (21.9 yd) shuttle run (Léger et al., 1988) and PACER (Progressive Aerobic Cardiovascular Endurance Run) are two good examples. The PACER test will be introduced in detail in the FitnessGram section later in this chapter.

Step Tests

Several step-test protocols are available for estimating aerobic capacity. These tests are based on the linear relationships among workload, heart rate, and $\dot{V}O_2$max discussed previously. Generally, each participant steps up and down to an up, up, down, down cadence until a specific workload, heart rate, or time is achieved. The aerobic capacity is then estimated from the heart rate response or the recovery heart rate. Participants with higher aerobic capacity will have a faster return to lower heart rates. The YMCA 3-Minute Step Test is one of the simplest step tests to administer and is useful for initial testing of participants who may not be fit.

YMCA 3-MINUTE STEP TEST

Purpose

To assess aerobic fitness in mass testing situations with adults.

Objective

To step up and down to a set cadence for 3 minutes and measure the resulting heart rate.

Equipment

- 12-inch (30.5 cm) step
- Metronome set at 96 beats per minute
- Watch or timer
- Stethoscope (carotid pulse can also be used)

Instructions

The participant listens to the metronome to become familiar with the cadence and begins when ready. The participant steps up, up, down, down to the cadence, which allows 24 steps per minute. This continues for 3 minutes. After the final step down, the participant sits down and the heart rate is immediately counted for 1 minute.

Scoring

The 1-minute recovery heart rate is the score for the test. Table 10.7 provides evaluative norms for test results.

Table 10.7 Male and Female Norms for Recovery Heart Rate After the 3-Minute Step Test (beats/min)

	AGE (YEARS)					
Male rating	**18-25**	**26-35**	**36-45**	**46-55**	**56-65**	**66+**
Excellent	50-76	51-76	49-76	56-82	60-77	59-81
Good	79-84	79-85	80-88	87-93	86-94	87-92
Above average	88-93	88-94	92-98	95-101	97-100	94-102
Average	95-100	96-102	100-105	103-111	103-109	104-110
Below average	102-107	104-110	108-113	113-119	111-117	114-118
Poor	111-119	114-121	116-124	121-126	119-128	121-126
Very poor	124-157	126-161	130-163	131-159	131-154	130-151
Female rating						
Excellent	52-81	58-80	51-84	63-91	60-92	70-92
Good	85-93	85-92	89-96	95-101	97-103	96-101
Above average	96-102	95-101	100-104	104-110	106-111	104-111
Average	104-110	104-110	107-112	113-118	113-118	116-121
Below average	113-120	113-119	115-120	120-124	119-127	123-126
Poor	122-131	122-129	124-132	126-132	129-135	128-133
Very poor	135-169	134-171	137-169	137-171	141-174	135-155

Data based on Golding (2000).

Rockport 1-Mile (1.6 km) Walk Test

Kline and colleagues (1987) presented another field method for estimating $\dot{V}O_2$max called the Rockport 1-Mile (1.6 km) Walk Test. The procedure involves using the time of a 1-mile walk, sex, age, body weight, and ending heart rate to estimate $\dot{V}O_2$max. The 1-mile walk requires participants to walk as fast as possible; their heart rates are taken immediately at the end of the walk. The 1-mile walk test was shown to be reliable ($r_{xx'} = .98$). The prediction equation (equation 10.1) produced a concurrent validity coefficient of .88 with a standard error of estimate of 5.0 ml · kg^{-1} · min^{-1}:

$$\dot{V}O_2max = 132.853 - (.0769) \times wt - (.3877) \times age + (6.315) \times sex - (3.2469)$$
$$\times 1\ mi\ walk\ time - (.1565) \times heart\ rate \tag{10.1}$$

where wt is weight in pounds, age is in years, sex is 0 for females and 1 for males, 1 mi (1.6 km) walk time is in minutes to the hundredths of a minute, heart rate is beats per minute at the end of the walk, and $\dot{V}O_2$max is ml · kg^{-1} · min^{-1}. Notice that this is a multiple regression equation (introduced in chapter 4).

The original study used a sample with an age range of 30 to 69 years. Further research supported the validity ($r_{xy} = .79$; $s_e = 5.68$ ml · kg^{-1} · min^{-1}) of the equation for adults aged 20 to 29 years (Coleman et al., 1987). As with any physical performance test, the 1-mile (1.6 km) walk can have improved reliability and validity if a practice trial is administered (Jackson, Solomon, and Stusek, 1992). Evaluative norms for people aged 30 to 69 years for the 1-mile (1.6 km) walk test and percentile norms for people aged 18 to 30 years are provided in tables 10.8 and 10.9, respectively.

The 6-minute walk test is commonly used for older adults and clinical populations and will be introduced in detail in the older adult functional fitness battery later in this chapter.

Table 10.8 Norms for the 1-Mile (1.6 km) Walk Test (Participants Aged 30 to 69 Years; min:sec)

Rating	Males (N = 151)	Females (N = 150)
Excellent	<10:12	<11:40
Good	10:13-11:42	11:41-13:08
High average	11:43-13:13	13:09-14:36
Low average	13:14-14:44	14:37-16:04
Fair	14:45-16:23	16:05-17:31
Poor	>16:24	>17:32

Based on Kline et al. (1987).

Table 10.9 Norms for the 1-Mile (1.6 km) Walk Test (Participants Aged 18 to 30 Years; min:sec)

Percentile	Males (N = 400)	Females (N = 426)
90	11:08	11:45
75	11:42	12:49
50	12:38	13:15
25	13:38	14:12
10	14:37	15:03

Adapted from Jackson et al. (1992).

Predicting $\dot{V}O_2$max Without Exercise

Jackson and colleagues (1990) developed an equation for estimating $\dot{V}O_2$max without an exercise test of any kind. The equations had reasonable validity coefficients (r_{xy} >.79) and standard errors of estimation (s_e <5.7 ml · kg^{-1} · min^{-1}); the latter is comparable to submaximal exercise test and field test standard errors of estimation. This, too, is a multiple regression equation. This technique allows for accurate estimation of aerobic capacity in situations in which large numbers of participants need to be evaluated, such as in epidemiologic research. The equation (equation 10.2) is as follows:

$$\dot{V}O_2\text{max} = 50.513 + 1.589 \times \text{self-reported physical activity} - 0.0289 \times \text{age in years} - 0.522 \times \text{percent body fat} + 5.863 \times \text{sex (female} = 0, \text{ male} = 1) \quad (10.2)$$

where self-reported physical activity was measured by a 0-7 rating scale for physical activity participation and fat% was measured by skinfolds.

Wier and colleagues (2006) demonstrated that body mass index (BMI), percent body fat, or waist girth could be used interchangeably in producing estimations of $\dot{V}O_2$max with essentially the same level of accuracy. Jurca et al. (2005) used data from three large international datasets and confirmed that cardiorespiratory fitness level could be estimated well (validities >.75) with a nonexercise test model, including sex, age, BMI, resting heart rate, and self-reported physical activity.

MEASURING BODY COMPOSITION

Obesity is a risk factor in the development of CVD, cancer, and adult-onset diabetes. As a consequence—and because the United States has a large percentage (>40%) of adults who are obese—measuring obesity in an accurate manner is an important goal.

The term *obesity* refers specifically to overfatness, not being overweight. A well-muscled athlete who is extremely fit may be considered overweight on a height and weight table but may actually be quite lean. In health-related fitness, the measurement of **body composition** involves estimating a person's percent body fat, which requires that his or her body density be determined. A good method for conceptualizing body composition is to divide the body into two compartments: *lean*, which includes muscle, bone, and organs and is of high density, and *fat*, which is of low density. For a fixed body weight, a leaner person with a lower percent body fat will have a higher body density than a fatter person of the same weight. In body density estimation, lean tissue is assumed to have an average density of 1.10 g/cm^3, whereas fat tissue is assumed to have an average density of 0.90 g/cm^3. This assumption leads to one of the errors in body composition measurement: Lean tissue and fat tissue do not have the same density, and types of lean tissue (e.g., bone vs. muscle) have different densities. This variable source of measurement error is present in the methods discussed next.

A variety of methods of body composition measurement are available, including the following:

- Dual-energy X-ray absorptiometry (DXA)
- Hydrostatic weighing
- Air displacement plethysmography
- Computed tomography scans and magnetic resonance imaging
- Isotopic dilution
- Ultrasound
- Bioelectrical impedance

- Total body electrical conductivity (TBEC)
- Near-infrared interactance
- Anthropometry (body mass index based on height and weight, skinfolds, and girths)

These methods are the result of the development of relatively new technologies. ACSM (2014b) provides a summary of the advantages and disadvantages of these techniques.

Laboratory Methods

As indicated earlier, there are a variety of laboratory procedures for assessing body composition. In the past, hydrostatic weighing was very popular, but it is being replaced by DXA techniques or air displacement plethysmography due to the practical inconvenience of hydrostatic weighing, which requires examinees to completely submerge into a water tank and expel all of the air in their lungs.

Dual-Energy X-Ray Absorptiometry (DXA)

DXA (previously called DEXA) use in clinical and research applications is growing as the equipment becomes affordable. The technique is based on a three-compartment model of body mineral stores, fat-free mass, and fat mass. DXA is an X-ray technology that passes rays at two energy levels through the body. The attenuation or changes of those rays as they pass through bone, organs, muscle, and fat provide estimates of bone mass, fat-free mass, and fat mass, which allows estimation of bone mineral density as well as percent body fat. Data on specific areas of the body, such as abdominal body fat, can also be gathered. DXA data can be used in research studies related to osteoporosis and body composition change in weight-loss or weight-control interventions. Estimates of total percent body fat have a reported standard error estimate (SEE) of less than 2% (ACSM, 2014b). Sources of measurement error are the lack of standardization among equipment manufacturers, variability between measurements from machines of the same manufacturer, and different software used by the same machine. Thus, for high reliability of measurements, it is strongly recommended that DXA scans be performed on the same instrument for repeated intrasubject measurements (ACSM, 2014b). Limitations include lack of availability (usually only hospitals and universities have the devices) and high cost to the public. In addition, pregnant women should avoid any X-ray-based test, although the radiation dose of DXA is small.

Hydrostatic Weighing and Air Displacement Plethysmography

Both hydrostatic and air displacement plethysmography are based on a two-compartment model of fat and fat-free mass. Hydrostatic weighing (underwater weighing), which is based on Archimedes' principle, was a popular method of laboratory assessment of body density (figure 10.8), which had been used as a criterion measure for validating such field methods as skinfold and girth measurements. The rationale behind underwater weighing is to determine one's body volume by determining the amount of water displaced during the procedure. To overcome water-related inconvenience in hydrostatic weighing, the new method of air displacement plethysmography was developed in which the examinee's body volume is estimated using compressed air while he or she is seated in a sealed chamber (figure 10.9). Although it has a few advantages over hydrostatic weighing (it is quicker, more comfortable, automated, and noninvasive), air displacement plethysmography is limited by its high cost and spandex-style clothing requirement. For more information on hydrostatic weighing and air displacement plethysmography, see Morrow, Jackson, Bradley, and Hartung (1986); Nieman (1995); and Fields, Goran, and McCrory (2002).

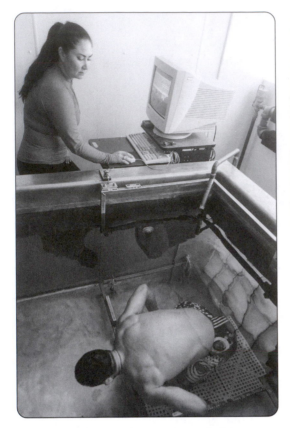

Figure 10.8 Hydrostatic weighing.
© Human Kinetics

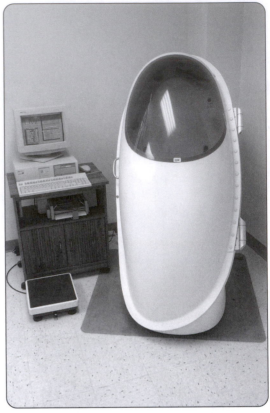

Figure 10.9 Air displacement plethysmography.
© Human Kinetics

Field Methods

Field methods for body composition assessment include bioelectrical impedance analysis (BIA), skinfold measurements, BMI, and the waist–hip girth ratio.

Bioelectrical Impedance Analysis (BIA)

Bioelectrical impedance analysis (BIA) measures the opposition of body tissues to the flow of a small (less than 1 mA) alternating current. Impedance is a function of two components: the resistance of the tissues themselves, and the additional opposition due to the capacitance of membranes, tissue interfaces, and nonionic tissues. BIA determines the electrical impedance of body tissues and provides an estimate of total body water, from which fat-free mass and body fat can be estimated. Applications of BIA often use multifrequency measurements, or a frequency spectrum, to evaluate differences in body composition caused by clinical and nutritional status. In 1994, the National Institutes of Health organized a conference on using BIA to measure body composition, and since then many validation studies have been completed, including low-cost, commercially available BIA devices, and applied to a variety of populations (Khalil et al., 2014; NIH Consensus Statement, 1994; Vasold et al., 2019). The advantages of BIA are that it is simple, quick, and noninvasive. Limitations include possible interruptions by other factors (e.g., skin condition and water consumption), prediction equation dependence, and possible high cost, which has recently been significantly improved by widely available low-cost BIA devices.

Skinfold Measurements

Determining body composition with hydrostatic weighing or DXA is necessary in research studies, but it is not a feasible method for body composition measurement in the field. One of the most feasible, reliable, valid, and popular methods of field estimation of body composition is the skinfold technique, which is the measurement of skinfold (actually, fatfold) thicknesses at specific body sites. The measurements are performed with skinfold calipers, such as those manufactured by Lange and Harpenden (figure 10.10).

Two research studies (Jackson and Pollock, 1978; Jackson, Pollock, and Ward, 1980) developed valid generalized equations for predicting body density from skinfold measurements at seven skinfold sites (chest, axilla, triceps, subscapular, abdomen, suprailium, and thigh) for males and females with an age range of 18 to 61 years. The equations were adapted for the YMCA (Golding, 2000) and provided predictions of percent body fat. Figure 10.11 illustrates the measures used for the three- and four-site prediction equations. The skinfolds were each highly correlated ($r > .76$) with hydrostatically determined body density. During their analysis, the researchers found that the skinfolds had a nonlinear, quadratic relationship to body density; age was also a useful predictor. Table 10.10 provides the relevant equations. Using skinfolds, sex, and age, the concurrent validity of skinfold equations for men exceeds .90 and for women is about .85.

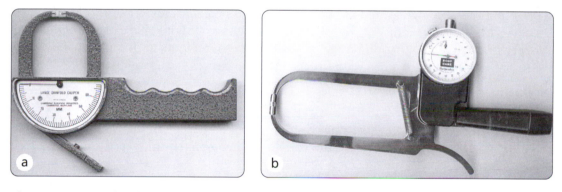

Figure 10.10 Skinfold calipers: *(a)* Lange and *(b)* Harpenden.

© James Morrow

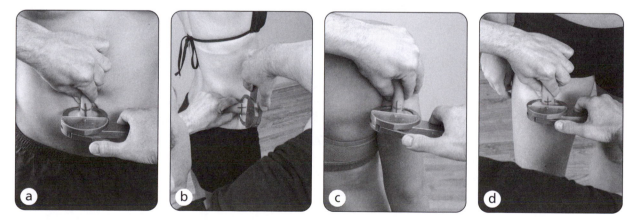

Figure 10.11 Where to measure skinfolds: *(a)* abdomen; *(b)* suprailium, ilium, or hip; *(c)* triceps; and *(d)* thigh.

© Human Kinetics

Table 10.10 YMCA Equations for Estimation of Percent Body Fat

FOUR SITES (.ABDOMEN, SUPRAILIUM, TRICEPS, AND THIGH)
Males
%fat = 0.29288 × (sum of 4 sites) − 0.0005 × (sum of 4 sites)^2 + 0.15845 × (age) − 5.76377 $r = .901$ $s_e = 3.49\%$
Females
%fat = 0.29669 × (sum of 4 sites) − 0.00043 × (sum of 4 sites)^2 + 0.02963 × (age) + 1.4072 $r = .846$ $s_e = 3.89\%$
THREE SITES (ABDOMEN, SUPRAILIUM, AND TRICEPS)
Males
%fat = 0.39287 × (sum of 3 sites) − 0.00105 × (sum of 3 sites)^2 + 0.15772 × (age) − 5.18845 $r = .893$ $s_e = 3.63\%$
Females
%fat = 0.41563 × (sum of 3 sites) − 0.00112 × (sum of 3 sites)^2 + 0.03661 × (age) + 4.03653 $r = .825$ $s_e = 3.98\%$

Data based on Golding (2000).

Skinfold measurements predict body density and percent body fat in a valid manner. However, to ensure reliability of your skinfold measures, you should have plenty of practice. Properly trained testers should be able to produce measurements with high reliability ($r_{xx'}$ >.90). The recommended steps for taking skinfold measures are as follows:

1. Lift skinfolds two or three times before placing the skinfold caliper and taking a measurement.

2. Place the calipers below the thumb and fingers and perpendicular to the fold so that the dial can be easily read, release the caliper grip completely, and read the dial 1 to 2 seconds later.

3. Repeat the process at least three times; the measures should not vary by more than 1 millimeter (0.04 in.). The median value should be used as the measure. An interval of at least 15 seconds should occur between each measurement to allow the site to return to normal. If you get inconsistent values, you should go to another site and then return to the difficult one. As many as 50 to 100 practice sessions on participants may be needed to develop reliable skinfold techniques; with proper preparation, you can achieve reliable measures with a variety of skinfold calipers (Morrow, Fridye, and Monaghen, 1986). For more specific rules for skinfold measurements, see Nieman (1995).

Properly taken skinfold measures are useful field estimates of body composition. Keep in mind that, as the data in table 10.10 show, there is a standard error of estimation (SEE) of up to 3.98% fat. When you report a participant's percent body fat, it is a good idea to inform the participant that it is an estimate and to state the potential error of that estimate—for example, "Your percent body fat is 15 with a potential range of 11 to 19." In dealing with very obese participants, you may not be able to take skinfolds and may need to use another technique.

YMCA SKINFOLD TEST

Purpose
To estimate a person's percent body fat.

Objective
To use a simple field method for estimating body composition characteristics.

Equipment
Skinfold calipers

Instructions
Take skinfold measurements at the abdomen, ilium, triceps, and thigh sites with the procedures described previously.

Scoring
Convert the skinfold measures to percent body fat using the equations in table 10.10. Compare the values to the recommended percent body fat levels and the evaluative norms provided in tables 10.11 and 10.12.

Table 10.11 ACSM-Recommended Levels of Percent Body Fat

				AGE (YEARS)		
	Essential	Minimal	Athletic	≤34	35-55	≥56
Male	3-5	5	5-13	8-22	10-25	10-25
Female	8-12	10-12	12-22	20-35	23-38	25-38

Data based on ACSM (2014a).

Table 10.12 Norms for Percent Body Fat in Males and Females

	AGE (YEARS)					
	18-25	26-35	36-45	46-55	56-65	66+
Male rating						
Excellent	3-7	4-10	5-13	8-16	11-17	12-18
Good	8-10	11-13	15-17	17-19	19-21	19-20
Above average	11-12	14-16	18-20	20-22	22-23	21-22
Average	13-15	17-19	21-22	23-24	24-25	23-24
Below average	16-18	20-22	23-25	25-27	26-27	25-26
Poor	19-21	23-26	26-28	28-30	28-29	27-29
Very poor	23-35	27-38	29-39	31-40	31-40	30-39
Female rating						
Excellent	9-17	7-16	9-18	12-21	12-22	11-20
Good	18-19	18-20	19-22	23-25	24-26	22-25
Above average	20-21	21-22	23-25	26-28	27-29	26-28
Average	22-23	23-25	26-28	29-30	30-32	29-31
Below average	24-26	26-28	29-31	31-33	33-35	32-34
Poor	27-30	29-32	32-35	34-37	36-38	35-37
Very poor	32-43	34-46	37-47	39-50	39-49	38-45

Data based on Golding (2000).

Body Mass Index (BMI)

The body mass index (BMI) is a simple measure expressing the relationship of weight to height. It is used in epidemiological research and has a moderately high correlation (r_{xy} = .69) with body density (fatness). It is easily calculated from the following formula:

$$BMI = \frac{Weight}{Height^2} \tag{10.3}$$

where weight is measured in kilograms and height in meters.

The ratings shown in table 10.13 have been applied to the BMI by the National Heart, Lung, and Blood Institute of the National Institutes of Health.

Table 10.13   Disease Risk Relative to Normal Weight and Waist Circumference

	BMI (kg/m²)	Obesity class	Waist circumference ≤102 cm (40 in.) (male) or ≤88 cm (35 in.) (female)	Waist circumference >102 cm (40 in.) (male) or >88 cm (35 in.) (female)
Underweight	<18.5		—	—
Normal weight	18.5-24.9		—	—
Overweight	25.0-29.9		Increased	High
Obesity	30.0-34.9	I	High	Very high
	35.0-39.9	II	Very high	Very high
Extreme obesity	40.0+	III	Extremely high	Extremely high

Data based on National Heart, Lung, and Blood Institute of the National Institutes of Health.

In the field, the BMI can serve as an acceptable substitute for skinfold measurements on very obese participants; however, do not use it on participants who are lean or of normal weight, for whom skinfolds are more accurate.

WWW! **Go to HK*Propel* to complete Student Activity 10.3.**

Distribution of Body Fat

It is well known that excessive body fat is a health risk, but another factor is the distribution of this body fat. People with excessive body fat on the trunk (android obesity) as compared with the lower body (gynoid obesity) have a higher risk of coronary heart disease (CHD). CHD is a component (along with stroke) of cardiovascular disease (CVD). As shown in table 10.13, men with a waist circumference greater than 102 centimeters (40 in.) and women with a waist circumference greater than 88 centimeters (35 in.) are considered to have increased risk for type 2 diabetes, hypertension, and CVD by the National Heart, Lung, and Blood Institute. A simple measure of this body composition risk factor is the **waist–hip girth ratio**, or the waist circumference divided by the hip circumference. Ratios greater than 1.0 for males and 0.80 for females are associated with a significantly increased risk of CHD (American Heart Association, 1994).

MEASURING MUSCULAR STRENGTH AND ENDURANCE

Many people, even professionals in exercise science, use the terms *strength*, *force*, *power*, *work*, *torque*, and *endurance* almost interchangeably. However, it is important for you as a measurement specialist to understand that each of the terms has a distinct meaning. **Work** is the result of the physical effort that is performed. It is defined by the following equation:

$$\text{work (W)} = \text{force (F)} \times \text{distance (D)} \tag{10.4}$$

For example, the equation 150 ft lb (~203.3 Nm) = 150 lb × 1 foot means that a weight of 150 pounds (68 kg) was moved 1 foot (0.3 m) in distance. Power is the amount of work performed in a fixed amount of time. It is defined by the following equation:

$$\text{power (P)} = (\text{F} \times \text{D}) / \text{time (T)} = \text{W} / \text{T} \tag{10.5}$$

A power value of 150 ft lb/s (~203.3 Nm/s) means that a weight of 150 pounds (68 kg) was moved 1 foot (0.3 m) in distance in 1 second. **Muscular strength** is the force that can be generated by the musculature that is contracting. **Torque** is the effectiveness of a force for producing rotation about an axis. Many computerized dynamometers report the torque produced during muscular contractions as well as the force or strength.

Muscular endurance is the physical ability to perform work. The pull-up test is often called a test of arm and shoulder strength, but in reality it is a measure of muscular endurance (if more than one is performed)—the maximum number of pull-ups performed is a measure of the amount of work that was completed. Muscular endurance can be categorized in terms of relative endurance or absolute endurance. **Relative endurance** is a measurement of repetitive performance related to maximum strength. **Absolute endurance** is a measurement of repetitive performance at a fixed resistance. For example, determining a participant's maximum strength on the bench press and then having him or her perform as many repetitions as possible at 75% of maximum strength is a relative endurance test. Performing the maximum number of repetitions at a fixed weight of 100 pounds (45.4 kg) would be a test of absolute endurance. Absolute endurance is highly correlated with maximum strength, but relative endurance has a low correlation with maximum strength.

Muscular actions are defined and categorized by specific terms, the most common of which are *concentric*, *eccentric*, *isometric*, *isotonic*, and *isokinetic*. The terms are defined as follows:

- *Concentric contraction.* The muscle generates force as it shortens.
- *Eccentric contraction.* The muscle generates force as it lengthens.
- *Isometric contraction.* The muscle generates force but remains static in length and causes no movement.
- *Isotonic contraction.* The muscle generates enough force to move a constant load at a variable speed through a full range of motion.
- *Isokinetic contraction.* The muscle generates force at a constant speed through a full range of motion.

In laboratory testing, researchers assess muscular performance generally by measuring the force, torque, work, and power generated in concentric, eccentric, isokinetic, and isometric contractions. In field situations, muscular performance is assessed with concentric, isotonic contractions. The term **repetition maximum (RM)** is used to indicate how many repetitions one can do. For example, 10RM is the maximum weight that can be lifted 10 times.

Laboratory Methods

Similar to laboratory measurements of aerobic capacity, measuring muscular strength and endurance in the laboratory requires expensive and sophisticated equipment and precise testing protocols. They are the most valid assessments but are difficult to administer.

Computerized Dynamometers

State-of-the-art muscular fitness measurement techniques involve using *computerized dynamometers*, which integrate mechanical ergometers, electronic sensors, computers, and sophisticated software. Computerized dynamometers allow for detailed measurements of force, work, torque, and power generated not only in terms of maximal values but of values throughout a range of motion. Isokinetic dynamometers allow the speed of movement to be controlled during testing. These devices are used by physicians, physical therapists, and athletic trainers in orthopedic clinics and in rehabilitation programs for patients recovering from orthopedic injury or surgery; researchers in exercise science also use them in strength and endurance studies. Strength is measured in terms of peak force and endurance in terms of fatigue rates in force production during a set of repetitions. Each of these devices has its own advantages and disadvantages. Because devices can be expensive, selection of a device requires an analysis of cost–benefit ratios.

The Biodex, a computerized isokinetic dynamometer (Biodex System 2 and Advantage Software 4.0, Biodex Medical Systems, Inc., Shirley, NY) is a device in widespread clinical use. Figure 10.12 illustrates graphic output of forces or torques that the system is able to generate.

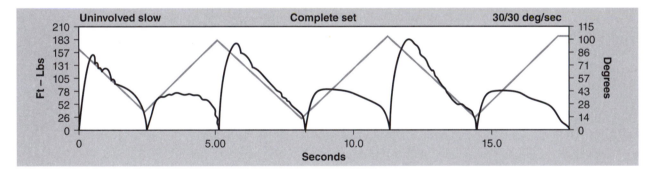

Figure 10.12   Graphic output of forces or torques produced on the Biodex dynamometer.

The Biodex can produce reliable ($r_{xx'}$ >.90) force measurements. However, as with any physical performance test, practice trials for both testers and test participants are needed to reduce measurement error. Mayhew and Rothstein (1985) provided a thorough discussion of the issues associated with dynamometer measurement. These devices must be calibrated regularly to ensure reliable and valid measurements. Computerized dynamometry for muscular strength and endurance assessment should be used to provide criterion measures for concurrent validity research on more feasible field tests of muscular fitness. This application of computerized dynamometry would be valuable, because most field assessments in muscular fitness have only content validity to support their use.

Back Extension Strength Test

Lower-back pain is a serious, prevalent health problem in the adult population that has been associated with lack of strength and endurance in the extensor muscles of the lower back (Suzuki and Endo, 1983). Graves and colleagues (1990) examined measurement issues associated with isometric strength of the back extensors (figure 10.13). Their test protocol has the participant begin at 72° of lumbar flexion and continue down to 0° of lumbar flexion. At each 12-degree interval, the participant exerts a maximal isometric contraction of the extensor muscles of the back. The maximal isometric torque is measured by the dynamometer's automated system. This protocol has produced reliable ($r_{xx'} > .78$) torque results. Average strength values for this test are provided in table 10.14.

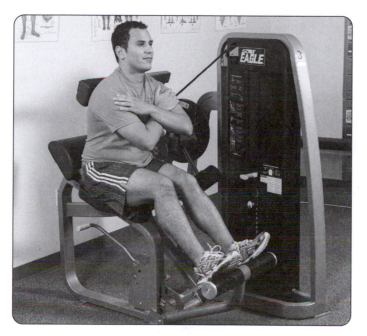

Figure 10.13   Back extension.
© Human Kinetics

Table 10.14   Mean Isometric Torque Values for Lumbar Extension ($N \cdot$ m/kg body weight)

	DEGREES OF LUMBAR FLEXION						
	0	12	24	36	48	60	72
Male	3.0	3.8	4.4	4.8	5.2	5.5	6.0
Female	2.2	2.7	3.0	3.1	3.3	3.5	3.9

Data based on Graves et al. (1990).

Handheld Dynamometers

Handheld dynamometers (HHDs) are also used to assess muscle strength. Although HHDs cannot provide as much information as isokinetic dynamometers, they have an advantage over isokinetic dynamometers in that they are portable and less expensive. These devices, which come in several shapes and sizes, provide an objective measurement of isometric force produced by a muscle or muscle group and can easily be used in the clinic, home, or office. The tester instructs the participant to perform the desired motion and then, holding the HHD in one hand and providing stabilization from unwanted motions with the other hand, keeps the HHD steady as the participant pushes or pulls against it. The tester must hold the HHD stable as the participant presses or pulls against it with as much force as he or she can generate. This type of test is known as a *make test*. The HHD's digital display will record the maximum force produced in kilograms or pounds so that the tester can easily read the display after the test is completed. The major disadvantage of HHD use is that the tester must be physically strong enough to effectively stabilize the device during test administration, which is most problematic during administration of strength tests of the larger muscles of the lower extremity. For this reason, many clinicians and scientists use straps to stabilize the HHD when testing muscles in the lower extremity whenever possible. The use of straps effectively removes the issue of tester strength and has been shown to dramatically improve reliability of testing lower-extremity muscles with use of an HHD. Research has demonstrated that HHD test results can be reliable and valid, but the force or torque results between HHDs of the same make and model and between HHDs of different makes and models can be inconsistent (Trudelle-Jackson et al., 1994). Thus, for reliable and valid results, practitioners should use a single HHD when comparing results across repeated measurements of any kind.

Manual Muscle Testing

A precursor to HHD testing was manual muscle testing (MMT) techniques, which are still used most often to assess muscle strength in clinical situations. MMT techniques originated in the 1940s; although the techniques have been refined over the years, they still retain the original basic testing principles. These techniques are used extensively in the clinic because they require no equipment and can be administered quickly. When using MMT, the tester asks a patient to move the joint through its available range of motion against gravity. If the person is able to move through the full range, the tester begins to apply resistance at the end of the range, then gradually increases the resistance and notes how well the person is able to hold the muscle contraction against the increasing resistance. After applying resistance for 4 or 5 seconds, the tester stops and rates the amount of resistance that was applied. The person's effort is graded on a scale of 0 to 5:

- Grade 5: Muscle contracts normally against full resistance.
- Grade 4: Muscular strength is reduced, but muscular contraction can still move the joint against resistance.
- Grade 3: Muscular strength is further reduced such that the joint can be moved only against gravity with the tester's resistance completely removed (e.g., the elbow can be moved from full extension to full flexion starting with the arm hanging down at the side).
- Grade 2: The muscle can move only if the resistance of gravity is removed (e.g., the elbow can be fully flexed only if the arm is maintained in a horizontal plane).
- Grade 1: Only a trace or flicker of movement is seen or felt in the muscle, or fasciculations are observed in the muscle.
- Grade 0: No movement is observed.

A system of pluses and minuses is often used to further sensitize the grading system. If a patient was not able to complete the motion against gravity, no resistance is applied and instead the patient is repositioned so that gravity is minimized. In this instance, the tester assesses whether that person can complete the joint range of motion when gravity is not opposing the motion. For example, when the quadriceps muscle is being assessed, the person is in a seated position and asked to fully extend the knee (e.g., against gravity). If the person completes the motion, then resistance is applied and graded accordingly. If the person is not able to fully extend the knee against gravity, then he or she is repositioned to a side-lying position with the knee flexed and asked to fully extend the knee from this position. Tester experience as well as tester strength plays a role when MMT techniques are used to assess muscular strength.

Field Methods

Field assessment of muscular strength and endurance involves lifting either external weights or body weight. The measure of maximum strength is the 1-repetition maximum (1RM), which is the maximum amount of weight a person can lift one time. Muscular endurance is assessed by performing either the maximum number of repetitions of a weightlifting exercise with a submaximal weight load or the maximum number of repetitions of a body-weight exercise such as sit-ups.

Upper- and Lower-Body Strength and Endurance

The ACSM (2014a) recommends 1RM values of the bench press and the leg press strength tests as valid measures of upper- and lower-body strength. The ACSM further suggests that the 1RM values be divided by a person's body weight to present a strength measure that is equitable across weight classes.

The following steps present a method for assessing the 1RM value for any given exercise. These steps were used to produce reliable ($r_{xx'}$ >.92) and valid measures of upper- and lower-body strength (Jackson, Watkins, and Patton, 1980).

1. The participant warms up with stretching and light lifting.
2. The participant performs a lift lighter than his or her estimated maximum. A practice session is extremely useful for novice participants.
3. To prevent fatigue, the participant rests at least 2 minutes between lifts.
4. The weight is increased by a small increment—5 or 10 pounds (2.3 or 4.5 kg), depending on the exercise and weight increments available.
5. The process is continued until the participant fails an attempt.
6. The last weight successfully completed is the 1RM weight.
7. Divide the 1RM by the participant's body weight.

If more than five repetitions are needed to determine the 1RM value, the participant is retested after a day's rest with a heavier starting weight.

Tables 10.15 and 10.16 provide norms for the bench press and leg press tests. These norms reflect measurements taken from the Universal Gym Weight Lifting Machine, a rack-mounted weight machine. These standards are not valid for testing done with free weights or another type of machine. One of the difficulties of strength measurement is that each strength-testing device—free weights, computerized dynamometers, or rack-mounted devices—produces different results. Use standards appropriate for your testing situation; it may be necessary for you to develop your own local standards.

Table 10.15 Bench Press Strength Ratings (1RM lb/lb [kg/kg] body weight)

	AGE (YEARS)				
	20-29	30-39	40-49	50-59	60+
Male rating					
Excellent	>1.26	>1.08	>0.97	>0.86	>0.78
Good	1.17-1.25	1.01-1.07	0.91-0.96	0.81-0.85	0.74-0.77
Average	0.97-1.16	0.86-1.00	0.78-0.90	0.70-0.80	0.64-0.73
Fair	0.88-0.96	0.79-0.85	0.72-0.77	0.65-0.69	0.60-0.63
Poor	<0.87	<0.78	<0.71	<0.64	<0.59
Female rating					
Excellent	>0.78	>0.66	>0.61	>0.54	>0.55
Good	0.72-0.77	0.62-0.65	0.57-0.60	0.51-0.53	0.51-0.54
Average	0.59-0.71	0.53-0.61	0.48-0.56	0.43-0.50	0.41-0.50
Fair	0.53-0.58	0.49-0.52	0.44-0.47	0.40-0.42	0.37-0.40
Poor	<0.52	<0.48	<0.43	<0.39	<0.36

Data based on The Cooper Institute (2002).

Table 10.16 Leg Press Strength Ratings (1RM lb/lb [kg/kg] body weight)

	AGE (YEARS)				
	20-29	30-39	40-49	50-59	60+
Male rating					
Excellent	>2.08	>1.88	>1.76	>1.66	>1.56
Good	2.00-2.07	1.80-1.87	1.70-1.75	1.60-1.65	1.50-1.55
Average	1.83-1.99	1.63-1.79	1.56-1.69	1.46-1.59	1.37-1.49
Fair	1.65-1.82	1.55-1.62	1.50-1.55	1.40-1.45	1.31-1.36
Poor	<1.64	<1.54	<1.49	<1.39	<1.30
Female rating					
Excellent	>1.63	>1.42	>1.32	>1.26	>1.15
Good	1.54-1.62	1.35-1.41	1.26-1.31	1.13-1.25	1.08-1.14
Average	1.35-1.53	1.20-1.34	1.12-1.25	0.99-1.12	0.92-1.07
Fair	1.26-1.34	1.13-1.19	1.06-1.11	0.86-0.98	0.85-0.91
Poor	<1.25	<1.12	<1.05	<0.85	<0.84

Based on The Cooper Institute (2002).

The YMCA Bench Press Test (Golding, 2000) is used to assess upper-body endurance. The Canadian Standardized Test of Fitness uses a push-up test to measure upper-body endurance. These tests are described in the following highlighted boxes.

Johnson and Nelson (1979) indicated that you can achieve a high reliability ($r_{xx'} = .93$) when these standardized test procedures are followed. Note: The results of this test will be negatively correlated with the body weight of the person tested.

YMCA BENCH PRESS TEST

Purpose
To assess the absolute endurance of the upper body.

Objective
To perform a set of bench press repetitions to exhaustion.

Equipment
Barbells (35 and 80 pounds [15.9 and 36.4 kg])

Metronome set at 60 beats per minute

Weightlifting bench

Instructions
The weight is 35 pounds (15.9 kg) for women and 80 pounds (36.4 kg) for men. The consecutive repetitions are performed to a cadence of 60 beats per minute, with each sound indicating a movement up or down.

Scoring
The test continues until exhaustion or until the participant can no longer maintain the required cadence. Table 10.17 provides norms for this test. Because this is an absolute endurance test, it is positively correlated with maximal strength and body weight or size.

Table 10.17 Male and Female Norms for the YMCA Bench Press Test (Number Completed)

	AGE (YEARS)					
	18-25	26-35	36-45	46-55	56-65	66+
Male rating						
Excellent	44-64	41-61	36-55	28-47	24-41	20-36
Good	34-41	30-37	26-32	21-25	17-21	12-16
Above average	29-33	26-29	22-25	16-20	12-14	10-10
Average	24-28	21-24	18-21	12-14	9-11	7-8
Below average	20-22	17-20	14-17	9-11	5-8	4-6
Poor	13-17	12-16	9-12	5-8	2-4	2-3
Very poor	0-10	0-9	0-6	0-2	0-1	0-1
Female rating						
Excellent	42-66	40-62	33-57	29-50	24-42	18-30
Good	30-38	29-34	26-30	20-24	17-21	12-16
Above average	25-28	24-28	21-24	14-18	12-14	8-10
Average	20-22	18-22	16-20	10-13	8-10	5-7
Below average	16-18	14-17	12-14	7-9	5-6	3-4
Poor	9-13	9-13	6-10	2-6	2-4	0-2
Very poor	0-6	0-6	0-4	0-1	0-1	0-0

Data based on Golding (2000).

CANADIAN STANDARDIZED TEST OF FITNESS—PUSH-UP TEST

Purpose
To assess upper-body endurance.

Objective
To perform push-ups to exhaustion.

Equipment
Mat

Instructions
Males perform the test with the toes touching the ground; females perform the test with the knees bent and touching the ground (figure 10.14).

Scoring
The number of correct repetitions is compared with the norms provided in table 10.18.

Figure 10.14 *(a)* Traditional and *(b)* modified push-ups.

© Human Kinetics

Table 10.18 Male and Female Norms for the Push-Up Test (Number Completed)

	AGE (YEARS)					
	15-19	20-29	30-39	40-49	50-59	60-69
Male rating						
Excellent	39+	36+	30+	22+	21+	18+
Above average	29-38	29-35	22-29	17-21	13-20	11-17
Average	23-28	22-28	17-21	13-16	10-12	8-10
Below average	18-22	17-21	12-16	10-12	7-9	5-7
Poor	0-17	0-16	0-11	0-9	0-6	0-4
Female rating						
Excellent	33+	30+	27+	24+	21+	17+
Above average	25-32	21-29	20-26	15-23	11-20	12-16
Average	18-24	15-20	13-19	11-14	7-10	5-11
Below average	12-17	10-14	8-12	5-10	2-6	1-4
Poor	0-11	0-9	0-7	0-4	0-1	0-1

Data based on *Canadian Standardized Test of Fitness: Operations Manual*, 3rd ed. (1986).

Trunk Endurance

The most universally used test of abdominal endurance is some type of a sit-up test. The YMCA's Half Sit-Up Test protocol is described here.

Reliability estimates for sit-up tests are typically ≥.68. Acceptable reliability requires the use of standardized procedures and practice trials. The participant's body weight is negatively correlated with the results of this test.

Robertson and Magnusdottir (1987) presented an alternative field test of abdominal endurance similar to the YMCA Half Sit-Up Test. Their curl-up test requires the participant to lift the head and upper back a distance that allows the extended arms and fingers to move 3 inches (7.6 cm) parallel to the ground rather than sit up completely. This requires less use of the hip flexors and more use of the abdominals than the standard sit-up test. The participant completes as many of these as possible in 60 seconds. The authors report excellent reliability ($r_{xx'} = .93$) for the test.

www ► **Go to HK***Propel* **to complete Student Activity 10.4.**

YMCA HALF SIT-UP TEST

Purpose

To assess abdominal endurance.

Objective

To perform the maximum number of half sit-ups in 1 minute.

Equipment

Stopwatch

Mat

Tape

Instructions

The participant lies face up on a mat or rug with the knees at a right angle (90°) and feet flat on the ground. The feet are not held down. The participant places his or her palms down on the mat with the fingers touching the first piece of tape (see figure 10.15a). The participant flattens the lower back to the mat and then sits up halfway so that the fingers move from the first piece of tape to the second, 3.5 inches (8.9 cm) away (figure 10.15b). The participant then returns the shoulders to the mat and repeats the movement as described. The head does not have to touch the surface. The lower back is to be kept flat on the mat during the movements; arching the back can cause injury. The number of half sit-ups performed in 1 minute are counted. Participants should be instructed to pace themselves so they can continue to do half sit-ups for the full minute.

Scoring

See norms in table 10.19.

(continued)

YMCA HALF SIT-UP TEST *(continued)*

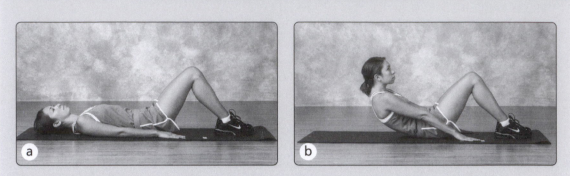

Figure 10.15 The sit-up test: *(a)* beginning position and *(b)* up position.
© Human Kinetics

Table 10.19 Male and Female Norms for the YMCA Half Sit-Up Test (Number Completed)

	AGE (YEARS)					
	18-25	26-35	36-45	46-55	56-65	66+
Male rating						
Excellent	77-99	62-80	60-79	61-78	56-77	50-66
Good	61-72	53-58	48-57	52-57	48-53	38-44
Above average	52-57	44-52	43-45	44-51	41-46	25-31
Average	43-49	37-41	33-39	36-41	33-39	26-30
Below average	37-41	33-36	29-32	29-33	28-32	22-24
Poor	29-35	26-32	24-28	21-25	21-25	15-21
Very poor	14-27	7-21	6-21	6-16	5-20	5-12
Female rating						
Excellent	68-91	54-70	54-74	48-73	44-63	34-54
Good	58-64	44-50	42-48	37-44	35-42	31-33
Above average	51-57	37-41	35-38	33-36	27-32	26-29
Average	41-48	33-36	30-32	30-32	23-25	21-25
Below average	34-38	28-32	23-28	25-28	18-22	16-20
Poor	28-33	22-26	19-22	19-23	11-15	10-13
Very poor	11-25	7-20	4-16	2-13	1-8	0-9

Based on Golding (2000).

MEASURING FLEXIBILITY

The measurement of **flexibility**, or range of motion of a joint or group of joints, is an important aspect of assessment of patients recovering from orthopedic surgery or orthopedic injuries. Because flexibility is specific to a joint and its surrounding tissues, there are no valid tests of general flexibility. For example, if you have flexible ankle joints, you may not have flexible shoulder joints.

Laboratory Methods

Miller (1985) summarized the measurement issues associated with the assessment of joint range of motion. He described clinical measurement techniques, including the following:

- Goniometry—manual, electric, and pendulum goniometers
- Visual estimation
- Radiography
- Photography
- Linear measurements
- Trigonometry

Miller asserted that radiography is the most reliable and valid method of measurement but that it has limited feasibility because of problems with radiation. Goniometry is the most feasible method of clinical assessment of flexibility; it can be reliable and valid if proper test procedures are followed. Reliability estimates for goniometric measurement of hamstring flexibility have exceeded .90 in past research (Jackson and Baker, 1986; Jackson and Langford, 1989). Clinical tests of flexibility should serve as criterion measures for validity studies of field tests of flexibility for health-related fitness. Such concurrent validity studies improve the validity of measurement of flexibility in the field.

Field Methods

As reported previously, the ACSM (2014a) indicates that lower-back pain is a prominent health problem in the United States. In theory, lack of flexibility in the lower back should be associated with lower-back pain, but valid research has yet to establish a relationship between the two (ACSM, 2014a). Flexibility of the lower back is measured with tests of trunk flexion and extension. Most adult fitness batteries include a test of trunk flexion that is a version of the sit-and-reach test.

Trunk Flexion

The sit-and-reach test is a universally used test of trunk flexion. Specifically, it was designed to measure the flexibility of the lower-back and hamstring muscles.

Keep in mind, however, that flexibility of the lower back has not been empirically related to lower-back pain. Research has shown that the sit-and-reach test is a valid measure of hamstring flexibility and is highly reliable ($r_{xx'}$ >.90) but is poorly correlated with a clinical measure of lower-back flexibility and probably is not a valid field test of lower-back flexibility (Jackson and Baker, 1986; Jackson and Langford, 1989). Further research (Jackson et al., 1998) documented no relationship between reported lower-back pain and performance on the sit-and-reach or sit-up tests. These observations provide little support for including sit-and-reach tests in health-related fitness assessments, but these tests are often present in both adult and youth fitness batteries. The YMCA adult fitness test version of the sit-and-reach test is described next.

YMCA ADULT TRUNK FLEXION (SIT-AND-REACH) TEST

Purpose

To assess trunk flexibility.

Objective

To reach as far forward as possible.

Equipment

Yardstick

Masking tape

Instructions

A yardstick is placed on the floor with an 18-inch (45.7 cm) piece of tape across the 15-inch (38.1 cm) mark on the yardstick. The tape should secure the yardstick to the floor. The participant sits with the 0 end of the yardstick between his or her legs. The heels should almost touch the tape at the 15-inch (38.1 cm) mark and heels should be positioned about 12 inches (30.5 cm) apart. With the legs held straight, the participant bends forward slowly and reaches with parallel hands as far as possible and touches the yardstick. The participant should hold this reach long enough for the distance to be recorded (figure 10.16).

Figure 10.16 The sit-and-reach test.
© Human Kinetics

Scoring

Perform three trials. The best score, recorded to the nearest quarter inch (0.6 cm), is compared with the norms supplied in table 10.20.

Table 10.20 Male and Female Norms for the YMCA Trunk Flexion Test (Inches)

	AGE (YEARS)					
	18-25	26-35	36-45	46-55	56-65	66+
Male rating						
Excellent	22-28	21-28	21-28	19-26	17-24	17-24
Good	20-21	19-19	18-19	16-18	15-16	14-16
Above average	18-19	17-17	16-17	14-15	13-13	12-13
Average	16-17	15-16	15-15	12-13	11-11	10-11
Below average	14-15	13-14	13-13	10-11	9-9	8-9
Poor	12-13	11-12	9-11	8-9	6-8	6-7
Very poor	2-11	2-9	1-7	1-6	1-5	0-4
Female rating						
Excellent	24-29	23-28	22-28	21-27	20-26	20-26
Good	22-22	21-22	20-21	19-20	18-19	18-19
Above average	20-21	20-20	18-19	17-18	16-17	17-17
Average	19-19	18-19	17-17	16-16	15-15	15-16
Below average	17-18	16-17	15-16	14-14	13-14	13-14
Poor	16-16	14-15	13-14	12-13	10-12	10-12
Very poor	7-14	5-13	4-12	3-10	2-9	1-9

Note: For metric measurements, refer to www.worldwidemetric.com/measurements.html

Data based on Golding (2000).

Trunk Extension

Measuring trunk flexion with the sit-and-reach test is prevalent in health-related fitness testing; however, trunk extension assessments are not generally included in health-related fitness testing, and little research has been conducted on developing a valid field test of this ability. Jensen and Hirst (1980) described a field test in which the participant lies prone with the hands clasped near the small of the back. The participant then raises the upper body off the floor as far as possible while an aide holds the legs down. The score is the distance from the suprasternal notch to the floor multiplied by 100 and divided by the trunk length. Safrit (1986) suggested determining the trunk length while the participant is seated; trunk length is then defined as the distance from the suprasternal notch to the seat. There is no reliability and validity information on this test. Furthermore, body weight and back extensor strength would be related to the scores on this test, which would limit it as a valid measure of flexibility. Developing a reliable and valid measure of lower-back extension requires additional research.

MEASURING BONE DENSITY

Bone density is a good indicator of bone health. In the laboratory or clinical setting, bone density may be tested using sophisticated equipment. In community and school settings, it can be estimated using field measures.

Laboratory Methods

A bone density test, known as *bone densitometry,* is used to measure bone mineral content and density. It may be done using X-rays, dual-energy X-ray absorptiometry (DXA), or CT scan. Because of its high accuracy, precision, and repeatability, DXA is considered the gold standard for measuring bone density (as well as body composition, mentioned earlier in this chapter).

A typical DXA report for bone mineral density (BMD) usually includes two norms—healthy young adults (T-score) and age-matched adults (z-score). The T-score here is different from the one we learned in chapter 3, where T-score has a mean of 50 and standard deviation (SD) of 10. Mathematically, the T-score here is a z-score, but the referenced mean and SD are from healthy 25- to 35-year-old adults—the age when bone density reaches its peak. Thus, T-score is determined when comparing an examinee's BMD to that of healthy 25- to 35-year-old adults of the same sex and ethnicity. As illustrated in figure 10.17, a positive T-score indicates that the bone is stronger than normal, whereas a negative T-score indicates the bone is weaker than normal:

- A T-score of –1 or above indicates a normal bone density.
- A T-score between –1 and –2.5 indicates low bone mass.
- A T-score of less than –2.5 indicates the presence of osteoporosis.

In general, the risk for bone fracture doubles with every SD below normal. Therefore, a person with a T-score of –1 (i.e., 1 SD below healthy normal) will have twice the risk for bone fracture than a person with normal BMD (T-score of 0).

The interpretation of z-score is the same as you learned in chapter 3, because a person's BMD is compared to an age-, sex-, race-, height-, and weight-matched norm (e.g., a person with a z-score of 1 indicates that his or her BMD is 1 SD above the mean BMD of his or her peers with the same age, sex, race, height, and weight).

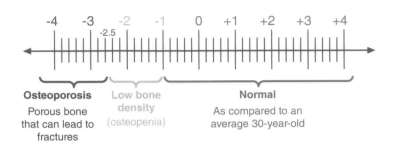

Figure 10.17 Clinical interpretation of T-score.

For children and youth less than 20 years old, only z-scores should be used. The diagnosis of osteoporosis in children should not be made on the basis of densitometric criteria alone. Rather than osteopenia or osteoporosis, the term *low bone density for chronological age* should be used if the z-score is below –2.0 for this population (Bachrach and Gordon, 2016).

Field Methods

In the community, the ultrasound method has been used to estimate bone density. In the school setting, jumping tests have been recently used and validated for this purpose.

Ultrasound

Ultrasound tests use sound waves to measure bone density, usually at an examinee's heel. It is quick and painless and it does not use potentially harmful radiation like X-rays. However, this screening test is only occasionally offered, often at events such as health fairs. If results from an ultrasound test find low bone density, DXA is advised to confirm the results. Another limitation of the ultrasound method is that it cannot measure the density of the bones in the hip and spine, the bones most likely to fracture from osteoporosis.

Jump and Strength Tests

Because participation in physical activity, especially strength training (Janz et al., 2015) and jumping (Fuchs, Bauer, and Snow, 2001), are positively related to BMD, efforts are being made to include these tests in health-related test batteries for youth (see FitnessGram later in this chapter for additional details).

HEALTH-RELATED FITNESS TEST BATTERIES

Several organizations have grouped health-related fitness test items together into test batteries. These test batteries and their documentation allow you to administer reliable and valid fitness tests, interpret the results, and convey fitness information to your program participants.

YMCA Physical Fitness Test Battery

Throughout this chapter we have highlighted the test battery used by the YMCA for its members' physical fitness assessments (Golding, 2000). The test is available and easily adaptable to many adult physical fitness testing situations. The entire test battery includes the following components:

- Height
- Weight
- Resting heart rate
- Resting blood pressure
- Body composition
- Cardiovascular evaluation
- Flexibility
- Muscular strength and endurance

We have previously described and provided norms for the 3-minute step test (cardiorespiratory endurance), skinfold measurements (body composition), the sit-and-reach test (flexibility), the bench press test (muscular strength), and the half sit-up test (muscular endurance).

Canadian Standardized Test of Fitness

The Canadian Standardized Test of Fitness was the result of the 1981 Canadian Fitness Survey, conducted on thousands of participants to develop an understanding of the fitness level of the Canadian population. The test battery includes the following components:

- Resting heart rate
- Resting blood pressure
- Body composition (skinfolds)
- Cardiorespiratory endurance (may use a variety of test results, including treadmill time, cycle ergometer results, and distance-run times)
- Flexibility (sit-and-reach test)
- Abdominal endurance (1-minute sit-up test)
- Upper-body strength and endurance (push-up test)

The President's Challenge: Adult Fitness Test

In May 2008, the President's Council on Fitness, Sports & Nutrition launched the President's Challenge Adult Fitness Test, an online adult fitness test battery designed to be self-administered (however, a partner is needed to complete the entire test battery). The test is designed to measure health-related physical fitness, including aerobic fitness, muscular strength and endurance, flexibility, and body composition. The battery includes the following tests:

- Aerobic fitness
  - 1-mile (1.6 km) walk test
  - 1.5-mile (2.4 km) run
- Body composition
  - Body mass index
  - Waist girth

- Muscular strength and endurance
    - Standard or modified push-up
    - Half sit-up test
- Flexibility
    - Sit-and-reach test

However, the President's Challenge Adult Fitness Test is now undergoing a review and is not currently available in its most recent format.

FITNESS TEST BATTERIES FOR OLDER ADULTS

The U.S. Census Bureau (2020) released estimates in June 2020 showing that the nation's older adult population (defined as aged 65 and over) has grown rapidly since 2010, driven by the aging of baby boomers born between 1946 and 1964. This population grew by over a third (34.2% or 13,787,044) during the past decade. Today, there are more than 46 million older adults living in the United States; by 2050, that number is expected to grow to almost 90 million. Between 2020 and 2030 alone, the time the last of the baby boom cohorts reach age 65, the number of older adults is projected to increase by almost 18 million. This means by 2030, 1 in 5 Americans is projected to be 65 or over. This increase is similar in other industrialized countries.

The health care costs for this aging population are a growing concern. Older adults have the highest rates of chronic diseases such as cardiovascular disease, cancer, diabetes, osteoporosis, and arthritis, and Medicare, the U.S. national health care program for older adults, is facing an uncertain financial future that may have grave implications for future generations of Americans. As professionals in human performance, we serve an important role in helping improve the physical activity and fitness levels of young, middle-aged, and older adults. Improved overall health of the population derived from increased physical activity and fitness would help alleviate the health care financing shortages facing the United States.

Aging is related to

- decreased sensations of taste, smell, vision, and hearing;
- decreased mental abilities (e.g., memory, judgment, speech);
- decreased function of the digestive system, urinary tract, liver, and kidneys;
- decreased bone mineral content and muscle mass, resulting in less lean body weight; and
- decreased physical fitness (e.g., cardiorespiratory endurance, strength, flexibility, muscular endurance, reaction and movement times, balance).

This last factor is of primary concern to anyone entering a health- and fitness-related profession. Studies show that older adults respond to appropriate endurance- and strength-training programs in a manner similar to younger adults. Hagberg and colleagues (1989) demonstrated a 22% increase in $\dot{V}O_2$max as the result of endurance training in 70- to 79-year-old males and females. Fiatarone and colleagues (1994) conducted a high-intensity program of strength training in older males and females with an average age of 87 years; participants were described as frail nursing home residents. Their study demonstrated strength gains of 113% and lean body mass gains of 3%. More important, the participants dramatically improved their walking speed and stair-climbing power, which are clinical measures of **functional capacity**, or the ability to perform the normal activities of daily living. These abilities are of prime importance for older adults if they are to maintain independent living status and a high quality of life (USDHHS, 1996). Understanding the relationships among physical activity, physical fitness, and functional capacity in older adults requires reliable and valid measurements of these metrics.

Senior Fitness Test

We have thus far defined health-related physical fitness in adults as including cardiorespiratory endurance, body composition, and musculoskeletal fitness. Although these factors are still important, an older person's health and fitness should also include motor fitness factors such as balance, reaction time, and movement time in order to focus on maintaining functional capacity, activities of daily living, and overall quality of life.

For example, falls are a major health problem among older adults. Many older adults suffer fractures from falls, and the risk for mortality increases after falls occur. Indeed, many older adults become fearful of falling again and restrict their movements and lifestyle. Strength and balance training in older adults improves fitness parameters that increase lean body and bone mass and lower the risk of falling and of suffering a fracture if one does fall (USDHHS, 1996).

In concert with this broader definition of health-related physical fitness in older adults, Rikli and Jones (1999a; 1999b) have developed a functional fitness battery for older adults called the Senior Fitness Test (Rikli and Jones, 2013b). The test battery incorporates measures of strength, flexibility, cardiorespiratory endurance, motor fitness, and body composition (table 10.21). Local and national advisory panels of health and fitness experts have assisted in selecting and validating the test items. Items were selected with a focus on delaying frailty and maintaining mobility in older adults, both of which are key factors in a healthy older population with a high quality of life. Test items were selected based on whether they met the following criteria: Does the item

- represent major functional fitness components (i.e., key physiologic parameters associated with the functions required for independent living);
- have acceptable test–retest reliability (>.80);
- have acceptable validity, with support for at least two to three types of validation (content-related, criterion-related, or construct-related);
- reflect usual age-related changes in physical performance; and
- require minimum equipment and space?

Table 10.21 Fitness Parameters and Items of the Senior Fitness Test

Physical fitness parameter	Test item
Lower-body strength	30 s chair stand
Upper-body strength	Arm curl
Lower-body flexibility	Chair sit-and-reach
Upper-body flexibility	Back scratch
Cardiorespiratory endurance	6 min walk or 2 min step-in-place
Motor fitness (composite measure of power, speed, agility, and balance)	8 ft (2.4 m) up-and-go
Body composition	Body mass index (BMI)

In addition, is the test item

- able to detect physical changes attributable to training or exercise;
- able to be assessed on a continuous scale across wide ranges of functional ability (frail to highly fit);
- easy to administer and score;
- capable of being self- or partner-administered in the home setting;
- safe to perform without medical release for the majority of community-residing older adults;
- socially acceptable and meaningful; and
- reasonably quick to administer, with individual testing time of no more than 30 to 45 minutes?

Reliability and validity for the test battery are acceptable. Test–retest reliability estimates exceeded .80 for all tests for older males and females. Criterion-related validity coefficients exceeded .70 for five of the seven performance tests for both males and females, and all seven performance tests demonstrated construct validity. Thus, the battery has been established with content validity and feasibility as guidelines. Further research has established sufficient reliability and validity for the battery.

In contrast to the belief that older adults may not be fit enough to be tested, Rikli and Jones (1999a) found that ambulatory, community-residing older adults with "no medical conditions that would be contraindicated for submaximal testing" (ACSM, 2014a) who had not been advised to refrain from exercise could be safely tested. Rikli and Jones (1999b) stated that older adults should not take the tests without physician approval if they

- have been advised by their doctors not to exercise because of a medical condition;
- have experienced chest pain, dizziness, or exertional angina (chest tightness, pressure, pain, heaviness) during exercise;
- have experienced congestive heart failure; or
- have uncontrolled high blood pressure (greater than 160/100).

Thus, with careful screening and caution, this test battery should be a valid and feasible tool for fitness assessment in older persons. The test items are presented in the pages that follow. Tables 10.22 and 10.23 (on pages 282-283) provide percentile norms for females and males, respectively, on the eight items on the Senior Fitness Test (Rikli and Jones, 1999b).

Rikli and Jones (2013a) extended their work with older adults by developing criterion-referenced fitness standards (recall chapter 7). For each of their tests, they identified a cut score that was associated with independent functioning for those between the ages of 60 and 94 (see table 10.24 on page 284). Reliabilities and validities for the criterion-referenced standards were between .79 and .97, indicating that the standards reflected good consistency (i.e., reliability) and predictability (i.e., validity). See Rikli and Jones (2013a; 2013b) for a complete list of these standards by fitness test and age groups.

30-SECOND CHAIR STAND

Purpose

To assess lower-body strength.

Objective

To complete as many stands from a sitting position as possible in 30 seconds.

Equipment

Stopwatch

Straight-back or folding chair (without arms) with seat height of approximately 17 inches (43.2 cm)

Instructions

For safety purposes, the chair is placed against a wall or stabilized in some other way to prevent it from moving during the test. The participant begins the test seated in the middle of the chair, back straight, and feet flat on the floor, arms crossed at the wrists and held against the chest (figure 10.18). On the go signal, the participant rises to a full stand and then returns to a seated position. The participant is encouraged to complete as many full stands as possible within 30 seconds. After a demonstration by the tester, the participant should do a practice trial of one or two repetitions as a check for proper form, followed by one 30-second test trial.

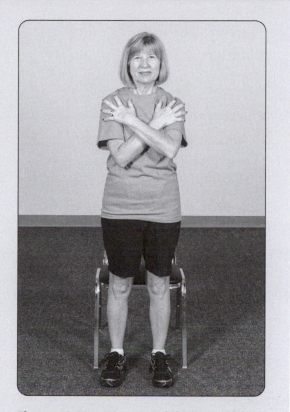

Figure 10.18   30-second chair stand test.

© Human Kinetics

Scoring

The score is the total number of stands executed correctly within 30 seconds. If the participant is more than halfway up at the end of 30 seconds, it counts as a full stand.

Adapted by permission from R.E. Rikli and C.J. Jones, "Development and Validation of a Functional Fitness Test for Community-Residing Older Adults," *Journal of Aging and Physical Activity* 7, no. 2 (1999): 129-161.

ARM CURL

Purpose

To assess upper-body strength.

Objective

To perform as many correctly executed arm curls as possible in 30 seconds.

Equipment

- Stopwatch
- Straight-back or folding chair
- Dumbbell hand weights (5 and 8 pounds [2.3 and 3.6 kg])

Instructions

The participant sits on the chair, back straight, feet flat on the floor, and with the dominant side of the body close to the edge of the chair. The participant holds the weight (5 lb [2.3 kg] for women and 8 lb [3.6 kg] for men) at the side in the dominant hand, using a handshake grip. The test begins with the participant's arm in the down position beside the chair, perpendicular to the floor. At the go signal, the participant turns the palm up while curling the arm through a full range of motion (figure 10.19), then returns the arm to a fully extended position. At the down position, the weight returns to the handshake grip position.

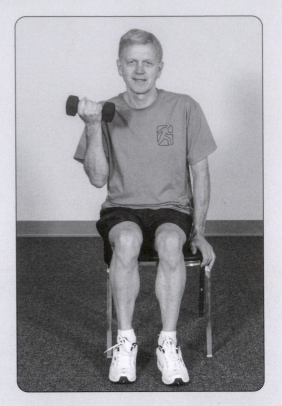

Figure 10.19   Arm curl test.
© Human Kinetics

The tester kneels (or sits in chair) next to the participant on the dominant arm side, placing his or her fingers on the person's midbiceps to stabilize the upper arm and to ensure that a full curl is made (participant's forearm should squeeze examiner's fingers). The participant's upper arm must remain still throughout the test.

The tester may also need to position his or her other hand behind the participant's elbow to help gauge when full extension has been reached and to prevent a backward swinging motion of the arm.

The participant is encouraged to execute as many curls as possible within the 30-second time limit. After a demonstration, give a practice trial of one or two repetitions to check for proper form, followed by one 30-second trial.

Scoring

The score is the total number of curls made correctly within 30 seconds. If the arm is more than halfway curled at the end of the 30 seconds, it counts as a curl.

Adapted by permission from R.E. Rikli and C.J. Jones, "Development and Validation of a Functional Fitness Test for Community-Residing Older Adults," *Journal of Aging and Physical Activity* 7, no. 2 (1999): 129-161.

6-MINUTE WALK TEST

Purpose

To assess aerobic endurance.

Objective

To walk as far as a participant can in 6 minutes along a 50-yard (45.7 m) course.

Equipment

Stopwatch

Long measuring tape (over 20 yards [18.3 m])

4 cones

20 to 25 Popsicle (craft) sticks for each participant

Adhesive name tags (for each participant)

Pen

Chalk or masking tape to mark the course

Instructions

A 50-yard (45.7 m) course is set up and marked in 5-yard (4.6 m) segments with chalk or tape (figure 10.20). The walking area needs to be well lit and have a nonslippery, level surface. For safety purposes, chairs are positioned at several points along the outside of the walkway.

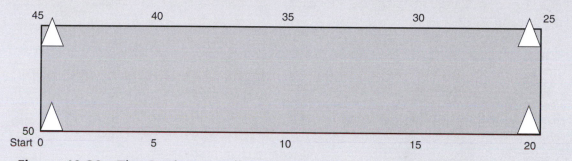

Figure 10.20 The 6-minute walk test uses a 50-yard (45.7 m) course marked in 5-yard (4.6 m) segments.

On the go signal, participants walk as quickly as possible (without running) around the course as many times as they can within the time limit. The tester gives a Popsicle stick (or other similar object) to the participant each time he or she completes one lap (alternatively, someone may tally laps). If two or more participants are tested at once, starting times are staggered 10 seconds apart so participants do not walk in clusters or pairs. Each participant is assigned a number to indicate the order of starting and stopping (self-adhesive name tags to number each participant can be used).

During the test, participants may stop and rest (sitting on the provided chairs) if necessary and then resume walking. The tester should move to the center of the course after all participants have started and call out elapsed time when participants are approximately half done, when 2 minutes are left, and when 1 minute is left. At the end of each participant's respective 6 minutes, participants are instructed to stop and move to the right where an assistant will record the score. A practice test is conducted before the test day to assist with proper pacing and to improve scoring accuracy.

The test is discontinued if at any time a participant shows signs of dizziness, pain, nausea, or undue fatigue. At the end of the test, participants should slowly walk around for about a minute to cool down.

Scoring

The score is the total number of yards (meters) walked (to the nearest 5 yards [4.6 m]) in 6 minutes.

Adapted by permission from R.E. Rikli and C.J. Jones, "Development and Validation of a Functional Fitness Test for Community-Residing Older Adults," *Journal of Aging and Physical Activity* 7, no. 2 (1999): 129-161.

2-MINUTE STEP-IN-PLACE

Purpose

An alternative test to assess aerobic endurance.

Objective

To walk in place as fast as possible for 2 minutes.

Equipment

- Stopwatch
- Tape measure or length of 30-inch (76.2 cm) cord
- Masking tape
- Mechanical counter (if possible) to ensure accurate counting of steps

Instructions

The proper (minimum) stepping height for each participant is at a level even with the midway point between the patella (kneecap) and the iliac crest (top of the hip bone). This point can be determined using a tape measure or by simply stretching a piece of cord from the patella to the iliac crest and then doubling the cord over to determine the midway point. To monitor

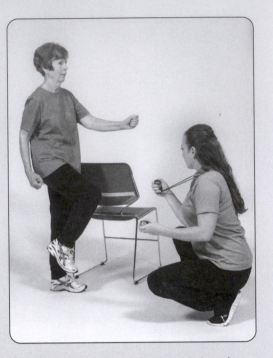

Figure 10.21 2-minute step-in-place.
© Human Kinetics

correct knee height when stepping, books can be stacked on an adjacent table, or a ruler can be attached to a chair or wall with masking tape to mark proper knee height.

On the go signal, the participant begins stepping (not running) in place, completing as many steps as possible within 2 minutes (figure 10.21). The tester counts the number of steps completed, serves as a spotter in case of loss of balance, and ensures that the participant maintains proper knee height. As soon as proper knee height cannot be maintained, the participant is asked to stop or to stop and rest until proper form can be regained. Stepping may be resumed if the 2-minute period has not elapsed. If necessary, the participant can place one hand on a table or chair to assist in maintaining balance. To assist with pacing, participants should be told when 1 minute has passed and when there are 30 seconds to go. At the end of the test, each participant should slowly walk around for about a minute to cool down.

A practice test before the test day can assist with proper pacing and improve scoring accuracy. On test day, the examiner should demonstrate the procedure and allow the participant to practice briefly.

Scoring

The score is the total number of steps taken within 2 minutes. Count only full steps (i.e., each time the knee reaches the point midway between the patella [kneecap] and iliac crest [top of the hip bone]).

Adapted by permission from R.E. Rikli and C.J. Jones, "Development and Validation of a Functional Fitness Test for Community-Residing Older Adults," *Journal of Aging and Physical Activity* 7, no. 2 (1999): 129-161.

CHAIR SIT-AND-REACH

Purpose

To assess lower-body (primarily hamstring) flexibility.

Objective

To sit in a chair and attempt to touch the toes with the fingers.

Equipment

Straight-back or folding chair with seat height of approximately 17 inches (43.2 cm)

Instructions

The participant sits in the chair and moves forward until sitting on the front edge. (For safety, the chair is placed against a wall and checked to see that it remains stable in this position.) The crease between the top of the leg and the buttocks should be even with the edge of the chair seat. Keeping one leg bent and the foot flat on the floor, the participant extends the other leg (the preferred leg) straight in front of the hip, with the heel flat on the floor and the foot flexed at approximately 90° (figure 10.22).

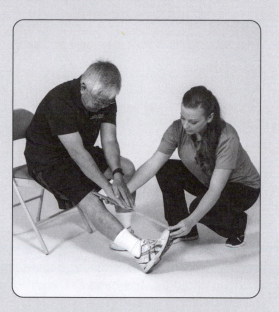

Figure 10.22   Chair sit-and-reach test.
© Human Kinetics

With the leg as straight as possible (but not hyperextended), the participant slowly bends forward at the hip joint, keeping the spine as straight as possible and the head in line with the spine (not tucked). The participant attempts to touch the toes by sliding the hands, one on top of the other with the tips of the middle fingers even, down the extended leg. The reach must be held for 2 seconds. If the extended knee starts to bend before scoring, the participant is asked to slowly sit back until the knee is straight. Participants are reminded to exhale as they bend forward; to avoid bouncing or rapid, forceful movements; and to never stretch to the point of pain.

After the tester provides a demonstration, participants are asked to determine the preferred leg—the leg that yields the better score. Then each participant is given two practice (stretching) trials on that leg, followed by two test trials.

Scoring

Using an 18-inch (45.7 cm) ruler, the tester records the number of inches a person is short of reaching the toe (minus score) or reaches beyond the toe (plus score). The middle of the toe at the end of the shoe represents a zero score. Both test scores are recorded to the nearest 0.5 inch (1.3 cm) and the best score is circled.

Adapted by permission from R.E. Rikli and C.J. Jones, "Development and Validation of a Functional Fitness Test for Community-Residing Older Adults," *Journal of Aging and Physical Activity* 7, no. 2 (1999): 129-161.

BACK SCRATCH

Purpose

To assess upper-body (shoulder) flexibility.

Objective

To reach behind the back with the hands to touch or overlap the fingers of both hands as much as possible.

Equipment

18-inch (45.7 cm) ruler (half of a yardstick)

Instructions

In a standing position, the participant places the preferred hand over the same shoulder, palm down and fingers extended, reaching down the middle of the back as far as possible with the elbow pointed up. The hand of the other arm is placed behind the back, palm up, reaching up as far as possible in an attempt to touch or overlap the extended middle fingers of both hands.

Without moving the participant's hands, the tester helps to see that the middle fingers of each hand are directed toward each other (figure 10.23). Participants are not allowed to grab their fingers together and pull.

After a demonstration, the participant determines the preferred hand and is given two stretching trials followed by two test trials.

Figure 10.23 Back scratch test.
© Human Kinetics

Scoring

The tester measures the distance of overlap or distance between the tips of the middle fingers to the nearest 0.5 inch (1.3 cm). Minus scores represent the distance short of touching middle fingers; plus scores represent the degree of overlap of middle fingers. The tester records both test scores and circles the best score. The best score is used to evaluate performance. (It is important to work on flexibility on both sides of the body, but only the better side has been used in developing norms.)

Adapted by permission from R.E. Rikli and C.J. Jones, "Development and Validation of a Functional Fitness Test for Community-Residing Older Adults," *Journal of Aging and Physical Activity* 7, no. 2 (1999): 129-161.

8-FOOT (2.4 M) UP-AND-GO

Purpose

To assess physical mobility (involves power, speed, agility, and dynamic balance).

Objective

To stand, walk 16 feet (4.9 m), and sit back down in the fastest possible time.

Equipment

Stopwatch

Tape measure

Cone (or similar marker)

Straight-back or folding chair with seat height of approximately 17 inches (43.2 cm)

Instructions

The chair is positioned against a wall or in some other way to secure it during testing. The chair needs to be in a clear, unobstructed area, facing a cone marker exactly 8 feet (2.4 m) away (measured from the front edge of the chair to the back of the marker). There should be at least 4 feet (1.2 m) of clearance beyond the cone to allow ample turning room for the participant.

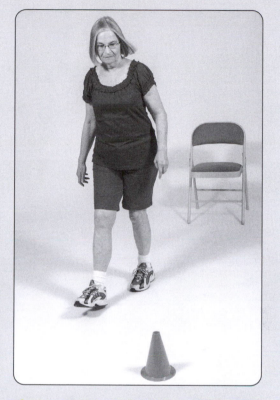

Figure 10.24 8-foot (2.4 m) up-and-go test.

© Human Kinetics

The tester reminds the participant that this is a timed test and that the objective is to walk as quickly as possible (without running) around the cone and back to the chair. The participant starts in a seated position in the chair with erect posture, hands on thighs, and feet flat on the floor with one foot slightly in front of the other. On the go signal, the participant gets up from the chair (may push off thighs or chair), walks as quickly as possible around the cone, and returns to the chair and sits down (figure 10.24). The tester should serve as a spotter, standing midway between the chair and the cone, ready to assist the participant in case of loss of balance. For reliable scoring, the tester must start the stopwatch on go, whether or not the person has started to move, and stop the stopwatch at the exact instant the person sits in the chair.

After a demonstration, the participant should walk through the test one time to practice and then perform two test trials. Participants are reminded that the stopwatch will not be stopped until a participant is fully seated in the chair.

Scoring

The score is the time elapsed from the go signal until the participant returns to a seated position on the chair. Both test scores are recorded to the nearest tenth of a second. The best score (lowest time) is circled and used to evaluate performance.

Adapted by permission from R.E. Rikli and C.J. Jones, "Development and Validation of a Functional Fitness Test for Community-Residing Older Adults," *Journal of Aging and Physical Activity* 7, no. 2 (1999): 129-161.

Table 10.22 Age-Group Percentile—Female

	30 s chair stand		60-64 (n = 595)	65-69 (n = 1027)	70-74 (n = 1240)	75-79 (n = 937)	80-84 (n = 502)	85-89 (n = 305)	90-94 (n = 141)
percentile	10th	no.	9	9	8	7	6	5	2
	25th		12	11	10	10	9	8	4
	50th		15	14	13	12	11	10	8
	75th		17	16	15	15	14	13	11
	90th		20	18	18	17	16	15	14
	Arm curl		60-64 (n = 598)	65-69 (n = 1034)	70-74 (n = 1258)	75-79 (n = 953)	80-84 (n = 519)	85-89 (n = 329)	90-94 (n = 146)
percentile	10th	no.	10	10	9	8	8	7	6
	25th		13	12	12	11	10	10	8
	50th		16	15	15	14	13	12	11
	75th		19	18	17	17	16	15	13
	90th		22	21	20	20	18	17	16
	6 min walk		60-64 (n = 356)	65-69 (n = 617)	70-74 (n = 728)	75-79 (n = 513)	80-84 (n = 276)	85-89 (n = 152)	90-94 (n = 79)
percentile	10th	yd	495	440	420	365	310	260	195
	25th		545	500	480	430	385	340	275
	50th		605	570	550	510	460	425	350
	75th		660	635	615	585	540	510	440
	90th		710	695	675	655	610	595	520
	2 min step-in-place		60-64 (n = 264)	65-69 (n = 491)	70-74 (n = 597)	75-79 (n = 489)	80-84 (n = 279)	85-89 (n = 167)	90-94 (n = 61)
percentile	10th	no.	60	57	53	52	46	42	31
	25th		75	73	68	68	60	55	44
	50th		91	90	84	84	75	70	58
	75th		107	107	101	100	91	85	72
	90th		122	123	116	115	104	98	85
	Chair sit-and-reach		60-64 (n = 591)	65-69 (n = 1037)	70-74 (n = 1250)	75-79 (n = 954)	80-84 (n = 514)	85-89 (n = 332)	90-94 (n = 151)
percentile	10th	in.	−3.0	−3.0	−3.5	−4.0	−4.5	−4.5	−7.0
	25th		−0.5	−0.5	−1.0	−1.5	−2.0	−2.5	−4.5
	50th		2.0	2.0	1.5	1.0	0.5	−0.5	−2.0
	75th		5.0	4.5	4.0	3.5	3.0	2.5	1.0
	90th		7.0	6.5	6.0	5.5	5.0	4.5	3.5
	Back scratch		60-64 (n = 592)	65-69 (n = 1030)	70-74 (n = 1246)	75-79 (n = 946)	80-84 (n = 517)	85-89 (n = 323)	90-94 (n = 148)
percentile	10th	in.	−5.5	−6.0	−6.5	−7.5	−8.0	−10.0	−11.5
	25th		−3.0	−3.5	−4.0	−5.0	−5.5	−7.0	−8.0
	50th		−0.5	−1.0	−1.5	−2.0	−2.5	−4.0	−4.5
	75th		1.5	1.5	1.0	0.5	0.0	−1.0	−1.0
	90th		4.0	3.5	3.0	3.0	2.5	2.0	2.0
	8 ft (2.4 m) up-and-go		60-64 (n = 594)	65-69 (n = 1033)	70-74 (n = 1244)	75-79 (n = 938)	80-84 (n = 497)	85-89 (n = 306)	90-94 (n = 142)
percentile	10th	s	6.7	7.1	8.0	8.3	10.0	11.1	13.5
	25th		6.0	6.4	7.1	7.4	8.7	9.6	11.5
	50th		5.2	5.6	6.0	6.3	7.2	7.9	9.4
	75th		4.4	4.8	4.9	5.2	5.7	6.2	7.3
	90th		3.7	4.1	4.0	4.3	4.4	5.1	5.3
	Body mass index		60-64 (n = 572)	65-69 (n = 1016)	70-74 (n = 1213)	75-79 (n = 916)	80-84 (n = 504)	85-89 (n = 337)	90-94 (n = 149)
percentile	10th	kg/m²	19.6	19.8	20.3	19.8	19.6	19.5	18.3
	25th		22.8	23.0	23.1	22.5	22.0	21.8	21.1
	50th		26.3	26.5	26.1	25.4	24.7	24.3	24.1
	75th		29.8	30.0	29.1	28.3	27.4	26.8	27.1
	90th		33.0	33.2	31.9	31.0	30.0	29.0	29.5

Note: For metric measurements, refer to www.worldwidemetric.com/measurements.html

Adapted by permission from R.E. Rikli and C.J. Jones, "Functional Fitness Normative Scores for Community-Residing Older Adults, Ages 60-94," *Journal of Aging and Physical Activity* 7, no. 2 (1999): 162-181.

Table 10.23 Age-Group Percentile—Male

			60-64 (n = 230)	65-69 (n = 460)	70-74 (n = 498)	75-79 (n = 434)	80-84 (n = 226)	85-89 (n = 108)	90-94 (n = 71)
	30 s chair stand	no.							
percentile	10th		11	9	9	8	7	6	5
	25th		14	12	12	11	10	8	7
	50th		16	15	15	14	12	11	10
	75th		19	18	17	17	15	14	12
	90th		22	21	20	19	18	17	15

			60-64 (n = 229)	65-69 (n = 458)	70-74 (n = 498)	75-79 (n = 440)	80-84 (n = 232)	85-89 (n = 113)	90-94 (n = 71)
	Arm curl	no.							
percentile	10th		13	12	11	10	10	8	7
	25th		16	15	14	13	13	11	10
	50th		19	18	17	16	16	14	12
	75th		22	21	21	19	19	17	14
	90th		25	25	24	22	21	19	17

			60-64 (n = 144)	65-69 (n = 281)	70-74 (n= 294)	75-79 (n = 230)	80-84 (n = 130)	85-89 (n = 60)	90-94 (n = 48)
	6 min walk	yd							
percentile	10th		555	500	480	395	370	295	215
	25th		610	560	545	470	445	380	305
	50th		675	630	610	555	525	475	405
	75th		735	700	680	640	605	570	500
	90th		790	765	745	715	680	660	590

			60-64 (n = 92)	65-69 (n = 211)	70-74 (n = 225)	75-79 (n = 226)	80-84 (n = 119)	85-89 (n = 50)	90-94 (n = 38)
	2 min step-in-place	no.							
percentile	10th		74	72	66	56	56	44	36
	25th		87	88	80	73	71	59	52
	50th		101	101	95	91	87	75	69
	75th		115	116	110	109	103	91	86
	90th		128	130	125	125	118	106	102

			60-64 (n = 228)	65-69 (n = 461)	70-74 (n = 494)	75-79 (n = 434)	80-84 (n = 231)	85-89 (n = 113)	90-94 (n = 74)
	Chair sit-and-reach	in.							
percentile	10th		−6.0	−6.0	−6.5	−7.0	−6.0	−8.0	−9.0
	25th		−2.5	−3.0	−3.5	−4.0	−5.5	−5.5	−6.5
	50th		0.5	0.0	−0.5	−1.0	−2.0	−2.5	−3.5
	75th		4.0	3.0	2.5	2.0	1.5	0.5	0.5
	90th		6.5	6.0	5.5	5.0	4.5	3.0	2.0

			60-64 (n = 228)	65-69 (n = 457)	70-74 (n = 489)	75-79 (n = 430)	80-84 (n = 226)	85-89 (n = 113)	90-94 (n = 73)
	Back scratch	in.							
percentile	10th		−10.0	−10.5	−11.0	−12.0	−12.5	−12.5	−13.5
	25th		−6.5	−7.5	−8.0	−9.0	−9.5	−10.0	−10.5
	50th		−3.5	−4.0	−4.5	−5.5	−5.5	−6.0	−7.0
	75th		0.0	−1.0	−1.0	−2.0	−2.0	−3.0	−4.0
	90th		2.5	2.0	2.0	1.0	1.0	0.0	−1.0

			60-64 (n = 229)	65-69 (n = 461)	70-74 (n = 492)	75-79 (n = 436)	80-84 (n = 227)	85-89 (n = 106)	90-94 (n = 72)
	8 ft (2.4 m) up-and-go	s							
percentile	10th		6.4	6.5	6.8	8.3	8.7	10.5	11.8
	25th		5.6	5.7	6.0	7.2	7.6	8.9	10.0
	50th		4.7	5.1	5.3	5.9	6.4	7.2	8.1
	75th		3.8	4.3	4.2	4.6	5.2	5.3	6.2
	90th		3.0	3.8	3.6	3.5	4.1	3.9	4.4

			60-64 (n = 228)	65-69 (n = 460)	70-74 (n = 491)	75-79 (n = 429)	80-84 (n = 230)	85-89 (n = 114)	90-94 (n = 69)
	Body mass index	kg/m²							
percentile	10th		22.0	22.1	21.6	21.4	21.7	21.8	20.2
	25th		24.6	24.7	24.0	23.8	23.8	23.3	22.4
	50th		27.4	27.5	26.6	26.4	26.1	24.9	24.9
	75th		30.2	30.3	29.2	29.0	28.4	26.5	27.4
	90th		32.8	32.9	31.6	31.4	30.5	28.0	29.6

Note: For metric measurements, refer to www.worldwidemetric.com/measurements.html

Adapted by permission from R.E. Rikli and C.J. Jones, "Functional Fitness Normative Scores for Community-Residing Older Adults, Ages 60-94," *Journal of Aging and Physical Activity* 7, no. 2 (1999): 162-181.

Table 10.24 Criterion-Referenced Fitness Standards for Maintaining Physical Independence in Older Adults

	AGE (YEARS)							% of decline reflected over 30 years
	60-64	65-69	70-74	75-79	80-84	85-89	90-94	
Lower-body strength (number of chair stands in 30 s)								
Female	15	15	14	13	12	11	9	40.0
Male	17	16	15	14	13	11	9	47.1
Upper-body strength (number of arm curls in 30 s)								
Female	17	17	16	15	14	13	11	35.3
Male	19	18	17	16	15	13	11	42.1
Aerobic endurance (yards [meters] walked in 6 min)								
Female	625	605	580	550	510	460	400	36.0
Male	680	650	620	580	530	470	400	41.2
Alternate aerobic endurance (number of steps in 2 min)								
Female	97	93	89	84	78	70	60	38.1
Male	106	101	95	88	80	71	60	43.4
Agility/dynamic balance (8 ft [2.4 m] up-and-go, s)								
Female	5.0	5.3	5.6	6.0	6.5	7.1	8.0	37.5
Male	4.8	5.1	5.5	5.9	6.4	7.1	8.0	40.0

Note: The proposed fitness standards were developed for use with the Senior Fitness Test (SFT) battery (Rikli and Jones, 2013a; 2013b). The standards are based on actual SFT scores obtained by moderate-functioning older adults in previously published cross-sectional database (Rikli and Jones, 1999b) with scores adjusted as appropriate to reflect other relevant information in the literature, including an increased rate of decline over the years when performance is tracked longitudinally versus cross-sectionally.

Vingren et al. (2014) indicate that the fundamental concepts important to assessing physical fitness and physical activity in older adults are reliability, objectivity, relevance, validity, generalizability, error, norm-referenced, criterion-referenced, and formative and summative evaluation. Many of these concepts are central to measurement and evaluation of human performance and used repeatedly throughout this text.

FITNESS TEST BATTERIES FOR CHILDREN AND YOUTH

As we have discussed, assessing physical fitness in youth has changed from a motor-fitness emphasis to a health-related emphasis in the United States and abroad. Youth fitness test batteries use criterion-referenced standards for achieving or not achieving fitness awards. For a complete guide to youth fitness testing, see Safrit and Pemberton (1995).

Table 10.25 lists two youth fitness batteries used to measure and evaluate physical fitness: FitnessGram and Eurofit. Three elements are present in the test batteries:

1. Health-related fitness items
2. Criterion-referenced standards for each test
3. Motivational awards

Contact information for these two fitness batteries is shown in the highlight box.

Table 10.25 Youth Fitness Test Batteries

	FitnessGram	Eurofit
Aerobic capacity	PACER (r) Mile run Walk test	Endurance shuttle run Bicycle ergometer test
Body composition	Skinfolds (r) Body mass index	Height Weight Skinfolds
Abdominal strength and endurance	Curl-ups (r) Plank	Sit-ups
Upper-body strength and endurance	Push-ups (r) Modified pull-ups Pull-ups Flexed arm hang Hand grip strength	Hand grip Bent-arm hang
Trunk extensor strength and endurance	Trunk lift (r)	
Flexibility	Back saver sit-and-reach Shoulder stretch	Sit-and-reach
Running speed and agility		Shuttle run
Speed of limb movement		Plate tapping
Power	Vertical jump	Standing long jump
Balance		Flamingo balance

Note: r = recommended.

FitnessGram, which uses criterion-referenced standards, is the most widely used health-related youth fitness test in the United States. FitnessGram is also used by the Presidential Youth Fitness Program, which is a comprehensive school-based program that promotes health and regular physical activity for America's youth.

The Eurofit physical fitness test battery, which includes a combination of health-related and motor-fitness components, is used on the European continent. Eurofit combines health- and performance-related items with additional motor ability items (e.g., plate tapping, power, balance).

FitnessGram

We highlight the FitnessGram physical fitness battery here because it has been validated, used in many states by millions of children, and includes health-related, criterion-referenced standards. These criterion-referenced standards identify up to three fitness zones—Healthy Fitness Zone (HFZ), Needs Improvement Zone (NI), and Needs Improvement Health Risk Zone (NI-Health Risk)—depending on the selected test. The program includes optional tests, computer software, and support. The intention of the original FitnessGram was to convey a message to parents about their children's fitness status. The FitnessGram report, which shows the zone a student is in and corresponding exercise recommendations, has been a unique future of this fitness educational program from the very beginning. Figure 10.25 illustrates the FitnessGram student report. Following are descriptions of eight items from the battery. The criterion-referenced standards for selected tests are provided in tables 10.26 and 10.27. For more updated and complete information, visit https://fitnessgram.net.

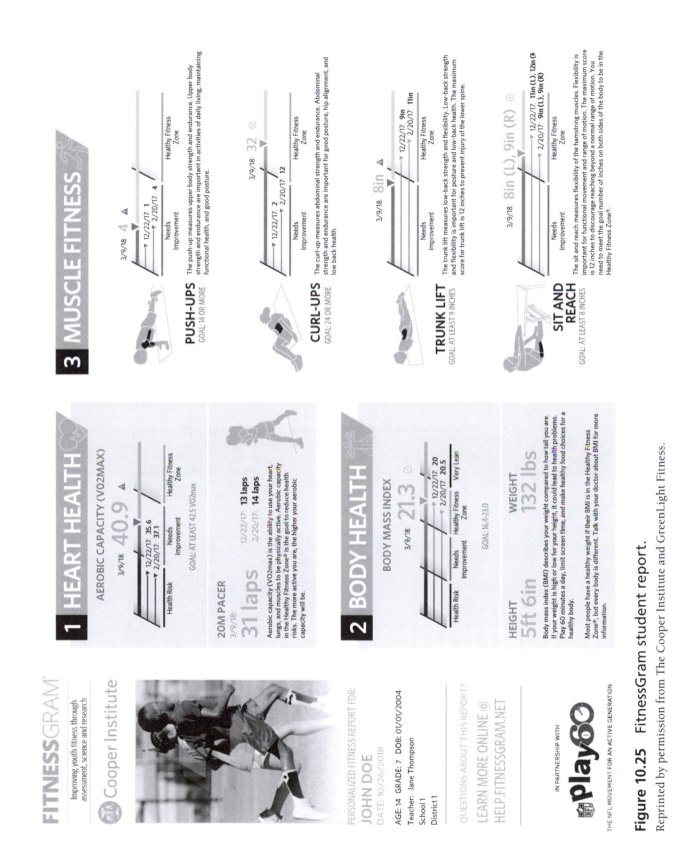

Figure 10.25 FitnessGram student report.

Reprinted by permission from The Cooper Institute and GreenLight Fitness.

Table 10.26 Selected FitnessGram Health-Related Criterion-Referenced Standards: Boys

Age	Aerobic capacity $\dot{V}O_2max$ (ml • kg^{-1} • min^{-1}) PACER, 1-mile (1.6 km) run, and walk test			Percent body fat				Body mass index			
	NI-Health Risk	NI	HFZ	Very lean	HFZ	NI	NI-Health Risk	Very lean	HFZ	NI	NI-Health Risk
5	Completion of test. Lap count or time standards not recommended.			≤8.8	8.9-18.8	18.9	≥27.0	≤13.8	13.9-16.8	16.9	≥18.1
6				≤8.4	8.5-18.8	18.9	≥27.0	≤13.7	13.8-17.1	17.2	≥18.8
7				≤8.2	8.3-18.8	18.9	≥27.0	≤13.7	13.8-17.6	17.7	≥19.6
8				≤8.3	8.4-18.8	18.9	≥27.0	≤13.9	14.0-18.2	18.3	≥20.6
9				≤8.6	8.7-20.6	20.7	≥30.1	≤14.1	14.2-18.9	19.0	≥21.6
10	≤37.3	37.4-40.1	≥40.2	≤8.8	8.9-22.4	22.5	≥33.2	≤14.4	14.5-19.7	19.8	≥22.7
11	≤37.3	37.4-40.1	≥40.2	≤8.7	8.8-23.6	23.7	≥35.4	≤14.8	14.9-20.5	20.6	≥23.7
12	≤37.6	37.4-40.2	≥40.3	≤8.3	8.4-23.6	23.7	≥35.9	≤15.2	15.3-21.3	21.4	≥24.7
13	≤38.6	38.7-41.0	≥41.1	≤7.7	7.8-22.8	22.9	≥35.0	≤15.7	15.8-22.2	22.3	≥25.6
14	≤39.6	39.7-42.4	≥42.5	≤7.0	7.1-21.3	21.4	≥33.2	≤16.3	16.4-23.0	23.1	≥26.5
15	≤40.6	40.7-43.5	≥43.6	≤6.5	6.6-20.1	20.2	≥31.5	≤16.8	16.9-23.7	23.8	≥27.2
16	≤41.0	41.1-44.0	≥44.1	≤6.4	6.5-20.1	20.2	≥31.6	≤17.4	17.5-24.5	24.4	≥27.9
17	≤41.2	41.3-44.1	≥44.2	≤6.6	6.7-20.9	21.0	≥33.0	≤18.0	18.1-24.9	25.0	≥28.6
>17	≤41.2	41.3-44.2	≥44.3	≤6.9	7.0-22.2	22.3	≥35.1	≤18.5	18.6-24.9	25.0	≥29.3

Age	Curl-up (no. completed)	Trunk lift (in.)	90° push-up (no. completed)	Modified pull-up (no. completed)	Flexed arm hang (s)	Back saver sit-and-reach (in.)	Shoulder stretch
5	≥2	6-12	≥3	≥2	≥2	8	Healthy Fitness Zone = touching fingertips together behind back on both the right and left sides.
6	≥2	6-12	≥3	≥2	≥2	8	
7	≥4	6-12	≥4	≥3	≥3	8	
8	≥6	6-12	≥5	≥4	≥3	8	
9	≥9	6-12	≥6	≥5	≥4	8	
10	≥12	9-12	≥7	≥5	≥4	8	
11	≥15	9-12	≥8	≥6	≥6	8	
12	≥18	9-12	≥10	≥7	≥10	8	
13	≥21	9-12	≥12	≥8	≥12	8	
14	≥24	9-12	≥14	≥9	≥15	8	
15	≥24	9-12	≥16	≥10	≥15	8	
16	≥24	9-12	≥18	≥12	≥15	8	
17	≥24	9-12	≥18	≥14	≥15	8	
>17	≥24	9-12	≥18	≥14	≥15	8	

Note: For metric measurements, refer to www.worldwidemetric.com/measurements.html

Adapted by permission from The Cooper Institute, *FITNESSGRAM/ACTIVITYGRAM Test Administration Manual*, updated 4th ed. (Champaign, IL: Human Kinetics, 2010), 65.

Table 10.27 Selected FitnessGram Health-Related Criterion-Referenced Standards: Girls

Age	Aerobic capacity $\dot{V}O_2$max (ml • kg^{-1} • min^{-1}) PACER, 1-mile (1.6 km) run, and walk test			Percent body fat				Body mass index			
	NI-Health Risk	NI	HFZ	Very lean	HFZ	NI	NI-Health Risk	Very lean	HFZ	NI	NI-Health Risk
5	Completion of test. Lap count or time standards not recommended.			≤9.7	9.8-20.8	20.9	≥28.4	≤13.5	13.6-16.8	16.9	≥18.5
6				≤9.8	9.9-20.8	20.9	≥28.4	≤13.4	13.5-17.2	17.3	≥19.2
7				≤10.0	10.1-20.8	20.9	≥28.4	≤13.5	13.6-17.9	18.0	≥20.2
8				≤10.4	10.5-20.8	20.9	≥28.4	≤13.6	13.7-18.6	18.7	≥21.2
9				≤10.9	11.0-22.6	22.7	≥30.8	≤13.9	14.0-19.4	19.5	≥22.4
10	≤37.3	37.4-40.1	≥40.2	≤115.5	11.6-24.3	24.4	≥33.0	≤14.2	14.3-20.3	20.4	≥23.6
11	≤37.3	37.4-40.1	≥40.2	≤12.1	12.2-25.7	25.8	≥34.5	≤14.6	14.7-21.2	21.3	≥24.7
12	≤37.0	37.1-40.0	≥40.1	≤12.6	12.7-26.7	26.8	≥335.5	≤15.1	15.2-22.1	22.2	≥25.8
13	≤36.6	36.7-39.6	≥39.7	≤13.3	13.4-27.7	27.8	≥36.3	≤15.6	15.7-22.9	23.0	≥26.8
14	≤36.3	36.4-39.3	≥39.4	≤13.9	14.0-28.5	28.6	≥36.8	≤16.1	16.2-23.6	23.7	≥27.7
15	≤36.0	36.1-39.0	≥39.1	≤14.5	14.6-29.1	29.2	≥37.1	≤16.6	16.7-24.3	24.4	≥28.5
16	≤35.8	35.9-38.8	≥38.9	≤15.2	15.3-29.7	29.8	≥37.4	≤17.0	17.1-24.8	24.9	≥29.3
17	≤35.7	35.8-38.7	≥38.8	≤15.8	15.9-30.4	30.5	≥37.9	≤17.4	17.5-24.9	25.0	≥30.0
>17	≤35.3	35.4-38.5	≥38.6	≤16.4	16.5-31.3	31.4	≥38.6	≤17.7	17.8-24.9	25.0	≥30.0

Age	Curl-up (no. completed)	Trunk lift (in.)	90° push-up (no. completed)	Modified pull-up (no. completed)	Flexed arm hang (s)	Back saver sit-and-reach (in.)	Shoulder stretch
5	≥2	6-12	≥3	≥2	≥2	9	Healthy Fitness Zone = touching fingertips together behind back on both the right and left sides.
6	≥2	6-12	≥3	≥2	≥2	9	
7	≥4	6-12	≥4	≥3	≥3	9	
8	≥6	6-12	≥5	≥4	≥3	9	
9	≥9	6-12	≥6	≥4	≥4	9	
10	≥12	9-12	≥7	≥4	≥4	9	
11	≥15	9-12	≥7	≥4	≥6	10	
12	≥18	9-12	≥7	≥4	≥7	10	
13	≥18	9-12	≥7	≥4	≥8	10	
14	≥18	9-12	≥7	≥4	≥8	10	
15	≥18	9-12	≥7	≥4	≥8	10	
16	≥18	9-12	≥7	≥4	≥8	10	
17	≥18	9-12	≥7	≥4	≥8	10	
>17	≥18	9-12	≥7	≥4	≥8	10	

Note: For metric measurements, refer to www.worldwidemetric.com/measurements.html

Adapted by permission from The Cooper Institute, *FITNESSGRAM/ACTIVITYGRAM Test Administration Manual*, updated 4th ed. (Champaign, IL: Human Kinetics, 2010), 66.

PACER

Purpose

To measure aerobic capacity.

Objective

To run back and forth across a 20-meter (21.9 yd) space at a specified pace that gets faster each minute for as long as possible.

Equipment

A flat, nonslippery surface at least 20 meters (21.9 yd) long that allows a width of 1.02 to 1.52 meters (40-60 in.) for each student

CD or cassette player with adequate volume

PACER CD or audiocassette

Measuring tape

8 or more cones

Pencil

Copies of score sheet

Note: Students should wear shoes with nonslip soles.

Instructions

Mark the 20-meter (21.9 yd) course with cones to divide lanes and a tape or chalk line at each end. If using the audiotape, calibrate it by timing the 1-minute test interval at the beginning of the tape. If the tape has stretched and the timing is off by more than 0.5 seconds, obtain another copy of the tape. Make copies of the score sheet for each group of students to be tested.

Before the test day, allow students to listen to several minutes of the tape so that they know what to expect. Students should then be allowed at least two practice sessions.

Allow students to select partners to record each others' laps. Have students who are being tested line up behind the start line.

The PACER CD has a version with music and one with only the beeps. The PACER tape has two music versions and one beep-only version. Each version of the test will give a 5-second countdown and tell students when to start. A single beep will sound at the end of the time for each lap. A triple beep sounds at the end of each minute. The triple beep serves the same function as the single beep and also alerts the runners that the pace will get faster.

Students should run across the 20-meter (21.9 yd) distance and touch the line with the foot by the time the beep sounds. At the sound of the beep, they turn around and run back to the other end. If some students get to the line before the beep, they must wait for the beep before running the other direction.

The first time a student does not reach the line by the beep, he or she reverses direction immediately. Allow a student to attempt to catch up with the pace. The test is completed for a student when he or she fails to reach the line by the beep for the second time. Students just completing the test should continue to walk and stretch in the cool-down area. Figure 10.26 provides diagrams of testing procedures.

Note: The PACER also has a 15-meter (16.4 yd) option for those with restricted areas not permitting a 20-meter (21.9 yd) test. The CD is adjusted for the reduced test length.

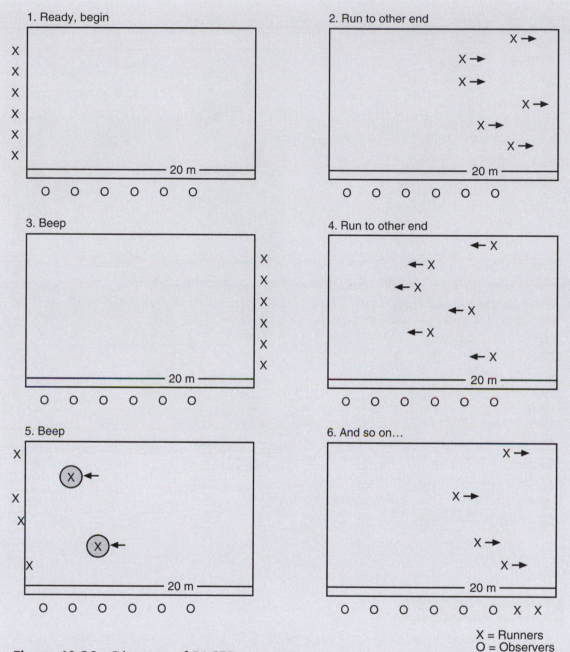

Figure 10.26 Diagram of PACER.

Adapted by permission from The Cooper Institute, *FITNESSGRAM/ACTIVITYGRAM Test Administration Manual*, updated 4th ed. (Champaign, IL: Human Kinetics, 2010), 31.

Scoring

In the PACER test, a lap is one 20-meter (21.9 yd) distance (from one end to the other). Have one student record the lap number (cross off each lap number on a PACER score sheet) while their partner runs. The recorded score is the total number of laps completed by a student. For ease in administration, count the first lap that a student does not reach the line by the beep. Be consistent with all students and classes.

Reprinted by permission from The Cooper Institute, 2010, *FITNESSGRAM/ACTIVITYGRAM test administration manual*, updated 4th ed. (Champaign, IL: Human Kinetics), 31.

SKINFOLD MEASUREMENTS

Purpose

To measure percent body fat.

Objective

The thickness of the skinfolds at the triceps and medial calf sites is measured and used to estimate percent body fat.

Equipment

Skinfold calipers (both expensive and inexpensive calipers have been found to provide reliable and valid measures)

Instructions

The triceps skinfold is taken on the right arm over the triceps muscle. The skinfold is vertical and midway between the acromion process of the scapula and the elbow (see figure 10.11c on page 253). The medial calf skinfold is a vertical skinfold taken at the level of maximal girth of the calf on the right leg (figure 10.27). The foot rests on a stool or other device so that the knee is at a 90° angle.

Scoring

Measure each skinfold three times and record the median value. Record the skinfold thickness to the nearest 0.5 mm.

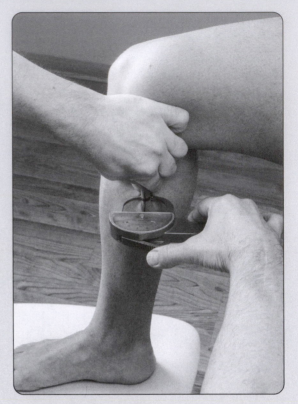

Figure 10.27 FitnessGram calf skinfold measurement.

© Human Kinetics

Mastery Item 10.4

Assume you want to estimate percentage of body fat with the skinfold measures. Identify two sources of error that might arise.

CURL-UP

Purpose

To measure abdominal strength and endurance.

Objective

To perform as many curl-ups as possible (maximum of 75).

Equipment

Gym mats

30 inch (76.2 cm) × 3 inch (7.6 cm) cardboard strips (for kindergarten to grade 4)

30 inch (76.2 cm) × 4.5 inch (11.4 cm) cardboard strips (for grades 5 and up)

Player and audio recording (for controlling cadence)

Instructions

Students perform this test in groups of three. One student performs the curl-ups, the second supports the head of the performer, and the third secures the strip so that it does not move. Each student being tested lies in a supine position on the mat with knees bent at an approximate 140° angle, legs apart, and arms straight and parallel to the trunk with the palms resting on the mat. The fingers are extended and the head is in the partner's hand resting on the mat. The cardboard strip is placed under the knees with the fingers touching the nearer edge. The third student stands on the strip so that it does not move during the test. The student being tested curls up so that the fingers slide to the other side of the strip (figure 10.28). That student performs as many curl-ups as possible while maintaining a cadence of 1 curl-up every 3 seconds, stopping at a maximum number of 75.

Figure 10.28 FitnessGram curl-up test.
© Human Kinetics

Scoring

Record the number of curl-ups to a maximum of 75.

TRUNK LIFT

Purpose
To measure trunk extensor strength and flexibility.

Objective
To lift the upper body from the ground using the muscles of the back and hold the position to allow an accurate measurement from chin to the ground.

Equipment
Gym mats

Yardstick or ruler

Instructions
The student being tested lies face down on a mat with toes pointed and hands placed under the thighs. The student lifts the upper body in a slow and controlled pace to a maximum of 12 inches (30.5 cm). The student holds the position until a ruler is placed in front of him or her and the measurement is taken from the floor to the chin (figure 10.29).

Figure 10.29 FitnessGram trunk lift test.
© Human Kinetics

Scoring
Perform two trials and record the highest score to the nearest inch (centimeter), up to a maximum of 12 inches (30.5 cm).

90° PUSH-UP

Purpose
To measure upper-body strength and endurance.

Objective
To complete as many 90° push-ups as possible.

Equipment
Metronome (for controlling cadence)

Instructions
Students work in pairs, with one counting while the other is tested. The student being tested lies face down on the ground with the hands under the shoulders; legs straight, parallel, and slightly apart; and the toes tucked under the feet. At a start command, the student pushes up with the arms until they are straight (figure 10.30). The legs and back should be kept straight throughout the test. The student lowers the body using the arms until the arms bend to a 90° angle and the upper arms are parallel to the floor. This action is repeated as many times as possible, following the cadence of one repetition every 3 seconds. The test is continued until the student cannot maintain the pace or demonstrates poor form.

Figure 10.30 FitnessGram push-up test.
© Human Kinetics

Scoring
Record the number of push-ups performed.

VERTICAL JUMP

Purpose

To estimate bone health by measuring lower-body muscular power.

Objective

To jump vertically as high as one can.

Equipment

Measuring tape or marked wall

Chalk for marking wall or vertical testing devices (e.g., Vertec or jump mat)

Recording sheets

Pen

Instructions

With both feet flat on the floor, the student stands with one side of the body by the wall or measuring device, takes a breath (inhalation is recommended), and reaches as high as possible with one arm. This is recorded to the nearest ½ inch as the standing reach. Standing with feet hip-width apart, the student performs the downward phase of a squat by bending the knees and swinging the arms back, then performs a vertical jump using an explosive movement off of both feet while reaching as high as possible with one arm. At the highest point of the jump, the student touches the wall or pushes the measuring device with the fingers of the reaching hand (see figure 10.31). The vertical jump height (result) is the difference between standing reach and highest point of jump to the nearest ½ inch. Each student will be allowed one practice jump followed by two test jumps, ultimately taking the highest outcome of the test jumps.

Note: The student's body weight is necessary to obtain health-related standards for this assessment.

Figure 10.31 Vertical jump: *(a)* standing reach, *(b)* squat, and *(c)* jump height.
© Human Kinetics

Scoring

Use the highest vertical jump score, which is the difference between standing reach and the highest point of jump to the nearest ½ inch.

PLANK

Purpose
To measure core strength and endurance.

Objective
To hold the plank position as long as possible (up to 3 minutes).

Equipment
Flat and clean surface

Stopwatch

Recording sheets

Pen

Instructions
Lying face down on the floor, the student raises the body with only forearms, hands (palms flat or fisted are acceptable), and toes touching the floor. The feet should be no wider than shoulder-width apart, the forearms in line with the shoulders, and both elbows at approximately 90°. Ensure that the student is breathing normally and the body is positioned in a straight line from the neck through the hips and to the heels; the line should not deviate up or down more than approximately 2 inches (see figure 10.32). Once the correct position is assumed, start the stopwatch and encourage the student to maintain the position as long as possible. Correction cues can be given during the test when form begins to falter. Terminate the test when consecutive cues are given that did not result in an adequate correction in form, or the student fatigues or voluntarily stops the test, or the student holds the plank position for 3 minutes.

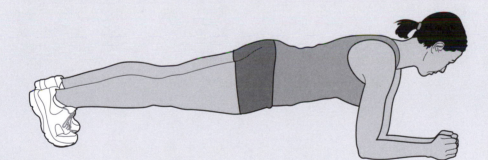

Figure 10.32 Test position for plank.

Scoring
Record the plank holding time in seconds (up to a maximum of 3 minutes).

HANDGRIP STRENGTH

Purpose
To measure isometric strength.

Objective
To squeeze the handgrip device as hard as possible.

Equipment
Handgrip device such as dynamometer

Instructions
Let the student select his or her dominant hand. Adjust the handle of the handgrip device until the student's index finger is bent at a 90° angle. The student stands with the feet hip-width apart and the handgrip device pointing toward the floor. The handgrip device should be in line with the student's body, but not touching the thigh or the lower half of torso (see figure 10.33). Instruct the student to squeeze the handgrip device as hard as possible for a duration of 2 seconds while keeping the arm at the side. Read the result from the handgrip device. The student should be given a practice trial to get used to the grip and then execute two test trials. The result (rounded to the nearest kilogram) is the highest of the two test trials for the dominant hand.

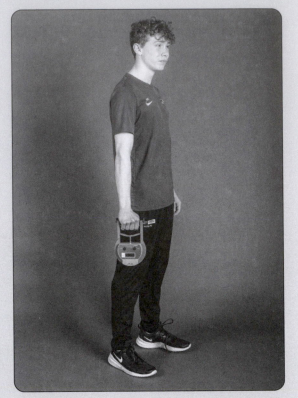

Figure 10.33   **Side view of handgrip test.**
Daniel Erman

Scoring
The highest score from the trials (rounded to the nearest kilogram) for each hand.

Enhancing Reliable and Valid Fitness Test Results With Children

The tests presented in the fitness batteries in table 10.25 have different levels of evidence to support their validity. Distance runs and skinfolds have demonstrated criterion-related validity in research (Lohman, 1989; Safrit et al., 1988). The tests of muscular fitness, pull-ups, push-ups, modified pull-ups, flexed arm hang, sit-ups, curl-ups, and the sit-and-reach have limited criterion-related validity support but are generally accepted on the basis of content or logical validity. Research has questioned the assumption of content validity for the sit-and-reach, pull-up, and sit-up tests (Engelman and Morrow, 1991; Hall, Hetzler, Perrin, and Weltman, 1992; Jackson and Baker, 1986). However, you can assume sufficient validity of your tests provided that you are using one of the fitness batteries presented in this chapter.

The practical issue is for you to take the appropriate steps during fitness test administration that will produce reliable test results. This is an important issue in youth fitness testing because research has shown that field tests of fitness in children can demonstrate poor reliability, especially in younger children (Rikli, Petray, and Baumgartner, 1992). To enhance reliability ($r_{xx'} > .80$), you can take a variety of practical steps to minimize measurement error, including the following:

1. Attain adequate knowledge of test descriptions.
2. Give proper demonstrations and instructions.
3. Develop effective student and teacher preparation through adequate practice trials.
4. Conduct reliability studies.

Of these steps, providing adequate practice trials is the most important. Children, especially younger ones, need several trials of a fitness test to learn how to take the test and to provide consistent results. For example, children need to practice the PACER several times to learn the appropriate pace and to provide a consistent time that is a reliable and valid representation of their cardiorespiratory endurance.

FITNESS TEST BATTERIES FOR SPECIAL POPULATIONS

Special populations include people with physical or mental disabilities or both. Lack of knowledge and understanding has clouded the history of society's care and attention to people with disabilities. Laws and changes in attitudes have provided an increasingly enlightened approach to interactions with people with disabilities; however, reliable and valid fitness assessment of adults with disabilities is not a well-researched or well-understood topic compared to that of other populations. In his book *Fitness in Special Populations*, Shephard (1990) provided a comprehensive and detailed presentation of exercise and fitness issues for people with disabilities. He stated that the assessment of fitness should include the following:

• Anaerobic capacity and power
• Aerobic capacity
• Electrocardiographic response to exercise
• Muscular fitness, including strength, endurance, and flexibility
• Body composition

Appropriate fitness assessment requires accurate classification of the person's disability and proper test and protocol selection (e.g., to test the aerobic capacity of a person with paraplegia, you might select a wheelchair ergometer or arm ergometer). Shephard (1990) provided fitness information for people with paraplegia and amputations; those who are blind or deaf; and those with intellectual disabilities or developmental disorders or who have autism, cerebral palsy, muscular dystrophy, or multiple sclerosis. Specific training and education are required for fitness measurement in these and other special populations.

Meeting the fitness needs of persons with disabilities is a challenge for the future and an area of professional opportunity. From a measurement and evaluation perspective, research on reliable and valid fitness tests in populations with specific disabilities is needed. Research is also needed to provide a better understanding of fitness levels and the necessary level of fitness for improved health in groups with disabilities.

Children With Disabilities

One of the biggest measurement challenges that you may confront as a professional in human performance is the assessment of physical fitness in children with disabilities. Keep in mind that the fitness test batteries discussed in this chapter would exclude or be biased against many children with **physical disabilities** (i.e., those having physical or organic limitations, such as cerebral palsy) as well as children with **mental disabilities** (i.e., those with mental or psychological limitations, such as autism) because of their specific disabilities. Before you can administer or evaluate fitness test results, you must consider the participants' physical and organic limitations; neural and emotional capacity; interfering reflexes; and acquisition of prerequisite functions, responses, and abilities (Seaman and DePauw, 1989). You can develop the basic knowledge and competence that you need for assessing the fitness and activity of children with disabilities from your instruction and learning experiences in **adapted physical education**—physical education adjusted to accommodate children with physical or mental limitations. The physical fitness tests selected should be appropriate for an individual student based on his or her disability as well as on the fitness capacity to be measured. Seaman and DePauw (1989) and Winnick and Short (1999; 2014) are excellent sources for detailed information on fitness assessment of children with special needs.

The Brockport Physical Fitness Test (Winnick and Short 1999; 2014), a health-related physical fitness test for youths aged 10 through 17 with various disabilities, was developed through a research study, Project Target, funded by the U.S. Department of Education. The test battery includes criterion-referenced standards for 25 tests. The test manual helps professionals consider each student's disability and select the most appropriate test and test protocol. Table 10.28 provides potential items for fitness assessment, the appropriate population with disabilities for the test, and reliability and validity comments for each test. The complete test kit includes a manual, a demonstration video, a fitness training guide, and Fitness Challenge software to help professionals administer the test and develop a database.

MEASUREMENT AND EVALUATION CHALLENGE

Jamal has done considerable thinking and investigative work since meeting with the executive director. He has familiarized himself with health-related fitness testing and the various options for testing available to him—specifically the YMCA Physical Fitness Test Battery, FitnessGram, and the Senior Fitness Test. Additionally, he realizes that the YMCA population may include individuals with special needs. Although Jamal's recommended fitness assessment concept will be health related, he knows that the specific test items he recommends will depend on the individual that is to be assessed.

Table 10.28 Brockport Physical Fitness Test Items

Test item	Intellectual disability	Blind with assistance	Cerebral palsy	Spinal cord injury	Congenital anomalies/ amputation	Reliability	Validity
PACER	R	R			O	Acceptable	Content concurrent
1-mile (1.6 km) run or walk		O			R	Acceptable	Concurrent
Target aerobic movement test	R		R	R	R	Acceptable	Content
Skinfolds	R	R	R	R	R	Acceptable	Concurrent
BMI	O	O	O			Acceptable	Concurrent
Reverse curl				R		Not reported	Content
Seated push-up			R	R	R	Not reported	Content
40-meter (43.7 yd) push or walk			R			Not reported	Content
Wheelchair ramp test			R/O			Not reported	Content
Bench press	O			O	R	Acceptable	Content
Dumbbell press			R/O	O	R/O	Acceptable	Content Concurrent
Extended arm hang	R					Acceptable	Content
Dominant grip strength	O		O	R	O	Acceptable	Construct
Isometric push-up	O					Acceptable	Content
Push-up		O				Acceptable	Content
Pull-up		O				Acceptable	
Modified pull-up		O				Acceptable	Content
Curl-up		R			R	Acceptable	Content
Modified curl-up		R				Acceptable	Content
Trunk lift	R	R			R	Acceptable	Content
Modified Appley test			R	R	R	Not reported	Content
Shoulder stretch	O	O				Acceptable	Content
Modified Thomas stretch			R	R		Not reported	Content
Back saver sit-and-reach test	R	R			R	Acceptable	Content
Target stretch test			R/O	R	R	Acceptable	Content Concurrent

Note: O = Option; R = Recommended.

SUMMARY

The material presented in this chapter should provide you with a sound basis for understanding the various factors involved in assessing health-related fitness reliably and validly. Mastering this material does not make you qualified to administer fitness tests in all populations or situations, but it is an important step toward achieving those qualifications. Health-related fitness testing, as with any other testing situation, requires appropriate test selection, preparation, practice trials, and attention to detail for sound measurement and evaluation of performance.

www! **Go to HK*Propel* for videos, homework assignments, and quizzes that will help you master this chapter's content.**

Assessment of Performance-Related Fitness

OBJECTIVES

After studying this chapter, you will be able to

- understand the differences between health-related and performance-related physical fitness;
- identify and define the components of performance-related fitness;
- evaluate and use available tests, including the Functional Movement Screen (FMS), to measure agility, balance, coordination, power, reaction time, speed, and other components; and
- understand critical issues related to administering performance-related fitness tests.

www **The lecture outline in HKPropel will help you identify the major concepts of the chapter.**

MEASUREMENT AND EVALUATION CHALLENGE

John, an athletic trainer, has just obtained an assistant coaching job for a local high school football team. The head coach was very excited to learn that John has had some training in assessing physical fitness and has given him the task of selecting an agility test for the team. John did an online search and found many agility tests that have been used for high school student athletes. Because John is not sure which agility test he should select, he decides to examine several on the basis of validity, reliability, practicability, and whether appropriate norms were developed to help him make a decision about which test to recommend.

Physical fitness is essential to any physically demanding performance and therefore has long been a focus in sports or in physically demanding jobs. As described in chapter 10, the concepts of health-related fitness and functional fitness were created in the 1970s to address the increase of chronic diseases, and the existing motor- or sport-centered fitness concept was recast as performance- or skill-related fitness. Performance-related fitness, then, is defined as the components of fitness that have a relationship with enhanced performance in sports and motor skills.

As described in chapter 10, the key components of health-related fitness are aerobic capacity, body composition, muscular strength and endurance, flexibility, and bone density. The components of performance-related fitness include such characteristics as agility, balance, coordination, power, reaction time, speed, and others (Corbin, Pangrazi, and Franks, 2000). It should be pointed out that health- and performance-related fitness, as well as functional fitness, are closely related and often interconnected. For example, people who possess performance-related fitness will be likely to engage in regular physical activities and therefore already have enhanced health-related fitness and a lower risk of hypokinetic diseases and conditions. Similarly, the performance of most sport or motor tasks involves most or all components of health-related fitness. Another good example is the fitness role of power, which, for many years, was considered a key component only for sport performance. Research studies (Reid and Fielding, 2012; Warburton et al., 2001), however, found that muscular power appears to decline with age more quickly than muscular strength, and loss of muscular power is strongly associated with decreases in functional ability and is a predictor for decreased mobility and premature mortality in older adults.

Meanwhile, some aspects of fitness may be more strongly related to a specific sport or type of activity than to health-related fitness. One example of this type of fitness is agility. An optimal amount of agility may be needed for an athlete to move from one point to another effectively, shift weight properly, and use speed optimally. Agility is highly task specific; thus, there is no valid measure of overall agility. Furthermore, agility does not appear to be highly correlated with health-related physical fitness. Nonetheless, a task-specific measure of agility might be important as a measure of performance-related physical fitness.

In this chapter, we will describe the assessment of several components of performance-related fitness, including both laboratory and field assessment methods. Many tests have been developed to assess each performance-related fitness component, but for this introduction we will only describe two or three tests for each component. Additional examples of tests can be found in *Kirby's Guide to Fitness and Motor Performance Tests* (1991) and the Topendsports website (www.topendsports.com/testing). At the end of this chapter, we will identify some critical issues related to administering and using performance-related fitness tests.

MEASURING AGILITY

Agility is the ability to rapidly change the position of the entire body in space with speed and accuracy. Agility is important in many sports activities, as exemplified by a match between two experienced badminton players or by the trampolinist executing a triple back somersault. In addition, agility is fundamental to skill in certain sports activities, such as volleyball (Mohr and Haverstick, 1956) and basketball (Hoskins, 1934). With the proper use of agility testing, coaches can determine which athletes on a team are most agile and which ones need to improve agility in order to improve overall performance.

Agility tests can be used by physical educators and coaches in several ways:

- To predict potential in different sports and activities
- To determine achievement, progress, and grades when agility is a specific objective in the teaching unit
- To evaluate the effectiveness of a specific unit of instruction to improve agility

Several specific features related to agility assessment should be noted:

• The surface area and the type of footwear could have an impact on the participant's scoring ability in certain tests, such as the side-step test. The use of a nonslip surface and the requirement that all students wear the same type of shoes should be considered.

• Considerable time is needed to administer certain agility tests to large groups. Two or more test stations are advised.

• Agility is often performance specific, and performance on one test (e.g., running) may not predict performance in other sport situations (e.g., rapid changes of body position).

• Although attempts have been made to develop a laboratory agility test, such as the light-based reactive agility test (Chelladurai et al., 1977), none has been considered the criterion measure of agility.

The following are a few agility test examples. Although some were developed long ago, they are still often used in practice.

SEMO AGILITY TEST

Purpose
To measure general agility when maneuvering forward, backward, and sideward.

Equipment

The free throw lane of a basketball court or any smooth area 12 feet (3.67 m) × 19 feet (5.79 m) with adequate running space around it

4 plastic cones (12 in. [30.38 cm] height)

Stopwatch

Instructions
The cones are placed squarely in each corner of the free throw lane, as shown in figure 11.1. The participant lines up outside the free throw lane (at A) with his or her back to the free throw line. At the go signal, the participant performs the following pattern:

(continued)

SEMO AGILITY TEST (continued)

1. Side step from A to B, passing outside of the corner cone

2. Backpedal from B to D, passing to the inside of the corner cone

3. Sprint forward from D to A, passing outside of the corner cone

4. Backpedal from A to C, passing to the inside of the corner cone

5. Sprint forward from C to B, passing outside of the corner cone

6. Side step from B to the finish line at A

A diagram of this pattern is shown in figure 11.1. Two trials are allowed. The following rules should be followed when administering the test:

1. When performing the side step, the feet should not cross over.

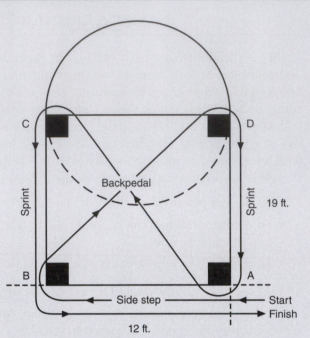

Figure 11.1 Diagram for the SEMO agility test.

2. When performing the backpedal, the participant must keep his or her back perpendicular to an imaginary line connecting the corner cones.

3. Incorrect procedure makes the trial invalid; the participant should be tested until one correct trial is completed.

4. At least one practice trial should be given.

Scoring

The better of two trials (recorded to the nearest tenth of a second) is recorded as the score. Norms are shown in table 11.1.

Table 11.1 Norms for SEMO Agility Test (College Students)

	SCORE (IN SECONDS)	
Performance level	Men	Women
Advanced	10.72 and below	12.19 and below
Advanced intermediate	10.73-11.49	12.20-12.99
Intermediate	11.50-13.02	13.00-13.90
Advanced beginner	13.03-13.79	13.91-14.49
Beginner	13.80 and above	14.50 and above

For more information on this test, see Kirby (1971).

ILLINOIS AGILITY RUN TEST

Purpose

To measure the ability to change positions of the whole body and then run a weaving pattern.

Equipment

8 cones or chairs

Floor tape

Stopwatch

Instructions

The course is marked by cones, with four center cones spaced 3.3 meters (10.8 ft) apart and four corner cones positioned 2.5 meters (8.2 feet) from the center cones (see figure 11.2). The participant assumes a prone position behind the starting line. At the go signal, the participant gets to his or her feet and runs forward to the first tape mark X. The participant must touch or cross the tape mark with his or her foot, turn and run back to the first center cone, then weave up and back through the four center cones. The participant then runs to the second tape mark X on the far line (again, touching or crossing the tape mark with the foot), and finally, runs across the finish line.

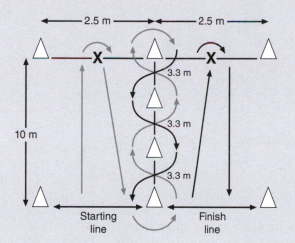

Figure 11.2 Diagram for the Illinois Agility Run Test.

Scoring

The time to complete each trial is recorded in seconds. Disqualification occurs if the participant fails to run the course as instructed, fails to complete the course, or moves any cones.

For more information on this test, see Cureton (1951).

SHUTTLE RUN TEST

Purpose

To measure speed and change of direction.

Equipment

An area equivalent to the width of a volleyball court

2 blocks of wood (2 in. × 2 in. × 4 in.)

Floor tape

Stopwatch

Instructions

Using tape, place two parallel lines on the floor 30 feet (9.1 m) apart. Place two wooden blocks behind one of the lines, as shown in figure 11.3. The student starts from behind the other line. On the go signal, the participant runs to the blocks, picks one up, runs back to the starting line, and places the block on the floor beyond the line. The participant runs back, picks up the other block, and runs across the finish line as fast as possible. Two trials are administered, with a rest in between.

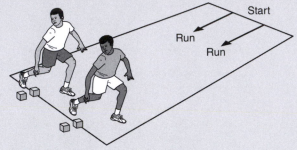

Figure 11.3 Shuttle run test.

Scoring

The time of the better of two trials, recorded to the nearest tenth of a second, is the score. Norms have been published for boys and girls, ages 6 through 17+ (Reiff et al., 1986).

For more information on this test, see Reiff et al. (1986).

MEASURING BALANCE

Balance is the ability to maintain equilibrium. Before a great deal of research was conducted on the nature of balance, it was deduced that two types of balance exist—static balance, the ability to maintain equilibrium in a stationary position, and dynamic balance, the ability to maintain equilibrium while in motion.

A coach, physical educator, exercise scientist, or clinician might be interested in measuring balance for a variety of reasons. For example, achievement of minimal balance standards might be a goal in an instructional unit, such as posture and body mechanics or gymnastics. Balance tests might also be used in assessing the movement patterns of young children and in evaluating the status of children in special education. The measurement of balance is useful in assessing potential in gymnastics, diving, and other forms of physical activity that require balance skills. This topic is also of great interest among older adults, for whom maintaining equilibrium is essential to prevent injury occurring as a result of falls. Tests are described for each type of balance.

Laboratory Methods

There are several populations for whom maintaining balance is a significant challenge (e.g., older adults, persons with cerebral palsy, and those with neural deficits leading to balance disorders). Recent advances in computer technology have led to the development of sophisticated rehabilitation tools for accurately monitoring, assessing, and training persons with balance deficits. One such device is the NeuroCom Balance Master, a dual force plate platform surrounded by an overhead safety bar and straps (https://balanceandmobilitytherapy.com/technology-BalanceMaster.html). This apparatus provides continuous visual feedback to participants regarding such attributes as center of gravity and postural alignment during static and dynamic balance testing. In addition, participant progress can be monitored during rehabilitative balance exercises under various conditions, such as with eyes open or eyes closed, and the sophisticated computer interface permits storage and analysis of patient demographic and test data and control over therapeutic training and assessment.

Field Methods

The following are some examples of field methods used to measure balance. The Bass stick test measures static balance and the Johnson modification measures dynamic balance.

BASS STICK TEST

Purpose

To measure the ability to balance in a stationary, upright position using a small base of support.

Equipment

Stick (1 in. × 1 in. × 12 in. [2.54 cm × 2.54 cm × 30.48 cm])

Stopwatch

Floor tape

Instructions

The test should be conducted on a smooth, flat floor. To begin, the participant places the dominant foot lengthwise on a special stick taped to the floor, as shown in figure 11.4. Both the ball of the foot and the heel should rest on the stick. The opposite foot is lifted from the floor and held in this position as long as possible. The stopwatch is started when the opposite foot leaves the floor and stopped when it touches the floor again, or when any part of the supporting foot touches the floor. Three trials are taken on each foot. If a participant loses balance during the first 3 seconds of the trial, the trial may be restarted.

Another version of this test is to put the foot crosswise on the stick.

Figure 11.4 Illustration of the Bass stick test.

© Human Kinetics

Scoring

The test score is the time for the best of three trials of each foot.

For more information on this test, see Bass (1939).

JOHNSON MODIFICATION OF THE BASS TEST

Purpose

To measure dynamic balance.

Equipment

Smooth floor area (50 in. × 180 in. [127 × 457 cm])

Stopwatch

Tape measure

Floor tape

Instructions

Using the floor plan shown in figure 11.5, the participant begins with the right foot on the starting mark and leaps to the first tape mark, landing on the left foot and balancing on the ball of the foot as long as possible, up to 5 seconds. The participant then leaps to the next square, lands on the right foot, and again tries to maintain balance on the ball of the foot for 5 seconds. This procedure is continued until the end of the floor pattern (see figure 11.5).

Scoring

For each successful landing on the tape mark, 5 points are scored; 1 additional point is scored for each second the participant remains balanced on the mark, up to 5 seconds. It is possible to score 10 points at each tape mark and 100 points on the total test. The participant is penalized 5 points for any of the following

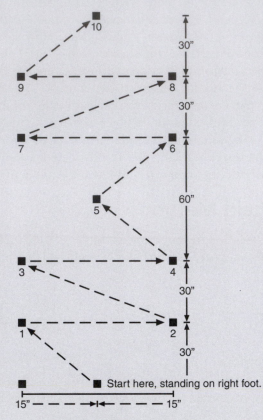

Figure 11.5 Floor pattern for Johnson modification of the Bass Test (1 in. = 2.54 cm).

landing errors: not stopping after landing on the tape mark, touching the floor with any body part other than the supporting foot, and not covering the tape mark with the ball of the foot. If the participant makes a landing error, the correct position can be resumed for the 5-second balance.

An additional 1-point penalty is given for the following errors during the 5-second balance: touching the floor with any body part other than the supporting foot and not holding the foot steady in the balance position. If balance is lost, the participant should return to the proper mark before leaping to the next mark.

For more information on this test, see Johnson and Leach (1968).

MEASURING COORDINATION

Coordination, known also as *motor coordination*, is the ability to use different parts of the body smoothly and efficiently in a sequence that preserves posture. Determination of the muscles to be used is based on the task to be done and the environment in which the task takes place (Cech and Martin, 2012).

Coordination is a complex ability that requires good levels of other fitness components such as balance, strength, and agility; an athlete who appears to be well coordinated may also be displaying these fitness components. It is often observed that people have better coordination on one side of the body than the other (e.g., few people can throw and catch equally well with both hands). Coordination is also a difficult ability to teach or coach; rather, it is something that is achieved through proper development throughout early life. As such, coordination tests are sometimes used in a test battery for monitoring a young person's physical development or lack thereof.

SODA POP TEST

Purpose

To measure manual dexterity and hand–eye coordination.

Equipment

Stopwatch

3 full 12 ounce (355 ml) soda pop cans

Cardboard template 32 inches (81.28 cm) high × 5 inches (12.7 cm) wide, with 6 circles 3.25 inches (8.26 cm) in diameter drawn in a straight line 1.5 inches (3.81 cm) apart (or a table with six taped paper markings as specified)

Instructions

The soda pop cans are placed in every other circle starting from the side of the hand being tested (see figure 11.6). The participant, seated at the table, begins the test by grasping the first can with the hand being tested, thumb pointing up and elbow joint bent between 100° and 120°. On the go signal, the stopwatch is started and the participant begins to turn cans of soda pop upside down in the following order:

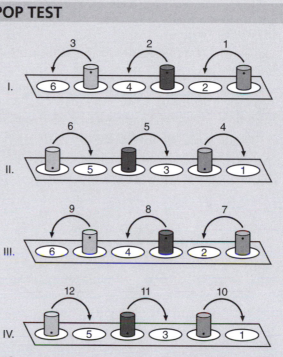

Figure 11.6 Illustration of the order of the soda pop test.

1. Place can 1 inside circle 2, followed by can 2 inside circle 4, and then can 3 inside circle 6.

2. Immediately return all three cans right side up to their original placements, starting with can 1, then can 2, then can 3. On this "return trip," grasp the cans with the hand in a thumb-down position.

3. The entire procedure is performed twice for one trial.

A video illustration of this test can be found at the following YouTube link: www.youtube.com/watch?v=VQQRbwGCdvA

Scoring

The time of each trial is recorded to the nearest tenth of a second. The participant is allowed two practice trials, then two actual trials. The best score is recorded.

For more information on this test, see Osness et al. (1990).

ALTERNATE-HAND WALL TOSS TEST

Purpose

To measure hand–eye coordination.

Equipment

- Tennis ball or baseball
- Wall tape
- Stopwatch

Instructions

Explain the test procedures to the participant and run an appropriate warm-up and practice. Put a mark a certain distance (e.g., 2 meters or 3 feet) from a smooth, solid wall. Standing behind the line facing the wall, the participant throws the ball from one hand in an underarm action against the wall and attempts to catch it with the opposite hand. The ball is then thrown back against the wall and caught with the initial hand (figure 11.7). The test can continue for a designated number of attempts or for a set time period (e.g., 30 seconds). By adding the constraint of a set time period, you also add the factor of working under pressure.

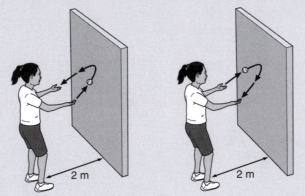

Figure 11.7 Alternate-hand wall toss test.

There are numerous variations that can be made to the procedures of this test depending on the desired outcomes: The size, weight, and shape of the object; the distance from the wall; the number of attempts; or the time period can all be varied.

Scoring

Table 11.2 lists general ratings for the alternate-hand wall toss test, based on the number of successful catches from a 2-meter distance in a 30-second period. These ratings may vary with differing variations listed above.

Table 11.2 Ratings for Alternate-Hand Wall Toss Test

Rating	Score
Excellent	>35
Good	30-35
Average	20-29
Fair	15-19
Poor	<15

Distance = 2 meters, time = 30 seconds

For more information on this test, see Harrison and Bradbeer (1982).

HARRE CIRCUIT TEST

Purpose

To assess participants' coordination abilities, perception of themselves in space, and dynamic total body coordination.

Equipment

10 meter × 10 meter (32 ft × 32 ft) space with smooth floor surface

1 mat

1 plastic cone

3 traffic cones or similar obstacles

Stopwatch

Instructions

Participants are instructed to complete the circuit shown in figure 11.8 as quickly as possible. The test initially requires the execution of forward rolls (only once after the start) and then three consecutive passages over and underneath three obstacles. The person administering the test measures the time and monitors the participant for errors (e.g., touching the obstacles or the cone placed in the middle of the circuit).

A video illustration of this test can be found at the following YouTube link: www.youtube.com/watch?v=yT44R4LCOY0

Scoring

The time employed to run the whole circuit is recorded to the nearest tenth of a second.

For more information on this test, see Harre (1982).

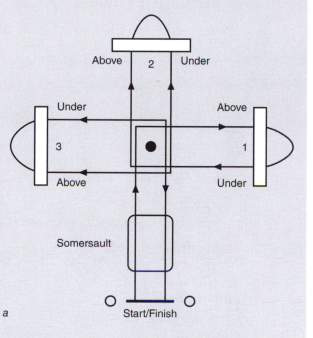

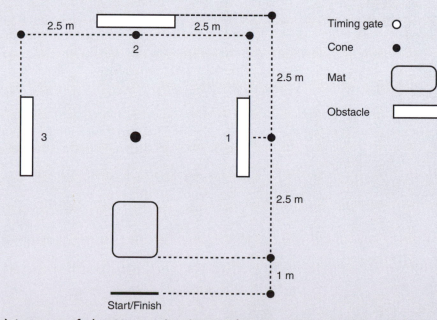

Figure 11.8 (a) Layout of the Harre circuit test (arrows and numbers indicate the test's route), and (b) schematic of distances among components and parts of the test.

MEASURING POWER

Power is the ability to generate maximum force in a minimum amount of time. From a mechanical standpoint, *power* is physical work divided by time, where work is defined as weight times the distance the weight is moved. Using this definition, Barlow (1970) and Considine (1970) examined the subdomain of *muscular power*, known also as *anaerobic power*, finding it to be unique to certain body regions, specifically the arms and legs. As a result, tests have been developed to specifically assess the power of arms (e.g., one-hand shot put, two-hand overhead shot put, medicine ball pitch, basketball throw) and legs (e.g., Margaria–Kalamen power test). A few commonly used tests are described below.

Laboratory Methods

Power can be accurately measured in the laboratory or clinical settings using cycle ergometers or similar equipment. A good example of this is the Wingate Anaerobic Test described next.

WINGATE ANAEROBIC TEST

The Wingate test (Bar-Or, 1987; Vandewalle et al., 1987), known also as the Wingate Anaerobic Test (WANT), is a cycle test of anaerobic leg power, conducted over 30 seconds. The test was developed at the Wingate Institute in Israel during the 1970s.

Purpose

To measure the anaerobic power of the lower body.

Equipment

Fleisch or a modified Monark cycle ergometer

Instructions

The participant should first perform a cycling warm-up of several minutes. The participant is instructed to pedal as fast as possible for 30 seconds. In the first few seconds, the resistance load is adjusted to the predetermined level, which is usually about 45 g/kg body weight (Fleisch) or 75 g/kg body weight (Monark) for adults. (The National Hockey League predraft testing, discussed in chapter 12, uses 90 g/kg.) Power athletes would generally use high resistances, whereas children and older adults may use lower resistances (figure 11.9).

A video illustration of this test can be found at the following YouTube link: www.youtube.com/watch?v=Hi2ZahzeLNY

Scoring

Some of the measures that can be gained from this test are mean and peak power, expressed in Watts (W) (ideally measured in the first 5-second interval of the test), relative peak power (determined by dividing peak power by body mass, expressed as W/kg), mean peak power, minimum peak power, and a fatigue index determined from the decline in power. The following formulas are used for computation:

Power output (kpm · min^{-1}) = [revs × resistance (kg) × distance (m) × 60] / time (sec),

Watts = kpm · min^{-1} / 6.123

W/kg = Watts / body weight (kg)

Fatigue index = [(peak power output − min power output) / peak power output] × 100

Figure 11.9   Wingate test.
© Human Kinetics

Field Methods

Simple, but valid and reliable, tests have also been developed to measure power, including the Margaria–Kalamen power test and standing long jump and vertical jump tests. Since the vertical jump test, which can be used for both health- and performance-related fitness, was introduced in chapter 10, only the Margaria–Kalamen power test and standing long jump test will be discussed here. Note, however, that when the vertical jump is used to measure performance-related fitness, the scores are often compared with a table of norms (e.g., see Hoffman, 2006).

MARGARIA–KALAMEN POWER TEST

Purpose
To test the power of the lower extremities.

Equipment
- Stopwatch
- Timing mats (optional)
- Tape measure
- Flight of 12 standard steps

Instructions
A starting line is marked 6 meters (19.6 ft) in front of the first step; the third, sixth, and ninth steps are also clearly marked. After a few practice runs to warm up, the participant stands ready at the starting line. On the go signal, the athlete sprints to and up the flight of steps as fast as possible, taking three steps at a time (stepping on the third, sixth, and ninth steps). The time to get from the third step to the ninth step is recorded (either using a stopwatch or using timing mats placed on those steps), starting when the foot first contacts the third step and stopping when the foot contacts the ninth step (see figure 11.10). Allow three trials of the test; use the best time for scoring.

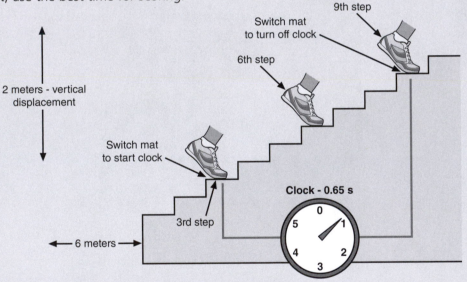

Figure 11.10 Illustration of Margaria–Kalamen test.

Scoring
Power is calculated using the following formula:

$$P = (M \times D) \times 9.8 / t$$

where P is power (Watts), M is mass (participant's weight in kilograms), D is the vertical distance between steps 3 and 9 (meters), t is time (seconds), and 9.8 is the constant of gravity.

For more information on this test, see Margaria et al. (1966) and Kalamen (1968).

www **Go to HK*Propel* to complete Student Activity 11.1.**

STANDING LONG JUMP

Purpose

To measure explosive leg power.

Equipment

Mat

Tape measure

Instructions

The participant stands with feet several inches apart and the toes behind the takeoff line. Prior to jumping, the participant swings the arms backward and bends the knees. The jump is accomplished by simultaneously extending the knees and swinging the arms forward. Measure from the takeoff line to the heel or any other part of the body that touches the floor nearest the takeoff line (see figure 11.11). Allow three trials.

Figure 11.11 Standing long jump.

AAHPERD Youth Fitness Test Manual, Revised by Paul Alfred Hunsicker and Guy G. Feiff. © 1976. American Alliance for Health, Physical Education, Recreation and Dance. Used with permission.

Scoring

Record the best of the three trials in feet and inches to the nearest inch. Performance standards from various tests are summarized in table 11.3 and the national norms have been published for boys and girls, ages 6 through 17+ (Reiff et al., 1986).

(continued)

STANDING LONG JUMP (continued)

Table 11.3 Standing Long Jump Performance Standards

	AGE IN YEARS (VALUES IN INCHES)											
	<6	7	8	9	10	11	12	13	14	15	16	17+
Boys												
AAHPERD 75th				64	64	67	71	75	80	86	90	93
PCFSN 75th	49	53	57	61	66	69	72	78	84	88	91	94
Chrysler Fund–AAU Fitness Test 80th	50	54	58	63	66	69	73	79	87	89	92	96
AAHPERD 25th				54	54	56	60	62	66	73	78	78
PCFSN 25th	39	42	47	50	53	57	60	64	69	73	78	82
Chrysler Fund–AAU Fitness Test 45th	43	48	51	55	59	61	64	70	76	79	83	87
Recommendation	42	46	50	53	56	60	63	68	73	77	81	85
Girls												
AAHPERD 75th				62	62	64	66	69	71	70	69	72
PCFSN 75th	45	48	52	55	60	64	67	69	70	71	71	72
Chrysler Fund–AAU Fitness Test 80th	48	51	54	59	61	64	69	71	72	73	73	74
AAHPERD 25th				49	49	52	54	57	58	59	57	59
PCFSN 25th	36	38	42	44	49	52	55	56	57	57	58	58
Chrysler Fund–AAU Fitness Test 45th	40	44	47	50	53	57	61	63	65	64	65	66
Recommendation	39	42	45	48	52	55	59	60	62	61	62	63

Note: For metric measurements, refer to www.worldwidemetric.com/measurements.html

For more information about this test, see AAHPERD Youth Fitness Test (1976).

MEASURING REACTION TIME

Quick reaction is essential for many sports (e.g., tennis, table tennis, basketball, boxing, and fencing) and for job performance (e.g., pilots and truck drivers). A person's quick action ability is often assessed by reaction time, which is the interval between the presentation of a stimulus and the initiation of the response. Although reaction time was initially thought to be a rather simple and easily measured phenomenon, it has been shown to be influenced by a number of variables, including the sense organ involved, the intensity of the stimulus, the preparatory set, general muscular tension, motivation, practice, the response required, fatigue, and one's general state of health. Nevertheless, some simple and easy reaction time tests were developed for physical education and sport. Very recently, apps for finger-touch reaction time tests have also been developed (e.g., Reaction Test Pro and LightsOut: Reflex Test).

NELSON HAND REACTION TEST

Purpose

To measure the speed of reaction of the hand in response to a visual stimulus.

Equipment

A reaction timer (see Mastery Item 11.1 to learn how to make the reaction timer)

Ruler

Table and chair

Instructions

The participant sits with the forearm and hand resting comfortably on the table. The tips of the thumb and index finger are held in a ready-to-pinch position about 3 or 4 inches beyond the edge of the table. The upper edges of the thumb and index finger should be in a horizontal position. The tester holds the stick-timer near the top, letting it hang between the participant's thumb and index finger. The baseline on the stick should be even with the upper surface of the participant's thumb. The participant is directed to look at the concentration zone (a black shaded zone between the 0.120- and 0.130-second lines on the stick) and to react by catching the stick between the thumb and index finger when it is released (figure 11.12). The participant should not look at the tester's hand nor move the hand up or down while attempting to catch the falling stick. Twenty trials are given. Each drop is preceded by a preparatory command of "ready." When the participant catches the stick, the score is read just above the upper edge of the thumb.

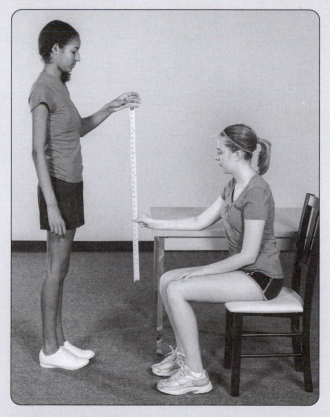

Figure 11.12 Nelson hand reaction test.

© Human Kinetics

(continued)

NELSON HAND REACTION TEST *(continued)*

Scoring

The five slowest and the five fastest trials are discarded and the middle 10 are averaged to get a final score. The recorded mean distance can be converted to time using table 11.4. The average reaction time for college men is approximately 0.21 seconds.

Table 11.4 Converting Distance to Time

Distance	Time
2 in. (~5 cm)	0.10 s (100 ms)
4 in. (~10 cm)	0.14 s (140 ms)
6 in. (~15 cm)	0.17 s (170 ms)
8 in. (~20 cm)	0.20 s (200 ms)
10 in. (~25.5 cm)	0.23 s (230 ms)
12 in. (~30.5 cm)	0.25 s (250 ms)
17 in. (~43 cm)	0.30 s (300 ms)
24 in. (~61 cm)	0.35 s (350 ms)
31 in. (~79 cm)	0.40 s (400 ms)
39 in. (~99 cm)	0.45 s (450 ms)
48 in. (~123 cm)	0.50 s (500 ms)
69 in. (~175 cm)	0.60 s (600 ms)

For more information on this test, see Johnson and Nelson (1986).

Mastery Item 11.1

Visit the following link to make a reaction timer: www.topendsports.com/testing/reaction make.htm

MEASURING SPEED

Speed, essential for successful performance in many motor activities, can be defined as the capacity to perform successive movements of the same pattern at a fast rate. Generally, when speed is discussed, one thinks of leg speed in running activities. But speed, like other performance-related fitness components, concerns many body parts and may vary from one part to another. In general, speed of muscle contraction would seem to be an innate quality, but speed of movement used when running sprints or in a game such as football can certainly be improved through training in proper technique and through continued practice in the coordination of movements.

50-YARD DASH

Purpose

To measure speed.

Equipment

Stopwatch (using two stopwatches or a split-second timer for simultaneous testing of two students is recommended)

Instructions

The 50-yard dash is usually administered outdoors, using any open area. The participant stands behind the starting line. Use the commands "Are you ready?" and "Go!," timing the latter signal with a downward sweep of the arm. The participant runs as fast as possible until he or she crosses the finish line. The stopwatch is started as the starter's arm reaches the downward position and is stopped as the finish line is crossed. Two trials are taken; the best score is used as the final score.

Scoring

Record the best time in seconds to the nearest tenth of a second. Norms have been published for boys and girls, ages 6 through 17+ (Reiff et al., 1986).

For more information about this test, see Reiff et al. (1986).

MEASURING OTHER PERFORMANCE-RELATED FITNESS COMPONENTS

Many other components can also be involved in sports performance, such as kinesthetic perception, or the ability to perceive the body's position in space (Singer, 1968). Although kinesthetic perception is by far the most difficult subdomain of human performance to measure in terms of reliability, validity, and objectivity, it is well accepted as an area that must be considered.

The increase in sports participation in recent decades has brought with it an increase in the risk of sustaining musculoskeletal injuries. It has long been thought that isolated muscle stretching would be an effective intervention to reduce muscle soreness or musculoskeletal injury; however, recent research has suggested that this may not be the case. To reduce injury risk, sports medicine professionals have begun to focus on improving movement patterns as opposed to focusing on rehabilitation of a specific joint. Current research suggests that tests assessing multiple domains of function (e.g., balance, strength, and range of motion) may simultaneously help identify athletes at risk for injury. Recently, Cook et al. (2006; 2014) devised a tool to assess fundamental movement called the Functional Movement Screen (FMS), detailed below.

Functional Movement Screen (FMS)

The FMS consists of seven fundamental movement pattern tests: deep squat, hurdle step, in-line lunge, shoulder mobility, active straight-leg raise, trunk stability push-up, and rotary stability. Each test is scored on a scale of 0 to 3, with the sum creating a composite score ranging from 0 to 21 points. It has been reported that a score of 13 or below indicates greater odds for sustaining an injury than scores of 14 or higher (Kollock et al., 2019). The FMS test kit includes a wooden measuring board (48 in. × 2 in. × 6 in. [122 cm × 5 cm × 15 cm]), hurdle, and measuring stick (figure 11.13). The assessment of these component tests is described below.

Figure 11.13 Functional Movement Screen test kit.

© Human Kinetics

DEEP SQUAT

Purpose

To assess bilateral, symmetrical, and functional mobility of the hips, knees, and ankles; shoulders; and thoracic spine.

Equipment

Measuring stick (dowel)

Measuring board

Instructions

The participant assumes the starting position with the feet aligned in the sagittal plane, approximately shoulder-width apart. The participant holds the measuring stick overhead with the elbows at a 90° angle. Next, the stick is pressed overhead with the shoulders flexed and abducted and the elbows extended (figure 11.14a). The participant then descends slowly into a squat position. The squat position should be assumed with the heels on the floor, head and chest facing forward, and the dowel maximally pressed overhead (figure 11.14b). As many as three repetitions may be performed. If any of the criteria for a score of three are not achieved, ask the participant to perform the test with the board from the FMS kit under the heels.

Figure 11.14 Frontal view of deep squat.

© Human Kinetics

Scoring

See table 11.5 for scoring criteria.

Table 11.5 Scoring for Deep Squat

Score	3	2	1
Criteria	• Upper torso is parallel with tibia or toward vertical • Femur below horizontal • Knees aligned over feet • Dowel aligned over feet	• Upper torso is parallel with tibia or toward vertical • Femur below horizontal • Knees aligned over feet • Dowel aligned over feet • Heels are elevated	• Tibia and upper torso are not parallel • Femur not below horizontal • Knees not aligned over feet • Lumbar flexion noted

A video illustration of this test can be found at the following YouTube link: www.youtube.com/watch?v=ewpuKzWEM04

HURDLE STEP

Purpose

To assess bilateral functional mobility and stability of the hips, knees, and ankles during the stepping motion as well as single-leg stance.

Equipment

Measuring stick (dowel)

Hurdle

Measuring board

Instructions

The participant places the feet together with the toes touching the base of the hurdle. The hurdle is then adjusted to the height of the tibial tuberosity, and the measuring stick is positioned across the shoulders below the neck (figure 11.15a). The participant then steps over the hurdle and touches the heel to the floor, maintaining the stance leg in an extended position (figure 11.5b), before returning to the starting position. The hurdle step should be performed slowly and as many as three times with each leg. If one repetition per leg is needed to meet the criteria provided, a 3 is given.

Figure 11.15 Frontal view of hurdle step.
© Human Kinetics

Scoring

See table 11.6 for scoring criteria.

Table 11.6 Scoring for Hurdle Step

Score	3	2	1
Criteria	• The hips, knees, and ankles remain aligned in the sagittal plane • Minimal movement in the lumbar spine • Dowel and hurdle remain parallel	• Alignment is lost between hips, knees, and ankles • Movement in the lumbar spine • Dowel and hurdle do not remain	• Contact with foot and hurdle • Loss of balance at any time

A video illustration of this test can be found at the following YouTube link: www.youtube.com/watch?v=U-n_Ar3RcLk

IN-LINE LUNGE

Purpose

To assess hip and ankle mobility and stability, quadriceps flexibility, and knee stability during rotational, decelerating, and lateral-type movements.

Equipment

Measuring stick (dowel)

Measuring board

Instructions

The tester obtains the participant's tibia length by measuring it from the floor to the tibial tuberosity. The participant places both feet on the board (or a tape measure taped to the floor) with the heel of back foot near the end of the board, and a distance equal to the tibia measurement is marked in front of the foot, applied from the end of the toes. The stick is placed behind the back, touching the head, thoracic spine, and sacrum. The hand opposite to the front foot should be the hand grasping the dowel at the cervical spine. The other hand grasps the dowel at the lumbar spine (figure 11.16a). The participant steps out on the board, placing the heel at the indicated mark, lowers the back knee enough to touch the ground behind the heel of the front foot (figure 11.16b), then returns to starting position. The lunge is performed up to three times bilaterally in a slow, controlled fashion. If one repetition is completed successfully, then a score of 3 is given for that extremity (right or left) and no other trial is necessary.

Figure 11.16 Side view of in-line lunge.

© Human Kinetics

Scoring

See table 11.7 for scoring criteria.

Table 11.7 Scoring for In-Line Lunge

Score	3	2	1
Criteria	• Minimal to no torso movement • Feet remain in sagittal plane on the board • Knee touches board behind the heel of front foot	• Movement noted in torso • Feet do not remain in sagittal plane on the board • Knee does not touch board behind the heel of front foot	• Loss of balance at any time

A video illustration of this test can be found at the following YouTube link: www.youtube.com/watch?v=gNm2Nkc4_o0

SHOULDER MOBILITY

Purpose

To assess bilateral shoulder range of motion, including internal rotation with adduction, external rotation with abduction, normal scapular mobility, and thoracic spine extension.

Equipment

Tape measure

Instructions

The tester first determines the hand length in inches by measuring the distance from the distal wrist crease to the tip of the third digit. Making a fist with each hand (with the thumb inside the fist), the participant assumes a maximally adducted, extended, and internally rotated position with one shoulder and a maximally abducted, flexed, and externally rotated position with the other, placing the fists on the back in one smooth motion. The tester then measures the distance between the two closest bony prominences (figure 11.17). The shoulder mobility test is performed as many as three times bilaterally.

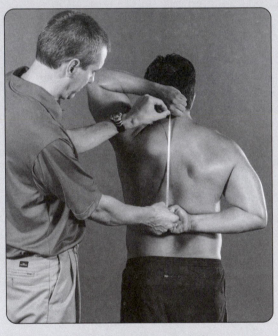

Figure 11.17   Shoulder mobility.

© Human Kinetics

Scoring

See table 11.8 for scoring criteria.

Table 11.8   Scoring for Shoulder Mobility

Score	3	2	1
Criteria	• Fists should be within one hand length	• Fists should be within one and a half hand lengths	• Fists fall greater than one and a half hand lengths

A video illustration of this test can be found at the following YouTube link: www.youtube.com/watch?v=UgEWbtvY44w

ACTIVE STRAIGHT-LEG RAISE

Purpose

To test the ability to disassociate the lower extremity from the trunk while maintaining stability in the torso and assess active hamstring and gastroc-soleus flexibility while maintaining a stable pelvis and active extension of the opposite leg.

Equipment

Measuring stick

Instructions

The participant lies supine with the arms by the sides, palms up and head flat on the floor. The tester then identifies the midpoint between the anterior superior iliac spine (ASIS) and midpoint of the patella and places the measuring stick at this position perpendicular to the ground (figure 11.18a). Next, the participant lifts the test leg as high as he or she can with a dorsiflexed ankle and an extended knee. The opposite knee should remain in contact with the ground, the toes should remain pointed upward, and the head should remain flat on the floor (figure 11.18b). Once the end range position is achieved, the malleolus is located and the score is recorded per the established criteria in table 11.9. The active straight-leg raise test should be performed as many as three times bilaterally.

Figure 11.18   Side view of active straight-leg raise.
© Human Kinetics

Scoring

See table 11.9 for scoring criteria.

Table 11.9   Scoring for Active Straight-Leg Raise

Score	3	2	1
Criteria	• Malleolus resides between midthigh and ASIS	• Malleolus resides between midthigh and midpatella	• Malleolus resides below midpatella

A video illustration of this test can be found at the following YouTube link: www.youtube.com/watch?v=r3mPcljUbms

TRUNK STABILITY PUSH-UP

Purpose

To test the ability to stabilize the spine in an anterior and posterior plane during a closed-chain upper-body movement and assess trunk stability in the sagittal plane while a symmetrical upper-extremity motion is performed.

Equipment

A clean, flat floor space

Instructions

The participant assumes a prone position with the feet together and the hands placed shoulder-width apart at the appropriate position (figure 11.19a). The knees are then fully extended and the ankles are dorsiflexed. The participant performs one push-up in this position. The body should be lifted as a unit; no lag should occur in the lumbar spine when performing this push-up (figure 11.19b). If the participant cannot perform a push-up in this position, the hands are lowered to the appropriate position.

Figure 11.19 Starting and finishing position view for trunk stability push-up.
© Human Kinetics

Scoring

See table 11.10 for scoring criteria.

Table 11.10 Scoring for Trunk Stability Push-Up

Score	3	2	1
Criteria	• Males perform 1 repetition with the thumbs above head • Females perform 1 repetition with thumbs in line with the chin	• Males perform 1 repetition with thumbs in line with the chin • Females perform 1 repetition with thumbs in line with the clavicle	• Males unable to perform 1 repetition with hands in line with chin • Females unable to perform 1 repetition with thumbs in line with clavicle

A video illustration of this test can be found at the following YouTube link: www.youtube.com/watch?v=QJ-gK66FkvA

ROTARY STABILITY

Purpose

To assess multiplane trunk stability during a combined upper- and lower-extremity motion.

Equipment

FMS kit board

Instructions

The participant gets into the quadruped position on the board with the shoulders, hips, and knees at 90° and the ankles dorsiflexed. The participant then flexes the shoulder and extends the same side hip and knee (figure 11.20a). The leg and hand are only raised enough to clear the floor by approximately 6 inches (15 cm). The same shoulder is then extended and the knee flexed enough for the elbow and knee to touch (figure 11.20b). This is performed bilaterally for up to three repetitions. If a score of 3 is not attained, then the participant performs a diagonal pattern using the opposite shoulder and hip in the same manner as described.

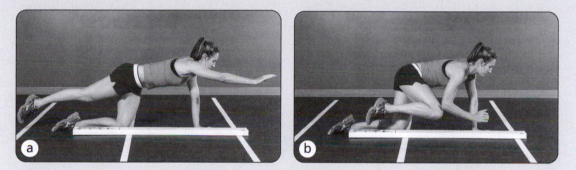

Figure 11.20 Starting and finishing position view for rotary stability.
© Human Kinetics

Scoring

See table 11.11 for scoring criteria.

Table 11.11 Scoring for Rotary Stability

Score	3	2	1
Criteria	• Performs 1 unilateral repetition while keeping torso parallel to board • Knee and elbow touch in line with the board	• Performs 1 diagonal repetition while keeping torso parallel to board • Knee and elbow touch in line with the board	• Unable to perform diagonal repetitions

A video illustration of this test can be found at the following YouTube link: www.youtube.com/watch?v=uFNFg6FbK6c

 **Go to HK*Propel* to complete Student Activity 11.2.**

ISSUES RELATED TO SELECTION, ADMINISTRATION, AND USE OF PERFORMANCE-RELATED FITNESS TESTS

Recall that, in the Measurement and Evaluation Challenge, John was asked by the head coach to select an agility test for their team. As we learned in chapters 6 and 7, the first two things to consider when we select a test should be validity (i.e., if a test measures what it is supposed to measure) and reliability (i.e., if the test is consistent). Though we ideally prefer a test with both validity and reliability coefficients of .80, these selection criteria could vary somewhat based on the nature of the performance test. If participants can easily demonstrate their maximal effort in a test (e.g., 50-yard dash), coefficients of .80 should be expected. If, on the other hand, significant skills are required to perform a test (e.g., Harre circuit test), a relatively low coefficient (e.g., .70) might be considered acceptable. Other factors that could have an impact on validity and reliability include sex, age, ability level, and familiarity with the test. The best way to learn about the validity and reliability of a test is to conduct a literature search. For example, by checking *Kirby's Guide to Fitness and Motor Performance Tests* mentioned earlier, John could easily determine that the SEMO Agility Test had a good reliability of $r = .88$ between trials 1 and 2 but relatively low validity from its correlations with dodging run ($r = .72$), shuttle run ($r = .63$), and side-step test ($r = .61$).

Besides validity and reliability, other secondary features related to practicality should also be considered. For example, is the test easy to administer? Will the test require expensive equipment? Can the test be administered to a large group and therefore be completed in a short time period? Can the test be used as an exercise in daily physical education classes? Can the test be administered by the teacher? Can the test be easily and objectively scored? With all these considerations and information collected together, the most appropriate test can be selected.

After selecting a test, the next step is to plan the administration of the test, which also results in consideration of complex functions. First, the participant must understand the test instructions and then be able to translate these instructions into the movements required by the test. Assume, for example, that John is going to run a pilot test of the shuttle run test before recommending it to the coach. On the first attempt the participant makes several errors—perhaps he or she threw the block across the line instead of placing it on the floor or forgot the number of times he or she should change direction. Did these errors occur because he or she did not understand the instructions or because he or she was unable to follow them? Before actual testing, the participant should see a demonstration of the test, first in slow motion, and then at full speed. Then he or she should be allowed to move through the test slowly while the test administrator gives cues on his or her movements. The test instructions can also be written on an index card or posted on a bulletin board. If the participant has difficulty with the instructions, a review of the written instructions might be helpful. Finally, the participant should have the opportunity to practice the test at maximum speed. Now he or she is ready to be tested. Remember that if the participant cannot perform the test properly, the trial must be thrown out and another taken. This is time consuming and inefficient. The time taken to prepare participants for the test is time well spent.

Preparation for Testing

When fitness tests are administered, testers expect the participant to perform at his or her maximum ability. If the participant has not received the preparation described previously, his or her true ability may not be measured—rather, the ability to perform a novel task would be measured, which is not the purpose of a fitness test. Although proper preparation for testing should be emphasized, however, it can be carried to extremes. Requiring the participants to practice a specific test day after day as part of the regular training program serves no purpose. If practicing the test increases the participants' overall performance-related fitness, some practice may be justifiable. Sometimes, other means of improving this specific component of performance-related fitness should be incorporated into the training program. This is because no developed test can measure all the components of fitness. Furthermore, it is more likely that some component of fitness is highly task-specific. This means that a performance on one test might not reflect how well (or poorly) one would perform on a different test even though both tests are meant to measure the same component. Yet, it is still very popular in the physical education and training fields to select a single test as a measure of a component of fitness, assuming that this test measures one's "general" ability of this component. This is not a fault of the test, but rather of the user's erroneous assumption. Other testing preparations include wearing appropriate clothes and shoes, performing the tests on nonslippery surfaces, and preparing methods for recording scores ahead of time.

Because fitness testing is usually conducted in large groups, assistants are often required. Students, teachers, parents, or other members of the community can be recruited as assistants. They should be thoroughly familiarized with the tests they will administer, which will not occur if they are simply given the test instructions to read. The following is a good rule of thumb: Preparing properly for testing always takes longer than you think it will! Do not assume anything. Bring assistants together for a practice session. Review and demonstrate the tests, giving each assistant experience in administering the tests or performing the assigned administrative task. Even when a test is administered year after year in a school, a brief review of procedures is essential.

Number of Trials

When a test is developed, information should always be included on the number of trials to be administered. The number of trials required is usually a function of determining how many trials are needed to obtain a reliable score. Let's suppose John is also interested in testing his players' dynamic balance ability and decided in advance that the test should have a reliability coefficient of at least 0.80. John decided to use the Johnson modification of the Bass Test described previously, but found that the reported reliability of this test was only .75. This is very close to the target reliability. With the addition of one or two trials, it should be possible to reach the target reliability. After experimenting with different numbers of trials (using the Spearman–Brown prophecy formula learned in chapter 6), John should be able to determine the number of trials needed to administer this test. Again, the main point is that the number of trials of a test should not be determined arbitrarily.

Use of Norms

It is very helpful if norms are available for the test you plan to administer, especially if you wish to determine the effectiveness of your program. To be used properly, norms should be available for the same age and sex as your participants. The norms should be used to determine the percentile scores of individual participants only. Using tables of norms to interpret group data, although appealing, is not appropriate, because the variability of individually based norms differs from the variability of group means. Caution should also be taken if outdated norms are used—in fact, many norms of commonly used performance-related fitness tests were developed decades ago (Hoffman, 2006).

MEASUREMENT AND EVALUATION CHALLENGE

After doing some research, John has decided to recommend the Illinois Agility Run Test because many validation studies have been conducted for this test, and many norm tables have been developed for a variety of age and sex groups. John will inform the coach that to be able to accurately measure players' agility using this test, some test preparations are needed. However, John is also aware that the agility measured by this test may not be able to predict agility in all football performance.

SUMMARY

In this chapter, performance-related physical fitness was introduced, and differences and similarities with health-related physical fitness were described. Key components of performance-related physical fitness and the selected tests for each component were introduced and illustrated. In addition, the Functional Movement Screen was described. Finally, critical issues related to administering and using fitness tests were discussed.

WWW! **Go to HK*Propel* for videos, homework assignments, and quizzes that will help you master this chapter's content.**

Assessment of Motor Abilities, Skills, and Performance

OUTLINE

OBJECTIVES

After studying this chapter, you will be able to

- differentiate between skills and abilities;
- apply sound testing procedures in ability and skill assessment;
- develop psychomotor tests with sufficient reliability and validity;
- differentiate among and be able to use various types of sport skills tests;
- develop tests and test batteries to select, classify, and diagnose athletes based on psychomotor tests;
- define and delineate the basic motor abilities; and
- properly use sport analytics, video analysis, and employment-related testing.

[www] The lecture outline in HK*Propel* will help you identify the major concepts of the chapter.

Motor performance, also known as *human performance* (Fleishman and Quaintance 1984), is the action or presentation of motor abilities and skills. These skills are essential to our daily lives (e.g., driving a car), as well as jobs (e.g., firefighting), sports (e.g., playing basketball), or emergency situations (e.g., carrying an injured person away from danger). Fleishman (1964) provided the modern foundation for work in this area, including the delineation between motor ability and **skill**:

Ability refers to a more general trait of the individual which has been inferred from certain response consistencies (e.g., correlations) on certain kinds of tasks. These are fairly enduring traits, which in the adult, are more difficult to change. Many of these abilities are, of course, themselves a product of learning, and develop at different rates, mainly during childhood and adolescence. . . . These abilities are related to performances in a variety of human tasks. For example, the fact that spatial-visualization has been found related to performance on such diverse tasks as aerial navigation and dentistry, makes this ability somehow more basic. . . . Skill refers to the level of proficiency on a specific task or limited group of tasks. As we use the term skill, it is task oriented. When we talk about proficiency in flying an airplane, in operating a turret lathe, or in playing basketball, we are talking about a specific skill. (p. 9)

Fleishman (1964) also classified ability as *psychomotor* or *perceptual-motor* ability according to its dimension. Battinelli (1984) further summarized this relationship in an article tracing the history of the debate over generality versus specificity of motor ability, stating the following:

The evident trend that emerges from the literature . . . over the years seems to demonstrate that the acquirement of motor abilities and motor skills through motor learning processes is dependent upon general as well as specific factors. The general components of motor ability (muscle strength, muscle endurance, power, speed, balance, flexibility, agility and cardiovascular endurance) have become the practical physical supports to motor learning, while skill specificities have been shown to be representative of the neural physiological processes exemplified in such learning. (p. 111)

The definition of the trait to be measured is important in determining the manner in which it should be measured. As noted in chapter 10, not only has the definition of physical fitness changed, but the manner in which it is measured has changed as well. In sport skills and motor performance testing, the distinction emphasized by Fleishman between ability and skill and later the advent of computers and advanced statistical techniques have also brought about changes. These changes, as well as the accepted practices associated with the measurement of motor abilities, skills, and performance, are presented in this chapter.

TESTING MOTOR ABILITIES

Forecasting in human performance and sport has long been a popular topic of debate. For example, is there such a thing as a natural athlete? What physical attributes are most important for high levels of athletic performance? Is it possible to measure athletic potential and predict future athletic success?

With the advent of computers and the application of multivariate statistical techniques to the analysis of human performance, researchers are now able to explore these questions in a manner that was not possible 40 years ago. Although the statistics involved are relatively complex, the objective is basic: to develop and use test batteries that can discriminate among performance levels.

Historical Overview

Early researchers operated on the theory that just as there were tests for assessing innate intelligence in the cognitive domain, there must also be a way to measure innate motor ability in the psychomotor domain. From the early 1920s through the early 1940s, these early researchers—Rogers, Brace, Cozens, McCloy, Scott, and others—concentrated on determining the physical components that are basic to and necessary for successful human performance.

One of the initial attempts was the development of classification indexes for categorizing students according to ability. This was to allow physical education classes to be formed homogeneously so that they could be taught with increased efficiency. The earliest classification indexes focused on predicting ability by age, height, and weight information. McCloy (1932) developed three classification indexes in his early work that could adequately group students:

Elementary: $(10 \times age) + weight$

Junior/senior high school: $(20 \times age) + (6 \times height) + weight$

College: $(6 \times height) + weight$

where age is in years, height is in inches, and weight is in pounds. By inspecting these formulas, McCloy found that at the college level, age was no longer an important factor in classification and that at the elementary level, height contributed little. This was one of the first attempts to predict performance, but at this point no motor performance tests were actually used.

Next, Neilson and Cozens (1934) developed a classification index based on the same principle; however, they used powers (exponents) of height in inches, age in years, and weight in pounds. The McCloy and the Neilson and Cozens classification systems were so similar that the correlation between them was .98.

At about the same time, researchers began classifying by motor ability testing. The term **general motor ability (GMA)**—referring to overall proficiency in performing a wide range of sport-related tasks—was introduced. Rogers and McCloy developed strength indexes that included some power tests, the standing long jump, and the vertical jump, which had been found to correlate moderately with ability in a variety of activities (Clarke and Bonesteel, 1935). To increase the accuracy of the prediction, test batteries were designed on the premise that certain motor abilities, such as agility, balance, coordination, endurance, power, speed, and strength, were the bases of physical performance.

An example of an early test of general motor ability is the Barrow Motor Ability Test (Barrow, 1954). Although the test was initially designed for college men, norms were developed later for both junior and senior high school boys. Originally, the test included 29 test items measuring eight theoretical factors: agility, arm–shoulder coordination, balance, flexibility, hand–eye coordination, power, speed, and strength. Barrow constructed an eight-item test battery to examine the validity of the factors and obtained a multiple correlation coefficient of .92 between the composite score for the eight-item battery and a composite score based on all 29 items. Test–retest reliability coefficients ranged from .79 to .89. From a measurement standpoint, however, the problem associated with this validation procedure is the composite criterion: Because both the predictor tests (a sub battery) and the criterion test (the full battery) included the same tests, Barrow was simply predicting the whole from a part, giving a spuriously high correlation of .92—a statistically invalid approach. The importance of this early motor ability test battery was Barrow's examination of the theoretical structure of the various components of athletic ability.

WWW **Go to HK*Propel* to complete Student Activity 12.1.**

Larson (1941) took another approach to the examination of motor ability. He analyzed the factors underlying 27 test items and six batteries and gathered additional statistical evidence on other test batteries related to measuring motor ability, finding reliability coefficients in excess of .86. This represented an attempt to examine the construct validity of motor ability tests through the use of a statistical technique called **factor analysis**. However, he also developed these batteries using the composite criterion approach, which reduced the validity of his findings.

The next step in the use of motor ability tests was the development of tests to measure **motor educability**—the ability to learn a variety of motor skills. These tests included track and field items as well as a number of novel stunts. One of the stunts involved having students grab one foot with the opposite hand and then jump to try to bring the other leg through the opening formed by the leg and the arm. Another item had the students jump up, spin around, and try to land facing exactly the same direction. Brace (1927) developed the Iowa Brace Test, involving 20 stunts, each scored on a pass/fail basis. McCloy later used the Iowa Brace Test as a starting point in developing a test for motor educability, ultimately selecting six combinations of 10 stunts to measure motor educability in the categories of upper elementary, junior high, and high school boys and girls. However, because they were all pass/fail items, it was an either–or evaluation, which tended to reduce reliability. Furthermore, the tests tended to correlate poorly with most sport performance measures, thus casting doubt on their validity for assessing motor educability in any realistic situation.

WWW **Go to HK*Propel* to complete Student Activity 12.2.**

The Sargent Jump was one of the first motor performance tests used to examine a person's potential for athletic performance (Sargent, 1921). Named after Dudley Sargent, the test was a vertical jump purported to be a measure of leg power. This test is widely used and measures a trait important to many power-based sports. Whereas this is a reliable test that is validly related to certain aspects of performance in a large number of sports, this single measure obviously gives an incomplete picture of overall athletic ability.

In the years that followed, the concept of **specificity**—motor abilities unique to individual psychomotor tasks—arose through the works of Henry (1956; 1958). Using correlational analysis, Henry stated that traits that have more than 50% shared variance ($r^2 > .50$; recall again the coefficient of determination from chapter 4), or, in other words, that have correlations above .70, were said to be general in nature. Any two tests that had correlations of .70 or less were said to be specific. Because of the magnitude of the correlations that Henry chose, most motor ability tests were found to be specific in nature. The cause-and-effect implication from this was that traits should be trained (and measured) specifically, and that anyone athletically proficient would have a high degree of many of these specific traits.

From the 1940s to the 1970s, other researchers, such as Seashore, Fleishman, Cumbee, Meyer, Peterson, and Guilford, continued to develop the notion that motor ability is specific rather than general in nature. Factors most often cited by these investigators included muscular strength and endurance, speed, power, cardiovascular endurance, flexibility, agility, and balance. In general, their theories were based on the correlations between physical factors. High correlation suggests that items have much in common, but low correlation suggests that items measure different traits. Thus, specificity of tasks can be viewed as a concurrent validity approach.

During this period, Fleishman (1964) was developing what he termed the **theory of basic abilities**. This theory serves as the basis for most of the subsequent scientific research that has been conducted in this area. Fleishman distinguished between skill and ability in the following manner: Abilities are more general and innate in nature, whereas skills are learned traits based on those abilities. For example, the tennis serve, badminton serve, and volleyball spike are all specific skills that involve similar overarm patterns; the overarm pattern is considered the ability. Fleishman was also one of the first to examine his theory using factor analysis. His work is classic in the field.

Measurement Aspects of the Domain of Motor Performance

A multitude of motor performance factors affect a person's ability to perform specific sport skills. These factors are the underlying bases for the domain of human performance. Using Fleishman's work as their basis, a number of other researchers, including Fleishman and Quaintance (1984), began examining the construct-related validity of these factors, improving on earlier works by expanding the taxonomies of human performance.

Examining the subdomains of human performance in order to understand the qualities necessary to perform various tasks has been the topic of research for a number of scholars in the field. It is important to examine these subdomains from a variety of different standpoints, because factors such as age and ability level could alter the structure of these domains. The primary subdomains of human performance are as follows:

- Muscular strength
- Speed
- Agility
- Anaerobic power
- Flexibility
- Balance
- Kinesthetic perception

These subdomains are components of health- and performance-related physical fitness, introduced in chapters 10 and 11.

GUIDELINES FOR MOTOR ABILITY, SKILLS, AND PERFORMANCE TESTS

When selecting or devising tests to measure motor ability, sport skills, or physical job performance, certain accepted testing procedures should be followed. Whether you use standardized tests or your own tests depends on your expertise and the specific use of the test. If you will be making comparisons with other groups, use some type of standardized test. If the information is simply for you, then you can modify a standard test or develop a new one to suit your purposes.

The American Alliance for Health, Physical Education, Recreation and Dance (AAH-PERD), now SHAPE America, provided guidelines for skills test development (Hensley, 1989); these are the basis for the development of the AAHPERD skills test series and also apply to the measurement of motor abilities. The guidelines state that skills tests should

- have at least minimally acceptable reliability and validity,
- be simple to administer and to take,
- have instructions that are easy to understand,
- require neither expensive nor extensive equipment,
- be reasonable in terms of preparation and administration time,
- encourage correct form and be gamelike but involve only one performer,
- be of suitable difficulty (neither so difficult that they cause discouragement nor so simple that they are not challenging),
- be interesting and meaningful to performers,
- exclude extraneous variables as much as possible,
- provide for accurate scoring by using the most precise and meaningful measure,
- follow specific guidelines if a target is used as the basis for scoring,
- require a sufficient number of trials to obtain a reasonable measure of performance (tests that have accuracy as a principal component require more trials than tests measuring other characteristics), and
- yield scores that provide for diagnostic interpretation whenever possible.

The guidelines also indicate that if a target is used as the basis for scoring, it should have the capacity to encompass 90% of the attempts; near misses may need to be awarded points. Determining target placement should be based on two principal factors: (1) the developmental level of each student (e.g., the height of a given target may be appropriate for a 17-year-old but not for a 10-year-old), and (2) the allocation of points for strategic aspects of the performance (e.g., the placement of a badminton serve to an opponent's backhand should score higher than an equally accurate placement to the opponent's forehand).

Because most tests of sport skills or motor abilities can be objectively measured, you can expect reliability and validity coefficients higher than those associated with written instruments. The AAHPERD guidelines suggest that reliability and validity coefficients exceed .70. However, although many reliability coefficients exceed this value, it is difficult to develop tests whose validity coefficients reach it. A validity coefficient of .70 would mean that nearly 50% of the variance of the test would be associated with the criterion (r^2 = .49; recall the coefficient of determination from chapter 4). When examining the validity of a test, consider not only its statistical relationship with the criterion, but also its practical relevance. Also, consider test feasibility as you select instruments to use.

EFFECTIVE TESTING PROCEDURES

The procedures for psychomotor testing are similar to those for written testing. These procedures may be classified as pretest, testing, and posttest duties. Thinking through all aspects of the testing procedures in detail is imperative so that you can gather consistent, accurate results. When testing in an academic setting, tests are normally administered as a pretest or as summative evaluation at the end of a unit. When testing athletes, it is important to remember that "effective testing can be done at any time during the training competition program depending on what you are looking for" (Goldsmith, 2005, p. 15).

Pretest Duties

Pretest planning is the first element of preparing to administer tests. The tester must be completely familiar with the test, the test items to be administered, the facilities, and the equipment. For a physical performance test, the students must be allowed time to practice so that they can learn how to pace themselves and how to perform the test in the most efficient manner. This way, the test will be an accurate measure of the students' actual learning, rather than of the students' abilities to perform in an unfamiliar situation.

With batteries of performance tests, there are other elements to consider. The first is the order of administration. If there are several physically taxing items in a test battery, spread them out across several days so that the test participants do not become unduly fatigued. Also, pair time-consuming items with items that do not take as long.

Consider whether you will need assistants to help record the data. This is important when you determine the type of score sheets to use. Two types of recording sheets are used when multiple testers and multiple stations are involved. With one form, all the data for a single test participant are recorded on one sheet (figure 12.1). One problem with this method is that the sheets can be misplaced or damaged as they go from station to station; however, this problem can be minimized if all forms for one group are placed on a single clipboard and the clipboard rotates from station to station. The second method is to provide each tester with a master list of all those taking the test; as the participants rotate through the stations, the testers simply record the data on the master list (figure 12.2). The problem with the second approach is the lengthy transcription process in the posttest recording stage. In addition, it is easy to write a student's scores in the wrong rows or blanks. If only one tester is available, we recommend using the master list method.

Testers themselves should be familiar with the test; you will need to develop and write standardized instructions so that each tester knows exactly what to do. The instructions should be read to the test participants so that each is given exactly the same information for the test. Consider giving each student a written copy of the instructions before testing. As stated, students should be familiar with the procedures and given an opportunity to practice the test in advance.

Soccer skills test score sheet

Name: _____

Class: _____

Speed dribble (# of touch violations in parentheses)

T1 _____ (   )

T2 _____ (   )

Control dribble

T1 _____

T2 _____

Passing		Left foot	Right foot	Shooting		
10 yd	1	_____	_____	Left target	1	_____
	2	_____	_____		2	_____
15 yd	1	_____	_____		3	_____
	2	_____	_____	Right target	1	_____
					2	_____
					3	_____

Figure 12.1   Sample individual score sheet.

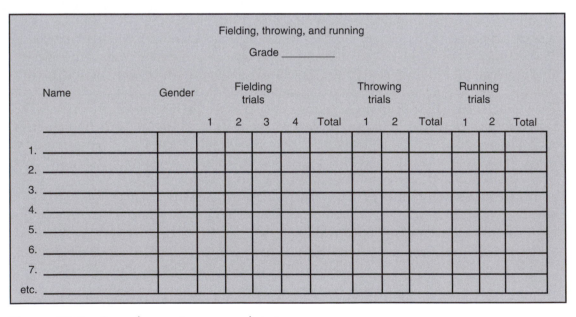

Figure 12.2   Sample master score sheet.

Testing Duties

The second phase is the actual testing. Prepare the test location as early as possible. Make sure that the surfaces are clean and address any safety concerns at this time, such as objects close to the testing area or unsafe surfaces or equipment. Give students the opportunity to warm up for physical performance tests. Present the instructions to the students and keep any hints or motivational comments consistent across all groups so that all students can do their best.

Posttest Duties

The final phase of the testing procedure involves transcribing the test results and analyzing the scores. The method of transcription will depend on how the data were actually gathered. The data need to be verified any time they are transferred from one medium to another; this is best done by having a helper call out scores. Each time the data are transcribed, they need to be proofed by another person.

The analysis you conduct depends on the purpose of the testing. If individual performance within the group is important, use norm-referenced analytical procedures. On the other hand, if the purpose of the test is to compare a person's performance with a standard, use criterion-referenced analysis.

Keep all test participants' scores as confidential as possible. If test participants help record the data, position them directly next to the tester. This can alleviate the problem of having to call out scores.

DEVELOPING MOTOR PERFORMANCE TESTS

Often, teachers, coaches, and researchers are interested in administering test batteries for a certain sport or motor performance area. In the analysis of athletic performance, tests of sport skills and motor performance are often used concurrently to give a teacher or coach additional information about prospective athletes. Strand and Wilson (1993) presented a flowchart for the construction of test batteries of this nature; we present a modified version of their flowchart in figure 12.3. Although the flowchart was developed primarily for sport skills, it is also applicable to motor ability testing. The steps are as follows:

• *Step 1.* Review the criteria for a good test. Basically, these criteria concern the statistical aspects of reliability, validity, and objectivity. The tester needs to be familiar with the equipment, personnel, space, and time available for testing. A test must be appropriate for the age and sex of the students and must be closely associated with the skill in question. Also, consider safety aspects at this time.

• *Step 2.* Analyze the sport to determine the skills or abilities to be measured. If you are trying to evaluate current performance, then skills tests are most appropriate; if you are trying to determine potential, then motor ability tests might be more useful. Combination batteries can also be useful, depending on the specific purpose of the test.

• *Step 3.* Review the literature. Once you have analyzed the sport for its salient areas, review previous test batteries and literature associated with the specific skills or motor performance areas. Consult experts (such as colleagues, textbooks, professionals, teachers, researchers) at this point.

• *Step 4.* Select or construct test items. Ensure that the test items are (a) representative of the performances to be analyzed, (b) administered with relative ease, (c) as closely associated with the actual performance as possible, and (d) practically important. Each test or

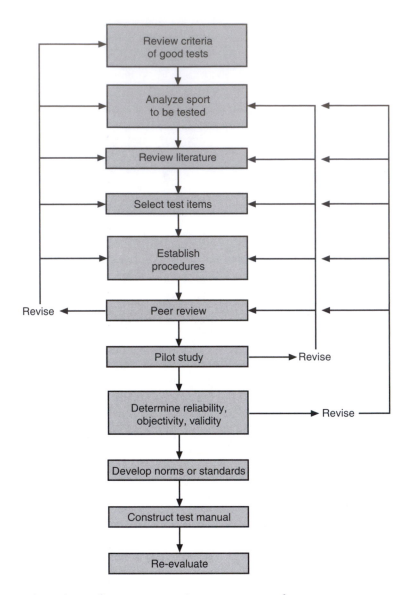

Figure 12.3 Flowchart for constructing motor performance tests.

item should measure an independent area. It is not an efficient use of time to have several items that measure the same basic skill or ability.

• *Step 5*. Establish the exact testing procedures. This includes selecting the number of trials necessary for each test item, the trials that will be used to establish the criterion score, and the order of the test items.

• *Step 6*. Include peer review. Have experts such as other teachers or coaches who are familiar with the activity examine the test battery.

• *Step 7*. Conduct pilot testing. If the selected test items are based on experts' opinions and a review of the literature, the pilot study analysis will help you determine the appropriateness of the test items. Pilot testing is an important step before the test item is finalized; it helps determine the total time of administration and the clarity of instructions and can uncover possible flaws in the test.

- *Step 8.* Determine the statistical qualities of each test item—its reliability, validity, and objectivity. The reliability coefficients are estimates and are accurate only for the groups tested. They can be group specific, especially with adolescents, so it is important that both sexes and all ages and skill levels included in the normative sample are tested. The methods for validating skill or motor ability tests are presented in chapters 6 and 7. Establishing content validity is important before the statistical evaluation of the tests. You can determine concurrent validity or construct validity at this time by performing the appropriate procedures. Remember that validity coefficients are estimates and are only appropriate for groups comparable to those tested. The stability of the motor performance tests is normally established with reliability coefficients that involve repeated measures. This is the one phase of motor performance testing that separates it from the preparation phases of many psychological or educational tests. Eliminate or modify potential items at this point if they have poor reliability or validity.

- *Step 9.* Establish norms for norm-referenced tests or determine standards for criterion-referenced tests.

- *Step 10.* Construct the test manual to fully describe the test, the scoring procedures, and its statistical qualities (reliability and validity). Follow the testing standards recommended by the American Educational Research Association, American Psychological Association, and National Council on Measurement in Education (2014).

- *Step 11.* Reevaluate the instrument from time to time. As time passes, your students may have different preparation levels; thus, norms and standards that were appropriate at one time may not be appropriate at another.

ISSUES IN SKILLS TESTING

The usual issues of reliability and validity are important in skills testing. However, there are two other issues that we highlight in this chapter. The first is feasibility. Because skills tests typically take time to administer, you may need to consider whether it is more important to spend time teaching the skills and having the students perform them or to spend the time evaluating the skills. The second issue is determining the best way to evaluate the skill. Skills testing often requires compromise between selecting a test that is extremely objective and reliable but not gamelike and selecting one that is highly valid but less objective and more time consuming. (For more on the latter approach, see chapter 15 for additional information about performance-based assessment.)

For example, consider Elijah's challenge at the beginning of this chapter. He might choose to measure several skills primarily associated with volleyball performance, along with conducting some motor performance tests. One such skill is the forearm pass. Assume he initially selected three tests to measure forearm passing: the self-pass, the wall pass, and the court pass. The self-pass and the wall pass are both simple tests that can be administered to large groups and require a minimal amount of space. They both produce consistently high degrees of reliability. In both cases, performers can practice the test by themselves and can administer a self-test. The tests are designed for group administration and partner scoring; they are feasible, involving minimal court markings and class time. However, these tests are not gamelike. For the self-pass, performers pass the ball repeatedly to themselves, the minimum criterion being that the ball must be passed at least 10 feet (3.0 m) off the ground on every repetition. For the wall pass, a performer passes the ball above a target line, again 10 feet (3.0 m) above the ground, while staying behind a restraining line 6 feet (1.8 m) from the wall. Most volleyball experts would agree that these tests may not readily transfer to game skills (e.g., the ability to pass a serve or a hard-driven spike).

The court pass test, on the other hand, places performers in a position in the backcourt, where they are most likely to receive the serve. They are asked to pass a ball that has been served or tossed by the tester to a target area. A point system is used to determine the accuracy and skill involved in the passing of 10 balls. This test has good reliability and is valid in terms of content validity. However, it requires much more administration time than the other two tests. Also, the testers must produce reasonably consistent serves or tosses. Elijah chose to go with the court pass to measure his team because of the small number of participants and their high skill level. If Elijah were testing beginners, either the self-pass or wall pass would test as effectively as the court pass.

Go to HK*Propel* to complete Student Activities 12.3 and 12.4.

AAHPERD's comprehensive testing program provides physical education instructors and coaches with a battery of reliable and valid tests that can be administered in a minimal amount of time with a minimal amount of court markings. The AAHPERD test selection guidelines indicate that skills tests should include the primary skills involved in performing a sport, usually not to exceed four. The tests of a battery should have acceptable reliability and validity and not have high intercorrelations. The battery in general should be able to validly discriminate among performance levels, thus demonstrating construct-related validity. The AAHPERD tests are designed primarily for the beginning-level performer. Table 12.1 includes the AAHPERD Basketball Skills Test (Hopkins, Schick, and Plack, 1984), Tennis Skills Test (Hensley, 1989), and Softball Skills Test (Rikli, 1991).

Table 12.1   AAHPERD Skills Test Batteries

Basketball	Tennis	Softball
Speed spot shooting	Ground stroke	Batting
Passing	Forehand and backhand	Fielding ground balls
Control dribble	Serve	Overhand throwing
Defensive movement	Volley	Base running

Both the softball and tennis skills test manuals contain rubrics for the skills that can be substituted for or used along with the objective tests.

Go to HK*Propel* to complete Student Activity 12.5.

SKILLS TEST CLASSIFICATION

When you construct skills tests, consider whether to use objective or subjective procedures. The AAHPERD skills test batteries include only objective tests; however, subjective ratings are included as alternate forms for the tennis and softball batteries. You must consider a number of factors, including time, facilities, number of testers, and number and type of tests to be used when deciding on a specific approach to skills testing. Also, you must be sure that you are measuring only one trait at a time.

Objective Tests

There are four primary classifications for objective skills tests:

1. Accuracy-based skills tests
2. Repetitive-performance tests

3. Total body movement tests

4. Distance or power performance tests

Some tests may be combinations of more than one classification. Each classification involves specific measurement issues.

Accuracy-Based Tests

Accuracy-based skills tests usually involve the skill of throwing or hitting an object, such as a volleyball, tennis ball, or badminton shuttlecock. They may also involve some other test of accuracy: throwing in football or baseball, free throws or other shots in basketball, or kicking goals in soccer. The primary measurement issue associated with accuracy tests is the development of a scoring system that will provide reliable yet valid results.

Consider a goal-scoring test in which the performer must kick a soccer ball into the goal from 12 yards (11.0 m) away. For a performer to get the maximum amount of points, the ball must go between the goal posts and a restraining rope 3 feet (0.9 m) inside them. A performer is allowed six kicks: three kicks to the left side of the goal and three kicks to the right side. Two points are awarded if the ball goes within the target area, 1 point if it simply goes into the goal. The problem with this test is that good shooters often try to put the ball in the target area but end up being just slightly off, for a score of 0, whereas less competent performers may shoot every shot for the middle of the goal to ensure that they receive at least 1 point. Good shooters may thus obtain lower scores than shooters who are not as skilled. This reduces the reliability and the validity of the test.

Mastery Item 12.1

Can you detect another weakness with a goal-scoring test that might affect its validity?

In volleyball tests, the court is often marked to delineate where the most difficult serves should be aimed. The problem with the goal-scoring test just mentioned is inherent in this test as well, because serves that fall outside the court are given a score of 0. To rectify this problem, slightly lower point values can be assigned to attempts falling within some area slightly outside the target. This can, however, create a problem of feasibility—that is, of marking the court for this area.

Another consideration with accuracy-based tests is the number of repetitions necessary to produce reliable scores. When the AAHPERD Tennis Skills Test was first developed, the serving component did not attain the reliability values necessary for inclusion in the battery. The Spearman–Brown prophecy formula (see chapter 6) indicated that to improve the reliability of the test significantly, the number of trials would need to be more than doubled, which would affect the feasibility of the test. Therefore, AAHPERD developed a modified scoring system that enabled the reliability to increase to an acceptable level. When using the multiple-trial approach in serving, testers must maximize reliability while minimizing the number of trials.

An example of an accuracy-based skills test is the North Carolina State University (NCSU) Volleyball Service Test (Bartlett, Smith, Davis, and Peel, 1991). Content validity was claimed because serving is a basic volleyball skill. An intraclass reliability coefficient of .65 for college students was reported. Procedures for the administration of this test are presented next.

NCSU VOLLEYBALL SERVICE TEST

Purpose

To evaluate serving in volleyball.

Equipment

Volleyballs

Marking tape

Rope

Scorecards or recording sheets

Pencils

Instructions

Regulation-sized volleyball courts are used for the serve test; the courts are prepared as shown in figure 12.4. The tester marks point values on the floor. The test participant stands in the serving area and serves 10 times either underhand or overhand.

Scoring

The scorer gives scores of 0 to balls that contact the net or land out of bounds and the higher point value to balls that land on a line.

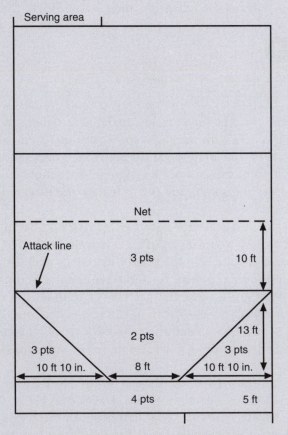

Figure 12.4 Floor plan for the NCSU Volleyball Service Test.

Reprinted from J. Bartlett, L. Smith, K. Davis, and J. Peel, "Development of a Valid Volleyball Skills Test Battery," *Journal of Physical Education, Recreation and Dance* 62, no. 2 (1991): 19-21. Reprinted by permission of Taylor & Francis (Taylor & Francis Ltd., www.tandfonline.com).

Repetitive-Performance Tests

Repetitive-performance tests, commonly called wall volleys or self-volleys, are tests that involve continuous performance of an activity for a specified time. They can be used to measure the strokes of racket sports, such as the forehand or backhand stroke in tennis, and volleying and passing in volleyball. Repetitive-performance tests usually have a high degree of reliability, but unless they are constructed carefully, they may not approximate the same type of action that is used in the game, in which case validity is reduced. Furthermore, because they are not gamelike, they may not transfer into actual game performance as well as some other court tests. Therefore, it is extremely important when using repetitive-performance tests that the testers make sure that the test participants use the correct form.

An example of a repetitive-performance test is the short wall volley test for racquetball (Hensley, East, and Stillwell, 1979). The test was originally administered to college students but is considered appropriate for junior and senior high school students as well. Test–retest reliability coefficients are .86 for women and .76 for men. A validity coefficient of .86 has been obtained by using the instructor's rating of students as the criterion measure. Procedures for the administration of this test are presented here.

SHORT WALL VOLLEY TEST

Purpose

To evaluate short wall volley skill.

Equipment

Rackets	Marking tape
Eye protection	Stopwatches
4 racquetballs per testing station	Scorecards or recording sheets
Measuring tape	Pencils

Instructions

The test participant stands behind the short service line, holding two racquetballs. An assistant located near the back wall holds two additional racquetballs. To begin, the participant drops a ball and volleys it against the front wall as many times as possible in 30 seconds. All legal hits must be made from behind the short line. The ball may be hit in the air after rebounding from the front wall or after bouncing on the floor. The ball may bounce as many times as the hitter wishes before it is volleyed back to the front wall. The participant may step into the front court to retrieve balls that fail to return past the short line but must return behind the short line for the next stroke. If a ball is missed, a second ball may be put into play in the same manner as the first. (The missed ball may be put back into play, or a new ball can be obtained from the assistant.) Each time a volley is interrupted, a new ball must be put into play by being bounced behind the short line. Any stroke may be used to keep the ball in play. The tester may be located either inside the court or in an adjacent viewing area.

Scoring

The 30-second count should begin the instant the student being tested drops the first ball. Trial 2 should begin immediately after trial 1. The score recorded is the sum of the two trials for balls legally hitting the front wall.

Mastery Item 12.2

What else could be done to increase the reliability of the short wall volley test?

Total Body Movement Tests

Total body movement tests (often called speed tests) assess the speed at which a performer completes a task that involves movement of the whole body in a restricted area, such as dribbling in basketball or soccer or base running in baseball and softball. These tests usually have a high degree of reliability because a large amount of interpersonal variability is associated with timed performances. These tests can be administered quickly, but they have two inherent problems. First, when this type of test is used, the course must validly approximate the skill used in a game, and in many cases flat-out speed of movement is not always required. For example, in basketball, even on the fast break there must be some degree of controlled speed to allow for control of the ball. Therefore, to measure basketball dribbling skill, AAHPERD has selected a control-dribble test involving dribbling around a course of cones (Hopkins, Schick, and Plack, 1984; see figure 12.5). Second, when such tests are used for evaluation, performance time is the valid criterion, but if you are interested in efficiency of performance, these tests would be highly related to a performer's speed. Obviously, a faster performer can complete the test in less time than a slower one, even if the slower performer has more skill in actual ball handling.

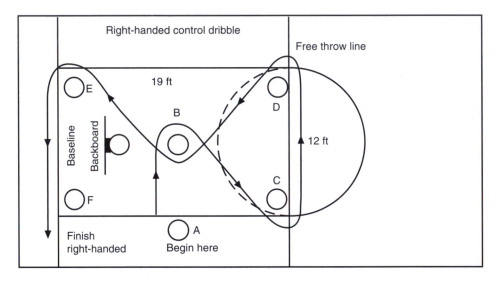

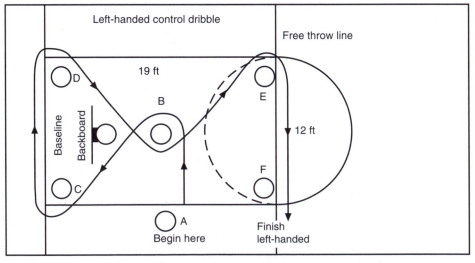

Figure 12.5 AAHPERD control-dribble test.

Reprinted by permission from D.R. Hopkins, J. Schick, and J.J. Plack, *Basketball for Boys and Girls: Skills Test Manual* (Reston, VA: AAHPERD, 1984).

A way to eliminate the speed problem inherent in total body movement tests is to create performance ratios. You create a **performance ratio** by dividing a performance time by a movement time (i.e., the total amount of time to complete the entire test) for the same participant. For example, you could determine a student's dribbling efficiency (a performance ratio) by comparing dribble time to total movement time. Performance ratios can be extremely effective motivational tools because performers are competing against themselves, trying to reduce the ratio to as close to 1 as possible (1 representing the best performance). However, because movement time is measured by this type of ratio, faster performers could be unduly penalized. Furthermore, performers might not give their maximum effort on the movement time test in order to achieve a better ratio score than their performance warrants. These ratios are not appropriate for use with a sports team because absolute performance is the primary measure. They are, however, an effective approach for measuring skill proficiency in a class setting.

Another example of a total body movement test is the defensive movement test from the AAHPERD Basketball Skills Test (Hopkins, Schick, and Plack, 1984). Intraclass reliability coefficients above .90 were reported, and concurrent validity established for the full test battery ranged from .65 to .95. The procedures for administering this test are presented next.

WWW Go to HK*Propel* to complete Student Activities 12.6 and 12.7.

DEFENSIVE MOVEMENT TEST

Purpose
To measure performance of basic defensive movement in basketball.

Equipment
Stopwatch
Standard basketball lane
Tape for marking change-of-direction points

Instructions
The tester marks the test perimeters as shown in figure 12.6. Only the middle line—rebound lane marker (C in figure 12.6)—is a target point for this test. The tester marks the additional spots outside the four corners of the area at points A, B, D, and E with tape. The test includes two trials, and the participant is allowed one practice trial beforehand. Each performer starts at A and faces away from the basket. On the go signal, the performer slides to the left without crossing the feet and continues to point B, touches the floor outside the lane with the left hand, executes a drop step, slides to point C, and touches the floor outside the lane with the right hand. The performer continues the course as diagrammed. The performer completes the course when both feet have crossed the finish line. Violations include foot faults (crossing of the feet during slides or turns and runs), failure of the hand to touch the floor outside the lane, and executing the drop step before the hand has touched the floor. If a performer violates the instructions, the trial stops and timing must begin again from the start.

(continued)

DEFENSIVE MOVEMENT TEST *(continued)*

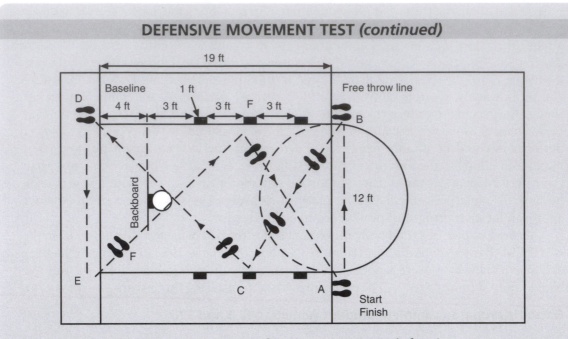

Figure 12.6 Perimeters and setup for the AAHPERD defensive movement test for basketball.

Reprinted by permission from D.R. Hopkins, J. Schick, and J.J. Plack, *Basketball for Boys and Girls: Skills Test Manual* (Reston, VA: AAHPERD, 1984).

Scoring

The score for each trial is the elapsed time required to legally complete the course. Record scores to the nearest tenth of a second for each trial; the criterion score is the sum of the two trials.

Distance or Power Tests

The final classification for objective skills tests is **distance** or **power performance tests**, which assess one's ability to project an object for maximum displacement or force. The badminton drive test for distance and the racquetball power serve test are examples of this type of test (Strand and Wilson, 1993), as are distance throws in softball and baseball and punt, pass, and kick competitions. A key consideration before using a test is whether the test requires or accounts for any correction for accuracy. For example, in the punt, pass, and kick contest, the distance off the line of projection is subtracted from the distance of projection. Because of this, a performer might hold back from using maximum force for fear of losing accuracy. In contrast, in track events such as the discus and shot put, these corrections are not made as long as the object is projected within a given area. In the volleyball arm power test using the basketball throw (Disch, 1978), accuracy is not considered an important component because arm power associated with the ability to spike a volleyball is being measured; therefore, absolute distance is the criterion. However, in other sport skills tests, such as throwing a football or baseball, accuracy may be important, so some correction needs to be used. The simplest correction is subtracting the distance off the line of projection from the total distance. As with all skills tests, it is important to ensure that these tests are performed in a gamelike manner.

An example of a distance or power performance test is the overhand throwing test from the AAHPERD Softball Skills Test (Rikli, 1991). Intraclass reliability coefficients were found to exceed .90 for all samples, and concurrent validity coefficients were found to range from .64 to .94. The procedures for administering this test are presented here.

OVERHAND THROWING TEST

Purpose

To measure the skill involved in throwing a softball for distance and accuracy.

Equipment

A smooth grass field that can be marked off in feet or meters

Measuring tape

Softballs

2 small cones or marking stakes

Instructions

The tester marks off a throwing line in feet or meters down the center of a large, open field area, with a restraining line marked at one end perpendicular to the throwing line. A back boundary line is marked off 10 feet (3.0 m) behind the restraining line. The participant stands between the restraining line and the back boundary line, back far enough to take one or more steps before throwing. After 3 to 4 minutes of short-throw warm-ups, the participant has two trials to throw the softball as far and as straight as possible down the throwing line without stepping over the restraining line. The tester or assistants are positioned in the field to indicate the spot where each ball first touches the ground using a cone or marking stake. If a participant steps on or over the restraining line before releasing the ball, the participant must repeat the trial.

Scoring

The net throwing score equals the throwing distance (the point on the throwing line perpendicular to the spot where the ball landed) minus the error distance (the distance the ball landed away from the throwing line). The player's score is the better of the two throws. The tester measures both the distance and the error score to the nearest foot or meter.

Mastery Item 12.3

What are some factors that could reduce the reliability of the overhand throwing test?

Subjective Ratings

Objective tests are attractive because they usually have a high degree of reliability, can be developed to produce good validity, and measure specific components of skill performance. However, **subjective ratings**—the value a rater places on a skill or performance based on personal observation—offer attractive alternatives for physical education teachers, coaches, and others interested in human performance analysis. Subjective ratings can be developed for process-oriented skills, in which the form of the skill is evaluated (e.g., diving and gymnastics). Performers are evaluated on their preliminary position, force production phase, and follow-through. Specific cues can be given to performers about where they may be losing efficiency in performance. Ellenbrand (1973) presented a classic example of a general rating scale for the performance of gymnastics stunts (table 12.2). This scale could easily be adapted to evaluate other process-oriented performances. The performance-based evaluation procedures presented in chapter 15 are closely aligned with subjective ratings; the same reliability, validity, fairness, and feasibility issues described there are important here.

Table 12.2 Ellenbrand Gymnastics Rating Scale

3 points	Correct performance. Proper mechanics. Executed in good form. Performer shows balance, control, and amplitude in movements.
2 points	Average performance. Errors evident in either mechanics or form. May show some lack of balance, control, or amplitude in movements.
1 point	Poor performance. Errors in both mechanics and form. Performer shows little balance, control, or amplitude in movement.
0 points	Improper or no performance. Incorrect mechanics or complete lack of form. No display of balance, control, or amplitude in movement.

The score for each item (event) is the product of a difficulty value and the execution rating. The sum of all test items in each event is the score for that event. The final test score is the sum of the scores for all events or all test items. There is no deduction for falls or repeated skills. However, a stunt that is executed with assistance receives a rating of 0.

Data from Ellenbrand (1973).

Another application of subjective ratings is to observe participants in the activity and give them a **global rating** based on their overall performance in a competitive situation. This allows the tester to evaluate several performers concurrently and possibly evaluate some intangible aspects of game performance that are not identifiable by performance of individual skills in a nongame setting. The problem with this approach lies in the number of observations possible. In one game, a student may have the opportunity to contact the ball several times, but in another game that student may make few or even no contacts. Other problems with subjective ratings are defining criteria and ensuring consistency among raters. In most physical education situations, a teacher is the only available person in class to do the rating, so the evaluation may be biased by preconceived notions about the performance or the specific performer; this is also true in many coaching situations. If several ratings could be obtained, reliability would be increased. However, this would also decrease the feasibility of the testing by either requiring more raters or increasing the number of viewing sessions necessary.

www Go to HK*Propel* to complete Student Activity 12.8.

Types of Rating Scales

Verducci (1980) delineated two basic types of scales: relative scales and absolute scales. **Relative scales** are scored by comparing performance to that of others in the same group. This normative approach has the virtue of distinguishing ability well within the group but creates problems if performers are to be compared with those in other groups. Relative scales are classified as follows: rank order, equal-appearing intervals, and paired comparisons.

In the widely used rank-order approach, all performers within a group are ranked on a given skill. If you are evaluating more than one skill, evaluate all the performers on one skill before moving on to the next. Rank-ordering forces differentiation among all performers, but it does not account for the degree of difference between them. Recall that the rankings result in ordinal numbers (see chapter 3).

The equal-appearing intervals method is often used when ranking groups of 20 or more participants. Using this technique, the rater places participants with similar performances into the same categories. For example, categories such as *best, good, average, poor,* or *worst* might be used. Typically, a larger percentage of participants fall into the middle categories than the extreme categories.

In the paired-comparison method, the rater compares each participant to every other participant and determines which of each pair is better than the other on the trait being assessed. When this is done for all possible pairs of participants, the results can be used to establish a relative ranking of all the participants in the group. This technique works well with groups of fewer than 10 participants.

With **absolute ratings**, a performer is evaluated on a fixed scale; performance is compared with a predetermined standard. This approach is not affected by the group in which a person is tested, and several people may end up with the same rating. Absolute scales can be classified into four types: numerical scales, descriptive scales, graphic scales, and checklists—the most widely used being numerical scales and checklists, which we will discuss in some detail. For discussions of less popular types of relative and absolute scales, see Verducci (1980; chapter 8).

The numerical scales in tables 12.2 and 12.3 describe the levels of performance needed to earn a certain number of points. The Ellenbrand gymnastics rating scale (table 12.2) ranges from 0 to 3, whereas the Hensley tennis rating scale (table 12.3) ranges from 1 to 5. In general, numerical scales range from 1 to 9 points; it is usually difficult to accurately discriminate more than nine performance levels. Numerical scales are most useful when performers can be classified into a limited number of ordered categories, and there is consistent agreement about the characteristics of each category.

Table 12.3 Tennis Rating Scale: Forehand and Backhand

5 = excellent	Proper grip, good balance, proper footwork, and near-perfect form. Demonstrates consistent stroke mechanics. Anticipates opponent's shots. Placement appropriate for opponent's weaknesses or position.
4 = good	Proper grip, good balance, adequate footwork, and acceptable but not perfect form. Demonstrates above-average consistency of stroke mechanics. Anticipates opponent's shots. Consistent placement within court area.
3 = average	Proper grip and acceptable balance, but footwork is poor. Form is somewhat erratic and inefficient, resulting in inconsistency in shot placement. Style of play may be defensive. Little anticipation of opponent's shots.
2 = fair	Uses improper grip at times, poor footwork, and basically incorrect form. Inconsistent stroke mechanics. Defensive style of play, merely trying to get ball over net. Little anticipation of opponent's shots. Unable to sustain a rally.
1 = poor	Incorrect grip, off balance, with poor footwork. Form is very poor and erratic. Inaccurate shot placement. No anticipation of opponent's shots. Experiences difficulty in getting ball over net.

Reprinted by permission from L.D. Hensley (ed.), *Tennis for Boys and Girls: Skills Test Manual* (Reston, VA: AAHPERD, 1989).

Checklists, which usually mark the absence or presence of a trait, are useful if both the process and the outcome are being evaluated. Each skill is checked when the task is demonstrated or explored. This criterion-referenced approach provides a concrete evaluation of given performance levels and gives specific feedback to each performer. It is easy for an instructor to determine which competencies are lacking and to provide extra practice in those areas. Table 12.4 presents the Level 2 Parent and Child Aquatics checklist from the *American Red Cross Water Safety Instructor's Manual* (American Red Cross, 2009).

Table 12.4 Parent and Child Aquatics Level 2

Skills	Completion goals
HOLDING AND SUPPORT TECHNIQUES	
Face-to-face positions	
• Hip support on front	Demonstrate
Back-to-chest position	
• Hip support on back	Demonstrate
• Back support	Demonstrate
• Arm stroke	Demonstrate
Side-to-side position	Demonstrate
Shoulder support	Demonstrate
WATER ADJUSTMENT, ENTRY, AND EXIT	
Water entry	
• Seated position	Demonstrate, with assistance
• Seated position—rolling over and sliding in	Demonstrate, with assistance
• Stepping or jumping in	Demonstrate, with assistance
• Using a ladder	Demonstrate
• Using stairs	Demonstrate
Exploring the pool	Explore, independently, in shallow water
Water exit	
• Using side of pool	Demonstrate
• Using a ladder	Demonstrate
BREATH CONTROL	
Underwater exploration	
• Opening eyes and retrieving objects below the surface	Explore, with support, in shallow water
• Opening eyes and retrieving submerged objects	Explore, with assistance, in shallow water
Bobbing	Demonstrate
BUOYANCY ON FRONT	
Front float	Demonstrate, with assistance
Front glide	Demonstrate, with support or assistance
Front glide to the wall	Demonstrate, with assistance
BUOYANCY ON BACK	
Back float	Demonstrate, with support or assistance
Back glide	Demonstrate, with support or assistance
CHANGING DIRECTION	
Roll from front to back	Demonstrate, with assistance
Roll from back to front	Demonstrate, with assistance
SWIM ON FRONT	
Passing between adults	Demonstrate, with assistance
Drafting with breathing	Demonstrate, with assistance
Leg action—alternating or simultaneous movements	Demonstrate, with assistance
Arm action—alternating or simultaneous movements	Demonstrate, with assistance
Combined arm and leg actions on front with breathing	Explore, with assistance
SWIM ON BACK	
Leg action—alternating or simultaneous movements	Demonstrate, with assistance
Arm action—alternating or simultaneous movements	Demonstrate, with support or assistance
Combined arm and leg actions on back	Explore, with support or assistance
WATER SAFETY	
Wearing a life jacket in the water	Discuss (parent) and demonstrate (child)
Reaching assists	Discuss or demonstrate (parent)
Basic water safety rules review	Discuss (parent)
Safety at the beach and at the water park	Discuss (parent)
Water toys and their limitations	Discuss (parent)

Data from American Red Cross (2009).

Common Errors in Rating Scales

Several common errors occur with rating scales. The most common of these, also discussed in chapter 9, is the **halo effect**, which is the tendency of a rater to elevate a person's score because of personal bias. Similarly, the rater may believe that the performance being evaluated is not indicative of a performer's normal level and rate that performer based on previous performances. The halo effect may also work in reverse; the rater may reduce a person's score because of negative bias.

Another common error, termed **standard error**, occurs when a rater has a standard different from that of other raters. Consider the judges' ratings in table 12.5. Inspection of the rating indicates that all three judges ordered the performances similarly; however, the numerical ratings from judge C were substantially below those from judges A and B, indicating that judge C was functioning with a standard vastly different from that used by judges A and B. This could be a major problem if all participants were not rated by all judges.

Table 12.5 Standard Error Exemplified by Three Judges' Ratings

	Judge A	Judge B	Judge C
Performer 1	9	8	4
Performer 2	8	9	4
Performer 3	7	7	3
Performer 4	5	6	1
Performer 5	5	5	1

A third error, called **central-tendency error**, reflects the hesitancy of raters to assign extreme ratings, either high or low. Assume that you are rating people on a scale of 1 to 5. There is a common tendency not to use the extreme categories, which reduces the effective scale to three categories (i.e., 2, 3, and 4). This not only causes scores to bunch around the mean, but also reduces the variability in the data, which can reduce reliability.

Suggestions for Improving Rating Scales

You can take several steps to alleviate many of the problems associated with rating scales.

1. Develop well-constructed scales. Here are several suggestions for doing this:
   a. State objectives in terms of observable behavior.
   b. Select traits that determine success.
   c. Define selected traits with observable behaviors.
   d. Select the degrees of success or attainment for each trait and define them with observable behaviors.
   e. Select and develop the appropriate scale for the rating instrument.
   f. Try out and revise the rating scale.
   g. Use the rating scale in an actual test situation.

2. Thoroughly train the raters. They should have a clear understanding of the traits measured and be able to thoroughly differentiate levels of performance.

3. Explain common rating errors to the raters. If they are aware of these pitfalls, they may avoid them.

4. Allow raters ample time to observe behaviors. This will increase the sampling unit of the performances.

5. Use multiple raters whenever possible. If this is not possible, then test several raters on a common group to check their objectivity. Finally, raters should rate one trait at a time, then move on to the next trait. Following this approach improves consistency.

Other Tests

In addition to objective tests and subjective ratings, other classifications of tests may be used for the measurement of human performance. Two such examples, performance-based testing and trials-to-criterion testing, are described briefly in the following paragraphs. Performance-based testing involves actual performance of the activity that is being assessed. Trials-to-criterion testing offers instructors a way to reduce the amount of time spent testing.

Performance-Based Testing

Differing from performance assessment, which obtains evidence for making inferences about a student's progress (see chapter 15 for additional information), performance-based testing results in a classification of skill based on the actual and specific performance that produces the score. In this situation, a concrete criterion exists. This occurs in sports such as archery, bowling, golf, and swimming. In archery, the score for a round of arrows indicates how well a performer did for that performance. Scores can be evaluated to examine the stability of performance at a given distance, or performance at different distances can be evaluated to determine the concurrent validity of shooting across distance.

Bowling and golf provide a unique situation: Although you can use total scores for assessment purposes, doing so does not evaluate specific elements of the game. Picking up a specific pin for a spare in bowling and combining the drive, short game, and putting in golf are elements that can be evaluated separately. These performances often focus on the process rather than the product, and rating scales often have a good degree of validity in these areas. The virtue of developing rating scales in these areas is that you have a concrete criterion against which to validate them. On the other hand, correlating a rating scale against an objective skills test may or may not represent validity. You may be simply getting a congruency of the two performances, neither of which may be valid.

© Human Kinetics

Performance-based testing, in which you can tell a performer's ability by looking at the total score, can be used in archery.

Trials-to-Criterion Testing

An alternative approach to the measurement of skills involves the use of **trials-to-criterion testing** (Shifflett and Shuman, 1982), in which a student performs a skill until he or she reaches a certain criterion performance. For example, consider a conventional free throw test involving 20 attempts. A class of 30 students would have 600 free throws to complete, which may not be feasible. This number can be reduced by using a trials-to-criterion approach, in which students would be given some number of successes to complete.

For example, instead of shooting 20 free throws and counting the number of successful shots made in 20 attempts, students might be instructed to shoot until they make 8 free throws. As soon as they make 8, they are to report the number of attempts it took to make the 8; the best score would be 8 attempts to make 8 shots. (At some point the test would have to be discontinued for those students who simply cannot make the 8 free throws, but for a given class, most of the shooters can probably make 8 out of 20 attempts.) If an average of 12 trials is needed to make 8 free throws, only 360 free throws would need to be attempted in a class of 30 students (rather than 600), thus saving a great amount of time. If the correlation between the trials-to-criterion test scores and the scores on the conventional free throw test (involving 20 attempts) is high, you would have an attractive way of reducing testing time. Furthermore, in a teaching situation, this would allow a teacher to spend additional time with the weaker shooters, while the stronger shooters can move on to other skills.

www Go to HK*Propel* to complete Student Activity 12.9.

PURPOSES OF MOTOR PERFORMANCE ANALYSIS

Motor performance analysis can be applied to several research questions. The primary purposes of human performance analysis are selection, classification, diagnosis, and prediction. (Note that some of these concepts were initially identified in chapter 1.) Analysis questions apply not only to athletic assessment, but also to job performance. *Selection* refers to the ability of a test (or test battery) to discriminate among levels of ability and thus allow someone to make choices. From an athletic standpoint, this could involve making cuts from or assigning players to teams. In job performance, selection is used in hiring. Firefighter and police recruits are often asked to complete physical performance tests that are related to the tasks they will perform in the line of duty. *Classification* involves clustering participants into groups for which they are best suited. In athletic situations, players are assigned to positions or events; job classification involves assignment to a task. *Diagnosis* concerns determining a person's weaknesses based on tests that are validly related to performance in a given area. Diagnosis is used in sport to design individualized training programs to help improve performance. In the working world, diagnostic tests could be used to examine job performance. Finally, although prediction overlaps somewhat with selection and possibly classification, it does provide a slightly different approach by examining the future potential of performers. See Disch and Disch (2005) for additional information and examples of selection, classification, diagnosis, and prediction.

In their work for the Canadian Association of Sport Science, MacDougall and Wenger (1991) discussed the following benefits that an athlete can derive from motor performance testing:

- Indicates athletes' strengths and weaknesses within the sport and provides baseline data for individualized training

- Provides feedback for the athlete and coach about effectiveness of the training
- Provides information about the athlete's current performance status
- Is an educational process to help the athlete and coach better monitor performance

They further stated that for testing to be effective, evaluators should follow these procedures:

- Include variables relevant to the sport, job, or task
- Select reliable and valid tests
- Develop sport-specific test protocols
- Control test administration rigidly
- Maintain the athlete's right for respect
- Repeat the testing periodically
- Report the results to both coach and athlete directly

There are two major statistical approaches to the analysis of the subdomains of human performance. The first is the correlational approach to examine the various relationships across groups. You can use multiple regression for this approach if only two groups are involved (correlation and regression are presented in chapter 4). The second approach is the divergent group approach using discriminant analysis, which examines groups (can be more than two) known to differ by using variables that are logically related to performance of a skill. Tests found to discriminate between performance levels in the divergent groups are then said to be predictors from a construct-related validity standpoint. The inferential statistical procedures presented in chapter 5 can also be used for this validation process.

The following sections present examples of analysis of human performance for the purposes of selection, classification, diagnosis, and prediction.

Selection

An interesting study of the use of predictive validity for selection was conducted by Grove (2001). He studied 74 male baseball players who were competing at the college level: junior college (JUCO) or Division I (D1). A third group comprised 16 players who were drafted by the professional league within 24 months of the testing (Pro). The players were measured on run times (30 and 60 yards [27.4 and 54.9 m]), throwing speed, vertical jump, and medicine ball throw. The data were analyzed by adjusting for age differences. Significant differences ($p < .001$) were found for vertical jump, medicine ball, and throwing speed. Additional analyses indicated that the JUCO group scored lower than the D1 and Pro groups on the vertical jump and medicine ball tests. The Pro group performed better on the throwing speed test than the other two groups. The means and standard deviations for all of the tests are presented in table 12.6. The battery was simple to administer, and with the exception of the vertical jump and medicine ball, the tests are widely used by baseball coaches and scouts at all levels. Grove concluded that the test battery used has merit as a cost-effective screening test for talent identification in baseball. He further stated that more pronounced differences may exist when players are grouped according to position as well as playing level.

Another example of selection is a predictive validity study of prospective National Hockey League (NHL) players by Tarter, Kirisci, Tartar, Weatherbee, Jamnik, McGuire, and Gledhill (2009). They used aggregate physical fitness indicators to predict success in the NHL. Their sample was 345 players invited to the annual combine conducted before the

Table 12.6 Descriptive Statistics and Subgroup Comparisons for Baseball Performance Tests

Measure	JUCO		D1		PRO	
	Mean	SD	Mean	SD	Mean	SD
30 yd [27.4 m] time (s)	4.04	.025	4.02	0.17	3.97	0.16
60 yd [54.9 m] time (s)	7.39	.045	7.34	0.28	7.27	0.24
Radar gun (mph)**	78.14	3.45	80.61	4.63	84.40	3.49
Vertical jump (cm)*	54.96	7.53	56.04	8.32	57.33	10.70
Medicine ball (kgm)*	14.50	1.61	16.47	1.37	16.18	1.80

Notes: Group sizes for the JUCO, D1, and Pro players were 32, 26, and 16, respectively. For measurement conversions, refer to a site such as www.worldwidemetric.com/measurements.html; kgm = product of the ball weight (in kilograms) and the distance throw (in meters).

*$p < .001$; **$p < .0005$

Data from Grove (2001).

NHL draft. The tests selected measured upper-body strength, lower-body power, aerobic fitness, anaerobic fitness, and body composition.

Tarter and colleagues (2009) used exploratory factor analysis (a statistical method to determine the components of the aspect being measured) and receiver operating curve analysis (a statistical method to set cutoff scores between performance levels) to derive an overall composite index. Their purpose was to determine the accuracy of this index for identifying players capable of transitioning into the NHL. Tarter and colleagues also hoped to establish the construct-related validity of the test battery used. The authors' criterion for success was those players who played a minimum of five games in the NHL within a four-year period.

Tarter and colleagues found that defensemen scoring in the 80th percentile on the composite physical fitness index (CPFI) had close to a 70% probability of meeting the criterion measure (i.e., transitioning into the NHL). Forwards scoring in the 80th percentile had a 50% chance. When success was examined at the 90th percentile, defensemen were found to have a 72% probability and forwards a 61% probability.

Based on these results, the authors developed a simplified assessment protocol that consisted of right- and left-hand grip strength, push-ups, long jump, body composition, and body weight. This simplified battery, although not as accurate as the full battery, was found to be acceptable for field testing situations. In their discussion, the authors further pointed out that physical fitness is just one ingredient required for success at the NHL level. Additional information coming from such sources as scouting reports, game statistics, and mental efficiency measures should lead to increased predictive validity (the use of this additional information will be explained in more detail later in this chapter). Tarter and colleagues (2009) concluded that using historical performance, skill observation, physical fitness, character, and neurocognitive measures together increased the accuracy of predicting success.

WWW! **Go to HK*Propel* to complete Student Activity 12.10.**

Classification

Leone, Lariviere, and Comtois (2002) provided an excellent example of classifying athletes based on anthropometric and biomotor variables. The participants in their study were elite adolescent female athletes with a mean age of 14.3 (standard deviation of 1.3 years). The athletes competed in tennis ($n = 15$), figure skating ($n = 46$), swimming ($n = 23$), and volleyball ($n = 16$). The descriptive values for the anthropometric and biomotor variables are presented in table 12.7.

Table 12.7 Physical Characteristics of the Athletes by Sport (Mean ± SD)

	Tennis (n = 15)	Skating (n = 46)	Swimming (n = 23)	Volleyball (n = 16)
Age (yr)	13.9 ± 1.3	14.7 ± 1.5	14.3 ± 1.3	13.8 ± 1.3
Body mass (kg)	50.6 ± 8.3	46.6 ± 8.0	54.3 ± 6.9	57.7 ± 8.3
Height (m)	1.61 ± 0.06	1.54 ± 0.07	1.62 ± 0.06	1.63 ± 0.05
Elbow (cm)	6.12 ± 0.30	5.87 ± 0.35	6.29 ± 0.26	6.40 ± 0.33
Knee (cm)	8.81 ± 0.43	8.63 ± 0.76	8.77 ± 0.34	9.31 ± 0.50
Biceps (cm)	25.5 ± 2.8	24.4 ± 2.3	27.8 ± 1.8	26.6 ± 2.2
Calf (cm)	34.0 ± 2.8	33.0 ± 2.7	34.4 ± 1.6	34.4 ± 2.2
Skinfolds (mm)	57.4 ± 17.8	47.7 ± 12.3	56.0 ± 15.0	63.1 ± 15.5
Push-ups (n)	57.8 ± 14.4	36.7 ± 13.5	62.1 ± 16.0	50.2 ± 13.5
Burpees (n)	46.1 ± 23.8	64.6 ± 33.2	52.5 ± 32.7	56.0 ± 28.4
Flexibility (cm)	37.3 ± 5.0	42.6 ± 5.1	41.0 ± 6.0	39.1 ± 6.9
$\dot{V}O_2$max (ml · kg^{-1} · min^{-1})	49.5 ± 4.4	48.3 ± 4.0	47.6 ± 3.1	48.9 ± 3.6

Data from Leone, Lariviere, and Comtois (2002).

Discriminant analysis of the tests revealed three significant functions ($p < .05$). Functions are essentially the underlying dimensions that are being measured by the tests. The maximum number of significant functions possible is $K - 1$, where K is the number of groups; therefore, this test battery maximally discriminated across the four groups. Inspection of the functions revealed that the first function (i.e., dimension) discriminated between the figure skaters and all of the other sports grouped together. The variables that accounted for this discrimination were body mass, height, push-ups, and biceps girth. The second function reflected differences between volleyball players and swimmers. The variables accounting for this discrimination were body mass, biceps girth, calf girth, and height. The third function differentiated the swimmers and tennis players. The variables responsible for this discrimination were body mass, biceps girth, calf girth, sum of skinfolds, and height. The significant discriminant functions were able to classify 88% of the players into their correct sports. The classification summary table is presented in table 12.8. The results of this study indicated that elite adolescent female athletes could be properly classified into their respective sports based on the selected test battery. The anthropometric variables were primarily responsible for most significant classifications. Of course, physical and anthropometric tests alone will not perfectly classify participants. Other factors, such as motivation and desire, could affect ultimate sport performance, and these factors are discussed in chapter 14.

Table 12.8 Classification for All Significant Discriminant Functions After Validation

Groups	N	Tennis	Skating	Swimming	Volleyball
		PREDICTED GROUP MEMBERSHIP, N (%)			
Tennis	15	11 (73.3)	0 (0.0)	3 (20.0)	1 (6.7)
Skating	46	0 (0.0)	46 (100)	0 (0.0)	0 (0.0)
Swimming	23	3 (13.0)	0 (0.0)	18 (78.3)	2 (8.7)
Volleyball	16	1 (6.3)	0 (0.0)	2 (12.6)	13 (81.3)

Diagnosis

A study by Doyle and Parfitt (1996), based on the principles of personal construct theory (Kelly, 1955), presented a unique quantitative performance profiling technique that involved not only motor performance factors, but also psychological parameters. Participants in the study were 39 track and field athletes (22 males and 17 females) with a mean age of 20.9 years (standard deviation = 2.26). The profiling technique examined how each athlete currently felt about his or her preparation for competition. Instead of actually completing performance tests, the athletes were asked to respond to questions rating themselves on the various parameters displayed in the profile (see figure 12.7). The athletes were asked to respond on a scale of 1 to 10, with 1 being not important at all and 10 being of crucial importance. Their reported scores were then correlated with their performance in three upcoming competitions. To establish a criterion for success, a person's performance was recorded as a percentage of his or her personal best time divided by his or her performance time. This allowed all the athletes to be compared across various events. Multiple correlations were calculated between the performance and the event competition scores. The results of the analysis indicated that the profiling technique could validly predict competition scores. Progressively stronger relationships were found between profile scores and performance measures from competition 1 to competition 3. It was concluded that there may be a learning process involved in the ability to rate the current state more precisely.

www Go to HK*Propel* to complete Student Activity 12.11.

Prediction

Although prediction is normally accomplished through regression analysis, an interesting example of construct validity is presented by Sierer, Battaglini, Mihalik, Shields, and Tomasini (2008). They studied college football players who participated in the 2004 and 2005 National Football League (NFL) combines to determine if selected tests used at the combine could discriminate between the drafted and nondrafted players. For the purpose of their study they characterized players into three groups: skilled, big skilled, and lineman. Skilled players were offensive and defensive backs and wide receivers. The big skilled players were linebackers, fullbacks, and tight ends. The linemen included both offensive and defensive linemen. Quarterbacks, punters, and place kickers were not included in the study. The variables included in the analysis were height, weight, 40-yard (36.6 m) dash times, bench press repetitions at 225 pounds (102 kg), vertical jump, standing long jump, pro-agility shuttle, and the results from the three-cone drill. To be included in the study, players had to have performed all of the tests. A total of 321 players met this criterion.

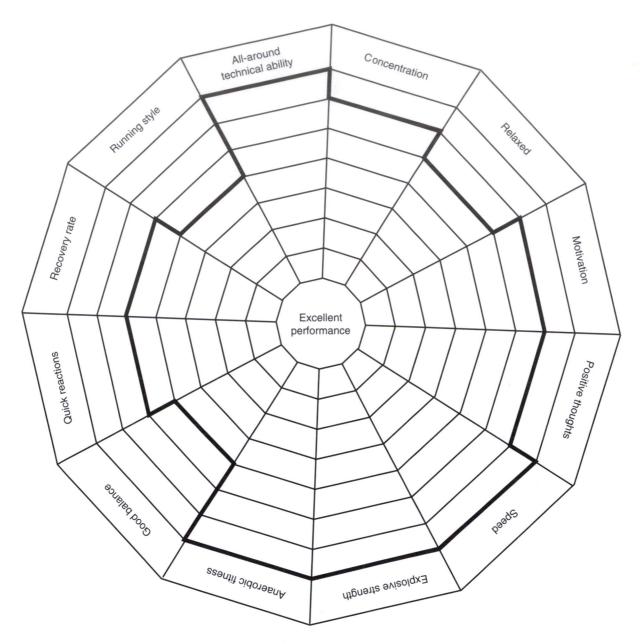

Figure 12.7 Sample performance profile.

The data were analyzed by groups using an independent t-test. The alpha level was chosen to be .05 but was divided by three to help control for the possibility of type I errors. If the drafted players outperformed the nondrafted players, the tests were considered to possess construct-related validity. For the skilled group, the 40-yard (36.6 m) dash time, vertical jump height, pro-agility shuttle time, and the results from the three-cone drill were all found to significantly distinguish between the groups. For the big skilled group, the 40-yard (36.6 m) dash and the three-cone drill were the only two tests that were found to be significant. For the linemen group, results from the 40-yard (36.6 m) dash, the 225-pound (102 kg) bench press, and the three-cone drill were all statistically significant.

Although a number of tests were not found to significantly differentiate between the drafted and nondrafted players, all of the tests that were found to be significant were in the direction of the drafted players. The fact that different tests were found to be significant for different groups indicates the specificity of physical abilities needed to perform tasks characteristic of the various groups. It would have been interesting to use a more sophisticated statistical technique than the independent t-test to analyze the groups and the drafted and nondrafted players at the same time. A 3 × 2 ANOVA would have allowed the authors to examine a possible interaction between the two independent variables.

Motor performance testing of athletes can be beneficial, but there are many factors that must be considered. Goldsmith (2005) listed 10 golden rules for testing competitive athletes. Those rules are as follows:

1. Test for things that make sense.
2. Test because you believe it will make a difference.
3. Test with a performance-focused goal.
4. Use a battery of tests and report the results in a profile. Don't combine into a single score unless you have an extremely good reason.
5. Try to provide test results as quickly as possible. This is both for your benefit and the athlete's.
6. Although tests should measure a single trait, be aware of possible effects on other aspects of performance (e.g., does fatigue as measured by one test have an effect on technique?).
7. Don't necessarily rely on previously developed tests; develop your own!
8. Keep records. These may be useful over time. Sometimes your memory may not perceive things as they actually were.
9. Remember the Max–Min–Con Principle: Use tests that maximize the experimental variance, minimize the error variance, and control for as many other factors as possible.
10. Educate athletes about testing. Impress upon them the importance of maximal effort for optimal results.

SPORT ANALYTICS

Sport analytics is a new term for statistical research primarily in the area of professional sports, but with applications to sport at all levels. It was popularized as a result of the book *Moneyball: The Art of Winning an Unfair Game* by Michael Lewis (2004), which recounts the work of Oakland A's general manager Billy Beane, who hired statistical consultant Paul DePodesta as his assistant general manager to examine players' performance. Although Bill James had been working in this area for a number of years and is widely accepted as the father of sabermetrics (the empirical analysis of baseball), Oakland was the first team to seriously apply statistical analysis to the drafting and acquisition of players.

As you've read in earlier sections of this text, most statistical analyses applied to testing athletes has been associated with physical or motor performance measures. The field of sport analytics examines various measures of in-game performance and also derives additional measures to help evaluate teams and players (e.g., placement, diagnosis, achievement confirmation, prediction, training evaluation, and motivation; Park and Kang, 2014). Beane was originally able to identify players who had been undervalued based on traditional baseball statistics: batting average, home runs, and runs batted in (RBIs). He and DePodesta began to use lesser-known measures—on-base percentage and slugging percentage—to draft players the Oakland A's could afford who could perform at the major-league level.

As a result of the Oakland A's strategy, sport analytics has become prominent in all areas of professional sport in the United States and is starting to be used by other sports and other organizations throughout the world. The use of traditional scouting departments, which was almost totally discounted in Lewis' *Moneyball*, has now become part of a more comprehensive model to provide a thorough picture of prospective players.

There are a number of considerations for teams that want to use the sport analytics approach, including finding people who have both the computer and statistical background to acquire and manipulate the data. Alamar (2013) stated that the goals of sport analytics are twofold: first, to save the team's decision makers time by providing all the relevant information necessary for evaluating players in an efficient manner, and second, to provide insight into novel ways of determining who is the best player suited for that team's system. Producing this result involves team personnel at all levels buying into the use of both quantitative and qualitative measures for decision making.

According to Alamar (2013), the three essential components of a sport analytics program are (1) data management, (2) predictive models, and (3) information systems. Some of the professional teams taking early advantage of analytics include the Portland Trail Blazers and the Boston Celtics of the NBA, the Philadelphia Eagles and New England Patriots of the NFL, and the St. Louis Cardinals and San Diego Padres, along with the Oakland A's in Major League Baseball. Alamar cites the Celtics' pick of Rajon Rondo in the 2006 NBA draft as an example of using analytics to identify a player who was undervalued, with Rondo's ability to rebound being an important contribution to the Celtics team.

The key to using analytics efficiently is to develop a system that works for a given team. Because of the massive amounts of data that may be gleaned from game statistics, videos, performance testing, psychological evaluations, medical testing, and the like, it is imperative that a comprehensive plan be developed to acquire and analyze this information. Both quantitative and qualitative sources must be blended into a report that provides decision makers with usable information. Analytics simply provides information; it does not make decisions.

A number of research studies have been conducted in the area of sport analytics, but much of the information is proprietary to the team that conducted the research. The criterion for a successful model is often dichotomous, which leads to a number of statistical approaches producing successful results. Most statistical analyses in the area of sport analytics involve the use of some type of multiple regression model. The criterion may be wins and losses when examining overall team performance, or it might be the probability of success when drafting or trading for players. The key to success in developing a valid analytical model is based on the approach developed by the specific team. Alamar (2013) states that there are five basic questions that must be answered before any analysis is attempted:

1. What was the thought process that led to the analysis?
2. What is the context of the result?
3. How much uncertainty is in the analysis?
4. How does the result inform the decision making process?
5. How can we further reduce the uncertainty? (p. 55)

Another sport analytics text, *Mathletics* (Winston, 2009), presents a number of analytical studies in baseball, football, basketball, and other sports. It gives examples of how a variety of questions can be approached from a statistical standpoint—using variables that are commonly available—for creating measures that provide novel insights into the analysis of success in sport. Topics include predicting team success from one year to the next, comparing various players to determine their actual worth for trade purposes or salary negotiations, and analyzing various lineups to determine the probability of success in varying situations.

We are just beginning to see the widespread use of analytics by virtually all professional sports teams. There is also a strong demand for trained sport analysts in a variety of settings, including media companies, collegiate sports teams, sports marketing agencies, sports betting companies, strength and conditioning clubs, and throughout the field of sport science. Although analytics may be used at any level, the need to acquire, manipulate, manage, analyze, and visualize extremely large datasets requires extensive computer and statistical skills. The days of using only qualitative recommendations and quantitative data acquired through performance testing for sport decision making are over.

Dataset Application

The chapter 12 large dataset in HK*Propel* provides sample data on high school football players. There are position, anthropometric, strength, endurance, and speed variables. Use SPSS to conduct the following:

1. Calculate descriptive statistics on the variables (chapter 3).

2. Calculate the correlations among the various variables (chapter 4).

3. Determine if various position players differ on measured variables (chapter 5).

4. Develop histograms graphing height and weight by player position (chapter 3).

5. Use the descriptive statistics you generated above and frequencies (chapter 3) to develop performance profiles for the treadmill, strength, and speed variables. Note that a lower time is better for the 30-yard (27.4 m) dash.

VIDEO ANALYSIS APPS

Video analysis, which has long been used to evaluate athletic performance, is the process of using any motion recording and gaining useful and actionable information from it—usually either through performance analysis or biomechanical analysis. Video analysis can help sport scientists, physiotherapists, sport coaches, team coaches, and the athletes themselves. (Even the simple review of recorded video, without doing actual analysis, can help athletes with biofeedback.) Performance analysis usually creates an actionable insight on how to get better from either modeling or normative data, whereas biomechanical analysis focuses on the kinesiology of movement. Coaches and teachers usually favor performance analysis; sports medicine professionals are interested in analyzing training techniques in athletic movement.

With the great progress in video and mobile technology, many new sport video analysis apps have been developed to record and accurately scan athletes' performances during training or competition. Recording a sport performance video and then playing it back to check the details can enhance and facilitate coaching, teaching, and self-improvement. For example, a slow-motion function can help the user analyze the movement of a baseball pitch or golf swing. Several analytical features are often included in these apps:

• *Side-by-side viewing.* Comparing technique, either from athlete to athlete or session to session, is the most common way to use this type of analysis. Simple side-by-side comparison is a straightforward method of distinguishing similarities and differences between elite and subelite performances or showing an athlete developing ability over time. This procedure is often seen in the television broadcasting of professional golf tournaments.

- *Overlay.* Showing the same athlete or different athletes at the same time is a great way to appreciate time and space. At times, overlaying a ghost view is useful to see changes in skill or technique as well. The Winter Olympics uses overlay views with skiers and the NFL uses overlay with the 40-yard dash.

- *Timers and chronometers.* Coaches use timers to assess and address the gap between the way athletes use available space and their speed of movement. Timers can help dissect a sporting action or divide an event into splits or milestones.

- *Measurement tools.* Distance with a known reference object on screen can provide excellent accuracy if the camera angle is square. Though not a perfect solution, reference markers allow accurate estimates within reasonable distances.

- *Joint angles.* Joint angles, and sometimes release angles, may be measured (within reason) if the camera is set up at the right point of action. For example, discerning the angle of the knee joint at the long jump board is possible if the camera is set at the right distance and height.

- *Composite picture sequence display.* Stromotion (or a similar tool) is popular because it creates a photo sequence for easy evaluation of movement. Because video is a sequence of images, the composite picture is both a favorite visual aid among coaches and a timeless teaching tool.

Not every app will include all these features, but they are essential tools for professionals who want to extract more from their video analysis. Most performance analysis will include actionable measurements for coaching.

The following are a few examples of sport analysis video apps commonly used in practice:

- *Dartfish Mobile (www.dartfish.com/mobile).* Dartfish designed its entry-level product, Dartfish Mobile, to assess fundamental activities such as light tagging and sharing. The chronometer is accurate enough to time a 10-meter dash within .01 seconds if the view is positioned correctly. Although this product is the lowest entry point in the company's product ecosystem, its purpose is to get people started with sharing and recording, and then move them to more robust options.

- *VIP & TEAM Coach's Eye.* Coach's Eye, the leading video platform for coaches and athletes, is available for Apple or Android devices. It allows side-by-side comparison, slow-motion playback, and options to add text to videos.

- *OnForm: Video Analysis App (www.getonform.com).* This app is free and includes a handful of tools that make it viable for simple tasks like annotating sporting actions qualitatively.

With the high-quality cameras of mobile phones and ever-improving technology, it is expected that sport analysis video apps will become an increasingly convenient tool for both assessing and coaching sport skills.

EMPLOYMENT-RELATED PERFORMANCE TESTING

To determine if an applicant is suited for a particular job, some employment-related assessments are essential. Although most tests today are cognitive, assessments of physical ability (e.g., lifting a certain amount of weight) and skill (e.g., operating complex machinery) are still often needed. In fact, for certain professionals (e.g., firefighters, police officers, military personnel, truck drivers), physical performance or skill testing is essential. Although most job performance tests are used for pre-employment screening or during recruitment processes, testing throughout employment is sometimes required.

There are three major focuses for employment-related performance testing: (1) worker safety, (2) worker productivity, and (3) legal requirements. Worker safety is to determine if a person is able to meet the physical demands of a job so that injury can be avoided. Worker productivity is to determine if a person has the endurance and skill to work efficiently. Finally, legal requirement is to help determine if a hire decision is appropriately made, because failing to hire a member of a protected group can result in discrimination litigation—for example, females are often a protected group for physically demanding jobs (Jackson, 2006b). This raises important considerations if the job to be completed involves strength or other physical abilities that often differ between males and females as a group.

Developing Employment Performance Tests

Although the key steps for developing a psychomotor test described earlier in this chapter are useful, developing an employment performance test also requires attention to the three focuses of an employment performance test—namely, if the test truly measures the performance required for a job, how the performance should be measured, and finally, how the evaluation standards are determined.

What to Measure

Determining what to measure is basically a question of content validity. Rather than consulting a group of content experts or stakeholders, as is usually done when determining the content validity of a cognitive test, the content validity of an employment performance test is usually determined through a job analysis, in which important work behaviors and key physical fitness, ability, and skill components are identified and prioritized. A number of methods have been used for the job analysis, including observing the job, recording job activities, interviewing workers and supervisors, and collecting job-related information using questionnaires (Gael, 1988). Jackson et al. (1998) used a questionnaire to determine tasks that oil production workers may encounter while performing their work. Below are a few tasks listed in their questionnaire:

- Climbing stairs while carrying loads weighing 50 to 75 pounds (23-34 kg)
- Holding and manipulating heavy fittings while in a crouched or awkward position
- Lifting loads of 75 to 100 pounds (34-45 kg) from the ground to waist height

Biomechanical, physiological, and psychological data are often collected at the same time as the job analysis so that specific physical demands from the job can be determined using scientifically validated measures (Jackson, 2006b).

How to Measure

There are generally two ways to measure job-related performance: work-sample or physical ability tests. A work-sample test consists of a set of job-related sample tasks (e.g., to lift up a specific weight or raise an extension ladder). One of the major advantages of work-sample tests is that validity is easy to confirm, since the tasks in the test are the actual tasks from the job. There are also many research studies to support using real-life tasks in a performance-related test; this is why many firefighter and police employment tests use the work-sample test (Arvey et al., 1992). There are, however, some limitations or disadvantages of the work-sample test. First, developing a realistic work-sample test may be expensive, and testing equipment is often difficult to move around or set up, making this type of testing impractical. Second, it may not be possible to measure candidates' maximal capacities. As a result, a candidate who barely passed the test using his or her maximal ability may not be an effective performer, and job-related injury risks could be high.

Physical ability tests are, in fact, the same performance-related physical fitness tests we learned about in chapter 11. The most commonly tested fitness components for job performance are strength and aerobic capacity. The key difference between physical ability tests and work-sample tests is that the former can measure test takers' maximal capacities. However, because physical ability tests are not job specific, some validation studies are needed to link the test with job-related performance (e.g., a strong person is not necessarily able to do a job-related lifting task effectively). Arnold et al. (1982) demonstrated that physical ability tests and work-sample tests can be highly correlated, but Jackson (2006b) reported that the correlations between isometric strength and work-sample tests ranged from .63 to .93. Jackson (2006b) further emphasized that physiological validation (Hodgdon and Jackson, 2000) is needed for the employment tests because they are not educational or psychological tests.

How to Set Performance Standards

Due to the nature of employment performance tests (determining whether a test taker has adequate ability to meet the job requirements), criterion-referenced evaluation decisions are appropriate, rather than a norm-referenced evaluation. This necessitates the development of a performance standard or cut score (for individual test items or the total test battery). A few additional factors should be considered when setting the standard (Jackson, 2006b):

• Adverse impact: If selection is unfair or biased against a protected group, including age (40 and over), sex, race, religion, disability status, and veteran status. Usually, when a performance standard is set higher, the chances increase of having an adverse impact on female applicants.

• Risk of injury: If the risk of injury to employees increases due to the hiring standard not being appropriately set (often set too low).

• Interpretation of the validation studies: If the validation studies provide sound physiological support for the test components selected and performance standards set by accepted scientific criteria.

• Environmental conditions: If the test conditions match the real-life working conditions and, if not, how the performance standards were adjusted.

When there is a physical demand or challenge in a job, three approaches are often used to avoid potential job-related injury and reduce workers' compensation costs: (1) ergonomics (i.e., redesign the work), (2) wellness and training (i.e., change the worker's fitness level or work habits), and (3) pre-employment testing (i.e., hire those with the physical ability to perform the work). Although pre-employment testing has been the most effective of the three, employers often decide to use the less effective and more expensive approaches of ergonomics along with wellness and training. As a result, physically demanding jobs are disappearing quickly and are unfortunately being replaced by regulated sedentary behavior (see Glover and Zhu [2017] for more information).

Pre-Employment Test Examples

We present two real-life pre-employment test examples: (1) the Candidate Physical Abilities Test (CPAT) for firefighter candidates, which is a work-sample test, and (2) the Army Combat Fitness Test (ACFT) for soldiers, which is a physical ability test.

Candidate Physical Abilities Test

The Candidate Physical Abilities Test (CPAT), developed by the International Association of Fire Fighters, is a tool to help fire departments select physically capable individuals to be trained as firefighters and perform essential job tasks at fire scenes. The CPAT consists of eight separate, sequential events (subtests), including the stair climb, hose drag, equipment carry, ladder raise and extension, forcible entry, search, rescue drag, and ceiling breach and pull. During completion of the CPAT, the candidate wears a 50-pound (22.68 kg) vest to simulate the weight of the protective clothing and self-contained breathing apparatus (SCBA). An additional 25 pounds (11.34 kg), using two 12.5-pound (5.67 kg) weights that simulate a high-rise pack (hose bundle), is added to the candidate's shoulders for the stair-climb event. Throughout all events, the candidate must wear long pants, a hard hat with chin strap, work gloves, and footwear with no open heel or toe. Watches and loose or restrictive jewelry are not permitted. The CPAT is a pass/fail test based on a validated maximum total time of 10 minutes and 20 seconds.

The hose drag is described in more detail below. For more information about other events, see *Candidate Physical Ability Test*, second edition (International Association of Fire Fighters, 2007).

HOSE DRAG

Purpose

To simulate the critical tasks of dragging an uncharged hoseline from the fire apparatus to the fire occupancy and pulling an uncharged hoseline around obstacles while remaining stationary.

Equipment

Uncharged fire hose with hoseline nozzle

Drum

Instructions

The candidate grasps a hoseline nozzle attached to 200 feet (60 m) of 1-3/4-inch (44 mm) hose and places the hoseline over the shoulder or across the chest, not exceeding the 8-foot (2.24 m) mark (the hoseline is marked to indicate the maximum amount of hose the candidates are permitted to drape across the shoulder or chest). The hoseline is also marked at 50 feet (15.24 m) past the coupling at the nozzle to indicate the amount of hoseline that the candidate must pull into a marked boundary box before completing the test.

The candidate drags the hose 75 feet (22.86 m) to a prepositioned drum, makes a 90° turn around the drum, and continues an additional 25 feet (7.62 m) (figure 12.8). Stopping within the marked 5- × 7-foot (1.52 m × 2.13 m) box, the candidate drops to at least one knee and pulls the hoseline until the hoseline's 50-foot (15.24 m) mark crosses the finish line. During the hose pull, the candidate must keep at least one knee in contact with the ground and the knee must remain within the marked boundary lines. This concludes the event. The participant then walks 85 feet (25.91 m) within the established walkway to the next event.

(continued)

HOSE DRAG *(continued)*

Figure 12.8   Hose drag testing.
© Human Kinetics

Scoring

The following practices are allowed:

- The candidate is permitted to run during the hose drag.
- The candidate is given one warning to keep at least one knee down.
- The candidate is given one warning to keep the knees in bounds.
- The candidate is given one warning for taking one step out of the box.

The following practices constitute a failure:

- The candidate fails to go around the drum.
- The candidate travels outside of the marked path.
- The candidate takes two steps out of the back of the box.
- The candidate commits a second infraction for not keeping at least one knee in contact with the ground.
- The candidate commits a second infraction for the knees being outside of the marked boundary.

Army Combat Fitness Test

Developed by the U.S. Army, the Army Combat Fitness Test (ACFT, 2018) has replaced the previous Army Physical Fitness Test. As part of military training, a candidate is expected to meet ACFT requirements regardless of age or sex. ACFT has six separate, sequential events (subtests) that assess the candidate's ability to perform physical tasks that could be encountered in combat conditions:

1. *Maximum deadlift (MDL).* The candidate performs a three-repetition maximum deadlift of a 60-pound hex bar and plates. The deadlifts replicate picking up ammunition boxes, a wounded soldier, supplies, or heavy equipment.

2. *Standing power throw (SPT).* The candidate tosses a 10-pound ball backward as far as possible to test the muscular explosive power that may be needed to lift yourself or a fellow soldier over an obstacle or to move rapidly across uneven terrain.

3. *Hand-release push-ups (HRP).* The candidate will have 2 minutes to do as many hand-release push-ups as possible. These are similar to traditional push-ups, but at the down position the candidate must lift the hands and arms from the ground and then reset to do another push-up.

4. *Sprint-drag-carry (SDC).* The candidate must run five times up and down a 25-meter lane, sprinting, dragging a sled weighing 90 pounds, and then carrying two 40-pound kettlebell weights. This is to simulate pulling a soldier out of harm's way, moving quickly to take cover, or carrying ammunition to a fighting position or vehicle.

5. *Plank (PLK).* The candidate must maintain a proper plank position for as long as possible. This event assesses the strength and endurance of the core muscles. Balance is a secondary component of fitness assessed by the PLK.

6. *2-mile run (2MR).* This is a timed run to assess endurance and cardiovascular strength.

Three levels of criterion-referenced standards were developed for the ACFT: *black* for soldiers in heavily physically demanding units or jobs; *gray* for soldiers in significantly physically demanding units or jobs; and *gold* for soldiers in moderately physically demanding units or jobs, which also represents the overall Army minimum standard for passing the ACFT. For more information, refer to www.army.mil/acft.

The corresponding points by level, as well as the minimum standards by event, are summarized as follows:

- *Black (70 points).* 180 pounds on the maximum deadlift; 8.5 meters for the power throw; 30 hand-release push-ups; 2 minutes, 9 seconds for the sprint/drag/carry; 5 leg tucks; and 18 minutes for the 2-mile run.

- *Gray (65 points).* 160 pounds on the maximum deadlift; 6.5 meters for the power throw; 20 hand-release push-ups; 2 minutes, 45 seconds for the sprint/drag/carry; 3 leg tucks; and 19 minutes for the 2-mile run.

- *Gold (60 points, Army minimum standard).* 140 pounds for the maximum deadlift; 4.6 meters for the power throw; 10 hand-release push-ups; 3 minutes, 35 seconds for the sprint/ drag/carry; 1 leg tuck; and 21 minutes, 7 seconds for the 2-mile run.

For more information, refer to the following website: https://armypubs.army.mil/epubs/ DR_pubs/DR_a/ARN33179-ATP_7-22.01-001-WEB-3.pdf

MEASUREMENT AND EVALUATION CHALLENGE

Elijah selected tests that measured basic abilities pertinent to the sport of volleyball. The information he gathered from administering these tests could be used as a diagnostic technique to help him develop individualized training programs for his athletes. He calculated percentiles for the data and generated performance profiles for all of the players. By examining the profiles, he could see which motor performance areas needed to be developed for each player.

An example of this type of profile examining three players is presented in table 12.9 (Disch and Disch, 1979). Player 1 had high percentile scores on all the motor performance tests and the anthropometric measures. This player was selected as an All-American setter at the U.S. Volleyball Association Collegiate Championships. The profile of player 2 includes high scores in the motor performance characteristics but lower scores on the anthropometric characteristics. This player was an excellent setter but did not quite reach the level of performance of player 1. Player 3 was found to have favorable anthropometric characteristics, but his motor performance profile was much below that of the first two players. The data indicate that player 3 needs to concentrate on improving his motor performance characteristics, which should improve his performance on the volleyball court.

Table 12.9 Men's Volleyball Performance Profile

Percen-tile	Weight (lb)	Height (in.)	Reach (in.)	Percent body fat	Vertical jump (in.)	Triple hop (in.)	Agility run (s)	20 yd dash (s)
99	200	78		5.70		344	7.7	2.5
95	189	77	100	5.94	29	341		2.7
90	188		99	6.15	27	333	7.8	
85	185	76	94.5	6.58	26	330		2.8
80	183	75.5	97	6.86	25	319	7.9	
75	182		96	6.99		313		
70	181	75		7.30	24	303	8.1	
65	180		95.5	7.41		302		2.9
60	179	74	95	7.55		297		
55	174		94.5	7.6	23	296	8.3	
50	172		94	7.74		295	8.5	
45	169	73		8.09		292		
40	162		93.5	8.21		287		
35	161		93	8.47		285		3.0
30	158	72	92.5	9.68		279	8.8	
25	157	71		9.88	22	276		
20	156	70.5		10.15				
15	154		90	10.88	21	272	8.9	3.1
10	151	70	89	11.63		266	9.4	3.3
5	136	69	88	11.63	20	254	9.5	3.4

Adapted from Disch and Disch (1979).

Note: Player 1: solid line; Player 2: dotted line; Player 3: dashed line.

SUMMARY

Reliable and valid measurement of sport skills and basic physical abilities has a prominent place in human performance testing. Assessment of psychomotor ability is an essential task you may undertake as a physical therapist, personal trainer, physiologist, physical education instructor, athletic coach, or exercise scientist. Whatever your position, a reliable and valid testing program will help you become a respected professional in human performance.

Skills testing comprises a variety of test methods, including objective procedures, subjective ratings, and direct performance assessment. An extensive presentation of a wide range of current sport skills tests is beyond the scope of this text. Those of you who are interested can find a thorough compilation of sport skills tests in Strand and Wilson (1993) and Collins and Hodges (2001). Motor ability testing, as you have read, has a long history in human performance and will take on increased importance in athletics and employment testing. The most important consideration is selecting valid tests that meet your test objectives and are feasible in terms of time and effort. The work of Kirby (1991) is a classic resource for descriptions and critiques of motor performance tests. Finally, many jobs or services require pre-employment physical tests, and it is essential for these tests to be developed and administered in ways that validly represent the work, as well as protect the test taker and the employer.

www | **Go to HK*Propel* for videos, homework assignments, and quizzes that will help you master this chapter's content.**

Assessment of Physical Activity and Energy Expenditure

OBJECTIVES

After studying this chapter, you will be able to

- define physical activity and energy expenditure;

- understand the relationship and difference between physical activity and physical fitness;

- understand key methods for measuring physical activities and their validity, reliability, advantages, and disadvantages;

- understand key methods for measuring energy expenditures and their validity, reliability, advantages, and disadvantages;

- determine how to select methods to measure physical activity and energy expenditure; and

- determine key methods for assessing physical activity and energy expenditure in children and youth.

www The lecture outline in HK*Propel* will help you identify the major concepts of the chapter.

MEASUREMENT AND EVALUATION CHALLENGE

The clinic in which Sofia is doing her summer internship is considering implementing a new service for overweight and obese children. Knowing that Sofia is a kinesiology major, the head physician asks Sofia to make a recommendation for which method would be best for monitoring children's physical activity. Although the topic of assessing children's physical activity was briefly covered in Sofia's exercise physiology class, she realizes that she needs to do more research so that she can provide an appropriate recommendation.

Physical activity is one of the most important topics in the fields of public health, physical education, and exercise science. Physical activity is most commonly defined as "any bodily movement produced by skeletal muscles that results in energy expenditure" (Caspersen, Powell, and Christenson, 1985, p. 126). Physical activity in daily life can be categorized based on its purpose, such as occupational, exercise and sport, household and family care, transportation, leisure, and other; it can also be categorized based on its intensity and energy expended, such as sleep, sedentary, light, moderate, and vigorous (figure 13.1). Exercise is a subset of physical activity that is planned and structured and typically involves repetitive bodily movement to improve or maintain one or more component of physical fitness. Overall, physical activity is a complex behavior, and accurately assessing it and the corresponding energy expenditure is challenging, especially under conditions of regular daily life, known also as *free-living conditions*. Given the epidemiological evidence supporting the role of physical activity in health promotion and chronic disease prevention, the reliable and valid measurement of physical activity is essential in determining the following:

- The amount of physical activity people do
- The amount of sedentary behavior in which individuals engage
- The role of physical activity in health status
- Factors that relate to physical activity behavior
- The effect of interventions to promote physical activity

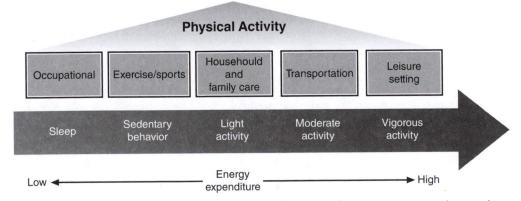

Figure 13.1 Classification of physical activity based on purpose or intensity.

After providing a brief chronological review of some significant events and publications related to physical activity measurement practice, this chapter will introduce the assessment of physical activity and energy expenditure, including key assessment methods, their validity and reliability, and their advantages and disadvantages. Thereafter, how to select a physical activity assessment method and how to determine the appropriate dose of physical activity will be described.

A CHRONOLOGICAL VIEW OF PHYSICAL ACTIVITY AND HEALTH ASSESSMENT

Compared to other areas, such as fitness and motor skills, assessing physical activities is relatively new, but has received great attention and made great progress since the 1990s. This section provides a brief introduction on related key events and studies.

1953: Morris et al. (1953) published a study of London transport workers and British civil servants that documented higher rates of coronary heart disease in men who were sedentary on the job.

1965-1966: The Yamasa Corporation of Japan produced an early pedometer embraced by Japanese walking clubs called "Manpo-Kei," which literally translated to "10,000-step meter" (Tudor-Locke et al., 2008).

1968: Kenneth Cooper published his landmark book *Aerobics*, in which a point system was included to quantify the amount of energy expended in various aerobic activities.

1991: Plenum Press published a book entitled *Activity Measurement in Psychology and Medicine*, by Warren Tryon. It was the first book outside of the field of exercise science that systematically introduced how to objectively assess physical activity.

1992-1994: The Second International Consensus Symposium on Physical Activity, Fitness, and Health was held in Toronto, Canada, in May 1992. Based on the symposium, the *Physical Activity, Fitness, and Health: International Proceedings and Consensus Statement* was published by Human Kinetics in 1994, containing a chapter by Ainsworth, Montoye, and Leon titled "Methods of Assessing Physical Activity During Leisure and Work."

1996: *Physical Activity and Health: A Report of the Surgeon General*, released in 1996 by the U.S. Department of Health and Human Services (USDHHS), was a landmark scientific presentation of the health benefits of physical activity and fitness. The report summarized the physiological and psychosocial benefits that people of all ages gain from a physically active lifestyle.

1996: Human Kinetics published *Measuring Physical Activity and Energy Expenditure* (Montoye et al., 1996), the first book in the field of exercise science to introduce methods of assessing physical activity and energy expenditure.

1997: *Medicine & Science in Sports & Exercise* (MSSE), the official journal of the American College of Sports Medicine(ACSM), published a supplement (Vol. 29, No. 6) containing a collection of physical activity questionnaires for health-related research.

1997: Under the leadership of Ted A. Baumgartner, the journal *Measurement in Physical Education and Exercise Science* was created; it has since published many research articles in the field of physical activity and energy expenditure.

1999-2002: Assisted by James R. Morrow, Jr., the Measurement and Evaluation Symposium of the American Association for Active Lifestyles and Fitness and the Cooper Institute jointly organized a conference entitled "Measurement of Physical Activity," which for the first time brought measurement and evaluation scholars together to consider critical measurement issues associated with assessing physical activity. As a result of the conference, 20 papers were published as a supplement in *Research Quarterly for Exercise and Sport* (Vol. 71, Suppl. 2, 2000), and Human Kinetics published a book entitled *Physical Activity Assessments for Health-Related Research* (Welk, 2002), which described major methods of assessing physical activity and related measurement issues.

2003-2006: Organized by Weimo Zhu and held at the University of Illinois at Urbana-Champaign (UIUC) in October 2003, the 10th Measurement and Evaluation Symposium, titled "Measurement Issues and Challenges in Aging Research," addressed issues related to assessing physical activity among older adults. The proceedings of the symposium, *Measurement Issues in Aging and Physical Activity* (Zhu and Chodzko-Zajko, 2006), were later published by Human Kinetics.

2003-2004: For the first time the National Health and Nutrition Examination Survey (NHANES, 2003-2004) used ActiGraph, an objective measure, to assess physical activity using a national study sample ($N = 4767$, including 2282 males and 2485 females).

2004-2005: In December 2004, a conference titled "Objective Measurement of Physical Activity: Closing the Gaps in the Science of Accelerometry" was jointly organized by the University of North Carolina and the National Cancer Institute. In 2005, the conference papers were published in an MSSE supplement (Vol. 37, Suppl. 11), describing the objective monitoring of physical activity in detail.

2007: ACSM and the American Medical Association launched the national initiative Exercise Is Medicine (EIM) to encourage health care providers to include exercise when designing treatment plans for patients. The basic assumption of the initiative is that exercise and physical activity are important in the treatment and prevention of a variety of diseases and should be assessed as part of routine medical care. An EIM goal is to have health care providers discuss physical activity behaviors each time they meet with patients, regardless of the purpose of the visit. A simple tool on how to assess the level of physical activity participation was included.

2007: In October 2007, assisted by James R. Morrow, Jr., and Weimo Zhu, the American Association for Active Lifestyles and Fitness and the Cooper Institute jointly organized the 11th Measurement and Evaluation Symposium to address the diversity issues in physical activity and health, including major measurement and evaluation challenges and possible solutions.

2008: In October 2008, the U.S. Department of Health and Human Services (USDHHS, 2008) released the *Physical Activity Guidelines for Americans* to provide physical activity recommendations for all Americans, which are consistent with the known research evidence relating physical activity to health.

2008: A symposium entitled "Walking for Health" was held at UIUC in October, resulting in an MSSE supplement (Vol. 40, Suppl. 7) entitled "Walking for Health: Measurement and Research Issues and Challenges."

2009-2012: In July 2009, a workshop "Objective Measurement of Physical Activity: Best Practice and Future Direction" was held at the National Institute of Cancer to update the best practice recommendations for the use of wearable monitors to assess physical activity. A set of papers from the workshop was published in an MSSE supplement (Vol. 44, Suppl. 1).

2010-2012: In July 2010, the National Cancer Institute, the U.S. Centers for Disease Control and Prevention (CDC), the National Institutes of Health Office of Disease

Prevention, the National Collaborative on Childhood Obesity Research, and the ACSM jointly organized a conference to explore the major challenges and opportunities for self-report methods in assessing active and sedentary behaviors. In 2012, a set of conference papers was published in a supplement entitled "Measurement of Active and Sedentary Behaviors: Closing the Gaps in Self-Report Methods" in the *Journal of Physical Activity & Health* (Vol. 9, Suppl. 1).

2012: In March, the 12th Measurement and Evaluation Symposium, titled "New Approaches in Measuring and Assessing Physical Activity," was held in Boston; it introduced the latest technologies and methods, as well as new statistical and analytical procedures.

2015: In April, the International Society for the Measurement of Physical Behavior (ISMPB) was established to promote and facilitate the study and application of objective measurement and quantification of free-living physical behaviors and their related constructs (e.g., energy expenditure, context) using wearable devices.

2015-2017: In October 2015, supported by the Society of Health and Physical Educators (SHAPE America) and the ACSM, Weimo Zhu organized a conference entitled "Sedentary Behavior and Health: Measurement Issues and Research Challenges" at UIUC. As a result of this conference, the book *Sedentary Behavior and Health: Concepts, Assessments, and Interventions* (Zhu and Owen, 2018) was later published by Human Kinetics.

2018: The second edition of the *Physical Activity Guidelines for Americans* was published by the Physical Activity Guidelines Advisory Committee, accompanied by a scientific report summarizing the scientific evidence on physical activity and health.

2018: *Journal for the Measurement of Physical Behavior*, the official journal of ISMPB, was established to promote the publication of high-quality research papers that employ or apply sensor-based measures of physical activity and research on movement disorders, sedentary behavior, and sleep.

PHYSICAL ACTIVITY AND PHYSICAL FITNESS

Healthy People 2030 and the *Physical Activity Guidelines for Americans* (USDHHS, 2018) have goals and objectives related to increasing physical activity but not specifically about the outcome of physical activity and health-related physical fitness (discussed in chapter 10). Higher levels of physical activity and fitness are both related to lower risks of morbidity (disease) and mortality (death). However, higher levels of physical fitness produce greater reductions in the risks of morbidity and mortality than higher levels of physical activity do (see figure 13.2 for an illustration of these relationships).

Two possibilities may explain this difference:

- Physical fitness may truly provide a greater reduction in the risk of chronic diseases and mortality than physical activity, or
- the measurement of the attribute of physical fitness tends to be more reliable and valid than the measurement of physical activity behavior.

Because the measurement of physical activity tends to include more error than the measurement of physical fitness, it is more difficult for research studies to determine accurate relationships between physical activity and health outcomes than between physical fitness and the same health outcomes.

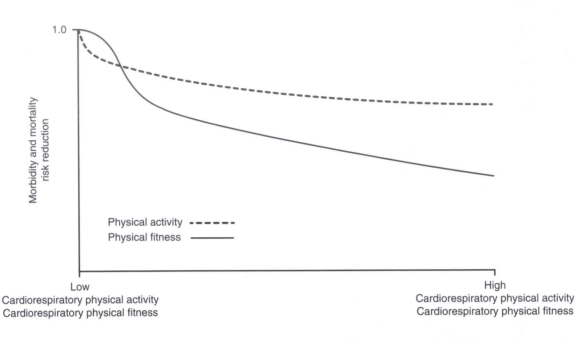

Figure 13.2 Impact of physical activity and physical fitness on morbidity and mortality risk reduction.

Mastery Item 13.1

Physical activity is negatively correlated with mortality. Name two other health variables that show negative relationships with physical activity and two that are positively correlated.

PHYSICAL ACTIVITY AND ENERGY EXPENDITURE

Though related, physical activity and energy expenditure are not identical. Physical activity, as introduced earlier, is a behavioral process in which body movement is produced by skeletal muscle contraction, whereas energy expenditure is a direct result of physical activity and reflects the net transfer of energy required to support the skeletal muscle contraction. Physical activity is typically defined in terms of mode (e.g., walking, jumping, dishwashing, chopping wood), frequency (number of sessions), duration (minutes per session), and intensity (*absolute* energy cost or *relative* effort as a percentage of maximal work capacity).

Energy expenditure, on the other hand, is used to quantify the total amount or *volume* of energy expended during a specified period of time. In practice, especially in weight control and management, energy expenditure is usually measured by kilocalorie (kcal), which is the amount of heat required to raise the temperature of a kilogram of water 1 °C and is often expressed simply as "calorie." As discussed in chapter 1 (and shown in figure 1.1), Pettee-Gabriel, Morrow, and Woolsey (2012) provided an excellent framework to summarize the relationship among physical activities (behaviors) and physical fitness and energy expenditure (attributes).

Commonly used terms related to energy expenditure measurement include the following:

- Basal metabolic rate (BMR), the energy expenditure used to maintain the body's basic biological processes in a completely resting state, which can be measured after a person has just woken up from an 8-hour sleep in a dark 70 °F room.

- Resting metabolic rate (RMR), the energy expenditure a person uses when sitting quietly in a chair, which is about 10% higher than BMR.
- Metabolic equivalents (METs), which represent the ratio of energy cost to RMR. For a 60-kilogram (132 lb) adult, RMR is assumed to be approximately 1 MET, which is equivalent to 3.5 mL $O_2 \cdot kg^{-1} \cdot min^{-1}$, or about 1 kcal $\cdot kg^{-1} \cdot hr^{-1}$.

By multiplying the MET level of a given physical activity by the duration (e.g., minutes) for which the activity is performed, one can compute the MET-minute (MET-min) value. This index quantifies the rate at which energy is expended for the duration an activity is performed while accounting for body size and resting metabolism. The *Compendium of Physical Activities* (Ainsworth et al., 1993; 2011) provides researchers and practitioners with a standardized linkage between specific activities, their purpose, and their estimated absolute energy cost expressed in METs. A sample entry from the compendium is listed below:

Code	METS	Major heading	Specific activities
17190	3.5	Walking	Walking, 2.8 to 3.2 mph, level, moderate pace, firm surface

Column 1 shows a five-digit code that indexes the general class or purpose of the activity. In this example, 17 refers to walking and 190 refers specifically to walking at a 2.8- to 3.2-mile (4.51-5.15 km) per hour pace. Column 2 shows the absolute energy cost of the activity in METs. Walking at this pace results in an energy expenditure of 3.5 METs, or 3.5 times the resting energy expenditure. Column 3 shows the general category of activity—walking—and column 4 shows the specific activity and conditions.

If a person completes a 30-minute walk at a 3.0 mph pace (4.83 km/hr), his or her energy expenditure for this particular activity would be 105 MET-min based on the following computation: frequency (1) × duration (30 min) × energy cost of 3.5 METs. The Physical Activity Guidelines mentioned earlier include a recommendation of 500 to 1000 MET-minutes per week for significant health benefits.

Mastery Item 13.2

Recall the frequency and the time you spent on specific activities in the past 24 hours and compute your energy expenditure in MET-min. Refer to the *Compendium of Physical Activities* website (https://sites.google.com/site/compendiumofphysicalactivities/home) for your specific activity METs.

METHODS OF ASSESSING PHYSICAL ACTIVITY

Many methods have been developed to measure and classify physical activity. Because there is no gold-standard criterion measure of physical activity in free-living conditions, and because some methods (e.g., observation) do not clearly fit into either self-report or objective measures, we will describe them here in terms of direct measures and indirect measures.

Direct Measures

Direct measures of physical activity are used to characterize the actual type, frequency, duration, and intensity of movement during specific observation periods. Physical activity records and motion sensors are used to collect information while an individual engages in various activities. Activity logs and checklists collect information at the end of a defined observation period (e.g., end of a day). The direct measures of physical activity are often used as validation criteria for indirect or subjective measures of physical activity.

Physical Activity Records

Physical activity records are ongoing accounts of activity patterns kept in diary format during a defined observation period. Usually, respondents record information about the type (e.g., sleep, walking, digging), purpose (e.g., transportation, occupation, exercise), duration (minutes), self-rated intensity (light, moderate, vigorous), and body position (e.g., reclining, sitting, standing, walking) for every activity completed during the specified observation period (e.g., 24 hours). Physical activity records are then scored using the physical activity compendium to link a MET intensity score with the type and purpose of each activity entered. After scoring, the physical activity records could provide a detailed account of the minutes spent doing various types, intensities, and patterns of physical activity. Activity-related energy expenditure (e.g., MET-min $\cdot$ d^{-1}) can then be quantified.

The validity and reliability of physical activity records have been examined for tracking the time spent in specific types and patterns (e.g., single continuous bouts, sporadic intermittent bouts) of physical activity that accounts for individual or population activity-related energy expenditure. For example, Conway et al. (2002) reported a difference of only 7.9 ± 3.2% between a 7-day physical activity record and doubly labeled water estimates (discussed later in this chapter) of free-living activity-related energy expenditure in men. Moderate to strong age-adjusted correlations (.35-.68) have been reported between total MET-min per day of energy expenditure from the physical activity record and an accelerometer for free-living men and women (Richardson et al., 1995). Physical activity records can be kept to obtain information on seasonal variations in habitual activity patterns (e.g., winter vs. summer). Having participants put entries into the activity record at the time a behavior is executed reduces the influence of recall bias.

Although physical activity records could provide a very detailed and comprehensive method of assessing free-living physical activity patterns, their feasibility is limited by cost, the potential for altered behavior, the need for multiple and extended observation periods to assess habitual activity patterns, and administrative burden on both the participant and administrator. For these reasons, physical activity records are often used only for individual activity assessments or as a criterion measure for validating simple physical activity measurements that are practical for use in the field.

Physical Activity Logs and 24-Hour Recalls

Physical activity logs are simplified physical activity records that are structured as a checklist of activities specific to the administrator's interests or the target population's usual activity pattern. For example, Ainsworth et al. (2000) present a modifiable list consisting of 20 to 50 activities originally developed for use in ethnically diverse populations of women. At the end of a specified observation cycle, respondents identify the type and duration of activities performed during that period of time. Activity-related energy expenditure (e.g., MET-min $\cdot$ d^{-1}) is computed by assigning intensity values from the compendium to each activity selected. The log takes only a few minutes to complete and can be scored quickly to provide information about the time spent in specific types or categories of physical activity and their resulting energy expenditures. Physical activity logs may be more convenient to complete and process than physical activity records. Alternatively, activity logs may underestimate actual physical activity levels and related energy expenditure if participants engage in activities other than those listed on the log. Because the log is completed at the end of the day, the degree of recall bias associated with this method is likely to be somewhat higher than for the physical activity record.

Test–retest reliability coefficients of physical activity logs ranged from $r_{xx'}$ = .22 to .58, and validity coefficients between total MET-hr $\cdot$ d^{-1} from the 24-hour recalls and an accelerometer were r = .74 and r = 0.32 for men and women, respectively (Matthews et al., 2000). Although physical activity logs or 24-hour recall may reduce administrator and participant burden, the potential for recall bias and alterations in physical activity patterns during assessment may mean that they are not suitable for populations with limited telephone access or who are unwilling to complete the phone interview, and may utilize a time frame that does not capture an individual's true habitual activity level.

Motion Sensors

Electronic motion sensors have become increasingly popular direct measures of free-living physical activity since the beginning of the 21st century (e.g., Freedson and Miller, 2000). Energy expenditure is also often extrapolated from this activity data under the assumption that movement (or acceleration) of the limbs and torso is closely related with whole body activity-related energy expenditure. There are several types of motion detectors that differ in cost, technology, and data output.

Pedometers

Pedometers are inexpensive devices used to quantify walking activity in terms of accumulated steps per unit of observation time (e.g., per day). Electronic pedometers are small battery-operated devices worn at the waist. The vertical forces of foot strike cause movement of a spring-suspended lever arm to open and close an electrical circuit, which registers as a "step." An estimate of distance walked is obtained by calibrating pedometer steps to an individual's stride length during usual walking pace over a known distance. In theory, step registration should reflect only the vertical forces of foot strike and hence walking activity; however, any vertical force through the hip area (e.g., sitting down hard onto a chair, riding on a bike or in a car over rough terrain) can trigger the device.

Pedometers have demonstrated reasonable precision for use in research and clinical settings where walking is the primary type of activity. Correlations of .48 to .93 have been reported between pedometer steps per day and activity data from accelerometers worn on the hip (Bassett et al., 2000). Although earlier studies have revealed that the precision of pedometers varied considerably according to manufacturer's brand and was lower for walking versus running activities (Bassett et al., 1996), more recent studies indicate that newer generation pedometers are accurate and robust with regard to their location on the body (Holbrook, Barreira, and Kang, 2009; Zhu and Lee, 2010). Currently, the function of recording steps has largely been taken over by smartphones or other wearable devices.

Accelerometers

Acceleration is the change in velocity over time and is usually expressed in gravitational force (g = 9.81 m/s^2). Degree of acceleration provides an index of movement because it indicates the rate at which distance is covered. Accelerometers, which are small battery-operated devices that measure the rate and magnitude of the body's displacement, are now the most popular physical activity measure. The key components of an accelerometer include a transducer, which converts movement to electronic signals, and a data acquisition system, which processes data and converts them from raw signals to desirable parameters. Movement can be measured in single or, now more commonly, multiple planes. Collected data are integrated and summed to the absolute value and frequency of acceleration forces over a defined observation period (e.g., every 15 or 60 seconds). Physical activity data, depending on the specific device, is output as an activity "count" or other unit. Statistical equations and cutoffs have been developed from controlled laboratory studies to estimate activity-related energy expenditure from the activity counts and determine the activity intensity in a specific time interval.

Many studies have examined the validity and reliability of accelerometry-based assessment of physical activity and energy expenditure. Their correlations with VO_2 during laboratory studies of treadmill walking and running ranged from moderately high to high, but were low with daily free-living activities (Ainsworth et al., 2000). Overall, accelerometers could provide information about the frequency, duration, intensity, and patterns of physical activity, including sleep; however, the specific type of physical activity sometimes cannot be determined. Accelerometers also tend to overestimate walking-related energy expenditure and underestimate lifestyle-related energy expenditure, and they do not account for energy expenditure owed to upper-body involvement or uphill walking. It is likely that these limitations reflect the accuracy of the regression equations used to predict activity-related energy expenditure rather than imprecision of motion detection by the accelerometer. Additional drawbacks of accelerometers include subject compliance issues, potentially altered physical activity patterns, and the cost of increasingly sophisticated instruments.

Indirect Measures

Indirect assessment methods provide surrogate measures of physical activity among free-living persons when direct methods would be too costly or impractical. Indirect measures are typically validated against objective direct measures of activity in controlled small-sample studies. The most common indirect method is the self-report physical activity questionnaire.

Physical Activity Questionnaires

Self-report questionnaires are the most frequently used method of assessing physical activity levels among individuals. Questionnaires are generally classified as global, recall, or quantitative based on their level of detail.

Global Questionnaires

Global activity questionnaires are short (e.g., 1-4 items) surveys that are self-administered to obtain a general index of physical activity. As a result, global questionnaires provide little detail on specific types and patterns of physical activity and do not allow for precise assessment of activity-related energy expenditure; they provide only simple classifications of activity status (e.g., active vs. inactive). Global questionnaires are typically used in health surveillance systems where sample sizes are very large, administrative time is limited, and the goal for assessment may be to merely track long-term population behavioral trends. Because of the brevity of the global questionnaire, the reliability is often acceptable, but validities are often poor due to lack of detail, which precludes comprehensive summaries of physical activity dimensions.

Recall Questionnaires

Recall questionnaires are typically 10- to 20-item instruments that detail the frequency, duration, and types of activities performed during a defined recall period (e.g., past month or year). These instruments can be interview based or self-administered. Scoring systems can be a simple ordinal scale (e.g., 1-5 representing low to high activity levels) or a summary score of continuous data (e.g., frequency or minutes of activity per week, MET · d^{-1}). The latter measure allows for quantification of time spent performing specific physical activities and their related energy expenditure, and therefore allows for the most precise interpretation of physical activity patterns and effects for application in various settings.

Recall surveys have demonstrated acceptable (though not high) levels of reliability and reproducibility. For example, 1-month test–retest correlations for activity-related energy expenditure computed from self-reported walking, stair climbing, and sport activity were .61 and .75 for men and women, respectively, who completed the College Alumnus questionnaire (Ainsworth, Leon, Richards, Jacobs, and Paffenbarger, 1993). Validity of the recall surveys, however, was relatively low, ranging from .20 to .58 (Richardson et al., 2001).

Recall surveys typically do a poor job of assessing nonoccupational, nonleisure, and nonsport activities, which may be particularly relevant sources of physical activity and energy expenditure among women and minorities (Jacobs et al., 1993). Activity information collected by recall survey is often subject to response bias (e.g., overestimated physical activities or intensity due to recall bias and social desirability) that may influence the precision of measurement. Bias appears to be highest for light and moderate intensity physical activities that are habitual behaviors (e.g., walking, housework) compared with vigorous sports and conditioning activities, which are planned and intentional behaviors. Therefore, one needs to be cautious when using self-report measures of physical activity. Troiano et al. (2008), for example, showed that approximately 50% of adults self-report that they meet national physical activity guidelines. However, when objective measures are obtained with accelerometry, less than 5% actually achieve the minimum moderate-to-vigorous physical activity (MVPA) guidelines.

Another limitation is that the structure of the survey may not include relevant population-specific activities, which would likely lead to an underrepresentation of an individual's actual physical activity level and related energy expenditure (LaMonte and Ainsworth, 2001; Warnecke et al., 1997). This issue may be particularly problematic when a questionnaire that has been validated for use in a population with specific age, sex, race, ethnicity, and sociocultural characteristics is used (or modified for use) in a population with distinct differences in these parameters. As with most measurements, a careful examination is required to determine if the survey has been validated for the targeted population and if the survey included the needed activity categories before employing it for the study or application.

Efforts have been made in this area to develop recall questionnaires to assess physical activity on an international scale. One such major undertaking is the development of the International Physical Activity Questionnaire (IPAQ) for cross-cultural use, which includes multiple language options in different formats (long, short, telephone, and self-administration). Kim, Park, and Kang (2013) conducted a meta-analysis study to examine the convergent validity evidence of the IPAQ. The results indicated small-to-medium effect size differences between IPAQ and objective measures (pedometer, accelerometer) across all physical activity categories, such as walking and MVPA. Visit the following link for more information on the IPAQ: https://sites.google.com/site/theipaq/.

Single-Response Scales

A simple self-report method using a single-response scale to assess physical activity behaviors has demonstrated excellent reliability and construct validity, comparable to those of more complicated assessment techniques (Jackson et al., 2007). Figure 13.3 depicts the five-level version of this scale, based on the stages of change in physical activity behavior. Test–retest reliability estimates have exceeded .90 for the scale, and the correlation (.57) between the scale responses and treadmill assessments of aerobic capacity was consistent with correlations for other self-report techniques.

| I don't exercise or walk regularly now, and I do not plan to start in the near future. |
| I don't exercise or walk regularly now, but I have been thinking of starting. |
| I am doing moderate physical activities fewer than 5 times a week or vigorous ones fewer than 3 times a week. |
| I have been doing moderate physical activities 5 or more times a week or vigorous ones at least 3 times a week for the last 1 to 6 months. |
| I have been doing moderate physical activities 5 or more times a week or vigorous exercise at least 3 times a week for 7 months or longer. |

Figure 13.3 The single-response physical activity scale (five levels).

Methods of Assessing Muscle-Strengthening Physical Activity

It should be pointed out that most of the physical activity assessment methods mentioned thus far focus on aerobic-related physical activity, such as walking, running, playing basketball, and the like. *Healthy People 2030* and the *Physical Activity Guidelines for Americans*, however, include objectives for muscle-strengthening activities. The health benefits of muscular strength and endurance were listed in table 10.2.

In the past, muscle-strengthening activities have been assessed by self-report questions. An example is the following question asked by the CDC's Behavioral Risk Factor Surveillance System telephone survey:

During the past month, how many times per week or per month did you do physical activities or exercises to STRENGTHEN your muscles? Do NOT count aerobic activities like walking, running, or bicycling. Count activities using your own body weight like yoga, sit-ups, or push-ups and those using weight machines, free weights, or elastic bands.

Times per week _____

Times per month _____

Never _____

Don't know / Not sure _____

Refused _____

These self-report questions provide little information about the respondent's actual muscle-strengthening behaviors. The following data are not obtained: What type of activities? What and how many muscle groups are involved? How many sets and repetitions are performed?

Very recently, new wearable devices, known as strength-training trackers, have been developed to measure and record users' strength-training doses—specifically, repetitions, resting time interval, time under tension, or velocity. There are, in general, four types of strength-training trackers:

1. *Inertial measurement unit (IMU) devices.* Placed on the wrist or waist, IMU devices (e.g., RecoFit) can count repetitions or recognize different exercises. These devices also include accelerometers and gyroscopes.

2. *Electromyography (EMG) devices.* These devices use a portable EMG attached to skin to detect muscle fiber activation.

3. *Force sensor devices.* These devices are embedded with a force sensor to detect tension or pressure on the device.

4. *Smartphone-based devices.* Smartphone-based trackers are devices controlled by a smartphone application or simply an application that uses a camera and an image-processing algorithm.

Practically, these trackers could be used to monitor some strength-training activities.

Methods of Assessing Sedentary Behavior

Sedentary behaviors and their effect on health have only recently begun to be fully investigated. Kang and Rowe (2015) present an overview of sedentary behavior measurement methods and a summary of the major challenges facing researchers, with a particular focus on sources of measurement error. For more information on concepts, assessment, and interventions related to sedentary behavior and health, see Zhu and Owen (2018).

METHODS OF ASSESSING ENERGY EXPENDITURE

Like assessing physical activity, many methods have been developed for assessing activity-related energy expenditure. In fact, some of the methods are used in both assessing physical activity and energy expenditure, with the former focused on the features of physical activity (e.g., intensity, duration, frequency, and mode) and the latter estimating energy expenditure spent during a specific time period. The methods that are precise enough to allow a direct measure of energy expenditure are called *criterion measures*; those that are less accurate, but can be used in large-scale studies with low cost, are called *field measures*. We will introduce these methods next.

Criterion Measures

Criterion measures are usually the best measure of energy expenditure. Although they are typically expensive and inconvenient to use, they are more accurate than field measures and thus used as a standard for field measures. In the studies of energy expenditure, direct calorimetry, doubly labeled water, indirect calorimetry, diary, and direct observation are frequently employed as criterion measures and they are described below.

Direct Calorimetry

Direct calorimetry is used to measure body heat production and is the most precise measure of energy expenditure. Calorimetry is performed under laboratory conditions, with the participant in a small airtight chamber that contains insulated pipes used to circulate water through the calorimeter. Based on the temperature and flow rate of the water as well as all sources of heat loss from the chamber, a participant's heat production, and thereby energy expenditure, can be measured within 1% to 2% error. Measurements are typically made over a 24-hour period and follow a 10- to 12-hour fast so that RMR can be accurately assessed. The size of the chamber is typically small and confined; therefore, it is not practical for assessing activity-related energy expenditure in a variety of normal daily activity patterns. A lag time between body heat release during exercise and calorimeter sampling may extend the periods of observation required for precise measurement. Issues with dissipating perspiration and altered mechanical efficiency during exercise may further bias the measurement away from a true representation of free-living energy expenditure. Cost and technical limitations make direct calorimetry generally infeasible for assessing activity-related energy expenditure outside of laboratory settings.

Doubly Labeled Water

Doubly labeled water (DLW), which is a biochemical procedure that tracks the rate of metabolic processes, is the most accurate method of measuring energy expenditure in free-living conditions. During the assessment, a person drinks a standardized amount of DLW ($^2H_2^{18}O$) and by measuring the elimination kinetics of ^2H and ^{18}O from a person's body via urine sample, total carbon dioxide (CO_2) production is determined for a specific period—ultimately, the estimation of energy expenditure in the period (Straling, 2002).

The validity evidence for DLW was conducted using a respiration chamber under controlled conditions. The difference between the two measures of energy expenditure was less than 5% (Montoye et al., 1996). Even though DLW is an accurate measure of energy expenditure and is not burdensome to subjects, its cost is high, and the specifics of physical activity, such as frequency, intensity, and duration of physical activity, cannot be measured (Welk, 2002).

Indirect Calorimetry

Indirect calorimetry has been the most frequently employed criterion measure and is considered an accurate measure of energy expenditure in free-living conditions (Montoye et al., 1996). This method uses the ratio of carbon dioxide production (VCO_2) to oxygen consumption (VO_2) to estimate energy expenditure. Because assessing physical activity-related energy expenditure in free-living conditions is the key interest, the portable indirect calorimetry system has proven popular to use, even though it is not as accurate as the direct calorimeter method. However, portable indirect calorimetry is still relatively expensive, unable to measure energy expenditure over the long term in free-living situations, requires some exercise physiology training to use, and may be too large to be practical for real-life performance testing.

Direct Observation

Direct observation was one of the earliest methods to assess physical activity. Observers recorded a variety of information about physical activity, such as behavioral information, pattern, type, frequency, time, and intensity. Simple observation usually includes behaviors such as sitting, lying, standing, walking, and running, and is especially useful for studying children's physical activity (McKenzie, 2002).

Direct observation has been validated by using heart rate monitors or motion sensors (e.g., McKenzie, Marshall, Sallis, and Conway, 2000). Interrater reliability was obtained by comparing different raters' coding scores of the same subject's activities, and intrarater reliability was obtained by assessing the difference between the first coding score (obtained from the real observation setting) and the second coding score (obtained from observation of videotaped activities of the same participants) from the same rater. Both inter- and intrarater reliability were high when measuring physical activity in general (McKenzie, 2002). Shortcomings of the observation method include expense and the time consumed to collect the data. Also, like the diary method, measurement error may occur when the raters classify physical activity using MET values (Kang, 2004; Welk, 2002).

Field Measures

Field measures are frequently employed to measure physical activity or to estimate energy expenditure in free-living conditions. Field measures are cheaper and more user-friendly and feasible than criterion measures; however, they are less accurate. Motion sensors, heart rate monitors, and self-report questionnaires are widely used field measures.

Motion Sensors

The components and functions of motion sensors were introduced earlier to assess physical activity. As mentioned, the information recorded may also be used to estimate activity-related energy expenditure. Unfortunately, according to a recent review on motion sensor prediction studies in which DLW was used as the criterion measure of energy expenditure, the bias at the individual level is very high even for these group-level validated motion sensors (Sardinha and Júdice, 2017). For children, only 13% of physical activity-related energy expenditure variance and 31% of total energy expenditure variance can be explained by the prediction based on accelerometers and for adults, 29% and 44%, respectively. The prediction improved slightly when heart rate measures were added.

Heart Rate Monitors

Heart rate is the earliest field measure used to assess physiological variables and has been widely used to estimate energy expenditure since 1907 (Montoye et al., 1996). To be able to estimate energy expenditure using heart rate, it has to be first calibrated to VO_2 consumption using indirect calorimetry in a laboratory setting. The derived prediction equation can be applied to estimate energy expenditure of physical activity in free-living conditions. The heart rate method is useful for large epidemiological studies because it is convenient and inexpensive. Using the flex heart rate (HR flex), which notes the highest pulse rate at rest and lowest during exercise (Ceesay et al., 1989), increases the accuracy of the estimated energy expenditure in free-living conditions. However, because heart rate is sensitive to many other factors, such as age, sex, temperature, humidity, fatigue, hydration, training status, and emotional stress (Rennie et al., 2000), energy expenditure estimated by heart rate can be biased by these factors. As with motion sensors, some significant errors were noticed when group-level predictions were applied at the individual level. With an additional individual calibration effort, the prediction can be improved (Lee et al., 2010), but this is difficult to implement in practice.

Multisensor Devices

Recent efforts have been made to improve the prediction of energy expenditure by combining motion sensors, heart rate monitors, and other biometric devices. One of the most successful devices in this category was the SenseWear Armband. Compared with a combination of accelerometry, heat flux, skin temperature, near-body ambient temperature, and galvanic skin response, the validity of the device was reported to predict physical activity-related energy expenditure and total energy expenditure with less than a 10% reduction in accuracy (Koehler and Drenowatz, 2017).

Questionnaires

Efforts have been made to use physical activity questionnaires to predict physical activity energy expenditure, but overall validity and reliability have been low (e.g., Sallis and Saelens, 2000); thus, use of questionnaires for this purpose is not recommended.

In short, assessing physical activity and related energy expenditure with sound validity and reliability remains challenging. As illustrated in table 13.1, each of the commonly used methods for assessing physical activity and energy expenditure has its own advantages and disadvantages, which should be considered when selecting a method.

Table 13.1 Advantages and Disadvantages of Commonly Used Methods of Measurement of Physical Activity and Energy Expenditure

Method	Advantages	Disadvantages
Doubly labeled water	Most accurate method; the gold standard for energy expenditure measurement Little subject burden	Very expensive Only measures total energy expenditure High requirement for technical expertise
Indirect calorimetry	Very accurate Portable Able to assess energy expenditure in real-life settings	Expensive Requires some physiological training Weight could be impractical for children or certain sports Not convenient for long-term daily tracking
Accelerometry	Objective; removes recall bias Easy to use Can conduct long-term tracking	Large variation in setting intensity cutoffs Cannot record certain activities well (e.g., swimming, cycling) Limited by wearing location Can be expensive
Heart rate monitors	Objective; used in both lab and field settings Relatively low cost Noninvasive	Can be affected by a variety of factors, including mood Inaccurate when measuring light and sedentary activities Inaccurate at wrist and arm sites
Pedometers	Useful for measuring walking, the most popular physical activity	Only measures one type of physical activity (walking) Could be impacted by other factors (e.g., age, speed, and settings)
Self-report methods	Can collect a large sample in a short period Can collect physical activity pattern information Low cost Low burden to subjects No impact on subject behaviors	Low validity and reliability Subjective measure Tend to overestimate subjects' physical activity participation Affected by recall errors

SELECTING A METHOD OF MEASUREMENT

With a long list of possible methods for measuring physical activity and energy expenditure, choosing a specific method best suited for a given application can be challenging. We offer you some specific considerations to help you make the best selection.

First, the method selected should be appropriate to the purpose of the measurement and the dimension of movement. For example, if a researcher is interested in measuring the prevalence of physical inactivity within a defined free-living population, a global physical activity questionnaire should be adequate. If the interest is in quantifying the activity-related energy expenditure during a weight-loss intervention research study, DLW or heart rate and accelerometer measures may be the methods of choice. If the purpose of the measurement is to help a patient identify his or her sedentary time so that a personalized intervention can be designed, a combination of an activity log or record and the heart rate-accelerometer method will be most effective.

The application setting should be another consideration. If the measurement is taken in a laboratory-based study, the most accurate methods should be considered. If, on the other hand, the measurement is to be taken in a regular daily setting, compliance should be the major consideration. If the method involves a wearable device, it should be practical and nonintrusive, with respect to an individual's daily lifestyle.

You also must consider the characteristics of the participants you intend to measure. This is particularly true when using physical activity questionnaires, but also applies to exercise testing methods and equations used to extrapolate accelerometer data to activity-related energy expenditure. Factors such as age; sex; education level; sociocultural, race, and ethnic status; and motivation to comply with measurement methodology must be considered. It would make little sense, for example, to survey habitual activity levels among end-stage heart failure patients using a physical activity questionnaire comprised primarily of occupational and sport activities. Similarly, using a physical activity survey or accelerometry regression equation that was validated in populations of college-aged white men to describe activity patterns and quantify activity-related energy expenditure among older Black women would be inappropriate. When possible, choose a measure that has demonstrated adequate psychometric properties among individuals similar to the target population. If such a measure does not exist, consider a pilot study to examine these issues prior to widespread use within the intended population.

It must be pointed out that not all measures of physical activity are appropriate for all populations. Table 13.2 provides a list of self-report and direct monitoring assessment procedures and the age groups for which they are suited. For children and youth, some additional considerations should be made, which will be described in detail in the following section.

Finally, financial costs related to instrumentation, data processing, and participant motivation are important when selecting a realistic measurement method. Methods that

Table 13.2 Physical Activity Assessment Methods for Specific Age Groups

Type	Instrument	Children	Adults	Older persons
Self-report	Task-specific diary	No	Yes	Yes
	Recall questionnaire	No	Yes	Yes
	Quantitative history	No	Yes	Yes
	Global self-report	No	Yes	Yes
Direct monitoring	Behavioral observation	Yes	Yes	Yes
	Job classification	No	Yes	No
	Heart rate monitor and motion sensor (ONE Tool)	Yes	Yes	Yes
	Heart rate monitor	Yes	Yes	Yes
	Electronic motion sensor	No	Yes	Yes
	Pedometer	No	Yes	Yes
	Gait assessment	Yes	Yes	Yes
	Accelerometers	Yes	Yes	Yes
	Horizontal time monitor	Yes	Yes	Yes
	Stabilometers	No	No	No
	Direct calorimetry	Yes	Yes	Yes
	Indirect calorimetry	No	Yes	Yes
	Doubly labeled water	Yes	Yes	Yes

Data based on Laporte, Montoye, and Caspersen (1985).

impose a substantial time burden on the administrator or participant may result in poor data quality. The most costly and burdensome measures are direct methods, and the least costly and burdensome methods are self-administered paper-and-pencil questionnaires. The trade-off, however, is in the precision of measurement.

Mastery Item 13.3

Identify three methods of measuring physical activity behaviors and the advantages and disadvantages associated with each method.

ASSESSING PHYSICAL ACTIVITY IN CHILDREN AND YOUTH

The concerns about youth fitness discussed in chapter 10 are matched by concerns about youth physical activity levels because childhood obesity continues to be a major public health problem. Public health officials and physical activity researchers need reliable and valid measures of physical activity in children and youth for effective research studies to promote increases in youth activity and fitness levels (Sallis et al., 1993). In 2018, Butte et al. developed the Youth Compendium of Physical Activities, which provides a list of 196 common activities in which youth often participate and estimates the energy expenditure associated with each activity. The details of the compendium can be found on the National Collaborative on Childhood Obesity Research website (www.nccor.org/tools-youthcompendium).

How many children complete the recommended 12,000 steps per day? Figure 13.4 shows the percentage of children and adolescents who complete 12,000 pedometer steps per day, as determined by a large-scale Canadian study (Craig, Cameron, and Tudor-Locke, 2013). The percentage of boys who complete the recommended steps is higher than the percentage of girls at every age; however, the percentage of both boys and girls drops considerably from ages 11 to 16.

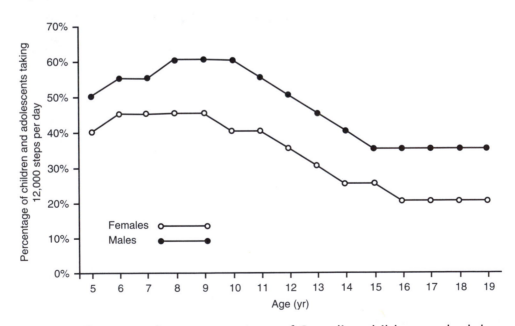

Figure 13.4 The approximate percentage of Canadian children and adolescents who take 12,000 steps per day.

This is a critical finding because physically inactive children tend to become inactive adults with increased risks for chronic diseases. For example, Dennison, Straus, Mellits, and Charney (1988) found that children with the poorest performances on distance-run tests had the highest risk of being physically inactive as young adults. Conversely, as Telama, Yang, Laasko, and Viikari (1997) reported, children and adolescents who are active tend to be more physically active as young adults than less active children. These kinds of longitudinal studies examining the relationship between youth physical activity and adult physical activity are impossible without accurate assessments of children's physical activity.

As discussed earlier in this chapter, assessing physical activity levels in children is most reliably and validly done via direct monitoring (e.g., accelerometers, pedometers, direct observation), but lack of feasibility limits applying such procedures in large-scale studies. In Sallis et al. (1993), three self-report instruments—a 7-day recall interview, a self-administered survey, and a simple activity rating—were examined for test–retest reliability and validity. Reliability ranged from .77 to .89 on the three self-report measures over all participants. However, as would be expected, reliability was higher for older children than for younger children. The 7-day recall had moderate concurrent validity (.44-.53) when related to a criterion of monitored heart rate; again, validity improved as the age of the children increased. Sallis and colleagues concluded that self-report techniques could be used with adolescents, but caution was necessary if used with younger children. Kang, Mahar, and Morrow (2016) described issues that often arise when measuring physical activity in children and adolescents and provided recommendations for assessing physical activity behaviors (see table 13.3).

Table 13.3 Considerations When Choosing Methods to Measure Physical Activity in Children

Considerations	Available tools
WHAT IS THE NATURE OF THE PHYSICAL ACTIVITY YOU WANT TO MEASURE?	
Walking behavior	Questionnaires Logs and diaries Pedometers Accelerometers
Minutes of physical activity at different intensities	Questionnaires Logs and diaries Heart rate monitoring Accelerometers Direct observation
Physical activity intensity, duration, and frequency	Questionnaires Logs and diaries Heart rate monitoring Accelerometers Direct observation
Total physical activity	Questionnaires Logs and diaries Heart rate monitoring Pedometers Accelerometers Direct observation

(continued)

Table 13.3 Considerations When Choosing Methods to Measure Physical Activity in Children *(continued)*

Considerations	Available tools
WHAT IS THE NATURE OF THE PHYSICAL ACTIVITY YOU WANT TO MEASURE? *(continued)*	
Energy expenditure	Questionnaires Logs and diaries Heart rate monitoring Accelerometers Doubly labeled water Direct observation
Mode or specific physical activities	Questionnaires Logs and diaries Direct observation
Context	Questionnaires Logs and diaries Direct observation
Sedentary behavior	Questionnaires Logs and diaries Accelerometers Direct observation
WHAT IS YOUR SAMPLE SIZE?	
Small (e.g., <100)	Questionnaires Logs and diaries Heart rate monitoring Pedometers Accelerometers Doubly labeled water Direct observation
Medium (e.g., 100-500)	Questionnaires Logs and diaries Heart rate monitoring Pedometers Accelerometers
Large (e.g., >500)	Questionnaires Pedometers Accelerometers
HOW LARGE IS YOUR BUDGET?	
Small	Questionnaires Logs and diaries Pedometers Direct observation
Medium	Heart rate monitoring Accelerometers Direct observation
Large	Doubly labeled water Direct observation
HOW MUCH BURDEN CAN PARTICIPANTS ENDURE?	
Low	Questionnaires Pedometers Direct observation
Moderate	Questionnaires Pedometers Accelerometers
High	Logs and diaries Heart rate monitoring Doubly labeled water

Considerations	Available tools
HOW MUCH TIME CAN YOU INVEST IN DATA ANALYSES?	
Little	Questionnaires Pedometers
Moderate	Logs and diaries Accelerometers
Can be detailed	Heart rate monitoring Accelerometers Doubly labeled water Direct observation
HOW MUCH TIME DO YOU HAVE TO COLLECT PHYSICAL ACTIVITY DATA?	
Not much	Questionnaires
Moderate	Logs and diaries Heart rate monitoring Pedometers Accelerometers Direct observation
Unlimited	Pedometers Accelerometers Doubly labeled water Direct observation
DO YOU WANT TO PROVIDE IMMEDIATE FEEDBACK TO THE PARTICIPANTS?	
No	Questionnaires Logs and diaries Heart rate monitoring Accelerometers Doubly labeled water Direct observation
Yes	Pedometers

Modified from Kang et al. (2016).

WWW Go to **HK***Propel* to complete Student Activity 13.1.

Dataset Application

A group of elementary school children wore pedometers for a week to measure their physical activity levels. The chapter 13 large dataset in HK*Propel* shows the daily number of steps calculated for each student. Use the data provided to do the following:

1. Determine the alpha coefficient for the 7 days. (Review chapter 6 if necessary.)

2. Determine if there is a significant difference in the mean number of steps taken by boys and girls for the entire week. (Hint: Use an independent t-test from chapter 5.)

3. Determine the correlation between the number of steps taken and the students' body weights (chapter 4).

4. Determine if there is a relationship between number of steps taken per week and the number of FitnessGram Healthy Fitness Zones achieved (chapter 4).

Figure 13.5 shows examples of questions used for middle school students as part of the Youth Risk Behavior Survey (YRBS) conducted by the CDC. Three of the questions are about physical activity behaviors and one is about sedentary behaviors (specifically, screen

39. During the past 7 days, on how many days were you physically active for a total of **at least 60 minutes per day**? (Add up all the time you spent in any kind of physical activity that increased your heart rate and made you breathe hard some of the time.)

    A. 0 days

    B. 1 day

    C. 2 days

    D. 3 days

    E. 4 days

    F. 5 days

    G. 6 days

    H. 7 days

40. On an average school day, how many hours do you spend in front of a TV, computer, smartphone, or other electronic device watching shows or videos, playing games, accessing the Internet, or using social media (also called "screen time")? (Do **not** count time spent doing schoolwork.)

    A. Less than 1 hour per day

    B. 1 hour per day

    C. 2 hours per day

    D. 3 hours per day

    E. 4 hours per day

    F. 5 or more hours per day

41. In an average week when you are in school, on how many days do you go to physical education (PE) classes?

    A. 0 days

    B. 1 day

    C. 2 days

    D. 3 days

    E. 4 days

    F. 5 days

42. During the past 12 months, on how many sports teams did you play? (Count any teams run by your school or community groups.)

    A. 0 teams

    B. 1 team

    C. 2 teams

    D. 3 or more teams

Figure 13.5 Middle school physical activity behavior questions from the 2021 Youth Risk Behavior Survey.

From www.cdc.gov/healthyyouth/data/yrbs/pdf/2021/2021-YRBS-Standard-MS-Questionnaire.pdf.

time). Research about physical activity and health has become increasingly focused on the amount of time children and youth regularly spend inactive. Television watching and computer use are examples of such sedentary behaviors that may be contributing to the increased overweight condition observed in children and adolescents.

It is important to note that self-report measures of physical activity behaviors in children and youth can be misleading. LeBlanc and Janssen (2010) report that 46% and 30% of boys and girls aged 12 to 19, respectively, self-report meeting the 60-minute daily physical activity guideline. However, the percentages drop to 4% (boys) and 1% (girls) when objective measures are obtained with accelerometers. Direct observation of physical activity behaviors can be expensive in terms of time and labor. However, new technologies have allowed reliable, valid, and feasible measurements using direct observation of physical activity. The observation instruments SOFIT, BEACHES, SOPLAY, and SOPARC were created by physical activity researchers at San Diego State University to assess physical activity behaviors in school, home, park, and recreation settings. They are summarized in table 13.4. Note the reliability and validation support based on information presented in chapters 6 and 7.

FitnessGram Methods

The FitnessGram physical fitness testing and education program also includes two approaches to assessing physical activity. The first is a brief questionnaire, based on items from the CDC's YRBS, assessing student participation in aerobic, muscular strength and endurance, and flexibility activities over the previous 7 days.

The second FitnessGram approach to physical activity assessment is ActivityGram, based on the Previous Day Physical Activity Recall (Weston, Petosa, and Pate, 1997). This segmented-day approach requires each participant to report activities in 30-minute blocks from 7:00 a.m. to 11:00 p.m. Students report the frequency, intensity, time, and type of physical activity, which provide content validity for the activity recall. This logging chart

Table 13.4 Instruments Used for Observation of Physical Activity Behaviors

Instrument	Purpose	Reliability	Validity
SOFIT (System for Observing Fitness Instruction Time) (McKenzie, Sallis, and Nader, 1991)	To obtain data on student activity levels, lesson content, and teacher interactions related to physical activity during physical education classes	Interobserver agreement ranges from 82% to 99%.	Correlations with various external criteria range from .42 to .99.
BEACHES (Behaviors of Eating and Activity for Children's Health: Evaluation System) (McKenzie, Sallis, Patterson, et al., 1991)	Allows an integrated assessment of eating and physical activity, including a wide range of potentially modifiable environmental and social factors	Interobserver agreement ranges from 90% to 99%. Observers continue training until agreement exceeds 80% to 85%.	BEACHES items correlate with external criteria (e.g., heart rate increased with each activity code increment).
SOPLAY (System for Observing Play and Leisure Activity in Youth) (McKenzie, Marshall, Sallis, and Conway, 2000)	To evaluate and measure the relationship between children's physical activity and their play environment (e.g., accessible, usable, organized, supervised, equipped)	Interobserver agreement ranges from .90 to .99. Intraclass correlations range from .75 to .98.	Content validation based on the SOFIT and BEACHES systems.
SOPARC (System for Observing Play and Active Recreation in Communities) (McKenzie, Cohen, Sehgal, Williamson, and Golinelli, 2006)	To obtain data on the characteristics of participants and their physical activity levels in park and recreational settings, also park characteristics	Interobserver reliabilities range from 88% to 99.8%.	Content validation based on the SOFIT and BEACHES systems.

is illustrated in figure 13.6. Once the data from the chart are entered into the FitnessGram computer software, an ActivityGram report is generated and given to each student. An example report is presented in figure 13.7.

WWW! **Go to HK*Propel* to complete Student Activity 13.2.**

Name_____ Teacher_____ Grade_____ Date_____

Record the main activity that you did during each 30-minute time period by writing the activity type and activity number in the appropriate box (types and numbers can be found in the box located at the bottom of the page). You may have done many things in each 30-minute time period, but try to pick the activity you did for most of the time. Then, check the box that describes how it felt (light/easy [L], moderate/medium [M], vigorous/hard [V]). Note: for all rest activities, use the **Rest** box and you can leave the L, M, or V columns blank. In the Time column, write the amount of time that the activity felt this hard or easy: **S** (some), **M** (most), or **A** (all).

Time	Type	Number	Rest	L	M	V	Time	Time	Type	Number	Rest	L	M	V	Time
7:00								3:00							
7:30								3:30							
8:00								4:00							
8:30								4:30							
9:00								5:00							
9:30								5:30							
10:00								6:00							
10:30								6:30							
11:00								7:00							
11:30								7:30							
12:00								8:00							
12:30								8:30							
1:00								9:00							
1:30								9:30							
2:00								10:00							
2:30								10:30							

Activity types and numbers

Lifestyle activity (LA)	Aerobic sports (AE)	Flexibility activity (FA)
1. Walk, bike, skate	11. Field sports	21. Martial arts
2. Housework/yardwork	12. Court sports	22. Stretching
3. Active games/play	13. Racket sports	23. Yoga
4. Active job	14. Aerobic sports–PE	24. Ballet dance
5. Other lifestyle activity	15. Other aerobic sports	25. Other flexibility
Aerobic activity (AA)	**Muscular activity (MA)**	**Resting (R)**
6. Aerobic class/dancing	16. Gymnastics	26. Schoolwork
7. Aerobic gym	17. Muscular sports	27. Computer/TV
8. Aerobic activity	18. Weightlifting	28. Eating/resting
9. Aerobic activity in PE	19. Wrestling	29. Sleeping
10. Other aerobic activity	20. Other muscular	30. Other rest

Figure 13.6 ActivityGram logging chart.

Reprinted by permission from The Cooper Institute, *FITNESSGRAM/ACTIVITYGRAM Test Administration Manual*, updated 4th ed. (Champaign, IL: Human Kinetics, 2010), 107.

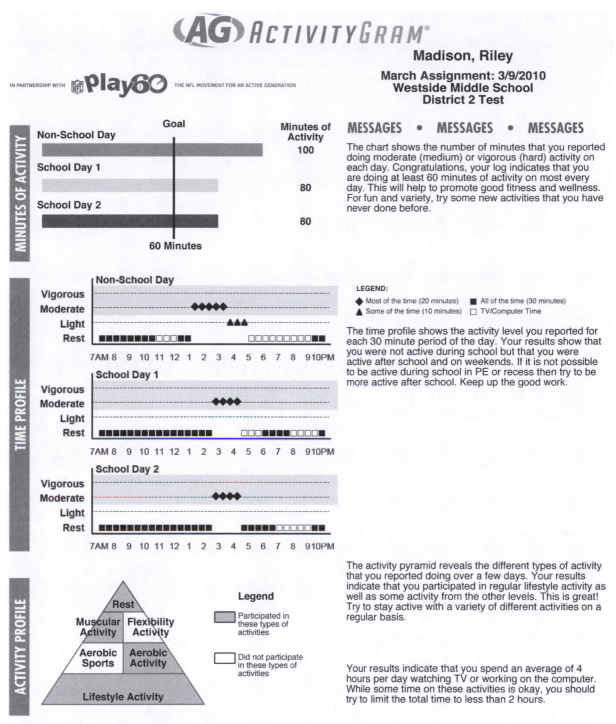

Figure 13.7 ActivityGram report.

Reprinted by permission from The Cooper Institute, *FITNESSGRAM/ACTIVITYGRAM Test Administration Manual*, updated 4th ed. (Champaign, IL: Human Kinetics, 2010), 82.

Dataset Application

Open the chapter 13 large dataset in HK*Propel*. Assume that the data represent the number of steps taken per day for each of 2 days and that you are interested in whether people take at least 7500 steps per day each day—a value that defines them as "somewhat active" (Tudor-Locke and Bassett, 2004). Do the following:

1. Calculate the descriptive statistics for each day (chapter 3). Do males and females differ in the number of steps they take (chapter 5)?

2. Correlate the results for the 2 days (chapter 4). This is an interclass reliability illustration (chapter 6).

3. Calculate the alpha coefficient for the 2 days (chapter 6). Variables have been created to indicate whether people have met the criterion for each day.

4. Use the procedures in chapter 7 to determine the percent agreement, phi coefficient, chi-square, and Kappa for meeting the criterion across these 2 days. Comment on the criterion-referenced reliability obtained.

DETERMINING DOSE OF PHYSICAL ACTIVITY AND ENERGY EXPENDITURE FOR HEALTH

The amount of physical activity and energy expenditure needed for good health has long been an interest in the field of exercise science (Blair, LaMonte, and Nichaman, 2004). A typical dose of physical activity and related energy expenditure is determined by the FITT formula, where F represents frequency (how often to exercise), I represents intensity (how hard to exercise), T represents time (how long to exercise), and T represents type (what kind of exercise to choose). In the 1970s, the aerobic exercise recommendation for developing and maintaining fitness in healthy adults was 3 to 5 days per week, at 50% to 85% $\dot{V}O_2$max (60%-90% maximum heart rate), for 15 to 60 minutes (Blair et al., 2004).

The ACSM suggests that a person must expend 150 kilocalories (kcal) per day (or 1000 kcal per week) on physical activity to achieve a health benefit from physical activity. The physical activity pyramid shown in figure 13.8 presents guidelines for physical activity. The basic concept is for people to develop physically active lifestyles that will produce sufficient levels of health-related fitness, promote physical and mental health, and limit sedentary behaviors. Figure 13.9 summarizes selected key guidelines from the *Physical Activity Guidelines for Americans* (USDHHS, 2018). From a health standpoint, the most important message is to limit the amount of time spent inactive, because physical inactivity is negatively related to all components of health-related physical fitness.

WWW! **Go to HK*Propel* to complete Student Activity 13.3.**

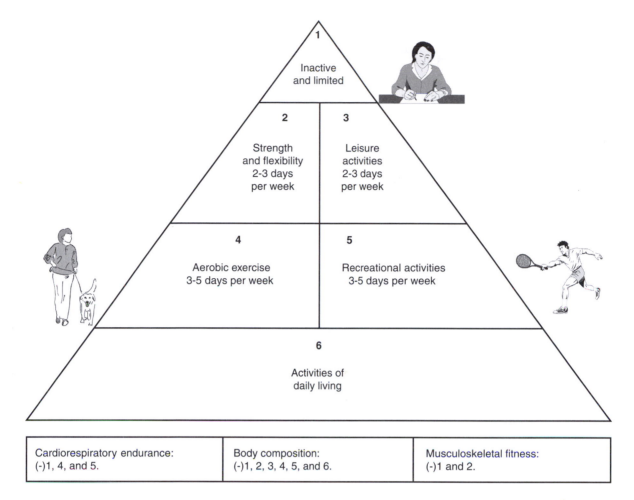

Figure 13.8 The relationship of health-related fitness to the physical activity pyramid.

Preschool-Aged Children

- Preschool-aged children (ages 3 through 5 years) should be physically active throughout the day to enhance growth and development.

- Adult caregivers of preschool-aged children should encourage active play that includes a variety of activity types.

Children and Adolescents

- It is important to provide young people opportunities and encouragement to participate in physical activities that are appropriate for their age, that are enjoyable, and that offer variety.

- Children and adolescents ages 6 through 17 years should do 60 minutes (1 hour) or more of moderate-to-vigorous physical activity daily:

(continued)

Figure 13.9 Selected key guidelines for physical activity.

Figure 13.9 *(continued)*

1. Aerobic: Most of the 60 minutes or more per day should be either moderate- or vigorous-intensity aerobic physical activity and should include vigorous-intensity physical activity on at least 3 days a week.

2. Muscle-strengthening: As part of their 60 minutes or more of daily physical activity, children and adolescents should include muscle-strengthening physical activity on at least 3 days a week.

3. Bone-strengthening: As part of their 60 minutes or more of daily physical activity, children and adolescents should include bone-strengthening physical activity on at least 3 days a week.

Adults

- Adults should move more and sit less throughout the day. Some physical activity is better than none. Adults who sit less and do any amount of moderate-to-vigorous physical activity gain some health benefits.

- For substantial health benefits, adults should do at least 150 minutes (2 hours and 30 minutes) to 300 minutes (5 hours) a week of moderate-intensity, or 75 minutes (1 hour and 15 minutes) to 150 minutes (2 hours and 30 minutes) a week of vigorous-intensity aerobic physical activity, or an equivalent combination of moderate- and vigorous-intensity aerobic activity. Preferably, aerobic activity should be spread throughout the week.

- Additional health benefits are gained by engaging in physical activity beyond the equivalent of 300 minutes (5 hours) of moderate-intensity physical activity a week.

- Adults should also do muscle-strengthening activities of moderate or greater intensity that involve all major muscle groups on 2 or more days a week, as these activities provide additional health benefits.

Older Adults

The key guidelines for adults also apply to older adults. In addition, the following key guidelines are just for older adults:

- As part of their weekly physical activity, older adults should do multicomponent physical activity that includes balance training as well as aerobic and muscle-strengthening activities.

- Older adults should determine their level of effort for physical activity relative to their level of fitness.

- Older adults with chronic conditions should understand whether and how their conditions affect their ability to do regular physical activity safely.

- When older adults cannot do 150 minutes of moderate-intensity aerobic activity a week because of chronic conditions, they should be as physically active as their abilities and conditions allow.

A variety of strategies are available for achieving a lifestyle that includes healthful amounts of physical activity. For example, hybrid pedometers, which combine some capacities of traditional pedometers and accelerometers, can be programmed to differentiate between walking and moderate or vigorous steps by adjusting a steps-per-minute function—thus, steps taken above a user-defined count are recorded as MVPA. These crosses between pedometers and accelerometers are relatively affordable and record steps, minutes of physical activity, and minutes of MVPA. The suggested number of steps per day suggested as an index of public health for adults is listed in chapter 1 (p. 7).

MEASUREMENT AND EVALUATION CHALLENGE

After some research and reading this chapter, Sofia has a good understanding of the current methods available to measure children's physical activity. She has decided to recommend the use of accelerometers and heart rate monitors to the head physician for several reasons:

1. Laboratory methods such as DLW and indirect calorimetry would be too expensive and impractical to use in a daily free-living setting.
2. The validity and reliability of self-report methods for children are too low to employ.
3. Although an observation method is good for children, it does not apply to this case because it cannot be used for monitoring free-living conditions.

Both accelerometers and heart rate monitors could overcome these limitations and have been demonstrated to be reasonably valid and reliable in tracking children's physical activity in a free-living environment. Sofia is confident that she can make a good contribution to the newly implemented program to help overweight and obese children.

SUMMARY

Because of the benefits of physical activity for health and prevention of chronic diseases, accurate measurement of physical activity and energy expenditure is critical for the fields of exercise science and public health. This chapter defined physical activity and energy expenditure and introduced the major methods to measure them, as well as the advantages and disadvantages of each method and how to select one appropriately. Given that a variety of methods are available, ranging from self-reports (e.g., diaries, logs, recall questionnaires) to direct monitoring (e.g., behavioral observation by a trained observer, electronic monitoring of heart rate or body motion, physiological monitoring using direct calorimetry, indirect calorimetry, or doubly labeled water), it is important to have an understanding of each.

WWW Go to HKPropel for videos, homework assignments, and quizzes that will help you master this chapter's content.

Psychological Measurements in Sport and Exercise

OUTLINE

OBJECTIVES

After studying this chapter, you will be able to

- define and identify the scope of the field of sport and exercise psychology;
- be aware of the performance enhancement and mental health aspects of sport psychology;
- recognize the benefits of physical activities and exercise psychology;
- differentiate between psychological states and traits;
- explain the differences between general and sport-specific psychological tests;
- discuss ethics and cautions in the use of psychological testing with athletes;
- describe the qualifications necessary for using and interpreting psychological tests;
- illustrate the feedback processes involved in psychological testing of athletes;
- discuss the use and abuse of psychological tests for team selection;
- identify factors related to the reliability and validity of general and sport-specific psychological inventories typically used in sport and exercise settings; and
- differentiate between the research and application perspectives of psychological inventories used in sport and exercise.

www **The lecture outline in HK*Propel* will help you identify the major concepts of the chapter.**

MEASUREMENT AND EVALUATION CHALLENGE

Rashad has just been hired to coach a team in the National Football League (NFL). As a football player in college, Rashad always believed that his mental skills and competitiveness really helped him achieve the high level he attained. Although he didn't have as much natural ability as many of his opponents or teammates, he was a confident, self-motivated athlete who was usually able to control his emotions, keep focused throughout the game, and stay positive even during a series of bad losses. Therefore, as he prepares to take over his first head coaching assignment, he believes that it's important not only to assess his players' physical abilities but also evaluate their mental skills.

However, Coach Rashad has little background in assessing players' personality and mental skills, and he has many questions. For example, what psychological inventories should he use to evaluate his players? Should he be administering and interpreting these tests, or should he hire a qualified sport psychologist to do so? When during the season should these psychological inventories be administered? Should he use interviews to find out about his players' mental skills? These are difficult questions, but if he can get the answers, Coach Rashad believes that the information derived from these psychological assessments will be invaluable to him and his players in understanding and improving their mental skills.

The purpose of this chapter is to provide you with an introduction to the evolving field of sport and exercise psychology and to highlight some of the measurement techniques and instruments typically used to assess psychological attitudes, states, and traits. In addition, we discuss issues related to the measurement and interpretation of psychological tests along with ethical considerations when using these tests with athletes and other sport and exercise participants.

SPORT PSYCHOLOGY: PERFORMANCE ENHANCEMENT AND MENTAL HEALTH

The field of sport psychology has developed so rapidly that many people do not have a clear understanding of what the field encompasses. Many people see the field of sport psychology as narrow, when in fact it has a wide scope and applications to many areas of our lives.

Most definitions of sport psychology clearly underscore two main areas: *performance enhancement* and *mental health*. The performance enhancement aspect of sport psychology refers to the effects of psychological factors on sport performance. These factors include anxiety, burnout, concentration, confidence, flow, motivation, mental preparation, and **personality** (the totality of a person's distinctive psychological traits). This aspect of sport psychology is not restricted to elite athletes. Rather, it spans a continuum from children participating in youth sports to older adults playing in recreational or competitive leagues. The key point is that the mind affects the body; therefore, the way we think and feel has an impact on how we physically perform. In competitive sports, participants' physical skills are often comparable, and so the key difference between winning and losing is often mental skills, not physical ones.

The other main focus of sport psychology is enhancing mental health and well-being through participation in sport. Our minds can have an important effect on our bodies; conversely, our bodies (i.e., the way we feel physically) can have an important effect on our minds. For example, research indicates that vigorous exercise is related to a reduction in **anxiety** (distress and tension caused by apprehension) as well as **depression** (a mental condition of gloom, sadness, or dejection). Sport participation has been related to increases in self-esteem and self-confidence. In essence, sport participation can increase our feelings of psychological well-being and thus exert a positive influence on our mental health and total well-being.

However, participating in competitive sports can also sometimes be frustrating and upsetting; losing and failing to meet expectations can lead to increases in anxiety, depression, and aggressiveness. In such cases, sport psychologists attempt to accentuate the positive aspects of sport participation so that participants will receive psychological benefits.

WWW **Go to HK*Propel* to complete Student Activity 14.1.**

Sport psychology researchers who study performance enhancement and those who study mental health have different objectives, so it is not surprising that their measurement objectives and the types of psychological tests they use differ considerably (although there is some overlap). For example, when studying performance

Sport psychology can help athletes manage frustrations and anxiety.

enhancement, researchers are typically interested in measuring psychological factors that influence performance, either positively or negatively. Their tests might measure attentional focus, confidence, flow, precompetitive anxiety, self-motivation, and imagery. For researchers interested in studying mental health, tests that measure anxiety, burnout, depression, self-esteem, self-concept, **mood**, and anger are appropriate.

EXERCISE PSYCHOLOGY: PSYCHOLOGICAL BENEFITS OF PHYSICAL ACTIVITY

Although the terms *sport psychology* and *exercise psychology* are often used interchangeably, the latter focuses on the psychological benefits gained from participating in exercise, which, as we learned in chapter 13, is a subset of physical activity that is planned and structured and typically involves repetitive bodily movement to improve or maintain physical fitness. There have been some significant findings in this area:

• *Academic performance.* Based on a meta-analysis, Sibley and Etnier (2003) reported that physically active students perform better on a variety of tasks than students who are not as physically active. In 2005, Grissom published a large-scale study showing a positive correlation between reading and mathematics scores on the Stanford Achievement Tests and physical fitness scores measured by FitnessGram for 884,715 fifth-, seventh-, and ninth-grade students in California. Since then, the relationship between academic performance and physical fitness and activity has been examined in many studies, and a review by the Centers for Disease Control and Prevention (CDC, 2010) concluded that there is substantial evidence to indicate that physical activity can help improve academic achievement, including grades and standardized test scores.

- *Stress management.* By increasing the production of the brain's "feel-good" neurotransmitters, such as endorphins, exercise participation has been associated with helping to reduce negative effects of stress, improve mood, and lower symptoms of mild depression and anxiety (Seaward, 2021). In addition, exercise participation can also improve sleep, which is often disrupted by stress, depression, and anxiety.

- *Cognitive function.* One of the most exciting areas in the field of exercise psychology is research showing that aerobics and resistance training can improve cognitive function in both older adults (Kramer et al., 1999) and youth (Hillman et al., 2008) by inducing structural and functional changes in the brain (Mandolesi et al., 2018).

- *Age-related psychological decline.* Exercise participation has also been demonstrated as one of the most effective means for preventing and treating age-related psychological decline, such as memory loss, mild cognitive impairment, and dementia, including Alzheimer's disease (Erickson et al., 2011; Gholamnezhad et al., 2020).

- *Psychological symptoms of chronic diseases.* Either as a result of illness or the side effect of treatment, many patients with chronic diseases suffer from negative psychological symptoms, such as anxiety, depression, and fatigue. Fortunately, exercise interventions have been effective for preventing or managing many of these symptoms (Campbell et al., 2019).

TRAIT VERSUS STATE MEASURES

When assessing personality and psychological variables in sport and exercise psychology, it is important to distinguish between trait and state measures. Trait psychology was the first to adopt the scientific approach in the study of personality. Trait psychology is based on the assumption that personality **traits**—the fundamental aspects of personality—are relatively stable, consistent attributes that shape the way people behave. In essence, the trait approach considers the general source of variability to reside within the person; it minimizes the role of situational or environmental factors. This means that a person who assesses at a high level for the trait of aggressiveness would tend to act more aggressively in most situations than someone who assesses at a lower level for the same trait.

Traits, or predispositions, may be acquired through learning or be genetically inherent. A well-known trait typology is extroversion–introversion, which relates to a person's general tendency to respond in either an outgoing or a shy manner, without regard for the situation. For example, an extrovert placed in a new situation in which he or she doesn't know anybody will likely be outgoing and try to meet people. In sport psychology, the traits that have been studied include anxiety, aggressiveness, self-motivation, confidence, and achievement motivation. In exercise psychology, self-efficacy, social physique anxiety, depression, self-confidence, and self-consciousness are the traits often studied.

Alternatively, in the situational, or state, approach, psychological **states** are viewed as a function of a particular situation or environment, and traits are given a subsidiary role in the explanation and prediction of behavior. Thus, behavior is expected to change from one situation to the next. In essence, psychological states are transitory, potentially changing rapidly as the situation changes.

For example, assume you are a reserve member of a basketball team and usually sit on the bench. Your team is playing in a championship game, but because you are not expected to play much, your anxiety right before the game is low. However, your coach tells you that one of the starters is ill and won't be able to play and that you're going to be starting. In a few seconds, your anxiety about how well you will perform in this important game has greatly elevated. The situation of starting on the team has caused a dramatic shift in your state anxiety level; this result has little to do with your trait anxiety.

www **Go to HKPropel to complete Student Activity 14.2.**

Although sport and exercise psychologists make the distinction between traits and states, you need to consider both when attempting to understand and predict behavior in sport and exercise settings. The idea that traits and states are codeterminants of behavior is known as the *interactionalist approach*; it is the approach most widely endorsed today. This approach to the study of personality and behavior suggests that one's personality, needs, interests, and goals (i.e., traits), as well as the particular constraints of the situation (e.g., win–loss record, attractiveness of the facility, crowd support), interact to determine behavior.

A study conducted by Sorrentino and Sheppard (1978) demonstrated the usefulness of the interactionalist approach. Their study tested male and female swimmers by having them perform a 200-yard (182.8 m) freestyle time trial, once swimming individually and once as part of a relay team. The situational factor assessed was whether each swimmer had a faster individual split time when he or she swam alone or as part of a relay team. In addition, the personality characteristic of affiliation motivation was assessed. This personality trait represents whether the swimmers were more approval oriented (viewing competing with others as positive) or rejection threatened (feeling threatened because they might let their teammates down in a relay situation). As the researchers predicted, the approval-oriented swimmers demonstrated faster times in relays than in individual events. In contrast, the rejection-threatened swimmers swam faster in individual events than in relays. Thus, the swimmers' race times involved an interaction between their personalities (affiliation motivation) and the situation (individual versus relay).

However, it should be noted that the relationship between states and traits may not always be distinct. In fact, Brose, Lindenberger, and Schmiedek (2013) found that current affective states influenced scores on trait responses. For example, participants who felt better at the time of measurement were more likely to report above-average appraisals of their trait affect and believe that current states were more relevant than general traits. How a person feels in a particular moment can influence his or her evaluation of his or her general affect (i.e., how he or she generally feels). These findings expose a prevalent methodological (validity) concern regarding single measures of traits, which may be corrected by taking multiple measurements.

An understanding of both trait and state measures can help a coach and athlete better prepare for the demands of competition.

© Human Kinetics

GENERAL VERSUS SPECIFIC MEASURES

For many years, almost all the trait and state measures of personality and other psychological attributes used in sport psychology came from general psychological inventories. These inventories measured general or global personality traits and states, with no specific reference to sport or physical activity. Examples of such inventories are the State–Trait Anxiety Inventory (Spielberger, Gorsuch, and Lushene, 1970), the Test of Attentional and Interpersonal Style (Nideffer, 1976), the Profile of Mood States (McNair, Lorr, and Droppleman, 1971; also see Morgan, 1980), the Self-Motivation Inventory (Dishman and Ickes, 1981), the Eysenck Personality Inventory (Eysenck and Eysenck, 1968), and the Locus of Control Scale (Rotter, 1966).

Psychologists have found that situation-specific measures provide a more accurate and reliable predictor of behavior in those situations than do general tests. For example, Sarason (1975) observed that some students were doing poorly on tests, not because they weren't smart or hadn't studied, but simply because they became too anxious and "froze up." These students were not particularly anxious in other situations, but taking exams made them extremely anxious. Sarason labeled such people *test anxious* and devised a situation-specific test called Test Anxiety that measures how anxious a person feels before taking exams. This situation-specific test was a better predictor of immediate pretest anxiety than a general test of trait anxiety. Clearly, we can better predict behavior when we have increased knowledge of the specific situation and how people tend to respond to it.

Along these lines, sport psychologists have begun to develop sport-specific tests to provide more reliable and valid measures of personality traits and states in sport, exercise, and physical activity contexts. For example, your coach might not be concerned about how anxious you are before giving a speech or before taking a test, but he or she would surely be interested in how anxious you are before competition (especially if excess anxiety is detrimental to your performance). A sport-specific test of anxiety would provide a more reliable and valid assessment of an athlete's precompetitive anxiety than a general anxiety test. Some examples of psychological inventories developed specifically for use in sport settings include the Sport Anxiety Scale (Smith, Smoll, and Schutz, 1990), the Task and Ego Orientation Sport Questionnaire (Duda, 1989), the Sport Motivation Scale (Briere, Vallerand, Blais, and Pelletier, 1995), the Physical Self-Perception Profile (Fox and Corbin, 1989), the Sport Imagery Questionnaire (Hall, Mack, Paivio, and Hausenblas, 1998), the Competitive State Anxiety Inventory-2 (Martens, Vealey, and Burton, 1990), the Group Environment Questionnaire (Widmeyer, Brawley, and Carron, 1985), the Trait Sport Confidence Inventory (Vealey, 1986), and the Perfectionism Inventory for Sport (Anshel, Weatherby, Kang, and Watson, 2009). Some sport psychologists have even gone a step further and developed tests for a specific sport rather than for sports in general, such as the Tennis Test of Attentional and Interpersonal Style (Van Schoyck and Grasha, 1981), the Anxiety Assessment for Wrestlers (Gould, Horn, and Spreeman, 1984), the Group Cohesion Instrument for Basketball (Yukelson, Weinberg, and Jackson, 1984), and the Australian Football Mental Toughness Questionnaire (Gucciardi, Gordon, and Dimmock, 2008).

To obtain improved predictions of behavior, a number of sport-specific multidimensional inventories have been developed that measure a variety of psychological skills believed important for performance success. The first of these scales to receive attention was the Psychological Skills Inventory for Sports (Mahoney, Gabriel, and Perkins, 1987). Although this scale was used a great deal after its development, it was later demonstrated that the psychometric properties were generally poor. Thus, other measures were developed, such as the Test of Performance Strategies (TOPS; Thomas, Murphy, and Hardy, 1999), which measures eight mental skills in competition and in practice (see figure 14.1); the Ottawa

Competition strategy

I talk positively to get the most out of competitions.

I perform without consciously thinking about it.

I visualize competition going exactly the way I want it to.

When the pressure is on I know how to relax.

I evaluate whether I achieve my competition goals.

Practice strategy

I use practice time to work on relaxation techniques.

My attention wanders while training.

I have difficulty increasing my energy level during workouts.

I talk positively to get the most out of practice.

I set goals to help me use practice time effectively.

Figure 14.1 Sample items from the Test of Performance Strategies (TOPS).

Items are scored on the following scale: 1 = never; 2 = rarely; 3 = sometimes; 4 = often; 5 = always. Data from Thomas, Murphy, and Hardy (1999).

TOPS has been updated by Varga (2020).

Mental Skills Assessment Tool-3 (Durand-Bush, Salmela, and Green-Demers, 2001) assessing 12 mental skills fitting under the three general categories of cognitive skills (e.g., imagery, competition planning, refocusing), foundation skills (e.g., goal setting, self-confidence, commitment), and psychosomatic skills (e.g., relaxation, activation, fear control); and the Athletic Coping Skills Inventory-28 (ACSI-28; Smith, Schutz, Smoll, and Ptacek, 1995), which has seven subscales (e.g., concentration, peaking under pressure, freedom from worry). These scales are becoming more popular because researchers often attempt to determine the effectiveness of a mental skills training program in part by showing that it can enhance a number of psychological skills. These types of scales have also been used in investigations of psychological characteristics of elite performers. For example, Taylor, Gould, and Roio (2008) used the TOPS to help discriminate between more and less successful U.S. Olympians at the 2000 Olympic Games in Sydney.

Over time, the construct of perfectionism has evolved from a general to a sport-specific measure. With the increased interest in perfectionism in sporting contexts, Dunn, Causgrove Dunn, and Syrotuik (2002) created the Sport Multidimensional Perfectionism Scale (Sport-MPS) to address the necessity to evaluate perfectionism specifically in a sport-domain context. A second version, the Sport Multidimensional Perfectionism Scale-2 (Sport-MPS-2; Gotwals and Dunn, 2009), updated and expanded the original scale to include additional subscales. Subsequent research has supported the psychometric properties of the instrument and demonstrated the scale's ability to differentiate global from sport perfectionism, as well as healthy from unhealthy perfectionist tendencies (Gotwals, Dunn, Causgrove Dunn, and Gamache, 2010). Anshel, Weatherby, Kang, and Watson (2009) attempted to measure sport perfectionism for competitive athletes based on a unidimensional approach. Their instrument, the Sport Perfectionism Scale, consists of 35 items with a 5-point Likert-type scale.

Similarly, efforts were also made in exercise psychology to develop specific tests or scales for exercise participation, fitness, and function. For example, Albert Bandura (1977) introduced self-efficacy, a construct defined as the confidence in one's abilities to successfully perform a particular behavior. Since then, a variety of physical activity and fitness-specific self-efficacy scales have been developed for particular populations, including Self-Efficacy and the Stages of Exercise Behavior Change (Marcus et al., 1992), Lifestyle Education for Activity (LEAP; Dishman et al., 2004), walking self-efficacy in women with diastolic heart failure (Gary, 2006), and self-efficacy related to physical activity and function in older adults (McAuley et al., 2011).

WWW **Go to HK*Propel* to complete Student Activity 14.3.**

QUANTITATIVE VERSUS QUALITATIVE MEASUREMENT

Now that you know something about the field of sport and exercise psychology, we discuss the two approaches (quantitative and qualitative methodologies) that sport and exercise psychologists use to gain a better understanding of the psychological factors involved in sport, exercise participation, and physical activity.

Quantitative research, numerical in nature and the more traditional of the two approaches, involves experimental and correlational designs that typically use precise measurements. There are usually objectively measured variables (often controlled in a laboratory setting), and psychological states and traits are assessed with reliable and valid psychological inventories. In quantitative research, the researcher tries to stay out of the data-gathering process by using laboratory measurements, questionnaires, and other objective instruments. The data are then statistically analyzed with computations performed by computers. The obtained scores are often interval or ratio in scale (see chapter 2).

Although **qualitative research**, which is textual in nature, is often depicted as the antithesis of the traditional quantitative methods, it should be seen as a complementary method. Qualitative research methods generally include field observations, ethnography, and interviews seeking to understand the meaning of a participant's experience in a particular setting and how the components mesh together as a whole. To this end, qualitative research focuses on the essence of the phenomenon and relies heavily on people's perceptions of their world. Hence, the objectives are primarily description, understanding, and meaning. Qualitative data are rich, because they provide depth and detail; they allow people to be understood in their own terms and in their natural settings (Patton, 1990). The researcher does not manipulate variables through experimental treatment but rather is interested more in the process than in the product. Through observations and interviews, relationships and theories are allowed to emerge from the data rather than being imposed on them; thus, induction is emphasized (in contrast to quantitative research, in which deduction is primary). Finally, in qualitative research, the researcher interacts with the subjects, and the sensitivity and perception of the researcher play crucial roles in procuring and processing the observations and responses.

A study by Holt and Sparkes (2001) provided an example of the use of ethnography to study group cohesion in a team over the course of a season. The researcher (one of the authors) spent a season with a soccer team as a player and coach and collected data via participant observation, formal and informal interviews, a field diary, and a reflective journal. Having the researcher embedded in a group for an entire season allowed a richness of data and understanding that could not occur through the use of quantitative data collection techniques alone.

Additionally, data gathered from qualitative methods often serve as a foundation for the development or refinement of quantitative instruments and operational definitions. For example, the concept of "mental toughness" has suffered from conceptual and definitional ambiguity and has consequently been the subject of much debate. To address this issue, Jones, Hanton, and Connaughton (2002) interviewed international performers to gather data regarding the athletes' perceptions of what attributes comprise mental toughness and how it should be defined. Not only did this study provide practical information for the evolving debate regarding mental toughness, but it also exploited the often-overlooked insights of the athletes themselves. Furthermore, the study inspired and provided procedural guidance for subsequent sport-specific research directed toward an understanding of mental toughness in soccer (Thelwell, Weston, and Greenlees, 2005), Australian football (Gucciardi, Gordon, and Dimmock, 2008), and cricket (Bull, Shambrook, James, and Brookes, 2005).

A mixed-methods approach (both qualitative and quantitative) may allow for precise measurement and statistical analyses while simultaneously including qualitative data to help explain and expand upon the numbers in a holistic and comprehensive manner. Such an approach can allow an investigator to consider the idiosyncrasies of each person's responses, which might otherwise go unexplored. An example of a mixed-methods design can be seen in a study by Gould, Dieffenbach, and Moffett (2002), in which Olympic champions gave interviews and completed a battery of psychological inventories. The researchers were able to evaluate a diverse range of psychological characteristics based on quantitative and qualitative data. Not surprisingly, the results provided an extensive amount of information that has helped sport psychologists better understand what underlying factors influence these elite athletes. For instance, the quantitative methods (questionnaires) found that the athletes scored high in confidence. Through interviews, it was determined that a sport program or organization that fostered, nurtured, and instilled positive attitudes and skills in athletes helped build their confidence. Similarly, athlete confidence was also found to be influenced by having an optimistic, achievement-oriented environment while growing up as well as by having coaches express confidence in an athlete's development.

It should be noted that qualitative studies have become increasingly popular (e.g., Bloom, Stevens, and Wickwire, 2003; Culver, Gilbert, and Trudel, 2003; Lamb et al., 2018; Sparkes, 1998; Strean, 1998; Stuart, 2003). For a review of the role of qualitative research in sport and exercise psychology, see Brewer, Vose, Raalte, and Petitpas (2011). Most assessments of psychological traits and states in sport psychology have taken place through inventories and questionnaires that have been carefully developed to provide high reliability and validity—a quantitative approach.

Quantitative Methods

As noted, most psychological assessments in the field of sport and exercise psychology use the traditional quantitative methodology of the questionnaire (see Tenenbaum, Eklund, and Kamata [2012] for a summary of advances in sport and exercise psychology measurement). There are a number of quantitative questionnaires to choose from; two of the most popular employ Likert scales and semantic differential scales. Each of these scales can help define the multidimensional nature of the construct being assessed.

Likert Scales

A Likert scale is a 5-, 7-, or 9-point scale that assumes equal intervals between responses. In the example provided here, the difference between *strongly agree* and *agree* is considered equivalent to the difference between *disagree* and *strongly disagree*. This type of scale is used to assess the degree of agreement or disagreement with statements and is widely used in attitude inventories. An example of an item using a Likert scale is as follows:

All high school students should be required to take 2 years of physical education classes.

Strongly agree	Agree	Neutral	Disagree	Strongly disagree
5	4	3	2	1

A principal advantage of scaled responses is that they permit a wider choice of expression than categorical responses, which are typically dichotomous—that is, offering such choices as *yes* and *no* or *true* and *false*. The five, seven, or more intervals also increase the reliability of the instrument. In addition, different response words can be used in scaled responses, as shown in figure 14.2.

Never	Sometimes	Often	Frequently	Always		
1	2	3	4	5		
Strongly agree	Agree	No Opinion	Disagree	Strongly disagree		
1	2	3	4	5		
Not important at all				Extremely important		
1	2	3	4	5		
Always						Never
1	2	3	4	5	6	7
Agree						Disagree
1	2	3	4	5	6	7	8	9	10

Figure 14.2 Examples of scaled responses.

Semantic Differential Scales

Another popular measuring technique involves using a semantic differential scale, which asks participants to respond to bipolar pairs of adjectives, or those with opposite meanings, such as *weak–strong*, *relaxed–tense*, *fast–slow*, and *good–bad* with scales anchored at the extremes. Respondents are asked to choose a point on the continuum that best describes how they feel about a specific concept. Refer to table 14.1 on attitudes toward physical activity; notice that you can choose any of seven points that best reflects your feelings about the concept.

The process of developing semantic differential scales involves defining the concept you want to evaluate and then selecting specific bipolar adjective pairs that best describe respondents' feelings and attitudes about the concept. Research has indicated that the semantic differential technique measures three major factors. By far, the most frequently used factor is evaluation—the degree of goodness you attribute to the concept or object being measured. Potency involves the strength of the concept being rated, and the activity factor uses adjectives that describe action. The following list shows some examples of bipolar adjective pairs that measure the evaluation components; this is an example of a differential scale for measuring attitudes toward competitive sports for children.

Table 14.1 Semantic Differential Scales for Measuring Attitudes Toward Physical Activity

Physical activity								
Good	____	____	____	____	____	____	____	Bad
Pleasant	____	____	____	____	____	____	____	Unpleasant
Relaxed	____	____	____	____	____	____	____	Tense
Hot	____	____	____	____	____	____	____	Cold
Healthy	____	____	____	____	____	____	____	Unhealthy
Nice	____	____	____	____	____	____	____	Awful
Delicate	____	____	____	____	____	____	____	Rugged
Active	____	____	____	____	____	____	____	Passive

Evaluation
- Pleasant–unpleasant
- Fair–unfair
- Honest–dishonest
- Good–bad
- Successful–unsuccessful
- Useful–useless

Potency
- Strong–weak
- Hard–soft
- Heavy–light
- Dominant–submissive
- Rugged–delicate
- Dirty–clean

Activity
- Steady–nervous
- Happy–sad
- Active–passive
- Dynamic–static
- Stationary–moving
- Fast–slow

WWW **Go to HK*Propel* to complete Student Activity 14.4.**

Qualitative Methods

As discussed earlier, qualitative methods are becoming popular in sport psychology research because they provide a richness of information that is often untapped when traditional questionnaires are used. The following are some common methods of qualitative research.

Interviews

The interview is undoubtedly the most common source of data in qualitative research. Interviews range from a highly structured style, in which questions are determined in advance, to open-ended interviews, which allow for free responses. The most popular mode of interview used in sport psychology research is the semistructured interview. Each participant responds to a general set of questions, but the test administrator uses different probes and follow-up questions depending on the nature of a participant's response. A good interviewer must first establish rapport with participants to allow them to open up and describe their true thoughts and feelings. It is also important that an interviewer remain nonjudgmental regardless of the participant's responses. Above all, an interviewer has to be a good listener.

Using a digital recorder is probably the most common method of recording interviews, because it preserves the entire interview for subsequent data analysis. Although a small percentage of participants are initially uncomfortable being recorded, this uneasiness typically disappears quickly. Taking notes during an interview is another frequently used method to record data; sometimes note-taking is used in conjunction with recording when an interviewer wants to highlight certain important points. One drawback with taking notes without recording is that it keeps the interviewer busy, thus interfering with his or her thoughts and observations of the participant.

A good example of the use of interviews to collect qualitative data in sport psychology is the work of Gould, Dieffenbach, and Moffett (2002) noted earlier. To better understand the psychological characteristics of Olympic athletes and their development, the researchers used qualitative methods (along with some quantitative psychological inventories) and an inductive analytic approach to research. They studied 10 former Olympic champions (winners of a combined 32 medals); 10 Olympic coaches; and 10 parents, guardians, or significant others (one for each athlete). The interviews focused on the mental skills needed to become an elite athlete as well as how these mental skills were developed during the early, middle, and later years of the athletes' careers. The interviews, transcribed verbatim from recordings, were analyzed by content analysis, a method that organizes the interview into increasingly complex themes and categories representing psychological characteristics and their development. Among the numerous findings, it was revealed that these Olympic athletes were characterized by the ability to focus and block out distractions, maintain their competitiveness, set and achieve goals, and control anxiety, along with mental toughness and confidence. In addition, coaches and parents were particularly important in assisting the development of these athletes (although others were also involved to a lesser or greater extent). Specifically, coaches, parents, and others provided support and encouragement, created an achievement environment, modeled appropriate behavior, emphasized high expectations, provided motivation, taught physical and psychological skills, and enhanced confidence through positive feedback. Such depth of information could not have been collected through the strict use of questionnaires and other psychological inventories.

Observation

Observation is second only to the interview as a means of qualitative data collection. Although most early studies relied on direct observation with note-taking and coding of certain categories of behavior, the current trend is to use video. Just as using a digital recorder to conduct an interview allows a researcher to review everything a participant said, video likewise captures a participant's behavior for future analysis. Because participants who know that they are being watched and recorded may change their behavior, it is important that the observer be unobtrusive to the participants under study. Observers can seem less obtrusive by watching for several days before actually starting to record their observations so that the novelty of their presence wears off and behavior can occur naturally.

A classic example of using observations in sport psychology is seen in the seminal work of Smith, Smoll, and Curtis (1979) on the relationship between coaches' behaviors and young athletes' reactions to these behaviors. In the first part of the investigation, observers were trained to watch Little League baseball coaches during practice and games and carefully note what the coaches did. The observers recorded these behaviors over several months. After compiling literally thousands of data points, the researchers attempted to collapse the behaviors into some common categories. The end result of this process categorized coaching behaviors into those initiated by a coach (spontaneous) versus those that were reactions to a player's behavior (reactive).

For example, a coach yelling at a player who made an error was showing a reactive behavior. However, a coach's instruction to players on how to slide was considered a spontaneous behavior. Within these categories of reactive and spontaneous behaviors were subcategories, such as positive reinforcement, negative reinforcement, general technical instruction, general encouragement, and mistake-contingent technical instruction. These subcategories of coaching behaviors resulted in the development of an instrument called the Coaching Behavior Assessment System, which allowed the researchers to conduct several studies investigating the relationship between specific coaching behaviors and players' evaluative reactions to these behaviors. For example, in one study, team members who played for coaches who gave predominantly positive (as opposed to negative) reinforcement liked their teammates more, wanted to continue playing next year, and saw their coaches as more knowledgeable and as better teachers than did team members who played for coaches who did not favor positive reinforcement.

Using observation to assess physical activity behavior is another good example. Although many objective methods are available to assess physical activity (see chapter 13), the systematic observation approach could provide rich contextual information related to the impact of social, cultural, and environmental factors on physical activities. In fact, a set of validated, low-inference observation tools with little participant burden has been developed for studying physical activity (McKenzie and van der Mars, 2015).

CAUTIONS WHEN USING PSYCHOLOGICAL TESTS

Psychological inventories are crucial to sport and exercise psychologists. Such tests help evaluate the accuracy of different psychological theories and provide practitioners with a tool for applying theory to practice. We focus on the use of psychological tests in applied settings because it is here that abuses of test results and misconceptions in analysis are more likely to occur. Who is qualified to administer psychological tests to athletes or clients? The American Psychological Association (APA) recommends that test administrators have the following knowledge:

• *An understanding of testing principles and the concept of measurement error.* A test administrator needs a clear understanding of statistical concepts such as correlation, measures of central tendency (mean, median, and mode), variance, and standard deviation. No test is perfectly reliable or valid. Tests work only in specific settings.

- *The ability and knowledge to evaluate a test's validity for the purposes for which it is employed.* A qualified test administrator will recognize that test results are neither absolute nor irrefutable and that there are potential sources of measurement error. He or she will do everything possible to eliminate or minimize such errors. For example, testers must be aware of the potential influences of situational factors as well as interpersonal factors that may alter the way test scores are interpreted. In addition, cultural, social, educational, and ethnic factors can all have a large impact on an athlete's test results.

- *Self-awareness of one's own qualifications and limitations.* Unfortunately, there have been cases in the field of sport and exercise psychology in which people were not aware of their own limitations, and thus used tests and interpreted results in a manner that was unethical and potentially damaging. For example, many psychological inventories are designed to measure psychopathology or abnormality. To interpret these results, a test administrator needs special training in psychological assessment and possibly in clinical psychology. Without this background, it is unethical to use such tests with clients. Thus, testers should make sure they have the appropriate training to administer and interpret the tests that they employ.

Some psychological tests are used specifically, but inappropriately, to determine whether an athlete should be drafted onto a team or to determine if an athlete has the right psychological profile for a certain position (such as a middle linebacker in U.S. football). This practice was particularly rampant in the 1960s and 1970s, though it seems to have abated. These unethical uses of psychological tests can cause an athlete to be hastily eliminated from a team or not drafted simply because he or she does not appear to be "mentally tough." In truth, however, it is difficult to predict athletic success from the results of these tests. An athlete's or a team's performance (often measured in terms of winning and losing) is a complicated issue affected by such factors as physical ability, experience, coach–player compatibility, ability of opponents, and teammate interactions. It would be naive to think that simply knowing something about an athlete's personality provides enough information to predict whether he or she will be successful.

What types of psychological tests should be given to athletes or clients, and what conditions should be established for test administration and feedback? Most importantly, before they actually take the tests, test takers should be told the purpose of the tests, what the tests measure, and how the tests are going to be used. In several cases, athletes were given psychological tests with no explanation as to why they were taking the test; furthermore, these athletes received no feedback about the results and interpretation of the tests. This is unethical and violates the rights of those taking the tests. In most cases, the tests should be used to help test administrators and test takers better understand the test takers' psychological strengths and weaknesses so that they can focus on increasing their strengths and improving their weaknesses. In addition, test takers should be provided with specific feedback on the results of the testing process. Feedback should be provided in a way that test takers can gain insight into themselves and understand what the test indicates. The results and feedback can then serve as a springboard to stimulate positive changes.

If test takers are not told the reason for the testing, they will typically become suspicious regarding what the test is going to be used for. In such cases it is not unusual for test takers to worry that a coach or a teacher will use the test to select the starters or cut undesirable players. Given these circumstances, test takers will be increasingly likely to attempt to exaggerate their strengths or positive behaviors and minimize their weaknesses or negative behaviors—for example, claiming to be calm, cool, and collected when he or she really feels nervous and tight in critical situations. This response style of "faking good" can distort the true results of the test and make its interpretation virtually useless. Thus, it is important that test takers be assured of confidentiality in whatever tests they take, because this increases the likelihood that they will answer truthfully. A sport or exercise psychology consultant who has formal training in psychological assessment and measurement is the best person to administer and interpret psychological tests.

A test not only has to be reliable and valid, but also needs to be validated for the particular sample and situation in which it is being conducted. For example, you might choose a test that was developed using adults and administer it to youth aged 13 to 15 years. However, the wording of the test might be such that the younger individuals do not fully understand the questions; thus, the results are not relevant. Similarly, a test might have been developed on a predominantly white population, and your athletes or clients happen to be mostly African American and Hispanic. Cultural differences might lead to problems in interpreting the results with different populations.

As explained earlier, it is often a mistake to compare a test taker's psychological test results with norms that were developed years ago for a different population. Rather, the more critical point is how a test taker feels in relation to him- or herself, which represents an intrapersonal approach. In essence, the information gleaned by the use of psychological tests should be used to help test takers improve their skills relative to their own standards rather than to compare themselves with others.

Finally, caution should also be made with regard to the quality of the tests to be selected. Many existing scales were developed several decades ago, using only college students during test development. Validity and reliability of these tests were often examined under an outdated framework, and other critical qualities (e.g., test fairness) were not examined at all (Zhu, 2012b). As a result, application of these tests may not generate accurate or appropriate information.

WWW Go to HK*Propel* to complete Student Activities 14.5 and 14.6.

NEW TECHNOLOGY FOR ASSESSMENT PRACTICES

There is no question that technology has changed the world, and so it should not be surprising that it has also changed the manner of assessment in sport and exercise psychology. One of the most exciting areas is examining how psychological functions and benefits from sports are related to the brain structure and its functional activities. Using magnetic resonance imaging (MRI), for example, Colcombe et al. (2006) found a significant increase in brain volume (in both gray and white matter regions) in older adults who participated in aerobic fitness training, but not for older adults who participated in stretching and toning (nonaerobic) activities. In another study, Hillman et al. (2009) used electroencephalographic (EEG) techniques to examine children's cognitive control while involved in differing levels of physical activity and found that even a single, acute bout of moderately intense aerobic exercise (i.e., walking) may improve attention in preadolescent children.

Another example is the study by Muzik et al. (2018) of Wim Hof, known as the "the Iceman," who holds several world records for prolonged resistance to cold exposure. During this study, Hof wore a specifically designed bodysuit that researchers infused with temperature-controlled water while acquiring imaging data to measure changes in his biology to cold exposure. Using functional magnetic resonance imaging (fMRI) to study his brain and positron emission tomography (PET) to study his body, the researchers compared Hof's results with a group of healthy comparison participants. They found that by practicing the breathing and meditation methods he invented, Hof was able to keep his skin temperature relatively invariant to cold exposure. The researchers attributed this to his increased sympathetic innervation and glucose consumption in intercostal muscles, which appeared to allow Hof to generate heat that dissipates to lung tissue and warms circulating blood in the pulmonary capillaries.

TESTS USED IN SPORT AND EXERCISE PSYCHOLOGY

As mentioned previously, many tests, inventories, and scales have been developed for the measurement practices of sport and exercise psychology. Some commonly used tests are summarized below and more detailed information of these tests can be found on HK*Propel*. For additional reviews of the tests used in sport and exercise psychology, see Anshel (1987), Duda (1998), and Ostrow (2002).

Selected Tests Used in Sport Psychology

Psychological skills training (PST) has grown rapidly, and with it, a need for precise measurement tools. A wide range of questionnaires have been developed to evaluate imagery, self-talk, concentration, goal setting, and a number of other skills. Many such scales have demonstrated satisfactory psychometric properties and enhanced a practitioner's ability to measure psychological skills and the effectiveness of PST programs. Table 14.2 provides examples of some of these instruments. Many of the related references provide information on reliability and validity, as well as norms (if available) and illustrations of their use in sport psychology settings.

WWW Go to HK*Propel* to complete Student Activity 14.7.

Selected Tests Used in Exercise Psychology

As introduced earlier, exercise psychology is primarily concerned with studying the psychological and emotional effects of exercise, as well as promoting physical activity and healthy behaviors. Like sport psychology, it is an interdisciplinary field that integrates knowledge from biological and social sciences, health psychology, neuroscience, and exercise science. Understandably, measurement in a discipline based on such diversity can be a challenging and multifaceted issue. However, the importance of understanding psychological determinants and consequences of exercise and physical activity has never been more important, because public health concerns have continued to escalate globally.

Questionnaires used in exercise psychology target a range of factors, including self-perceptions, mood, motivation, attitudes, and emotions regarding exercise. Because exercise psychology is a multidimensional and interdisciplinary field, a combination of questionnaires may provide a more comprehensive view of the participants. Table 14.3 illustrates selected tests in exercise psychology.

Table 14.2   Selected Sport Psychology Tests and References

Content area	Examples of well-validated instruments	Related references
Competitive anxiety		
	Sport Competition Anxiety Test	Martens (1977)
	Sport Anxiety Scale-2	Smith, Smoll, Cumming, and Grossbard (2006)
	Competitive State Anxiety Inventory-2	Craft, Magyar, Becker, and Feltz (2003) Cox, Martens, and Russell (2003)
Psychological skills		
	Ottawa Mental Skills Assessment Tool Visual-Motor Behavior Rehearsal	Greenspan and Feltz (1989) Weinberg and Comar (1994) Fournier, Calmels, Durand-Bush, and Salmela (2005) Brown and Fletcher (2013)
Imagery		
	Vividness of Movement Imagery Questionnaire-2	Driskell, Copper, and Moran (1994) Isaac, Marks, and Russell (1986) Murphy and Jowdy (1992)
Self-talk		
	Automatic Self-Talk Questionnaire for Sports	Zourbanos and colleagues (2009)
Concentration		
	Test of Attentional and Interpersonal Style	Nideffer (1976) Van Schoyck and Grasha (1981) Albrecht and Feltz (1987)
Goal setting		
	Task and Ego Orientation in Sport Questionnaire	Hanrahan and Biddle (2002) Duda (1989)
Confidence		
	Self-Efficacy Mental Toughness	Nelson and Furst (1972) Weinberg, Gould, and Jackson (1979) Weinberg, Gould, Yukelson, and Jackson (1981) Clough and Strycharczyk (2012) Weinberg, Butt, and Culp (2011)
Trait and state sport confidence		
	Trait Sport Confidence Inventory State Sport Confidence Inventory	Vealey (1986)

(continued)

Table 14.3 Selected Exercise Psychology Tests and References *(continued)*

Content area	Examples of well-validated instruments	Related references
Motivation		
	Personal Incentives for Exercise Questionnaire Exercise Motivation Scale	Markland and Hardy (1993) Li (1999)
Behavioral regulation		
	Behavioral Regulation in Exercise Questionnaire	Mullan, Markland, and Ingledew (1997)
Stages of change		
	Stages of Change Scale for Exercise and Physical Activity	Marcus, Selby, Niaura, and Rossi (1992) Prochaska and DiClemente (1983) Velicer and Prochaska (1997) Marcus and colleagues (1994)

Selected General Psychological Tests Used in Sport and Exercise

Although the use of sport- and setting-specific psychological tests is the current preferred method, the use of general psychological tests in competitive sport and physical activity settings has added a great deal to the literature, in both the performance enhancement and psychological well-being domains. Table 14.4 illustrates selected general scales used in sport and exercise psychology.

Table 14.4 Selected General Psychological Instruments and References

Content area	Examples of well-validated instruments	Related references
Self-motivation		
	Self-Motivation Inventory	Dishman and Ickes (1981)
Mood		
	Profile of Mood States (POMS)	McNair, Lorr, and Droppleman (1971) Terry (2000) Rowley, Landers, Kyllo, and Etnier (1995)
Attention		
	Test of Attentional and Interpersonal Style	Nideffer (1976)

Dataset Application

The chapter 14 large dataset in HK*Propel* consists of data on 200 boys and girls aged 12. The variables include the following:

- Gender: 1 (boy), 2 (girl)
- FitnessGram PACER test results
- Healthy Fitness Zone (HFZ) achievement: 0 (did not achieve HFZ), 1 (did achieve HFZ)

- Body mass index (BMI)
- Student responses to 12 body satisfaction questions: 6-point Likert scale, ranging from 1 (extremely dissatisfied) to 6 (extremely satisfied)
- Student responses to 6 endurance self-efficacy questions: 6-point Likert scale, ranging from 1 (very false of me) to 6 (very true of me)
- Total score for the body satisfaction and endurance self-efficacy scales

Determine the following:

1. What is the alpha coefficient for the 12 body satisfaction variables (chapter 6)?

2. What is the alpha coefficient for the 6 endurance self-efficacy variables (chapter 6)?

3. What is the Pearson product-moment correlation between the body satisfaction scale and the endurance self-efficacy scale (chapter 4)?

4. What percentage of students are in the FitnessGram PACER Healthy Fitness Zone (HFZ) (chapters 3, 7, and 10)?

5. What is the Pearson product-moment correlation between the BMI and PACER laps (chapter 4)?

6. Do those achieving the FitnessGram HFZ differ from those not achieving the HFZ on body satisfaction, endurance self-efficacy, BMI, and PACER laps (chapters 3 and 5)?

7. Answer some other interesting questions you might have for this dataset.

MEASUREMENT AND EVALUATION CHALLENGE

After reading this chapter, Coach Rashad has learned about the types of psychological inventories at his disposal. He knows about the guidelines for using psychological tests with athletes, and he knows that he can get the best prediction if he uses tests that are sport specific. Armed with this information, Rashad decides on the following approach:

1. He will hire a qualified sport psychologist to coordinate the psychological evaluations and to interpret all psychological inventories.

2. He and the test administrator will tell his athletes specifically what the inventories are being used for and will provide feedback about their results.

3. He will use a few trait inventories before the season starts, such as the Sport Competition Anxiety Test (SCAT) and the Trait Sport Confidence Inventory (TSCI), as well as a multidimensional scale (ACSI-28), to increase his understanding of his athletes' general psychological profiles.

4. He will assess his athletes just before competition using state measures such as the Competitive State Anxiety Inventory-2 (CSAI-2) and the State Sport Confidence Inventory (SSCI), which are found online, to determine how the athletes are feeling just before games.

With this information, Coach Rashad, in conjunction with the sport psychologist, will devise a mental training program to help his athletes practice and develop their mental skills. By combining this psychological testing and training with his work on the physical aspects of training, Coach Rashad hopes that his football team will be ready to perform at their optimal level.

SUMMARY

The field of sport and exercise psychology has been expanding rapidly, with the two major areas being performance enhancement and mental health. Part of this expansion has involved developing and refining the measurement of psychological traits and states. However, these advances have also highlighted issues involving the use and abuse of psychological testing in sport and exercise settings. The American Psychological Association provides guidelines for the use of psychological testing to ensure that athletes are treated in an ethical manner and that test administration and feedback are conducted responsibly.

Although much early work in sport psychology used standardized scales to assess personality and other psychological constructs, more qualitative methods have been developed and used, including in-depth interviews and observation. These methods, along with the development of sport-specific psychological inventories, have resulted in more accurate and reliable prediction of behaviors in sport and exercise settings. Although sport-specific tests are currently preferred, several general psychological inventories have also been used extensively in sport and exercise settings; these have added to our understanding of sport-related behavior. Finally, the rapid development of new technology is greatly improving our abilities in psychological assessment and our abilities to serve athletes and the public.

Remember that most instruments consist of several scales. This is because psychological measures need to reflect the multidimensionality of personality, perception, and other psychological factors. Personality and its associated constructs are multifactorial in nature; subscales help researchers get at these various factors. In fact, there has been a trend in sport and exercise psychology to develop scales that assess a variety of psychological characteristics or skills and are by definition multidimensional. The result is that a number of factors are assessed; however, each factor is typically made up of only a few items and thus may not be as reliable as a scale focusing on only that mental skill (e.g., anxiety, confidence).

www Go to HK*Propel* for videos, homework assignments, and quizzes that will help you master this chapter's content.

Performance-Based Assessment: Alternative Ways to Assess Student Learning

OBJECTIVES

After studying this chapter, you will be able to

- define performance-based assessment and distinguish it from traditional standardized testing;
- discuss current trends in assessment practices in education;
- identify various types of performance-based assessments;
- identify criteria for judging the quality of performance when using performance-based assessments;
- explain the advantages and disadvantages of performance-based assessment; and
- identify guidelines for developing and using performance-based assessment.

WWW **The lecture outline in HK*Propel* will help you identify the major concepts of the chapter.**

The authors acknowledge the contributions of Dr. Jacalyn L. Lund (Georgia State University) for her earlier contributions to the text in this chapter.

MEASUREMENT AND EVALUATION CHALLENGE

Mariko is a middle school physical education teacher who is planning to teach a soccer unit. After completing the unit, her students will show their knowledge and skills in a small-sided (four vs. four players) soccer game. Mariko also wants to assess the students' affective domain behaviors when evaluating game play. During the unit, Mariko plans to use several formative assessments to help her diagnose student problems and offer feedback about how to improve their soccer playing skills. Because she is anxious to encourage students to use higher-level thinking skills, Mariko will use several types of assessment. Her goal is to have her students play soccer to demonstrate competence on several standards from the *National Standards and Grade-Level Outcomes for K-12 Physical Education* (SHAPE America, 2014). Knowing that one of the characteristics of performance-based assessment is that students must know the criteria by which they will be assessed before the start of instruction, Mariko plans to develop her assessments before beginning the unit. What assessments should Mariko use to assess each student's understanding and performance of the various skills, knowledge, and attitudes needed to play soccer?

Assessment and accountability are two of the most frequently discussed topics in education. An increasing interest in using assessment to enhance student learning has coincided with the educational reform movement that led to the development of standards and standards-based instruction. Increased interest in assessment and having students use higher-level thinking skills has led to a focus on performance-based assessment techniques that align with outcome-based content standards. To distinguish from the traditional multiple-choice testing format, a variety of terms such as *alternative assessment*, *authentic test*, *performance assessment*, *portfolio*, and *true test* were used to describe the new testing format in its early development.

The term **assessment** has been popularized by educators over the last three decades and is now used more often in education than either of the terms *measurement* or *evaluation*. Recall from chapter 1 that evaluation is the judgment made about the quality of a product, whereas assessment is "the gathering of evidence about a student's level of achievement and making inferences on student progress based on that evidence" (SHAPE America, 2014, p. 90).

But what is performance-based assessment, and how is it used in physical education? How can it be used to assess learning in all domains and thus be used to determine whether a student is physically educated? The next section will address these and other related questions.

IMPETUS FOR DEVELOPING AN ALTERNATIVE TYPE OF ASSESSMENT

There has long been widespread concern among parents, business leaders, government officials, and educators about school effectiveness. The educational reform movement that began in the mid-1980s called for a dramatic shift in the way schools delivered education, as well as for increased demands on learning that would require students to demonstrate higher levels of learning and mastery of content. Standardized tests, which typically featured the types of selected-response questions used to measure knowledge (the lowest form of learning according to Bloom's taxonomy), were deemed insufficient for measuring student learning. Measuring students' abilities to analyze, synthesize, and evaluate knowledge, rather than just acquire it, required a different approach to testing.

As demands for accountability and dissatisfaction with the traditional forms of assessment increased, Wiggins (1989) called for a new type of assessment that would assess higher levels of student thinking and be part of the learning process. The use of the word *assessment* instead of *testing* signaled a shift from focusing on a test at the end of instruction to the use of assessment to enhance student instruction. The word *assessment* comes from the French word *assidere,* which means "to sit beside." The shift implied that teachers were no longer gatekeepers who judged whether students had learned, but rather coaches who were responsible for increasing student learning. The "gotcha" mentality was replaced by the belief that assessment methods should facilitate teaching, enhance learning, and result in greater student achievement.

Along with the change in the purpose for assessment, Wiggins also advocated for assessments that would have meaning for students. He argued that students should demonstrate learning in more ways than by penciling in answers on a standardized test. When Wiggins introduced performance assessment, the measurement practice often dictated how the performance should be designed and administered. Feuer and Fulton (1993) later identified several performance-based assessments appropriate for use in education, including essays, exhibitions, portfolios, and oral discourse. These assessments were complex and represented meaningful activities that might be done by professionals in the field. In fact, performance assessment is not new to physical assessment. According to a history review by Madaus and O'Dwyer (1999), it can be traced back to the Sung dynasty (960-1279 C.E.), when performance tests were used in military examinations (the candidates were scored on three arrows shot from a horse at a man-shaped target).

Ultimately, this educational reform movement led to the development of content standards that were written in terms of what students should know and be able to do. Several disciplines (e.g., math, science, language arts, social science) have national standards defining essential skills and knowledge for that content area. The National Association for Sport and Physical Education (NASPE) first announced content standards for physical education in 1995 in the publication *Moving Into the Future—National Standards for Physical Education: A Guide to Content and Assessment* (later revised in 2004). In 2014, new national standards, along with a curriculum framework, were published in *National Standards and Grade-Level Outcomes for K-12 Physical Education.* The new standards (SHAPE America 2014, p. 12) are as follows:

- Standard 1: The physically literate individual demonstrates competency in a variety of motor skills and movement patterns.
- Standard 2: The physically literate individual applies knowledge of concepts, principles, strategies, and tactics related to movement and performance.
- Standard 3: The physically literate individual demonstrates the knowledge and skills to achieve and maintain a health-enhancing level of physical activity and fitness.
- Standard 4: The physically literate individual exhibits responsible personal and social behavior that respects self and others.
- Standard 5: The physically literate individual recognizes the value of physical activity for health, enjoyment, challenge, self-expression, and/or social interaction.

To meet the assessment needs of these standards, a performance assessment tool called PE Metrics was developed (NASPE, 2008) and is now in its third edition (SHAPE America, 2019).

Clarification of Terminology

Although Wiggins originally used the terms *true test* and **authentic assessment**, the terms **performance-based assessment** and **alternative assessment** also started appearing in the literature. Some people like to distinguish between the terms, whereas others (Herman, Aschbacher, and Winters, 1992; Lund and Kirk, 2020; Zhu, 1997) use the terms interchangeably.

Using any adjective is problematic if it causes confusion. For example, one physical educator may argue that a tennis forehand test delivered with a ball machine is not authentic, but that if a player is rallying with another player, it would be an authentic assessment. In fact, while both are perfectly good skill tests, neither is an example of the complex assessment to which *performance-based assessment* alludes. With this type of assessment, authenticity is important only because it tends to increase the assessment's meaningfulness and relevance to students. The more authentic one can make an assessment—meaning that it represents something that a professional in the field might do or value—the greater the chance that students will see the relevance of the assessment and get excited about completing it.

For the purpose of this chapter, the term *performance-based assessment* will be used to indicate the complex assessments used to determine student learning, or "the variety of tasks and situations in which students are given opportunities to demonstrate their understanding and to thoughtfully apply knowledge, skills, and habits of mind in a variety of contexts" (Marzano, Pickering, and McTighe, 1993, p. 13).

In physical education, the term *performance-based assessment* should not be confused with tests of performance, such as skills tests and fitness tests. Skills and fitness tests are not complex and meaningful as entities (e.g., a real tennis game is not simply players hitting forehand shots back and forth over the net), and so are not considered performance-based assessments.

Because data serve as an integral part of the instructional process, diagnostic assessments are used before the start of a unit to inform teachers about prior student knowledge (Lund and Veal, 2013). **Formative assessment** is the term used to describe assessment done during the learning process; this can be self- or peer assessment or administered by a teacher. The results do not contribute to student grades but rather are used to provide feedback to students to correct or improve the final performance. Students can use these results to set goals and monitor their progress toward the final learning goal or objective. Teachers can also use the data to establish goals for student learning, map progress toward final learning goals, or plan future instruction. Although performance-based assessments are useful as part of the coaching and teaching process, traditional assessments such as skills tests can be used as formative assessments as well.

Summative assessments are those given at the conclusion of instruction. They represent the final level of student achievement and are associated with grading, evaluation, or both. Summative assessments are administered by a teacher or someone with authority for the class.

As discussed in chapter 1, the terms *formative* and *summative* refer to the purpose of the assessment rather than a certain format or type of assessment. Formative assessments might be used throughout a unit for feedback and teacher planning but then become a summative assessment if used at the conclusion of a unit as a final assessment of student learning. Although students can improve their performance with formative assessments, there is no opportunity to improve a grade or score on a summative assessment.

Performance-based assessment uses nontraditional techniques to ensure students learn and retain physical activity knowledge.

Mastery Item 15.1

How does the learning measured by a traditional test (selected response) differ from that measured by performance-based assessments?

How does performance-based assessment differ from traditional assessment (i.e., multiple-choice or true–false tests, sport skills tests, or physical fitness tests)? Herman, Aschbacher, and Winters (1992) identified several characteristics to describe performance-based assessments that help explain what makes performance-based assessments different from traditional testing formats.

• *Students are required to perform, create, produce, or solve something.* Several types of assessments allow students to demonstrate their knowledge of a topic or their ability to do something through the creation of a product or performance. Creating a jump rope routine or a dance or playing lacrosse requires a different type of learning that traditional tests would not measure. With performance-based assessments, students can do the work individually or as part of a group. The resulting work typically takes several days to complete, and the process used to complete it is as important as the final product or performance. Note that student expectations here align with the highest level of Anderson and Krathwohl's (2001) revision of Bloom's taxonomy: creating.

• *Students are required to use higher-level thinking and problem-solving skills.* It is difficult to measure student ability to analyze, synthesize, or evaluate with a test consisting of selected-response questions. When measuring higher levels of thinking, teachers can give students a problem and have them solve it. In physical education classes, written tests often ask questions about the dimensions of a court, when a sport was first played, who invented the sport, or how to define terms associated with the activity. These are all examples of knowledge or comprehension, according to Bloom's taxonomy. Assessments of game play or the ability to choreograph a dance require students to use higher levels of thinking and problem-solving skills, thus making them more challenging for students. Additionally, these assessments often reveal true levels of student understanding or misunderstanding.

• *Tasks are used that represent meaningful instructional activities.* Students rarely get excited about the chance to take a written test. Performance-based assessments provide students the opportunity to demonstrate learning in a variety of ways. For example, students could demonstrate their knowledge of rules while officiating a game or serving as the game announcer, which may be far more meaningful and relevant than responding to questions about rules on a written test. Teachers might ask students in a dance class to write a critique of a dance performance to demonstrate their knowledge of choreography. These are examples of meaningful tasks that allow students to demonstrate what they have learned.

• *When possible, real-world applications are used.* Because the preceding examples are all activities that occur in the real world of sport, students are more likely to see the relevance of the task. Some assessment books recommend that student performance should have an audience when possible. Playing a game for others or distributing a publication gives the assessment an additional level of accountability simply because someone other than a teacher will have an opportunity to view it.

• *Professional judgment is used to score the assessment results.* Performance-based assessments are administered and scored by professionals who have training and experience in the subject or content being evaluated, using a list of criteria previously identified as significant. When an external panel (i.e., not a teacher) is used to do the scoring, training on using the scoring rubrics is typically done to ensure that the scorers understand the criteria and expectations and that assessments are scored with consistency.

• *Students are coached throughout the learning process.* This is one of the most exciting aspects of performance-based learning. Students are provided with guidance and feedback throughout the process so that the final performance or product is the best that it can be. Because students know the criteria in advance, they are capable of comparing their own performance to the criteria (self-assessment) or can assist other students by providing peer feedback. Because the final goal is outcome based, various people can provide knowledge and assistance to help each student achieve that goal.

TYPES OF PERFORMANCE-BASED ASSESSMENT

As previously mentioned, there are many types of performance-based assessments, each of which has strengths and deficiencies. The goal of assessment is to accurately determine whether students have learned the materials or information taught and reveal whether they have complete mastery of the content with no misunderstandings. Just as researchers use multiple data sources to determine the truthfulness of the results, teachers can use multiple types of assessment to evaluate student learning. The real challenge comes in selecting or developing performance-based assessments that complement both each other and more traditional assessments to equitably assess students in physical education and human performance. Because assessments involve the gathering of data or information, some type of product, performance, or recording sheet must be generated. The following are some examples of various types of performance-based assessments used in physical education.

Observation

Human performance provides many opportunities for students to exhibit behaviors that may be directly observed by others, a unique advantage of working in the psychomotor domain. Wiggins (1998) used physical activity when providing examples to illustrate complex assessment concepts, because they are easier to visualize than would be the case with a cognitive example. The nature of performing a motor skill makes assessment through observational analysis a logical choice for many physical education teachers. In fact, investigations of measurement practices of physical educators have consistently shown a reliance on observation and related assessment methods (Hensley, 1997; Matanin and Tannehill, 1994; Mintah, 2003). In addition, as described in chapter 13, there are now observation software and apps available to assess physical education and activity, such as SOFIT (McKenzie, Sallis, and Nader, 1991), iSOFIT, SOPLAY (McKenzie et al., 2000), and SOPARC (McKenzie et al., 2006) (see also https://thomckenzie.com/useful-tools/observation/ for more observation-based assessment tools).

Observation is a skill used during instruction by physical education teachers to provide students with feedback to improve performance. However, without some way to record results, feedback is not an assessment. A checklist of critical elements or some type of recording method to tally the number of times a behavior occurred may be used. Keeping game play statistics is another example of recording data during observation. Peers can also assess each other or students can analyze their own performances using criteria provided on a checklist or a game play rubric. Figure 15.1 shows an example of a recording form that could be used for peer assessment.

BEHAVIOR	OBSERVED	NOT OBSERVED
Feet shoulder-width apart		
Bat in air ready to swing		
Draws or cocks the bat backward in preparation to swing		
Steps, putting total body behind the ball		
Swings through the ball		
Rolls wrists to generate power		
Follows through across body		

Figure 15.1 Peer assessment checklist for softball batting.

When using peer assessment, it is best to have the assessor only observing. When the person recording the assessment results is also expected to take part in the assessment (e.g., tossing the ball to the person being assessed), he or she cannot both participate and do an accurate observation. In the case of large classes, teachers might even use groups of four, in which one student is being evaluated, a second student is feeding the ball, a third student is doing the observation, and a fourth student is recording the results that the third student provides orally.

[WWW] **Go to HK*Propel* to complete Student Activity 15.1.**

Individual or Group Projects

Projects have long been used in education to assess a student's understanding of a particular topic. Projects typically require students to apply their knowledge and skills to complete a prescribed task, which often calls for creativity, critical thinking, analysis, and synthesis. Examples of student projects used in physical education include the following:

- Demonstrating knowledge of invasion game strategies by designing a new game
- Doing research on obesity and then developing a brochure for people in the community that presents ideas for developing a physically active lifestyle
- Demonstrating knowledge of fitness components and how to stay fit by designing one's own fitness program using personal fitness test results
- Choreographing and performing a videotaped dance
- Doing research on childhood games and teaching children from a local elementary school how to play them

Criteria for assessing the projects are developed and the results of the project are recorded.

Group projects involve a number of students working together on a complex problem that requires planning, research, internal discussion, and presentation. Group projects should include a component that each student completes individually to avoid having a student receive credit for work that others perform. Another way to avoid this issue is for each member of the group to award "paychecks" to the other group members (e.g., split a $10,000 check) and provide justifications for the amount given to each person. To encourage reflections on the contributions of others, students are not allowed to give an equal amount to everyone. These "checks," along with the rationales for allocation of funds to members of the group, are confidential and submitted directly to the teacher in an envelope so that others in the group do not see the amount that other group members "paid" them.

The highlight box shows an example of a project designed for middle school, high school, or college students that involves a research component, analysis and synthesis of information, problem solving, and effective communication.

EXAMPLE OF A DANCE PROJECT

The purpose of this assessment is for students to demonstrate their ability to read and understand the directions for an aerobic, folk, or other type of structured dance and then teach that dance to peers. They must demonstrate the ability to teach the steps, count to the music, and coach students to use the music and steps in harmony to perform the dance. Students are evaluated based on their ability to read a set of dance directions and interpret them, teach the dance to others, and competently perform the dances that they are taught by other students in the group.

Instructions

Working in groups of six, you will demonstrate the ability to perform six structured dances to music. Each student will be required to read the directions for a dance, learn those steps, and then perform the dance to the music. (If you do not know how to do the step that the routine calls for, you may need to do some research.) After successfully doing the first step, each student will teach the dance to the other five students in the group. All dances will be performed by the group and videotaped. Additionally, each student will be videotaped while teaching others the dance steps. Each student must demonstrate the ability to explain the steps to others, count to the music, and coach others to fix mistakes.

Portfolios

Portfolios are systematic, purposeful, and meaningful collections of a student's work designed to document learning over time. The type of **portfolio**, its format, and the general contents are usually prescribed by the teacher. The guidelines used to format a portfolio will be based on the type of learning that the portfolio is used to document. Teachers often have students use working portfolios as repositories of documents or artifacts accumulated over a defined time period. Other types of process information could also be included, such as drafts of student work, records of student achievement, or documentation of progress over time. Portfolio artifacts may also include input provided by parents, peers, administrators, or others.

A showcase portfolio contains the artifacts submitted for assessment consisting of examples of the student's best work. Typically, teachers specify three to five documents or artifacts that all students must submit and then allow students to choose two or three optional artifacts. For the self-selected items, each student should evaluate his or her work and choose those products that best represent the type of learning identified for this assessment. Each artifact included in a showcase portfolio is accompanied by a reflection, in which the student explains the significance of the item and the type of learning it represents.

It's a good idea to limit the portfolio to a certain number of artifacts to prevent the portfolio from becoming a scrapbook that has little meaning to a student and to avoid giving teachers a monumental task when scoring them. This also requires students to exercise some judgment about which artifacts best fulfill the requirements and exemplify their level of achievement. The portfolio itself is usually a file or folder that contains a student's collected work. The contents could include items such as a training log, student journal or diary, written reports, photographs or sketches, letters, charts or graphs, maps, copies of certificates, computer disks or computer-generated products, completed rating scales, fitness test results, game statistics, training plans, report of dietary analyses, and even video or audio recordings. Collectively, the artifacts selected provide evidence of student growth and learning over time as well as current levels of achievement. The potential items that could become portfolio artifacts are almost limitless. A list of possible portfolio ideas that may be useful for physical activity settings is shown in the highlight box (Kirk, 1997). A teacher would never require that a portfolio contain all of these items; the list is only offered as a way to generate ideas.

A rubric should be used to assess portfolios in much the same manner as any other product or performance. Providing a rubric to students along with the assignment allows them to self-assess their work and thus be more likely to produce a high-quality portfolio. Because they are designed to show growth and improvement in student learning, portfolios are assessed holistically. The reflections that describe each artifact and why the artifact was selected for inclusion in the portfolio provide insights into levels of student learning and achievement. Teachers should remember that format is less important than content; the rubric should be weighted to reflect this. Table 15.1 illustrates a qualitative analytic rubric for assessing a portfolio.

With the availability of new technology tools like wikis and blogs, an electronic portfolio (e-portfolio) can make the portfolio presentation effective and powerful. Evidence by e-portfolio may include writing samples, photos, videos, hands-on projects, skill demonstrations, observations by mentors and peers, and reflective thinking. The key aspect of an e-portfolio, as with a regular portfolio, is the reflection on why each piece of evidence was chosen and what was learned from the process (Butler, 2006).

For additional information about portfolio assessments, see Lund and Kirk (2010); for a suggested scoring scale for portfolios, see Kirk (1997). Melograno's *Portfolio Assessment for K-12 Physical Education* (2000) and Cambridge's *Eportfolios for Lifelong Learning and Assessment* (2010) also contain helpful information.

POTENTIAL PORTFOLIO PROJECTS FOR PHYSICAL ACTIVITY

- Write a self-assessment of current skill level and playing ability, with individual goals for improvement.
- Conduct ongoing self- and peer assessments of skill performance and playing performance (process–product checklists, rating scales, criterion-referenced tasks, task sheets, game play statistics).
- Prepare a graph or chart that shows and explains performance of particular skills or strategies across time.
- Analyze your game play performance (application of skills and strategies) by collecting and studying your individual game statistics (e.g., shooting percentage, assists, successful passes, tackles, steals).
- Apply your knowledge and skills to create and perform an aerobic dance, step aerobics, or gymnastics routine (a routine script or videotape of the performance may go in your portfolio).
- Document participation in practice, informal game play, or organized competition outside of class.
- Keep a daily physical activity journal in which you set daily goals; record successes, setbacks, and progress; and analyze situations to make recommendations for present and future work.
- Using a self-analysis or diagnostic assessment, select or design an appropriate practice program and complete schedule. Record results.
- Set up, conduct, and participate in a class tournament. Keep group and individual records and statistics.
- Demonstrate your knowledge of the game by writing a newspaper article as if you were a sports journalist reporting on the class tournament or a game.
- Develop and edit a class sport or fitness magazine.
- Complete and record a play-by-play and color commentary of a class tournament game as if you were a radio or television announcer.
- Interview a successful competitor about his or her process of development as an athlete and his or her current training techniques and schedule (audiotape or videotape may go in your portfolio).
- Interview an athlete with a disability about his or her experience of overcoming adversity. Apply what you have learned to your situation (audiotape, videotape, or a written reflection may go in your portfolio).
- Write an essay on the subject of what you learned and accomplished during a specific activity unit and what you learned about yourself in the process.

Table 15.1 Qualitative Scoring Rubric for Complete Portfolio

	Format and design	Content items for each section	Reflection on portfolio tasks
Bull's-eye	Follows prescribed format with no errors. Design is attractive. Shows creativity. Well organized. Strong alignment with instructional goals.	A variety of artifacts shows both breadth and depth of learning. Items are selected that document growth and evidence of student learning and achievement. Comprehensive and complete coverage of desired student learning.	Reflections are thoughtful and document depth of understanding. Reflection provides insight about student learning and mastery of the desired goals for learning.
On target	Follows prescribed format with few errors. Design is neat with some creativity. Format is organized and easy to follow. Effective instructional alignment with learning goals.	Artifacts selected demonstrate competence in the subject. Some breadth of understanding and few if any misconceptions. Student is able to demonstrate growth and understanding of the subject.	Reflections provide a clear rationale for including the artifact. The intent of the reflection is easy to determine. Clearly written and easy to understand.
Getting close	Follows prescribed format for the most part. Little evidence of creativity or imagination. Some grammar and spelling errors. Format may be messy or lack organization in some parts.	Basic information presents an acceptable level of understanding. Artifacts help document student learning and growth. One or more artifacts may not align with instructional goals.	Reflections are to the point and direct, conveying reasons that the artifacts demonstrate learning. Some are short and fail to capture a depth of understanding.
Missed the mark	Does not follow prescribed format. Numerous errors. Inappropriate design. Disorganized and not aligned with the goals of learning. Gives the sense that it was completed at the last minute.	Many inappropriate or unrelated items included, indicating a lack of basic knowledge. Poor-quality work. Little variety or depth.	Little evidence of personal reflection on the tasks. Vague and repetitive. Little thought or rationale for including the various artifacts.

EXAMPLE OF A PORTFOLIO PROJECT

The following is an example of a portfolio project for a middle school or high school activity unit. This portfolio project could be used with a variety of sports, such as volleyball, basketball, tennis, badminton, or soccer.

Instructions

To demonstrate your progress, you will submit up to eight artifacts completed during the sport education unit that demonstrate your knowledge of the skills, rules, and strategies of the sport and your ability to analyze the game. You must also document your contributions as a member of the team. The following three items are required:

1. An article reporting the class tournament game that you observed, including game statistics and an analysis of the game based on team or individual performances.

2. The results from a game play assessment done on your team that demonstrates your skill, competence, and ability to play the sport.

3. An example of how you fulfilled a role on a "duty team" that also documents your knowledge of the game. This might include an example of your ability to officiate, announce, or keep game statistics. (Note: "Duty team" members are not playing games but are the referees, announcers, scorekeepers, line judges, etc.)

Select two to five additional artifacts of your own choosing. Each artifact must be accompanied by a reflection that explains the type of knowledge demonstrated by the artifact and a justification as to why you selected this item for inclusion in your portfolio.

Mastery Item 15.2

What if your instructor for this class required you to submit a portfolio that represented your learning in the class? Make a list of the important things that you have learned in this class (the learning that you wish to document) and then identify possible artifacts that you could use to document this growth, including three or four key artifacts that your instructor might require from everyone. Some of your initial work might contain errors (remember, you are trying to document your growth), whereas others represent high levels of achievement. Write a reflection for one of the assignments that explains why you chose it. (Note: If this were an actual portfolio assignment, reflections for each artifact would be written.) Describe how this portfolio would allow you to show your growth and competence in measurement and evaluation.

Mastery Item 15.3

How would the learning demonstrated in a portfolio differ from that demonstrated on a final exam (e.g., multiple-choice or essay questions)?

Performances

Student performances can be used as culminating assessments at the completion of an instructional unit. For example, teachers might organize a gymnastics or track and field meet at the conclusion of a unit to allow students to demonstrate the skills and knowledge that they gained during instruction. Game play during a tournament is also considered a student performance. Students might also demonstrate their skills and learning in one of the following ways:

- Performing an aerobic dance routine for a school assembly
- Organizing and performing a jump rope show at the halftime of a basketball game
- Performing in a folk dance festival at a county fair
- Demonstrating wushu (a Chinese martial art) at the local shopping mall
- Training for and participating in a local road race or cycling competition

Although performances do not produce a written product, there are several ways to gather data for assessment purposes. A score sheet can be used to record student performance using the criteria from a game play rubric, which can be written so that students are assessed on all three learning domains (psychomotor, cognitive, and affective). Game play statistics are another example of a way to document performance. Performances can also be videotaped to provide evidence of learning.

In some cases teachers might want to use event tasks, or performances that are completed in a single class period. Students might demonstrate their knowledge of net or wall game tactics by playing a scripted game as it is being recorded. The ability to create movement sequences or a dance that uses different levels, effort, or relationships could be demonstrated during a single class period with an event task. Many adventure education activities that demonstrate affective domain attributes can also be assessed using event tasks.

Mastery Item 15.4

After graduation, many exercise science students go on to teach clients as personal trainers, physical therapists, and so on. How could you document your ability to teach using a performance or an exhibition?

Student Logs

Documenting student participation in physical activity is often difficult. Teachers can assess participation in an activity or skill completed outside of class using a log, which records behavior over time (see figure 15.2). Practice trials during class that demonstrate student effort can also be documented with logs. Often the information recorded shows changes in behavior, trends in performance, results of participation, progress, or the regularity of physical activity. A student log is an excellent artifact for use in a portfolio. Because logs are usually self-recorded documents, they are not used for summative assessments unless as artifacts in a portfolio or for a project. If teachers want to increase the importance placed on a log, a method of verification by an adult or someone in authority should be added.

| Name _____ | Teacher _____ | |
| Class _____ | | |

Activity	Date	Length of time participated

Figure 15.2   Activity log.

Journals

Journals can be used to record student feelings, thoughts, perceptions, or reflections—both positive and negative—about events or results, and may be used to document the personal meaning associated with one's participation in physical activity. Journal entries would not be an appropriate summative assessment by themselves, but might be included as an artifact in a portfolio. Journal entries are an excellent way for teachers to take the pulse of a class and determine whether students value what they are learning. Teachers must be careful not to assess journal entries for the actual content, because doing so may cause students to write what teachers want to hear (or give credit for) instead of their genuine feelings. Instead, teachers could hold students accountable for completing journal entries. Some teachers also use journals as a way to log participation over time.

Mastery Item 15.5

Which type of performance-based assessment would you use to best represent the knowledge you have gained about measurement and evaluation from this class?

ESTABLISHING CRITERIA FOR PERFORMANCE-BASED ASSESSMENTS

One of the characteristics of a performance-based assessment is that scoring criteria are provided when the assessment task is first presented to students. Some teachers make the mistake of not determining scoring criteria when developing assessments, but without an effective scoring standard, the assessment task is nothing more than an instructional activity. Teachers must always clarify their expectations to students regarding assessments, and these expectations are the criteria that accompany the assessment.

Setting criteria for performance-based assessments is different from setting criteria for traditional assessments because a numerical score isn't always used to judge the performance, but the fundamental principle remains the same. When students are given the criteria in advance, they no longer need to play the guessing game of *What's on the test?* The concept of giving the scoring criteria with the assessment should not be new to a physical education major; it is common to many physical education tests. For example, FitnessGram provides a range of scores needed to reach the Healthy Fitness Zone. Similarly, teachers often tell students the score they must achieve on a skills test to receive a certain grade or require students to reach a certain level of expertise before allowing them to play a game. With both fitness and skills tests, teachers can inform students of the criteria that they must strive to reach.

There are several types of criteria that teachers can use to denote the standards for performance on an assessment. **Process criteria** look at the quality of performance for students completing the assessment, or how students complete a task. In physical education, the term *process criteria* is used to denote the critical elements needed for a student to demonstrate correct form. For example, the peer assessment form in figure 15.1 includes the critical elements that are important for softball batting. Process criteria are critical for students to focus on correct form when they are first learning a skill. In many assessments for elementary school children, process criteria are typically used to assess student performance. For regular classroom teachers, the term *process criteria* is used in reference to class participation, homework completion, or student effort.

When the criteria are stated in terms of the outcome of student performance, they are called **product criteria**. Examples of product criteria are the number of times a student must serve the volleyball or the number of consecutive forward rolls students must do in a tumbling sequence. A rule of thumb is that teachers should not specify product criteria without also requiring correct form (process criteria).

When they want to measure student improvement, teachers may choose to use **progress criteria**, which measure the amount of improvement rather than levels of achievement (Guskey and Bailey, 2001). Some teachers like progress criteria, especially with students who have little or no experience with a sport or activity. However, a problem with using progress criteria is that, because there is no established score that everyone must reach, they create a moving target, meaning that some students may be required to reach a higher level of performance than others to receive a similar grade. For example, Taylor, who had never played tennis, was unable to do a forehand rally and received a score of zero on the pretest. Chris, who was more experienced, was able to hit three consecutive forehands during the pretest. Using the data from the pretest, the teacher determined that for each student to receive an A, Taylor would need to get six hits on the continuous rally test, whereas Chris would need to get 10. At the conclusion of the unit, both students were able to do six hits on the continuous rally test. Taylor received an A for the test, and Chris did not. An additional problem with using progress criteria is that some students might deliberately perform at a level below what they are capable of so that they can demonstrate improvement with relatively little personal skill development.

Simple Scoring Guides

The simplest type of scoring guide for performance-based assessments is a checklist. Checklists are often used by teachers to evaluate process criteria for sport skills (e.g., forearm pass, overhead pass, spike) or fundamental motor skills (e.g., hop, skip, leap). With a checklist, all traits or characteristics are of equal worth, and no decision is made about the degree to which the characteristic is present or about the quality of the performance: The trait is either demonstrated or it is not. Teachers indicate whether the element is present by marking a recording sheet. Some teachers use a single blank in front of the element, whereas others use two columns of blanks so that the evaluator can check either "Yes" or "No." The latter method is preferred when checklists are used as peer or self-assessments to ensure that students have observed each item on the list. Checklists are relatively simple to create and are useful as a written product to document an observation. When checklists are used, teachers must not make them too long, especially for younger children. It is difficult for even an experienced teacher to observe six or seven things simultaneously. Younger children should look at no more than two or three items, depending on their age. When using checklists containing multiple characteristics for peer assessments, teachers should require students to perform a sufficient number of trials so that the observer has an adequate opportunity to accurately observe the skill (see figure 15.3). Peers must be taught the purpose of checklists and how to use them if they are to be effective peer assessments.

Volleyball forearm pass	Trial 1		Trial 2		Trial 3	
Hands clasped, knees bent	Yes	No	Yes	No	Yes	No
Contact on flat surface of the arms	Yes	No	Yes	No	Yes	No
Uses legs to provide force without swinging arms	Yes	No	Yes	No	Yes	No
Platform directed to target on follow through	Yes	No	Yes	No	Yes	No

Figure 15.3 Repeated trials of volleyball assessment.

The **point system scoring guide** is another type of performance record with a listing of traits or characteristics considered important for the performance. It differs from a checklist in that it has points listed for each characteristic so that items considered to have greater value or importance are given more points. As with checklists, when using a point system scoring guide, the observer makes no judgment of quality—the trait is either present or is not. After the presence or absence of the traits or elements has been determined, the number of points can be added and a total score assigned. Teachers should not award partial credit for an item that is partially present, because this practice can affect the reliability of the rubric. If an element can be broken down into subelements, then each of the subelements should be assigned a point value so that partial credit for the category is awarded with consistency. Figure 15.4 shows an example of a point system scoring guide.

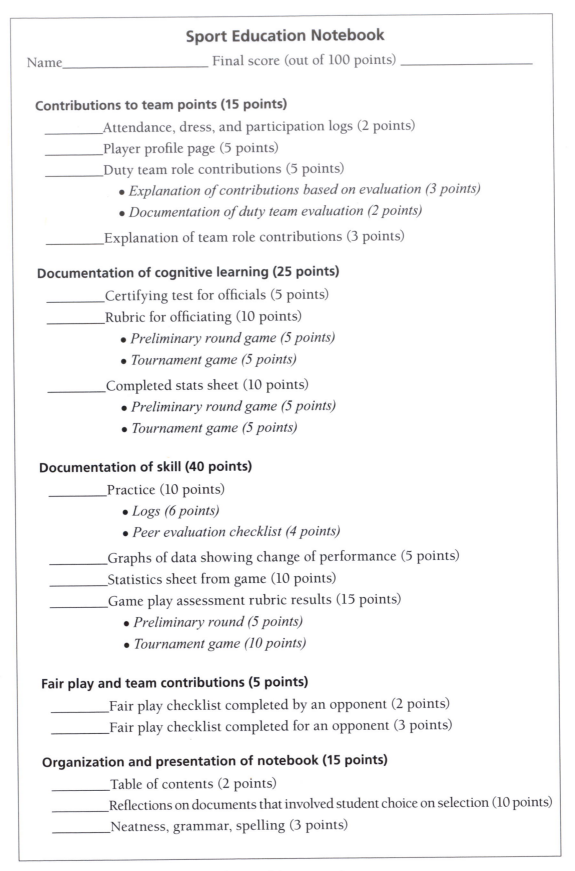

Sport Education Notebook

Name_____ Final score (out of 100 points) _____

Contributions to team points (15 points)

_____Attendance, dress, and participation logs (2 points)

_____Player profile page (5 points)

_____Duty team role contributions (5 points)

• *Explanation of contributions based on evaluation (3 points)*

• *Documentation of duty team evaluation (2 points)*

_____Explanation of team role contributions (3 points)

Documentation of cognitive learning (25 points)

_____Certifying test for officials (5 points)

_____Rubric for officiating (10 points)

• *Preliminary round game (5 points)*

• *Tournament game (5 points)*

_____Completed stats sheet (10 points)

• *Preliminary round game (5 points)*

• *Tournament game (5 points)*

Documentation of skill (40 points)

_____Practice (10 points)

• *Logs (6 points)*

• *Peer evaluation checklist (4 points)*

_____Graphs of data showing change of performance (5 points)

_____Statistics sheet from game (10 points)

_____Game play assessment rubric results (15 points)

• *Preliminary round (5 points)*

• *Tournament game (10 points)*

Fair play and team contributions (5 points)

_____Fair play checklist completed by an opponent (2 points)

_____Fair play checklist completed for an opponent (3 points)

Organization and presentation of notebook (15 points)

_____Table of contents (2 points)

_____Reflections on documents that involved student choice on selection (10 points)

_____Neatness, grammar, spelling (3 points)

Figure 15.4 Point system scoring guide example.

Rubrics

Rubrics are used to assess complex assessments (e.g., game play or dances, student projects, portfolios). When using a rubric, assessors are required to make a judgment about the level of quality of student performance. Because making a judgment about the quality of the performance requires more time than simply deciding whether a characteristic is present, fewer traits or characteristics are typically evaluated on rubrics. There are two types of rubrics—analytic and holistic. **Analytic rubrics** are used to evaluate individual traits or characteristics of the performance separately. Analytic rubrics are useful for formative assessments so students can see how well they performed or how they were scored on the various traits or descriptors and use the results to make improvements.

There are two types of analytic rubrics. **Quantitative rubrics** are useful for assessing sports with sequential skills where consistency of form affects the quality of performance, such as target games (e.g., archery, bowling) and individualized activities (e.g., diving, shot put). Numerical scores given are based on whether a student performs the various phases in the performance sequence consistently (e.g., 6 out of 6) or just occasionally (e.g., 2 out of 6), or somewhere between those two levels. When developing a quantitative rubric, a teacher might identify four or five phases of the performance and then assess each of these for consistency (see table 15.2). A **qualitative rubric** has written explanations that describe the quality of the performance for the various levels of the traits or characteristics being evaluated. These rubrics are used to assess complex performances that involve multiple domains or dimensions of cognitive learning, such as game play or dance performances. Some qualitative rubrics identify errors commonly seen at certain levels of performance. These descriptions help communicate what the evaluator should or should not be looking for with regard to the trait. Table 15.3 shows a qualitative rubric for softball game play.

Table 15.2 Quantitative Rubric for Shot Put: Class Recording Sheet

Steps	Starting position	Start of glide	Turn	Explosion	Landing
Critical elements	Shot is cradled against the neck. Holds the shot on finger pads. Body is leaning forward. Weight is on throwing side foot and knees are bent.	Thinker position. Kicks backward with nonsupporting leg as the supporting leg straightens. Back is facing the direction of the throw. Supporting foot starts the turn.	Plants nonsupporting leg and opens hips. Supporting leg starts to straighten. Body starts to straighten.	Pushes shot, extending arm. Weight shifts to nonthrowing leg. Releases the shot at a 45° angle while jumping into the air.	Lands on throwing side leg with side of body facing the target. Avoids falling out of the circle when releasing the shot. Exits by the back of the ring.
Name	Score	Score	Score	Score	Score

Directions: Observe each athlete and rate each of the steps on a scale using the following definitions:

4 = exhibits the behaviors consistently or almost always (3 × or more in a row)

3 = usually exhibits the behaviors (4 out of 6 trials)

2 = exhibits behaviors about half the time

1 = exhibits behaviors occasionally (2 × or fewer out of 6 trials)

0 = does not complete the assignment or does not participate

Table 15.3 Qualitative Rubric for Softball Game Play

Characteristic	Rookie	Junior varsity	Varsity	All star
Base running	Ignores base coach while running. Runs on every hit regardless of the situation. Jogs slowly around the bases.	Hustles to reach the base. Watches the base coach to keep track of whether to stay or advance.	Knows own abilities and does not over-estimate the ability to safely advance. Rounds the bases using the left foot for greatest efficiency. Occasionally slides to avoid a tag.	Stretches a hit to extra bases when possible. Slides to avoid being tagged or to slow the approach to the bag. Follows the base coach's advice but is capable of making accurate decisions.
Batting	Chops when swinging the bat. Swings at every pitch or fails to swing. Often makes mistakes about whether the pitch is good enough to swing at.	Good form. Swings through the ball; knows which pitch is good enough to swing at.	Typically makes contact with the ball when batting. Good judge of the strike zone.	Consistently hits the ball even when it is thrown with speed. Rarely strikes out. Is able to place the ball depending on the situation.
Fielding and throwing	Fails to judge many of the balls, which causes errors; over-runs the ball. Waits for ground balls to come to the player.	Can make a play and throw a runner out. Struggles to throw the ball for a distance longer than 40 ft (12.2 m). Ball is thrown in general direction of the target—usually throws a catchable ball.	Can throw from third to first to make the play. Knows when to make the play on the hop but usually tries to catch the ball on the fly. Charges ground balls.	Judges a batted ball accurately. Charges ground balls and can make an accurate throw on the run. Is able to leap and make difficult catches. Covers the field and plays the position well.
Knowledge of rules	As a base runner, does not know whether to advance or stay when the ball is hit. Knows basic rules (e.g., 3 strikes is an out; 4 balls is a walk; 3 outs per inning).	Knows the difference between a force out and a tag out. Knows enough of the rules to keep the scorebook.	Good knowledge of the rules. Knows more complicated rules such as infield fly rule. Can coach players while playing defense.	Can accurately officiate a game. Can answer most questions about rules correctly.
Strategies	Knows some strategies but applies them inconsistently during a game. Does not back up other players.	Knows what to do with runners on various bases with various outs.	Backs up the play. Uses offensive and defensive strategies to improve chances to win. Changes positions on the field depending on the game situation (runners and outs).	Reads the players and moves around position according to the game situation (runners and outs) and who is up to bat.
Fair play	May berate others when they make a mistake. Does not compliment others. Cares about own play with little regard to team success.	Ignores the mistakes of others and congratulates good play of others on the team.	Talks to teammates when they make a mistake to calm them down. Enthusiastic about the play of others.	Congratulates others on good play, even players on the opposing team. Is willing to sacrifice hit for the team. Will sit the bench if it benefits the team.

In contrast to an analytic rubric, which scores each characteristic individually, **holistic rubrics** are used to evaluate the entire performance with a single score. A holistic rubric describes all the characteristics of a level of performance in a single paragraph. Holistic rubrics are most commonly used with summative assessments when students are not attempting to improve their performance based on feedback. Table 15.4 is an example of a holistic rubric.

Table 15.4 Holistic Rubric for Tennis Playing Ability

5—Excellent	Consistently executes all strokes with little or no conscious effort, resulting in few unforced errors. Uses spin to add difficulty to shots and for strategic advantage. Uses the rules to gain an advantage. Anticipates opponent's shots and can control shots to employ effective strategies that lead to winning the point. First serve is strong and usually lands in bounds; difficult to return. Calls shots fairly and accurately.
4—Good	Demonstrates competency and ability to perform basic tennis skills while making few errors. Demonstrates an understanding of the rules by self-officiating a game correctly. When asked, provides a strategy appropriate for an upcoming game. Usually selects appropriate strategy and shot selection for situation and generally displays consistent performance. Is able to serve consistently and accurately.
3—Satisfactory	Displays basic understanding of tennis and is able to perform fundamental skills adequately to play a game. Performance is steady, relying mostly on ground strokes with some net play. Understands basic strategies but lacks ability to effectively employ them. Often relies on second serve during a game. Knows the rules well enough to play a basic game of tennis.
2—Fair	Primarily uses the forehand and some backhand shots to keep the game going. May run around his or her backhand to avoid using the shot. Has difficulty placing shots away from an opponent. Uses little strategy and frequently returns the ball to an opponent without making the player(s) move. Is able to keep score but gets confused and cannot answer questions. Serve is inconsistent.
1—Poor	Rarely, if ever, performs skills well enough to sustain a rally. Fails to return most serves. Is able to get the serve in play less than half of the time. Makes mistakes on basic rules such as where to serve, line calls, etc. Is not skilled enough to adjust to the performance of the opponent.

One drawback of using holistic rubrics is that student performance is rarely at the same level for every characteristic. This situation can be addressed in one of two ways. If a student demonstrates some characteristics at one level and the rest of the characteristics at another level, then the lowest level at which the characteristics were demonstrated is the score assigned to that student's work. A second way to address multiple levels of performance is to weight the characteristics listed on the rubric; the score on those items of highest importance determine the score given to the performance or product. The decision about how to apply the rubric must be decided before scoring, and all raters must use the same method for consistent results. When using holistic scoring, evaluators are often given examples of past student work called *exemplars*. Because a holistic score is an overall impression of the total work, these exemplars help evaluators calibrate the intent of the rubric. When scoring student work, it is helpful to compare items to the exemplars to maintain high levels of reliability. Exemplars for written work are fairly easy to provide; however, when used for student performances, video recordings are needed.

Developmental rubrics are used for students at all levels of proficiency, from beginning to advanced players. Players can see where they are on a continuum of skill development and how much they still need to master. Table 15.5 is an example of a simple developmental rubric for catching and throwing. Sports such as gymnastics or diving use a developmental scoring system; all performers are judged using the same criteria, regardless of age.

Table 15.5 Developmental Rubric for Catching and Throwing

Developmental level	Catching	Throwing
6	Can catch an object thrown with increased velocity or catch an object while moving; can sequence catching and throwing skills smoothly	Can throw with velocity and accuracy; is able to hit a target from a distance of 25 ft (7.6 m)
5	Can transfer catching skills to a game situation	Can transfer throwing skills to a game situation
4	Can catch a variety of objects at different levels with a partner	Shows trunk rotation and accuracy
3	Can catch a variety of self-tossed objects	Follows through toward target
2	Can catch a bounced ball from a partner	Shows opposition when stepping
1	Extends arms toward thrower, shows avoidance reaction; traps the ball rather than catching it	Shows limited body movement; arm dominates the throw

Adapted from the Wichita Public Schools, Kansas.

When creating rubrics, teachers can write a **task-specific rubric** for a single sport or activity (refer to the softball rubric in table 15.3), or a **generalized rubric**, which can be used for several related sports or activities. Consider level 5 from the tennis rubric found in table 15.4. If the terminology concerning the serve were changed slightly to read "Service is strong and usually lands in bounds," then the rubric could be used for serving with several net games. Removing the phrase about spin would be necessary for badminton games. These are examples of how a task-specific rubric could be modified to a generalized rubric. In many instances, games within a category (e.g., invasion, net or wall, target, field) or similar activities (e.g., different dance and rhythm styles such as folk, square dance, hip hop, jazz, social dance) could be assessed with a single rubric that would capture the essence of learning for a variety of sports and activities within the classification.

Teachers should avoid using hypergeneral rubrics for assessment, or those that fail to include any meaningful benchmarks, which leaves too much room for subjective decisions. Table 15.6 is an example of a hypergeneral rubric that provides little guidance.

Table 15.6 Example of a Hypergeneral Rubric

Level	Knowledge
4	Demonstrates a thorough understanding of the important concepts or generalizations and provides new insights into some aspect of that information
3	Displays a complete and accurate understanding of the important concepts or generalizations
2	Displays an incomplete understanding of the important concepts and generalizations and has notable misconceptions
1	Demonstrates little understanding of the concepts and generalizations and has several misconceptions

Adapted from the Mid-Continent Research for Education and Learning (McREL), Aurora, CO.

Developing a Rubric

The rubric often determines whether the person using the assessment is able to make a valid inference about student learning. Most performance-based assessment tasks have strong content or face validity—they represent what a teacher wants students to know and be able to do. For example, when teaching a badminton unit, the ultimate goal is to teach students to play the game of badminton at a reasonable level of competence; therefore, requiring students to play a game allows for the assessment of that ability. However, to properly assess this ability, the rubric must include the right items. For example, knowing the rules is an important part of being able to play a game. If the rubric failed to include knowledge of rules as a characteristic to be evaluated, there would be no way to determine whether students had learned them. Furthermore, a competent player will use a variety of shots to move the opponent around the court and try to place the shuttlecock where the opponent cannot return it; therefore, this ability should be an item included on the rubric. If the rubric fails to allow a teacher to gather adequate information about students' ability, the decision about their game play competence will not be based on good data. Selecting the right characteristics to use on a rubric is an essential part of the process used to develop an assessment.

Lund and Veal (2013) identified seven basic steps to follow when writing qualitative rubrics:

1. Identify what needs to be assessed.
2. Brainstorm ideas for the behaviors representative of what you are trying to achieve.
3. Identify the descriptors.
4. Decide how many levels are needed.
5. Write levels for the rubric.
6. Pilot the assessment.
7. Revise the rubric.

Writing a qualitative rubric is difficult unless one has a product or performance to use when creating the written descriptions for the various levels. Doing an actual assessment reveals levels of student misunderstandings and common errors, as well as essential elements for the various levels of performance. A holistic rubric is fairly easy to develop from a qualitative rubric because the descriptors for a given level can be combined into a single paragraph.

As stated earlier, quantitative rubrics are often selected for sports containing a single skill for which consistency of performance is critical to success, such as target games (see the example rubric in table 15.2, written for the shot put). Use the following steps when developing a quantitative rubric:

1. Identify the content.
2. Decide on the phases or steps important for consistent performance.
3. Identify the critical elements for each phase (no more than two or three).
4. Determine how many levels you wish to use.
5. Define the levels by indicating the level of consistency related to that level.
6. Develop a recording sheet.
7. Pilot the rubric.
8. Revise the assessment.

See Lund and Veal (2013) for a detailed discussion of how to develop rubrics.

Mastery Item 15.6

Develop a rubric that could be used with a game play assessment for a sport or a rubric for a physical activity (e.g., dance, gymnastics, in-line skating).

WWW **Go to HK*Propel* to complete Student Activities 15.2 and 15.3.**

SUBJECTIVITY: A CRITICISM OF PERFORMANCE-BASED ASSESSMENTS

Performance-based assessments are often criticized because of the subjectivity associated with the scoring process. However, as Danielson (1997) points out, there is a degree of subjectivity with all testing and assessment, because the decision about what knowledge to assess is quite subjective. Think about classes in which professors use two tests to evaluate whether students had mastered all the material covered in a course lasting an entire semester. Did instructors use questions from the book, lectures, or a mixture of the two? Have you ever had a teacher who asked questions about the captions under the pictures in the textbook? Some instructors choose to use questions from the instructor's guide, which were written by someone who has never met the instructor or attended the class. Some multiple-choice questions actually have two plausible answers (depending on how one interprets the stem) and only one opportunity to select the right response. Many of these issues with subjectivity can be mitigated by following the guidelines for test development provided in chapter 9.

The point of this discussion is to recognize that subjectivity occurs in most types of testing and assessment; the difference is *when* it occurs. With traditional written tests, the chance for subjectivity occurs with the selection of content, the writing of questions, and the selection of the correct response. With performance-based assessments, the chance for subjectivity occurs when the assessment is evaluated. As Danielson (1997) notes, most teachers who use performance-based assessments are competent professionals—rather than referring to their judgments of quality as *subjective*, she states that teachers are using professional judgment to make decisions about student learning. A physical educator or coach who is knowledgeable about a sport can observe players and assess their level of performance with a high degree of accuracy. Performance-based assessments formalize this process and let students know the criteria used to assess the performance while allowing teachers to focus on those elements identified as being important. Performance-based assessments allow teachers to coach students and thus maximize student achievement.

SELECTING APPROPRIATE PERFORMANCE-BASED ASSESSMENTS

Generally, the overall purpose of assessment is to provide good information for making valid decisions about learning. Good assessment information provides an accurate indication of a student's performance and enables a teacher, coach, or other decision maker to make appropriate decisions. But what constitutes good assessment information? What determines the quality of an assessment? Herman, Aschbacher, and Winters (1992) suggest that when considering the quality of an assessment, we are really asking the following:

1. Does an assessment provide accurate information for decision making?
2. Do the results permit accurate and fair conclusions about student or athlete performance?

3. Does using the results contribute to sound decisions?

To be able to answer yes to these questions, three criteria must be met: reliability, validity, and fairness.

Reliability

As previously defined, reliability relates to the consistency of results. Assessments that are scored inconsistently are essentially useless because they do not provide meaningful information. Inasmuch as performance-based assessment depends heavily on the professional judgment of teachers (or other assessors) for scoring and interpretation, one needs to be particularly concerned about interrater reliability or objectivity. Similar to the concept presented in chapter 6, this means that if more than one person is doing an evaluation, a student's score should be the same regardless of who did the evaluation. Intrarater reliability is also important. A teacher evaluating students in the morning must apply the criteria in the same way when using the assessment in the afternoon. For example, a teacher may discover that a rubric is insufficient and begin to mentally add criteria to the rubric to clarify it. When teachers apply criteria differently as the evaluation process continues, they drift from the original intent of the assessment, thus affecting reliability. Extraneous variables such as the halo effect (discussed in chapter 12) also must not enter into the decision. In short, an evaluator must strive to minimize inconsistency (i.e., error) in scoring to have confidence that the judgment is a result of the actual performance.

Besides the inconsistency caused by raters, the variability of tasks (e.g., teachers may use different tasks for the assessment due to their school's curriculum differences) and facilities (e.g., some schools have a standard indoor gym whereas others may have to use a multipurpose space for their physical education instruction) can result in inconsistencies. When designing a performance-based assessment, especially if the assessment is used for a high-stakes test, every effort should be made to eliminate or control these inconsistencies.

© Human Kinetics

It's important that an instructor create assessment guidelines that are reliable, consistent, and fair to all students, regardless of their gender, ability, or economic status.

A good performance scoring guide or rubric is essential for consistent scoring. When using a rubric, an evaluator must be able to distinguish between levels, and the words used to describe the levels must create a clear picture of what the performance or product should demonstrate. When developing performance scoring guides, teachers must select the most important traits or characteristics; judging a performance is difficult when too many characteristics are identified. When developing rubrics, the more levels one uses, the lower the reliability (Guskey and Bailey 2001). It is easier to make distinctions between two levels of performance (acceptable and unacceptable) than to make distinctions among 10 levels. Common sense tells one that as the distinctions between the levels become smaller, agreement between evaluators will decrease. The purpose of the assessment should determine the number of levels used.

To ensure that scoring guides or rubrics are thorough and accurate but not unwieldy, always pilot test them before using them for an important assessment. When multiple people will be using the same assessment and rubric (as with a state- or districtwide common assessment), training is essential. With training, scorers can reach high levels of reliability (Herman, Aschbacher, and Winters, 1992). The use of exemplars helps to ensure that the scoring stays consistent.

Validity

As you know from chapter 6, validity is an indication of how well an assessment actually measures what it is supposed to measure. Although reliability is necessary, it is not a sufficient condition for validity. An assessment could be perfectly reliable but not relevant to the decision for which it is intended. If the assessment result is not related to the characteristic reportedly being measured, it may jeopardize accurate conclusions and subsequent decisions. Broadly speaking, validity relates to the meaning and consequences attached to the test scores (Messick, 1995). Popham (2003) suggests that the question is not whether an assessment is valid, but rather whether the assessment allows users to make valid judgments about the results gathered from the process.

The primary form of validity reported for performance-based assessments, particularly those developed by teachers for classroom use, is content-related (face) validity, based on the assumed relationship between instruction and assessment. Although this linkage is important, content-related validity alone should not be accepted as sufficient evidence for the use of an assessment method. Logically, the most accomplished students should receive the best scores. The value of performance-based assessments is increased if there is corroborating evidence that they allow teachers to reach valid conclusions. However, inasmuch as most performance-based assessments are substantially different from standardized testing, traditional validation procedures may be inappropriate (Miller and Legg, 1993). Measurement and evaluation experts must work with teachers to resolve this issue and develop ways to ensure validity for performance-based assessments.

To make a valid judgment about student learning, teachers must use assessment tasks that have good face validity (have the potential to measure the type of learning that is targeted), use rubrics with characteristics that identify those behaviors and appropriate levels of student performance, and use multiple data sources to gather evidence about student learning.

Fairness

Although fairness is not a psychometric property in the same sense as reliability and validity, it is critically important in all forms of assessment, whether traditional or alternative.

Fairness simply means that an assessment allows all students, regardless of gender, ethnicity, or background, equal opportunity to do well. Although there is tremendous diversity within our society and not all students come to school with the same background, motivation, or values, all students should have an equal opportunity to demonstrate the skills and knowledge being assessed. Fairness should be evident in the development or selection of the assessment task as well as in the criteria used for judging the performance or product.

Is the assessment task fair and free from bias? Does the assessment task favor boys or girls, students from a particular ethnic group, students who have lived in a particular area, or those whose families have greater financial resources? To be fair, the task should reflect knowledge, values, and experiences that are familiar to and appropriate for all students, and should seek to measure the knowledge and skills that all students have had adequate time to acquire. Making sure the scoring procedures and criteria for assessment are free from bias helps to ensure that the ratings of a performance reflect an examinee's true capability.

It is commonly thought that traditional evaluation procedures tend to demonstrate greater reliability, objectivity, and validity than performance-based assessment, largely because of the increased objective nature of the traditional assessment. However, many of the more objective, valid, and reliable assessments provide only indirect measures of student achievement. If teachers use only traditional testing formats, they are left to evaluate student ability to play a game through a written test measuring one's knowledge of the rules and various assessments of individual skills. Students' abilities to apply knowledge of rules and strategy, to use skills in an open environment, and to play fairly simply cannot be captured with traditional assessments. Performance-based assessments allow a **holistic assessment** of student learning. The challenge for practitioners who use these assessments is to carefully consider the issues of validity, reliability, and fairness to ensure that any assessment used leads to appropriate decisions about student learning.

ISSUES TO CONSIDER WHEN DEVELOPING PERFORMANCE-BASED ASSESSMENTS

It should be obvious by now that the nature of performance-based assessments precludes the publication of assessments that have universal appeal and usefulness. Clearly, one size does not fit all when it comes to performance-based assessments. This does not mean that one cannot adapt and enhance ideas obtained from other sources, but most professionals will need to create their own assessments that meet their specific needs. When developing any type of assessment, remember that it consists of both a task and performance criteria—that is, to create an assessment, one needs both to create a task and identify the criteria used to evaluate it.

To be most useful in an instructional setting, a performance-based assessment should consider both context (situation–task) and performance (construct–skill); in other words, the assessment task should represent a completed performance with contextualized meaning that is directly related to the eventual use of the skill (Siedentop, 1996). Skill assessments are useful for allowing students to practice skills in a closed environment as well as providing an opportunity for teachers to observe students performing the skill, which doesn't always happen during game play. Because basic individual skills serve as the foundation for future activity or performance, assessing both the individual skills and the application of these skills in game performance is necessary. Too often students are able to perform the skills on a test but are unable to use them in a game. Performance-based assessments provide teachers with a way to assess the application of skills in a meaningful context.

The following are some things to consider when developing assessments.

Determining the Purpose

The critical first step in the process of developing assessments is determining the purpose of the assessment. For instance, is the purpose to diagnose initial skill levels to know where to start instruction? To determine deficiencies in a student's performance or the product and give feedback (formative assessment)? To ascertain achievement of specified objectives to award a grade (summative assessment)? To obtain information for evaluating a physical education program or to comply with state reporting requirements (curricular assessment)? The purpose of doing the assessment will determine the types of assessments that are selected.

Deciding How to Document Student Learning

Next, you must decide what type of evidence will be used to document student learning. Some teachers might want to document fitness improvement by requiring students to complete logs of their activities. Others might simply pre- and posttest students using heart rate monitors and the information from the FitnessGram PACER item. If the students are at a beginning skill learning level, process variables (correct form) are important to emphasize; if, however, students are at more advanced levels, adventure education activities may allow them to demonstrate responsible personal and social behavior. Note that self-assessments are good for documenting the learning process but are not appropriate evidence for a summative assessment.

Selecting the Appropriate Task

As stated, an assessment task must allow a teacher to gather the evidence needed to make the correct decision about student learning, or else it serves no useful purpose, and the results are not meaningful for decision making. One must ask if the assessment task elicits the desired performance or work. Have students had the opportunity to acquire the knowledge and skills needed for the task? If not, consider assessing something else. Also keep in mind that not all targets or desired outcomes are best measured using performance-based assessments.

The following performance-based assessments are useful for assessing process or facilitating student learning:

- Logs
- Journals
- Self-evaluations
- Peer assessments
- Interviews
- Performance checklists

For assessing product or outcome measures, the following performance-based assessments could be used:

- Projects
- Performances (e.g., game play, event tasks)
- Portfolios

Setting Performance Criteria

Previous parts of this chapter have discussed the importance of **performance criteria**. Remember: A task alone does not constitute an assessment—criteria are necessary for judging the performance or product and telling you how good is good enough. Some teachers and professionals, particularly those just starting out with performance-based assessment, tend to focus only on the task, to the exclusion of the criteria. How will you know if a student has accomplished the learning goal or met the target? How will you know when the performance is good enough? How will you know if the portfolio is acceptable? How will you be able to diagnose a student's performance and provide meaningful feedback? Each of these questions is answered by the criteria, not the task.

Determining Time Constraints

Time is a critical factor in selecting assessments. Teachers should select assessments that will allow them to accurately evaluate student learning, and performance-based assessments can be particularly time consuming and labor intensive. However, assessments should be carefully scheduled so that teachers can realistically manage them—they should not be a burden.

In large classes, teachers struggle to assess students. Using performance-based assessments as formative assessments can help bridge the gap between wanting to assess student learning and lack of time to do assessments. When checking for student learning, it is sufficient to sample students with a range of abilities and not assess each student in the class. Peer and self-assessments can also improve cognitive learning and provide feedback.

Teachers must also allocate adequate time to teach a unit. Some teachers allocate 1 to 2 weeks per unit or activity. This is an insufficient amount of time to adequately teach a skill. If students don't have the opportunity to learn the skills and how to participate in the activity, there really is nothing to assess. Typically, with large classes, more time is spent on managerial issues and less time on instruction; therefore, more time must be allocated if student learning is to be optimal and meaningful.

IMPROVING ASSESSMENT PRACTICES IN PHYSICAL EDUCATION SETTINGS

The purpose of this chapter is not to convince people that performance-based assessments are the solution to the assessment dilemmas that teachers in kindergarten through grade 12 schools face—rather, they expand the possibilities for assessment and are probably best used along with other traditional forms of assessment. Consider the following example that uses both types of assessment:

Mrs. Gaylor is an experienced teacher who has selected pickleball to teach net and wall tactics. She wants her students to be able to play pickleball at the conclusion of the unit and has decided that she will use a rubric to assess students' overall abilities during a class doubles tournament. The rubric used for assessing game play will require students to know the rules, work with a partner, demonstrate positive sport behavior, use correct form when executing shots, strategically place the ball away from the opponent, and use the serve to gain an offensive advantage. She will not assess individual skills during the game because form is often sacrificed as players make an attempt at an errant ball. Instead, skills tests will be used for evaluating the volley shot, the serve, and a continuous rally. All of the skills tests are performed against a wall so that a student does not need to depend on another student to demonstrate his or her own skillfulness. Although knowledge of rules will be one of the categories on the game play rubric, an additional written test (comprising multiple-choice and short essay questions) will be given so that lower-skilled students who might know the rules but not be able to demonstrate them during game play will have the opportunity to demonstrate that knowledge.

On the first day of the unit, Mrs. Gaylor informs students of her expectations, demonstrates the skills tests, and provides students with the game play rubric, which is also posted on the wall so that students can refer to it when needed. During the unit, Mrs. Gaylor incorporates the skills tests into her teaching progressions. Students can take the skills tests multiple times—the goal is to reach the criterion scores that Mrs. Gaylor gave when explaining the tests. Students may take the skills tests before and after the instructional part of class and during the class tournament when they are not playing a game. At the start of class, students pick up a 3- × 5-inch (7.6 × 12.7 cm) note card to record practice trials and results, which will serve as a participation log. By looking at the practice logs, Mrs. Gaylor can see which skills need more work and use this information while planning her lessons. Additionally, Mrs. Gaylor is keeping track of those students passing their skills tests and knows which students need more assistance and which need additional challenges because they have achieved the basic level of competence.

The tasks used for instruction are designed to teach students the skills and tactics needed to play pickleball. The content of the lessons is guided by the information (data) that Mrs. Gaylor is getting from her formative assessments. Before the start of game play, a written test is given to ensure that all students have cognitive knowledge of the rules. When students start to play games, she will observe them multiple times using the game play rubric. The areas of lower performance will be addressed in future lessons. Classes conclude with students completing exit slips that require them to answer questions about the content of the day's lesson. On the exit slip, students also have an opportunity to ask any questions about things from the class that they didn't understand, request additional instruction on an area that is proving difficult for them to learn, or ask for challenges such as additional skills or game play tactics to help them continue to improve.

Mastery Item 15.7

Identify the types of student learning that Mrs. Gaylor can document using the procedure just explained. How do traditional skill assessments work with performance-based assessments to enhance student learning?

As shown in the previous example, there is a need for both performance-based assessment and traditional techniques, depending on the purpose of the assessment. Regardless of the approach taken, teachers should use meaningful assessment in physical education class or other physical activity settings. The design and incorporation of clear, developmentally appropriate, and explicitly defined scoring rubrics are essential to ensure valid, consistent, and fair inferences about learning.

Stiggins (1987) suggested that the most important element in designing performance-based assessments is the explicit definition of the performance criteria. Moreover, Herman, Aschbacher, and Winters (1992) stated that criteria for judging student performance lie at the heart of performance-based assessment. If performance-based assessments are to realize their promise and live up to expectations, then it is essential that high-quality assessments be accompanied by clear, meaningful, and credible scoring criteria.

The following guidelines (adapted from Gronlund, 1993) suggest ways to improve the credibility and usefulness of performance-based assessment in physical education.

1. Ensure that assessments are congruent with the intended outcomes and instructional practices of the class.
2. Recognize that, together, observation and informed judgment with written results comprise a legitimate and meaningful method of assessment.
3. Use an assessment procedure that will provide the information needed to make a judgment about the intended student learning.
4. Use authentic tasks in a realistic setting, thus providing contextualized meaning to the assessment.
5. Design and provide clear, explicitly defined scoring rubrics with the assessment.
6. Be as objective as possible in observing, judging, and recording performance.
7. Record assessment results during the observation.
8. Use multiple assessments whenever possible.
9. Use assessment to enhance student learning.

A balanced approach to assessment is the prudent path to follow. The issue is not whether one form of assessment is intrinsically better than another; no assessment model is suited for every purpose. The real issue is determining what type of performance indicator best serves the purpose of the assessment and then choosing the assessment that will best provide this type of information.

Dataset Application

Using the chapter 15 large dataset, which is based on figure 15.3, in HK*Propel*, do the following:
- Confirm the total scores for each of the three trials (chapter 3).
- Determine the alpha reliability coefficient (chapter 6) for the reliability *across* total scores for the three trials. Comment on the reliability.
- Calculate the correlation coefficients (chapter 4) among the four measures taken during the first trial. Comment on these correlation coefficients.
- Determine whether boys and girls differ significantly (chapter 5) for the sum across all four measures across all three trials.

MEASUREMENT AND EVALUATION CHALLENGE

Let's return to Mariko: She has decided to use soccer to teach invasion games concepts; her primary purpose is for her students to be able to play a game of small-sided soccer. Psychomotor assessments will be emphasized in the unit. Mariko has decided to develop a game play rubric for soccer that includes maintaining possession of the ball and advancing the ball toward the goal (dribbling and passing), scoring (shooting or kicking to target), and preventing scoring (defense). Her rubric also includes elements of fair play (National Standard 4, SHAPE America, 2014). Skills tests for dribbling, passing, and shooting are used as formative assessments. Mariko has also developed some log sheets that allow students to document their efforts to practice skills outside of class. By looking at whether students have improved, Mariko has an indirect way to determine whether students are practicing as indicated. She also wants to observe student knowledge of the rules during game play and whether her students understand invasion game tactics. To document knowledge about tactics, she has created a project that requires students to develop a playbook with game play tactics and strategies. Students will demonstrate their knowledge of rules by officiating the small-sided soccer games during the class tournament. A rubric was also created to assess officiating knowledge.

SUMMARY

Performance-based assessment conducted in authentic settings in order to enhance learning and achievement was a foundation of educational reform. Performance-based assessments take many forms, but those typically used in physical education generally include individual or group projects, portfolios, performances, and student logs or journals. A method for recording learning is also essential so that teachers can use the information as they plan future lessons. A guiding principle for the development and use of any type of performance-based assessment is that it consists of both task and performance criteria. Furthermore, we have emphasized that consistency, validity, and fairness are just as important with performance-based assessment as they are with traditional standardized testing. For those who are interested in more information about performance-based assessments, see resource texts by Lund and Kirk (2020) and Lund and Veal (2013). In addition, assessment packages such as PE Metrics are designed to provide a collection of current, appropriate, and realistic assessment tools for physical educators.

www! **Go to HK*Propel* for videos, homework assignments, and quizzes that will help you master this chapter's content.**

APPENDIX
Microsoft Excel Applications

As we indicated in the textbook's preface, many of the procedures used to conduct the analyses associated with human performance measurement and evaluation decisions can be rather time consuming. However, the use of computers greatly facilitates these analyses. We have used SPSS (IBM SPSS Statistics) throughout the book to illustrate step-by-step procedures for analyzing data. SPSS is a sophisticated software package used for research, business, and educational decisions, and is widely available at universities, businesses, education and governmental agencies, and research centers. Less expensive student versions of SPSS are available; however, they are generally limited in the types of procedures that can be completed and in the number of participants or variables (or both) that can be included.

We are aware that some students may have difficulty accessing SPSS. Although not created to be an analytical package, Microsoft Excel is a widely available spreadsheet that can be also used to conduct all of the procedures that are presented in your textbook. Many of the analyses related to reliability, validity, measurement, and evaluation decisions will involve more steps than SPSS. Its advantages are its universal availability and relatively low cost.

All of the data tables and large datasets in each chapter are included in HK*Propel*. These tables are presented in both SPSS and Excel formats.

In this appendix, we provide you with directions and screen captures (from PC computers) that illustrate Excel use. We illustrate the procedures presented in chapters 2 through 7. Chapters 8 through 15 use the procedures introduced in chapters 2 through 7. It is not our intent to teach you how to become an Excel expert. Rather, as with SPSS, it is our intent to illustrate the specific tools that you will encounter when making measurement and evaluation decisions in kinesiology and human performance. As with SPSS, once you have learned the procedures in chapters 2 through 7, you are expected to be able to generalize these analyses to the remaining text chapters.

Some of the Excel procedures for analyses used in chapters 5 and 7 are quite cumbersome, particularly the chi-square (χ^2); thus, we have included templates for these analyses in chapters 5 and 7 of the HK*Propel* resource. Similarly, a template for the epidemiologic analyses is included in HK*Propel* in chapter 7. With the templates, you simply enter values into the cells provided and the results are calculated for you. A PowerPoint presentation illustrating Excel can be found in chapter 2 of HK*Propel*.

Many of the procedures listed in this appendix use Excel's Analysis ToolPak. The following descriptions are for activating the PC Analysis ToolPak if it is not already activated in your PC version of Excel 2019.

1. Open the Excel File Menu and click on Options (bottom left).

2. Under Excel Options, click on Add-ins.

3. Highlight Analysis ToolPak – VBA and click Go at the bottom of the screen.

4. Check Analysis ToolPak and then click OK. Data Analysis should then appear on the far right of your Data menu.

The Analysis ToolPak that allows you to perform many of the analyses is no longer available for recent versions of Excel for the Mac. The program StatPlus:mac LE is available at no cost from AnalystSoft. It may be downloaded from www.analystsoft.com/en/products/statplusmacle. Excellent documentation for Mac users is available along with this program. For more information about using Excel for statistical analysis, please see *Excel Statistics: A Quick Guide*, 3rd ed. (Salkind, 2015).

CHAPTER 2: USING TECHNOLOGY IN MEASUREMENT AND EVALUATION

The commands presented are from Excel 2019 (PC version). In the following tasks, you'll work with table 2.1, presented again here, as an introduction to working in Excel.

Table A.1 Sample Database (Data Matrix)

id	gender	age	weightkg	heightcm	stepsperday
1	0	20	50	165	5000
2	0	24	51	160	6000
3	0	21	62	173	7000
4	0	19	59	178	6500
5	0	23	43	145	4500
6	1	22	86	193	4800
7	1	25	65	183	4000
8	1	24	61	178	4200
9	1	28	75	173	3900
10	1	20	70	178	3500

Task 1

Input and verify the data. Do NOT enter the Mean and Stdevs lines.

Task 2

Compute BMI—steps.

1. Type "BMI" in column G, row 1.
2. Create BMI formula:
   a. Type "=" in cell G2.
   b. Click on cell D2 and then enter "/" in cell G2.
   c. Enter "(" and click on cell E2.
   d. Enter "*" and click on cell E2 again.
   e. Enter "/10000)".
   f. Click OK.
3. Use Excel to compute BMI for all 10 people.
   a. Place your cursor in cell G2. Click and hold the little box in the lower right corner of the cell.
   b. Drag the cursor down to cell G11 to provide BMIs for all participants.
   c. Use the Number box on the Home tab to round to two or three decimal places to make the display more aesthetically pleasing.

Task 3

Compute selected descriptive statistics (using the Functions menu).

1. Type "Mean" in cell A12.
2. Move your cursor to cell C12.
3. Go to the Formulas menu, click on More Functions, then Statistical, then AVERAGE.
   a. Place your cursor in cell C2 and scroll in all of the data for Age.
      Note: Excel will read blanks as missing values!
   b. Press Return.
4. To calculate all other means, keep your cursor in cell C12.
   a. Place crosshairs over the small rectangle in the lower right corner of cell C12.
   b. Drag the small rectangle with your cursor to cell G12—doing this should calculate all means.
   c. Round as you like!
5. To calculate standard deviation, type "Stdevs" in cell A13.
   a. Move your cursor to cell C13.
   b. Go to the Formulas menu, then More Functions, then Statistical and click on STDEV.S in the Statistical Functions menu.
   c. Scroll down to enter the numbers from C2 through C11. Do NOT include C12, which is the mean. Click OK.
   d. Place your cursor over the small rectangle in the lower right corner of the box and drag the cursor to cell G13 as you did in step 4.b for the means—this should calculate all standard deviations.
   e. Round to one decimal using the Number box on the Home tab.
      Note: We did not include gender when calculating the mean and standard deviation because gender is not a continuous variable.

Tasks 4 and 5

Compute maximum (use MAX) and minimum (use MIN) as just explained by using the Statistical Functions menu as well. Place the labels in cells A14 and A15, respectively. Compute in the same manner as for mean (AVERAGE) or standard deviation (STDEV.S). Be sure to include only data from row 2 through row 11 each time.

	A	B	C	D	E	F	G
1	id	gender	age	weightkg	heightcm	stepsperday	BMI
2	1	0	20	50	165	5000	18.37
3	2	0	24	51	160	6000	19.92
4	3	0	21	62	173	7000	20.72
5	4	0	19	59	178	6500	18.62
6	5	0	23	43	145	4500	20.45
7	6	1	22	86	193	4800	23.09
8	7	1	25	65	183	4000	19.41
9	8	1	24	61	178	4200	19.25
10	9	1	28	75	173	3900	25.06
11	10	1	20	70	178	3500	22.09
12	Mean		22.6	62.2	172.6	4940	20.69786
13	Stdevs		2.76	12.71	13.29	1183.4	2.1
14	Maximum			28	86	193	7000	25.05930703
15	Minimum			19	43	145	3500	18.36547291

CHAPTER 3: DESCRIPTIVE STATISTICS AND THE NORMAL DISTRIBUTIONS

Table 2.1

In the following steps, you'll use the Data Analysis ToolPak to compute descriptive statistics for the data in table 2.1.

1. Go to the Data menu in Excel and open Data Analysis.

2. You should get the following screen:

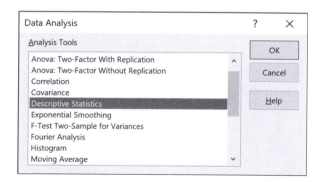

3. Select Descriptive Statistics, click OK, and then follow these instructions:

   a. Scroll in all of the data, including variable names (ONLY the data, NOT the mean, stdevs, maximum, and minimum).

   b. Check the Labels in First Row box.

   c. Leave New Worksheet Ply selected for Output options.

   d. Check the Summary statistics box.

4. Click OK, and your output should look like this:

5. Clean up the output by removing the redundant columns, rounding entries, and so on. The cleaned-up output should look like this:

Table 3.1

To create a histogram for the data in table 3.1, use the following steps:

1. Go to the Home menu and perform a descending sort (the commands will vary based on the version of Excel and the type of computer you are using). An example is shown in the next screen capture.

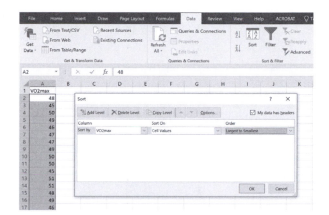

2. Scroll in all of your data.
3. Open the Sort function and perform a descending sort on VO2.
4. Be sure to click Descending and Header rows.
5. Click OK to sort your scores.
6. To create a histogram, you now have to create what Excel calls Bins, which are used to set up the frequency distribution for the histogram.

   a. Type the word "Bins" in the cell next to the label VO2 for your sorted data (cell B1).
   b. Type a column of numbers in the Bins column that corresponds to the range of the scores from highest to lowest (in this case, from 55 to 41).
   c. Now open the histogram program under Data and the Data Analysis. Scroll in your labels and scores in the Input Range and then scroll your Bins in where it asks for Bin Range.
   d. Be sure to check the Labels and Chart Output boxes.
   e. Leave New Worksheet Ply selected for Output options.
   f. Click OK.

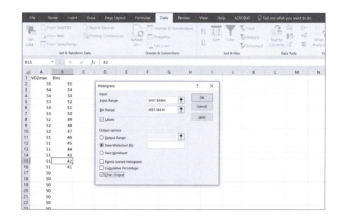

7. To clean up your output, click on the histogram chart (which usually comes out too flat). A box will appear around it.

8. Place your cursor over the lower right corner and drag the corner down and to the right to create an aesthetically pleasing graph, as shown below:

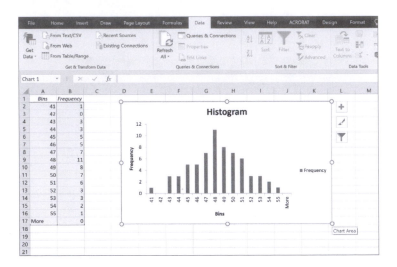

To calculate percentiles with Excel, use the Rank and Percentile program. Be on the Excel sheet where your data are located.

9. Select the Rank and Percentile program from the Data Analysis menu.

10. Input your data by placing your cursor over the variable name (VO2) and scroll in all of your data.

11. Be sure the Labels in First Row box is checked.

12. Select New Worksheet Ply for Output options.
13. Click OK, and your output should look like the output shown here.

	File	Home	Insert	Draw	Page Layout	Formulas	Data	Review

A1					Point			

	A	B	C	D	E	F	G	H
1	Point	VO2max	Rank	Percent				
2	1	55	1	100.00%				
3	2	54	2	96.80%				
4	3	54	2	96.80%				
5	4	53	4	92.10%				
6	5	53	4	92.10%				
7	6	53	4	92.10%				
8	7	52	7	87.50%				
9	8	52	7	87.50%				
10	9	52	7	87.50%				
11	10	51	10	78.10%				
12	11	51	10	78.10%				
13	12	51	10	78.10%				
14	13	51	10	78.10%				
15	14	51	10	78.10%				
16	15	51	10	78.10%				
17	16	50	16	67.10%				
18	17	50	16	67.10%				
19	18	50	16	67.10%				
20	19	50	16	67.10%				
21	20	50	16	67.10%				
22	21	50	16	67.10%				
23	22	50	16	67.10%				
24	23	49	23	54.60%				
25	24	49	23	54.60%				
26	25	49	23	54.60%				
27	26	49	23	54.60%				
28	27	49	23	54.60%				
29	28	49	23	54.60%				
30	29	49	23	54.60%				
31	30	49	23	54.60%				

Note that the ranks use the highest rank for tied ranks and the percent column is interpreted as percentiles, similar to SPSS.

CHAPTER 4: CORRELATION AND PREDICTION

Correlation

Excel allows you to calculate pairwise correlations or an entire correlation matrix.
Open Table 4.1 in HK*Propel*. Table 4.1 provides the scores for 10 students on three measures: body weight, chin-ups, and pull-ups.

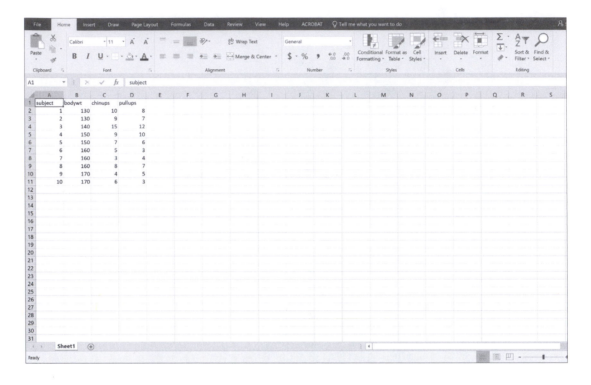

To calculate the individual correlations (pairwise), use the Formulas menu. Go to More Functions and Statistical to get to CORREL. It does not matter which variable you put in, array 1 or array 2. As an example, array 1 represents body weight (bodywt) and array 2 represents chin-ups (chinups). Be *certain* to remember what you did and label it correctly. The correlation will appear in the cell (B13) where you have placed your cursor.

Using Data Analysis, you can create an entire correlation matrix for the three variables. For practice, calculate the correlations for body weight and pull-ups and for chin-ups and pull-ups.

1. Select the Correlation program and click OK.

2. Scroll in all of the data, including variable names.

3. Check the Labels in First Row box.

4. Select New Worksheet Ply for Output options and click OK. The screen should look like this:

5. Click OK. You will get the following output:

6. To clean up, scroll in all of your output values and reduce the number of decimals to three with the formatting palette. The final matrix should look like this:

Regression

The same data in table 4.1 may be used to demonstrate prediction (called regression in Excel). Single-variable prediction is computed using Data Analysis: Regression. For this example, calculate pull-ups predicted from body weight.

1. Scroll in pull-up data, including the name, for Input Y Range.
2. Scroll in body weight data, including the name, for Input X Range.
3. Be sure the Labels box is checked.
4. Leave New Worksheet Ply selected for Output options.

5. Click OK.
6. On Summary Output, reduce the number of decimal places to three with the formatting palette.

CHAPTER 5: INFERENTIAL STATISTICS

Excel will calculate all of the statistics included in chapter 5; however, some of the options in SPSS are not available. Other limitations of Excel will be explained as they are presented.

Table 5.3

Because the calculation of chi-square (χ^2) is quite cumbersome in Excel, we have provided you with a template in HK*Propel*. You simply change variable names and value labels and enter values into the cells provided, and χ^2 and associated statistics are calculated for you. The screen capture for χ^2, proportion of agreement, Kappa, and phi is presented here:

Table 5.5

To use Excel for a t-test for two independent groups, input the data as shown below. Note the data are formatted differently than with SPSS.

1. From the Data Analysis programs, select t-test: Two-Sample Assuming Equal Variances.

2. Scroll Varsity Data (group1) in for Variable 1 Range.
3. Scroll Subvarsity Data (group2) in for Variable 2 Range.
4. Check the Labels box.
5. Select New Worksheet Ply for Output options.
   Note: The direction of the t-test will be based on which variable is entered first.

6. Click OK.

Table 5.6

1. From the Data Analysis programs, select t-test: Paired Two Sample for Means.

1	Preseason	Postseason
2	18	20
3	20	24
4	17	20
5	16	19
6	15	20
7	18	22
8	19	21
9	17	21

Data Analysis

Analysis Tools

Histogram
Moving Average
Random Number Generation
Rank and Percentile
Regression
Sampling
t-Test: Paired Two Sample for Means
t-Test: Two-Sample Assuming Equal Variances
t-Test: Two-Sample Assuming Unequal Variances
z-Test: Two Sample for Means

OK
Cancel
Help

2. Scroll Postseason Data (postseason) in for Variable 1 Range.
3. Scroll Preseason Data (preseason) in for Variable 2 Range.
4. Check the Labels box.
5. Select New Worksheet Ply for Output options.

t-Test: Paired Two Sample for Means

Input
Variable 1 Range: $B$1:$B$9
Variable 2 Range: $A$1:$A$9

Hypothesized Mean Difference:

☑ Labels

Alpha: 0.05

Output options
◯ Output Range:
◉ New Worksheet Ply:
◯ New Workbook

OK
Cancel
Help

6. Click OK.

	A	B	C	D	E
1	t-Test: Paired Two Sample for Means				
2					
3		Postseason	Preseason		
4	Mean	20.875	17.5		
5	Variance	2.4107	2.5714		
6	Observations	8	8		
7	Pearson Correlation	0.7746			
8	Hypothesized Mean Difference	0			
9	df	7			
10	t Stat	9			
11	P(T<=t) one-tail	2.1329E-05			
12	t Critical one-tail	1.8946E+00			
13	P(T<=t) two-tail	4.2657E-05			
14	t Critical two-tail	2.3646E+00			

Table 5.7

To use Excel for one-way ANOVA, you need to have equal N and input data as shown here. Note the data are formatted differently than with SPSS.

1. From the Data Analysis programs, select ANOVA: Single Factor.
2. Scroll in all of the data, including group labels.
3. Check the Labels in First Row box.
4. Select New Worksheet Ply for Output options.

5. Click OK.

SUMMARY

Groups	Count	Sum	Average	Variance
group1	5	941	188.2	61.7
group2	5	823	164.6	92.8
group3	5	765	153	127

ANOVA

Source of Variation	SS	df	MS	F	P-value	F crit
Between Groups	3217.6	2	1608.8000	17.1453	0.0003	3.8853
Within Groups	1126	12	93.8333			
Total	4343.6	14				

CHAPTER 6: RELIABILITY AND VALIDITY

There is no program in Excel to compute the alpha coefficient directly, but it can be computed using the Functions menu and the formula. The steps presented here use table 6.6 as an example.

As an alternative, you could calculate all of the variances associated with the trials and the total; you would then enter these into the alpha coefficient formula and complete the calculations by hand.

1. Type the word "VARA" in the cell below under "Participant" (cell A7).
2. Position your cursor in the cell next to "VARA" (cell B7).
3. Go to Formulas, then More Functions, then Statistical, and find VARA.
4. Because you have no missing data, Excel will compute the sample variance for the five participants (i.e., value1 ranges from B2:B6).
5. Now place your cursor in the lower right corner of the variance for trial 1 and drag the cursor across—that will produce variances for each trial as well as the total variance.

6. Once you have the variances for the trials and total, you can easily calculate alpha by hand. Notice we have included the alpha coefficient formula in cell B10 (i.e., "=(3/2)*(1-(B7+C7+D7)/E7)").

CHAPTER 7: CRITERION-REFERENCED TESTS: CUT SCORES, RELIABILITY, AND VALIDITY

The calculated statistics associated with criterion-referenced reliability and validity are based on the chi-square (χ^2) examples presented in chapter 5. Because the calculation of these statistics is quite cumbersome in Excel, we have provided you with a template in HK*Propel* for chapter 7. You simply enter values into the cells provided, and χ^2 and associated statistics are calculated for you. The screen capture for χ^2, proportion of agreement, Kappa, and phi is presented here:

Relative Risk

Spreadsheet: Relative Risk

Menu bar: File | Home | Insert | Draw | Page Layout | Formulas | Data | Review | View | Help | ACROBAT | Tell me what you want to do

Cell reference: B14   fx   MI Death

	A	B	C	D	E	F	G	H
1	**CRT measures for Epidemiology coefficients**							
2								
3								
4								
5	**Relative Risk**	**(RR)**						
6	**A measure of the strength of the association**							
7								
8	RR = IR exposed/IR unexposed							
9	(IR is incidence rate)							
10								
11	RR = (a/(a+b))/(c/(c+d))							
12								
13	Risk	Heart Disease						
14	Factor	MI Death	No MI Death	Total				
15	High CHOL	25	31	56				
16	No high CHOL	7	37	44				
17	Totals	32	68	100				
18								
19	IR exposed =	0.446						
20	IR not exposed =	0.159						
21								
22	RR =	2.806						
23								
24	Attributable Risk	0.6436364						
25								
26								
27	These data are from Table 7.10							
28								

Cross-Sectional Study

Spreadsheet: Cross-Sectional Study

Menu bar: File | Home | Insert | Draw | Page Layout | Formulas | Data | Review | View | Help | ACROBAT | Tell me what you want to do

Cell reference: C8   fx   37

	A	B	C	D	E	F	G	H	I
1	**Cross Sectional Study or Case Control**								
2									
3									
4									
5	Stimulus	Outcome Variable							
6	Variable	MI Death	No MI Death	Total					
7	High CHOL	25	31	56					
8	No High CHOL	7	37	44					
9	Total	32	68	100					
10									
11									
12	**Odds Ratio**	4.263							
13									
14									
15	These data are from Table 7.10								
16									
17									
18									
19									
20									
21									
22									
23									
24									
25									
26									
27									
28									

CHAPTERS 8 THROUGH 15

The data, examples, and measurement and evaluation analyses presented in chapters 8 through 15 use the procedures that you learned in chapters 2 through 7. When you are analyzing the large datasets or working through problems in the remaining chapters, simply return to the analytical procedures learned earlier and use the appropriate SPSS or Excel procedure.

Glossary

absolute endurance—A measure of repetitive performance against a fixed resistance (e.g., number of repetitions of a 100-pound [45.4 kg] bench press).

absolute rating—Evaluating performance on a fixed scale.

accuracy-based skills test—A skills test that assesses one's ability to serve or project an object (e.g., ball, shuttlecock) into a prescribed area or for distance and accuracy.

achievement test—In the cognitive domain, a test designed to measure the extent to which an examinee comprehends some body of knowledge.

adapted physical education—Physical education adjusted to accommodate students with physical or mental disabilities.

aerobic power—The body's ability to supply oxygen to the working muscles during physical activity; often called $\dot{V}O_2$max.

affective domain—Involves attitudes, perceptions, and psychological characteristics.

alpha level (α)—The probability of falsely rejecting a null hypothesis or making a type I error (i.e., claiming that there is a relation between the variables when, in fact, there is not). Also called the *significance level*.

alpha reliability—See *intraclass reliability*.

alternative assessment—An assessment technique that is different from traditional, standardized testing. Also called *authentic assessment*.

alternative hypothesis (H_1)—The alternative hypothesis is a statement that is the opposite of the null hypothesis.

analytic rubric—A way to score an assessment that lists the characteristics or traits important to the completion or performance of the task so the rater can evaluate the degree to which these characteristics or traits were met.

analytic scoring—A method of scoring responses to essay questions that involves identifying specific facts, points, or ideas in the answer and awarding credit for each one.

anxiety—Distress and tension caused by apprehension.

assessment—The process of collecting information and judging what it means.

authentic assessment—Assessment designed to take place in a real-life setting that provides authenticity and contextualized meaning.

behavioral objectives—Goals with specific measurable steps for their achievement.

body composition—The physical makeup of the body, including weight, lean weight, and fat weight.

cardiorespiratory endurance—The body's ability to extract and use oxygen in a manner that permits continuous exercise, physical work, or physical activities.

central tendency—Statistics that are located near the center of a set of scores.

central-tendency error—The type of rating scale error that results from the hesitancy of raters to assign extreme ratings.

checklist—A typically dichotomous rating of a trait.

coefficient of determination—A measure of variation in common between two variables. Interpreted as a percent, it is the square of the correlation (r^2).

cognitive domain—Involves knowledge and mental achievement.

composite score—A total score developed from the scores of a set of separate tests or performances.

concurrent validity—The relationship between a test (a surrogate measure) and a criterion when the two measures are taken relatively close together in time. It is based on the Pearson product-moment (PPM) correlation coefficient between the test and criterion.

content objectives—The specific course objectives determined by the instructor.

contingency table—A table used for cross-referencing two nominal variables.

convergent evidence—Validity evidence that supports relationships between test scores and other measures intended to assess similar constructs.

correlation—A measure of the relationship between two variables.

correlation coefficient—An index of the linear relationship between two variables, indicating the magnitude (i.e., amount) and the direction of the relationship.

criterion-referenced standard—A specific, predetermined level of achievement.

criterion-referenced test (CRT)—A test with specific, predetermined performance or health standards.

curvilinear relationship—An association among the variables that can best be depicted by a curve.

cut scores—Scores that establish identifiable groups or levels of performance; typically used with criterion-referenced testing.

dependent variable—The variable used as a criterion or that you are trying to predict (Y). Also called *outcome variable*.

depression—A mental condition of gloom, sadness, or dejection.

descriptive statistics—Mathematics used to organize, summarize, and describe data.

developmental rubric—A system used to judge performance across all levels from that of a beginner to that of an expert.

direct relationship—A positive relationship between two variables such that higher scores on one variable are associated with higher scores on the other variable. Similarly, lower scores on one variable are associated with lower scores on the other variable.

discriminant evidence—Validity evidence that supports relationships between test scores and other measures intended to assess different constructs.

distance or power performance test—A skills test that assesses one's ability to project an object for maximum displacement or force.

distractor—An incorrect option for a multiple-choice test item. Also called a *foil*.

educational objectives—General educational goals defined by various pedagogical experts.

epidemiology—The study of the incidence, distribution, and frequency of diseases (e.g., the study of the effects of physical inactivity on coronary heart disease).

error score—An unobservable but theoretically existent score that contributes to an inaccurate estimate of individual differences.

evaluation—A dynamic decision-making process that places a value judgment on the quality of what has been measured (e.g., a test score, physical performance).

extrinsic ambiguity—The characteristic of a test item that appears ambiguous to a respondent who does not understand the concept that the item is testing.

factor analysis—A statistical method using correlational techniques to describe the underlying unobserved dimensions (called factors) from measured variables.

fairness—A characteristic of an assessment, meaning the absence of bias, that allows all participants an equal opportunity to perform to the best of their capabilities.

fitness battery—A group of fitness tests to provide an overall assessment of physical fitness (e.g., FitnessGram).

flexibility—The range of motion of a joint or group of joints.

foil—An incorrect response to a multiple-choice test item. Also called a *distractor*.

formative assessment—Assessment that takes place during instruction and provides feedback so that students can make adjustments to improve their performance and enhance learning.

formative evaluation—An evaluation conducted during (as opposed to at the end of) an instructional or training program.

frequency distribution—A list of the observed scores and their frequency of occurrence.

functional capacity—The ability to perform the normal activities of daily living.

general motor ability (GMA)—The overall proficiency to perform a variety of psychomotor tasks.

generalized rubric—An evaluation system that can be used for multiple assessments that are related conceptually or by content.

global rating—Evaluating overall performance rather than assessing individual components of the action.

global scoring—A method of assessing an answer to an essay question that involves converting the general impression obtained from reading the answer into a score.

group differences method—A method to establish convergent or discriminant validity evidence in which two or three groups are compared to verify the construct being measured.

halo effect—The tendency to elevate a person's score because of positive bias. May also work in reverse (reduce a score because of negative bias).

health-related physical fitness—Attributes of fitness related to health outcomes; cardiorespiratory, body composition, and muscular fitness.

histogram—A graph using vertical bars to present the frequency distribution of the observed scores.

holistic assessment—Analysis based on the overall quality of a performance or product.

holistic rubric—A way to evaluate a performance with a single score using paragraph descriptions of the desired levels of performance for the various characteristics or traits considered important. Typically used with summative assessments.

hypothesis—A statement of relation between at least two variables.

independent variable—The variable related to a dependent variable. Often used as a predictor (X).

index of difficulty—A mathematical expression used in item analysis to estimate the percentage of examinees answering a test item correctly.

index of discrimination—A mathematical expression used in item analysis to estimate how well a test item discriminates among examinees who have been categorized by some criterion (typically the total score on the examination).

index of reliability—The theoretical correlation between observed score and true score; the square root of the reliability coefficient.

inferential statistics—Statistics used to test a hypothesis within a small group (sample) to draw conclusions about a larger group (population).

interclass reliability—A reliability coefficient computed with the Pearson product-moment correlation coefficient.

interrater reliability—Consistency between two or more independent judgments of the same performance.

intraclass reliability—A type of internal consistency reliability, based on ANOVA, where there may be unlimited trials. Alpha, KR_{20}, and KR_{21} are intraclass reliability estimates.

intrarater reliability—Consistency in scoring when a single rater scores the same test or performance two or more times.

intrinsic ambiguity—The characteristic of a test item that is truly ambiguous, even to the respondent who understands the concept that the item is testing.

inverse relationship—A negative relationship between two variables such that higher scores on one variable are associated with lower scores on the other variable. Similarly, lower scores on one variable are associated with higher scores on the other variable.

item analysis—A prescribed process used to examine the quality (e.g., the difficulty and the discrimination) of individual items on a written test.

Kappa (K)—A measure of agreement or association between categorical variables that is adjusted for chance.

keyed response—The correct response to a multiple-choice test item.

kurtosis—An indication of the shape of a distribution that specifies how flat or peaked the distribution is.

linear relationship—An association between two variables that can best be depicted by a straight line.

marginal—The sum of observations across a specific row or column of a contingency table.

mastery test—In the cognitive domain, a test designed to measure whether an examinee has achieved enough knowledge to satisfy a prescribed standard or criterion. Typically used with criterion-referenced testing.

maximal exercise test—An aerobic fitness test that requires the subject to exercise until reaching exhaustion (e.g., treadmill stress test).

maximal oxygen consumption ($\dot{V}O_2max$)—The criterion measure of aerobic capacity. Also called *maximal oxygen uptake*.

measurement—The act of assessing (e.g., assessing a knowledge or psychomotor test score or one's attitude toward physical activity).

mental disabilities—Having mental or psychological limitations (e.g., autism).

Microsoft Excel—A spreadsheet product of Microsoft that results in rows of subjects and columns of variables.

mood—An emotional state of mind, feeling, inclination, or disposition.

motor educability—The ability to learn motor skills.

multiple correlation—The relationship between one outcome (dependent) variable and multiple predictor (independent) variables.

muscular endurance—The physical ability to perform continuous muscular work.

muscular strength—The force that can be generated by the musculature that is contracting.

negative correlation—A correlation in which high scores on one variable are associated with low scores on another variable and low scores are associated with high scores; also called *inverse correlation*.

Net D—An index of discrimination for written test items that indicates the proportion of good discriminations remaining after neutral and bad discriminations are removed.

normal distribution—A bell-shaped, symmetric probability distribution.

norm-referenced standard—A level of achievement relative to a clearly defined subgroup, such as all males or females your age.

null hypothesis—A statement indicating there is no relationship between variables X (independent) and Y (dependent).

objectivity—The degree of interrater reliability; the ability of two or more raters to equivalently score a test.

observed score—One's score on a test. The observed score is the sum of a person's true score and error score.

parameter—A measure in the population of interest (e.g., the population mean).

Pearson product-moment (PPM) correlation coefficient—See *correlation coefficient*. The calculated Pearson product-moment must be between –1.00 and +1.00.

perceived exertion—The mental perception of the intensity of physical work.

percentile—The percentage of observations occurring at or below a given score.

percentile rank (PR)—The percentage of cases falling at or below a specific score in a distribution.

performance-based assessment—Testing method that requires a participant to create a product or performance that demonstrates his or her knowledge or skills.

performance criteria—Standards for judging a performance or product.

performance ratio—Dividing a performance score by another measure to better compare performance between subjects (e.g., weight, speed).

personality—The totality of a person's distinctive psychological traits.

phi coefficient—The Pearson product-moment correlation between two dichotomously scored variables, each of which can take on the value of 0 or 1.

physical activity—The act of bodily movement that requires the contraction of muscles and the expenditure of energy.

physical disabilities—Having physical or organic limitations (e.g., cerebral palsy).

physical fitness—A set of attributes that people have or achieve that relates to the ability to perform physical activity.

point system scoring guide—A list of characteristics used to judge a performance or product with points assigned to each item so that the characteristics are weighted in importance.

population—The target group of people or observations for which research findings are to be inferred.

portfolio—A systematic, purposeful, and meaningful collection of one's work that is assembled over time.

power—The amount of work performed in a fixed amount of time.

prediction—The ability to estimate the value of one variable from one or more other variables.

predictive validity—The relationship between a test (a surrogate measure) and a criterion when the criterion is measured in the future. It is based on the PPM correlation coefficient between the test and the criterion.

process criteria—Standards used to evaluate the quality of performance or how a student completes a task.

product criteria—Standards used to measure the outcome of a performance.

progress criteria—Standards used to measure student improvement during the course of instruction.

proportion of agreement (P)—Percentage of agreement between two measures.

psychomotor domain—Involves physiological and physical performance.

qualitative research—Perceptive measurement, typically textual in nature.

qualitative rubric—Written descriptions that provide qualitative information about the performance being assessed.

quantitative research—Precise measurement, typically numerical in nature.

quantitative rubric—Numbers (quantities) are assigned to indicate quality of the performance being assessed.

range—A measure of variability obtained by subtracting the lowest score from the highest score.

relationship—The statistical association between two or more variables.

relative endurance—A measurement of repetitive performance related to maximum strength (e.g., number of repetitions completed at 50% of 1RM).

relative risk—The risk of mortality (death) or morbidity (disease) associated with one group compared with another (e.g., smokers vs. nonsmokers; physically active vs. not active).

relative scale—Evaluating performance relative to others in a specific group.

relative scoring—A method of scoring responses to essay questions that involves reading all the answers to one question and arranging the papers in order according to degree of adequacy.

relevance—The degree to which a test pertains to the objectives of the measurement.

reliability—The degree to which repeated measurements of the same trait are reproducible under the same conditions; consistency.

repetition maximum (RM)—The maximum number of repeated movements that can be conducted (e.g., 1RM is the maximum weight that can be lifted one time).

repetitive-performance test—A skills test that involves continuous performance of an activity for a specified period of time (e.g., volleying).

research hypothesis—The researcher's hypothesis of the relationship between the independent variable (X) and the dependent variable (Y).

sample—A subgroup of the population with which scientific research is conducted.

scatterplot—A graphic representation of the relationship or correlation between two variables.

scientific method—A method of inquiry that requires the development of a hypothesis and the subsequent statistical testing of its plausibility.

significance—A probability of rejecting the null hypothesis when it is true (alpha). See *type I error*.

simple linear prediction—Using one variable (X) to predict the criterion (Y).

skewness—An indication of the shape of a distribution that specifies a lack of symmetry.

skill—A learned trait based on a person's abilities.

specific determiner—A word or a phrase in a written test item that provides an unintended clue to the correct answer.

specificity—Refers to motor abilities or skills that are unique to individual psychomotor tasks.

SPSS (IBM SPSS Statistics)—Data analysis software. From 2009 to 2010 the same software was referred to as PASW (Predictive Analytics Software).

stability reliability—Consistency of measures across time.

standard deviation—A linear measure of variability that takes into account every score in the distribution; the square root of the variance.

standard error—The type of rating scale error that results from differences in the standards of evaluation applied by raters of the same performance. Results from different raters having different standards.

standard error of estimate (SEE)—An indication of the amount of error when predicting Y from X; the standard deviation of the errors of prediction. Also called *standard error (SE)* or *standard error of prediction (SEP)*. A validity statistic.

standard error of measurement (SEM)—A value reflecting the degree to which a person's observed score fluctuates as a result of errors of measurement; it is interpreted in the same way as a standard deviation. A reliability statistic.

standard score—A score resulting from the conversion of observed values into a score with a given mean and a standard deviation. z-scores and T-scores are standard scores.

state—A psychological attribute that is related to situational changes.

statistic—A numerical value calculated in the sample to estimate a population parameter (e.g., a sample mean).

subjective rating—A value that an instructor places on a skill or performance based on personal observation.

submaximal exercise test—A fitness test that requires the subject to put forth less than maximal effort (e.g., using submaximal cycle ergometry to assess aerobic capacity).

summative assessment—Assessment that occurs at the conclusion of instruction and is designed to measure how much a student learned.

summative evaluation—A final, comprehensive evaluation conducted near or at the end of an instruction or training program.

surrogate measure—A test used to estimate the criterion (e.g., skinfold measurement is a surrogate measure of the criterion percent body fat that is obtained by DXA or underwater weighing).

table of specifications—A written test blueprint that indicates the proportion of test items dealing with each combination of content objective and educational objective.

task-specific rubric—A rubric and accompanying criteria written specifically for a single assessment task.

taxonomy—A classification system based on common characteristics (e.g., the cognitive, affective, and psychomotor domains).

test—An instrument (e.g., written test, performance test, or a wide variety of other instruments) used to make a particular measurement.

theory of basic abilities—Fleishman's theory that skills are based on a number of basic psychomotor abilities that are combined to perform the skill.

torque—A force producing rotation about an axis.

total body movement test—A test that assesses the speed at which a performer completes a task that involves movement of the whole body in a restricted area (e.g., basketball defensive shuffle test).

trait—A relatively stable, common, consistent psychological attribute.

trials-to-criterion testing—The performance of a skill until a standard of achievement is reached.

true score—An unobservable but theoretically existent score that contributes to one's observed test score; it contributes to an accurate estimate of individual differences.

type I error—A false rejection of the null hypothesis. Deciding there is a relation between variables when there actually is none. Expressed as a probability (α).

type II error—A false rejection of the alternative or research hypothesis. Failing to discern a relation between variables when a relation actually exists. Expressed as a probability (β).

validity—The degree of truthfulness in a test.

variability—The spread, or dispersion, of scores in a set of data.

variance (s^2)—A measure of variability; a measure of the spread of a set of scores based on the average squared deviation of each score from the mean.

waist–hip girth ratio—The waist circumference divided by the hip circumference. This measure provides an estimate of body fat distribution, which is a risk factor for cardiovascular disease.

work—The result of the physical effort that is performed; the product of the amount of force applied and the distance over which it is applied.

zero correlation—An indication that there is no linear relationship between two variables ($r = 0$).

References

Ainsworth, B.E., D.R. Bassett Jr., S.J. Strath, A.M. Swartz, W.L. O'Brien, R.W. Thompson, D.A. Jones, C.A. Macera, and C.D. Kimsey. 2000 Sep. Comparison of three methods for measuring the time spent in physical activity. *Medicine & Science in Sports & Exercise* 32(9 Suppl):S457-S464. doi:10.1097/00005768-200009001-00004

Ainsworth, B.E., W.L. Haskell, S.D. Herrmann, N. Meckes, D.R. Bassett Jr., C. Tudor-Locke, J.L. Greer, J. Vezina, M.C. Whitt-Glover, and A.S. Leon. 2011 Aug. Compendium of physical activities: A second update of codes and MET values. *Medicine & Science in Sports & Exercise* 43(8):1575-1581. doi:10.1249/MSS.0b013e31821ece12

Ainsworth, B., W. Haskell, A. Leon, D. Jacobs, H. Montoye, J. Sallis, and R. Paffenbarger. 1993. Compendium of physical activities: Classification of energy costs of human physical activities. *Medicine & Science in Sports & Exercise* 25:71-80.

Ainsworth, B.E., W.L. Haskell, M.C. Whitt, M.L. Irwin, A.M. Swartz, S.J. Strath, W.L. O'Brien, D.R. Bassett Jr., K.H. Schmitz, P.O. Emplaincourt, D.R. Jacobs Jr., and A.S. Leon. 2000. Compendium of physical activities: An update of activity codes and MET intensities. *Medicine & Science in Sports & Exercise* 32(Suppl.):498-504.

Ainsworth, B.E., A.S. Leon, M.T. Richardson, D.R. Jacobs, and R.S. Paffenbarger. 1993. Accuracy of the College Alumnus physical activity questionnaire. *Journal of Clinical Epidemiology* 46:1403-1411.

Ainsworth, B.E., and C.E. Matthews. 2001. Descriptive research in physical activity epidemiology. In *Research methods in physical activity,* 4th ed., ed. J.R. Thomas and J.K. Nelson, 291-308. Champaign, IL: Human Kinetics.

Ainsworth, B.E., H.J. Montoye, and A.S. Leon. 1994. Methods of assessing physical activity during leisure and work. In *Physical activity, fitness, and health: International proceedings and consensus statement,* ed. C. Bouchard, R.J. Shephard, and T. Stephens, 146-159. Champaign, IL: Human Kinetics.

Alamar, B. 2013. *Sports analytics: A guide for coaches, managers, and other decision makers.* New York: Columbia University Press.

Albrecht, R.R., and D.L. Feltz. 1987. Generality and specificity of attention related to competitive anxiety and sport performance. *Journal of Sport Psychology* 9, 231-248.

American Alliance for Health, Physical Education, Recreation and Dance (AAHPERD). 1976. *AAHPERD youth fitness test manual* (Rev. 1976 edition). Reston, VA: AAHPERD.

———. 1980. *Health-related physical fitness test manual.* Reston, VA: AAHPERD.

———. 1985. *Norms for college students: Health-related physical fitness test.* Reston, VA: AAHPERD.

———. 1988. *Physical best.* Reston, VA: AAHPERD.

American College of Sports Medicine (ACSM). 2010. *ACSM's health-related physical fitness assessment manual.* 3rd ed. Philadelphia: Lippincott, Williams & Wilkins.

———. 2014a. *ACSM's guidelines for exercise testing and prescription.* 9th ed. Philadelphia: Lippincott, Williams & Wilkins.

———. 2014b. *ACSM's resource manual for guidelines for exercise testing and prescription.* 7th ed. Philadelphia: Lippincott, Williams & Wilkins.

———. 2022. *ACSM's guidelines for exercise testing and prescription.* 11th ed. Philadelphia: Lippincott, Williams & Wilkins.

American College of Sports Medicine, Liguori, G., Feito, Y., Fountaine, C., and B. A. Roy. 2022. *ACSM's guidelines for exercise testing and prescription* (11th ed.). Philadelphia: Wolters Kluwer.

American College of Sports Medicine, D. Riebe, J.K. Ehrman, G. Liguori, and M. Magal. 2018. *ACSM's guidelines for exercise testing and prescription,* 10th ed. Philadelphia: Wolters Kluwer.

American Educational Research Association, American Psychological Association, and National Council on Measurement and Education. 1985. *Standards for educational and psychological testing.* Washington, DC: Author.

———. 1999. *Standards for educational and psychological testing.* Washington, DC: AERA.

———. 2014. *Standards for educational and psychological testing.* Washington, DC: AERA.

American Heart Association. 1994. *Heart and stroke facts.* Dallas: American Heart Association.

American Psychological Association. 1954. Technical recommendations for psychological tests and diagnostic techniques. *Psychological Bulletin* 51:201-238.

———. 1966. *Standards for educational and psychological tests and manuals.* Washington, DC: Author.

American Psychological Association, American Educational Research Association, and National Council on Measurement in Education. 1974. *Standards for educational and psychological tests.* Washington, DC: American Psychological Association.

American Red Cross. 2009. *American Red Cross water safety instructor's manual.* St. Louis: Mosby Lifeline.

Anderson, L.W., and D.R. Krathwohl, eds. 2001. *A taxonomy for learning, teaching and assessing: A revision of Bloom's taxonomy of educational objectives.* Complete ed. New York: Longman.

Anshel, M. 1987. Psychological inventories used in sport psychology research. *Sport Psychologist* 1:331-349.

Anshel, M.H., N.L. Weatherby, M. Kang, and T. Watson. 2009. Rasch calibration of a unidimensional perfectionism inventory for sport. *Psychology of Sport & Exercise* 10:210-216.

Arnold, J.D., J.M. Rauschenberger, W.G. Soubel, and R.M. Guion. 1982. Validation and utility of a strength test for selecting steelworkers. *Journal of Applied Psychology* 67:588-604.

Arvey, R., Nutting, S., and T. Landon. 1992. Validation strategies for physical ability testing in police and fire settings. *Public Personnel Management* 21(3):301-312.

Åstrand, P., and I. Rhyming. 1954. A nomogram for calculation of aerobic capacity (physical fitness) for pulse rate during submaximal work. *Journal of Applied Physiology* 7:218-221.

Bachrach, L.K., and C.M. Gordon. 2016. Bone densitometry in children and adolescents. *Pediatrics* 138(4):e20162398. doi:10.1542/peds.2016-2398

Bandura, A. 1977. Self-efficacy: Toward a unifying theory of behavioral change. *Psychological Review* 84(2):191-215. doi:10.1037/0033-295X.84.2.191

Barlow, D.A. 1970. Relation between power and selected variables in the vertical jump. In *Selected topics on biomechanics,* ed. J.M. Cooper, 233-241. Chicago: Athletic Institute.

Bar-Or, O. 1987. The Wingate anaerobic test: An update on methodology, reliability and validity. *Sports Medicine* 4(6):381-394. doi:10.2165/00007256-198704060-00001

Barrow, H.M. 1954. Test of motor ability for college men. *Research Quarterly* 25:253-260.

Bartlett, J., L. Smith, K. Davis, and J. Peel. 1991. Development of a valid volleyball skills test battery. *Journal of Physical Education and Dance* 62(2):19-21.

Bass, R.I. 1939. An analysis of the components of tests of semicircular canal function and static and dynamic balance. *Research Quarterly* 2:33-52.

Bassett, D.R., B.E. Ainsworth, S.R. Leggett, C.A. Mathien, J.A. Main, D.C. Hunter, and G.E. Duncan. 1996. Accuracy of five electronic pedometers for measuring distance walked. *Medicine & Science in Sports & Exercise* 28:1071-1077.

Bassett, D.R., B.E. Ainsworth, A.M. Swartz, S.J. Strath, W.L. O'Brien, and G.A. King. 2000. Validity of four motion sensors in measuring moderate intensity physical activity. *Medicine & Science in Sports & Exercise* 32(9 Suppl):S471-S480.

Battinelli, T. 1984. From motor ability to motor learning: The generality-specificity connection. *Physical Educator* 41(3):108-113.

Baumgartner, T.A., A.S. Jackson, M.T. Mahar, and D.A. Rowe. 2016. *Measurement for evaluation in kinesiology.* 9th ed. Burlington, MA: Jones & Bartlett Learning.

Berk, R.A., ed. 1980a. *A guide to criterion-referenced test construction.* Baltimore, MD: Johns Hopkins University Press.

———. 1980b. *Criterion-referenced measurement: The state of the art.* Baltimore, MD: Johns Hopkins University Press.

Bingham, W.V. 1937. *Aptitudes and aptitude testing.* New York: Harper.

Blair, S., W. Kannel, H. Kohl, and N. Goodyear. 1989. Surrogate measures of physical activity and physical fitness: Evidence for sedentary traits of resting tachycardia, obesity, and low vital capacity. *American Journal of Epidemiology* 129:1145-1156.

Blair, S., H. Kohl, R. Paffenbarger, D. Clark, K. Cooper, and L. Gibbons. 1989. Physical fitness and all-cause mortality: A prospective study of healthy men and women. *Journal of the American Medical Association* 262:2395-2401.

Blair, S.N., M.J. LaMonte, and M.Z. Nichaman. 2004 May. The evolution of physical activity recommendations: How much is enough? *American Journal of Clinical Nutrition* 79(5):913S-920S. doi:10.1093/ajcn/79.5.913S

Bland, J.M., and D.G. Altman. 1986. Statistical methods for assessing agreement between two methods of clinical measurement. *Lancet* 327:307-310.

Bloom, B.S., ed. 1956. *Taxonomy of educational objectives: Cognitive domain.* New York: McKay.

Bloom, G., D. Stevens, and T. Wickwire. 2003. Expert coaches' perceptions of team building. *Journal of Applied Sport Psychology* 15:129-143.

Booth, M.L., A. Okely, T. Chey, and A. Bauman. 2002. The reliability and validity of the Adolescent Physical Activity Recall Questionnaire. *Medicine & Science in Sports & Exercise* 34:1986-1995.

Borg, G. 1962. *Physical performance and perceived exertion.* Lund, Sweden: Gleerup.

———. 1998. *Borg's perceived exertion and pain scales.* Champaign, IL: Human Kinetics.

Brace, D.K. 1927. *Measuring motor ability.* New York: Barnes.

Brewer, B., J. Vose, J. Raalte, and A. Petitpas. 2011. Metaqualitative reflections in sport and exercise psychology. *Qualitative Research in Sport, Exercise and Health* 3(3):329-334.

Briere, N., R. Vallerand, M. Blais, and L. Pelletier. 1995. Development and validation of the French form of the Sport Motivation Scale. *International Journal of Sport Psychology* 26:465-489.

Brose, A., U. Lindenberger, and F. Schmiedek. 2013.

Affective states contribute to trait reports of affective well-being. *Emotion* 13(5): 940-948.

Brown, D., and D. Fletcher. 2013. *Does sport psychology work? A systematic and meta-analytic review of the effects of psychosocial interventions on sport performance.* Paper presented at the Annual Meeting of the Association for Applied Sport Psychology, New Orleans, LA.

Bull, S.J., C.J. Shambrook, W. James, and J. Brooks. 2005. Towards an understanding of mental toughness in elite English cricketers. *Journal of Applied Sport Psychology* 17(3):209-227.

Bungum, T.J., D.L. Peaslee, A.W. Jackson, and M.A. Perez. 2000. Exercise during pregnancy and type of delivery in nulliparae. *Journal of Obstetric, Gynecologic, and Neonatal Nursing* 29:258-264.

Butler, P. 2006. A review of the literature on portfolios and electronic portfolios, 1-23. New Zealand: Massey University College of Education.

Butte, N.F., Watson, K.B., Ridley, K., Zakeri, I.F., McMurray, R.G., Pfeiffer, K.A., Crouter, S.E., Herrmann, S.D., Bassett, D.R., Long, A., Berhane, Z., Trost, S.G., Ainsworth, B.E., Berrigan, D., and J.E. Fulton. 2018. A youth compendium of physical activities: Activity codes and metabolic intensities. *Medicine and Science in Sports and Exercise* 50(2):246-256. https://doi.org/10.1249/MSS.0000000000001430

Cambridge, D. 2010. *Eportfolios for lifelong learning and assessment.* San Francisco: Jossey-Bass.

Campbell, K.L., K.M. Winters-Stone, J. Wiskemann, A.M. May, A.L. Schwartz, K.S. Courneya, D.S. Zucker, C.E. Matthews, J.A. Ligibel, L.H. Gerber, G.S. Morris, A.V. Patel, T.F. Hue, F.M. Perna, and K.H. Schmitz. 2019. Exercise guidelines for cancer survivors: Consensus statement from international multidisciplinary roundtable. *Medicine & Science in Sports & Exercise* 51(11):2375-2390. doi:10.1249/MSS.0000000000002116

Canadian standardized test of fitness (CSTF): Operations manual. 3rd ed. 1986. Ottawa: Government of Canada, Fitness and Amateur Sport.

Caspersen, C. 1989. Physical activity epidemiology: Concepts, methods, and applications to exercise science. In *Exercise and sport science reviews,* ed. K. Pandolph, 423-473. Baltimore: Williams & Wilkins.

Caspersen, C.J., K.E. Powell, and G.M. Christenson. 1985. Physical activity, exercise, and physical fitness: Definitions and distinctions for health-related research. *Public Health Reports* 100(2):126-131.

Cech, D.J., and S.C. Martin. 2012. *Functional movement development across the life span.* 3rd ed. Philadelphia: W.B. Saunders Company.

Ceesay, S.M., A.M. Prentice, K.C. Day, P.R. Murgatroyd, G.R. Goldberg, W. Scott, and G.B. Spurr. 1989. The use of heart rate monitoring in the estimation of energy expenditure. A validation study using indirect whole-body calorimetry. *The British Journal of Nutrition* 61(2):175-186.

Centers for Disease Control and Prevention (CDC). 2010. *The association between school-based physical activity, including physical education, and academic performance.* Atlanta: U.S. Department of Health and Human Services.

Chelladurai, P., M.S. Yuhasz, and R. Sipura. 1977. The reactive agility test. *Perceptual and Motor Skills* 44:1319-1324.

Chen, P. Y., and P. M. Popovich. 2002. *Correlation: Parametric and nonparametric measures.* Thousand Oaks, CA: Sage.

Clarke, H.H., and H.A. Bonesteel. 1935. Equalizing the ability of intramural teams at a small high school. *Research Quarterly Supplement* 6(March):193-196.

Clough, P., and D. Strycharczyk. 2012. *Developing mental toughness: Improving performance, wellbeing and positive behaviour in others.* Philadelphia, PA: Kogan Page.

Cohen, J. 1988. *Statistical power analysis for the behavioral sciences.* 2nd ed. Hillsdale, New Jersey: Lawrence Erlbaum Associates.

Colcombe, S.J., K.I. Erickson, P.E. Scalf, J.S. Kim, R. Prakash, E. McAuley, S. Elavsky, D.X. Marquez, L. Hu, and A.F. Kramer. 2006. Aerobic exercise training increases brain volume in aging humans. *The Journals of Gerontology: Series A* 61(11):1166-1170. doi:10.1093/gerona/61.11.1166

Coleman, R., S. Wilkie, L. Viscio, S. O'Hanley, J. Porcari, G. Kline, B. Keller, S. Hsieh, P. Freedson, and J. Rippe. 1987. Validation of 1-mile walk test for estimating $\dot{V}O_2$max in 20-29 year olds [Abstract]. *Medicine & Science in Sports & Exercise* 19(Suppl. 2):S29.

Collins, D.R., and P.B. Hodges. 2001. *A comprehensive guide to sports skills tests and measurement.* 2nd ed. Lanham, MD: Rowman & Littlefield.

Committee on Fitness Measures and Health Outcomes in Youth, Food and Nutrition Board, and Institute of Medicine. 2012. *Fitness measures and health outcomes in youth,* ed. R. Pate, M. Oria, and L. Pillsbury. Washington, DC: National Academies Press. doi:10.17226/13483. www.ncbi.nlm.nih.gov/books/NBK241315/

Considine, W.J. 1970. A validity analysis of selected leg power tests utilizing a force platform. In *Selected topics on biomechanics,* ed. J.M. Cooper, 243-250. Chicago: Athletic Institute.

Conway, J.M., J.L. Seale, D.R. Jacobs, M.L. Irwin, and B.E. Ainsworth. 2002. Comparison of energy expenditure estimates from doubly labeled water, a physical activity questionnaire, and physical activity records. *American Journal of Clinical Nutrition* 75:519-525.

Cook, G., L. Burton, and B. Hoogenboom. 2006. Pre-participation screening: The use of fundamental movements as an assessment of function—Part 1. *North American Journal of Sports Physical Therapy* 1(2):62-72.

Cook, G., L. Burton, B. Hoogenboom, and M. Voight. 2014. Functional movement screening: The use of fundamental movements as an assessment of function - Part 2. *International Journal of Sports Physical Therapy*, 9(4):549-563.

Cooper, K. 1968. A means for assessing maximal oxygen intake. *Journal of the American Medical Association* 203:201-204.

Cooper Institute. 1999. *FitnessGram test administration manual*. 2nd ed. Champaign, IL: Human Kinetics.

———. 2010. *FitnessGram & ActivityGram test administration manual*. 4th ed. Champaign, IL: Human Kinetics.

Cooper Institute for Aerobics Research. 1987. *FitnessGram*. Dallas: Cooper Institute for Aerobics Research.

———. 1992. *FitnessGram*. Dallas: Cooper Institute for Aerobics Research.

Corbin, C.B., R.P. Pangrazi, and B.D. Franks. 2000. Definitions: Health, fitness, and physical activity. *President's Council on Physical Fitness and Sports Research Digest* 3(9):1-11.

Cox, J., and K. Cox. 2008. *Your opinion, please! How to build the best questionnaires in the field of education*. Thousand Oaks, CA: Sage.

Cox, R.H., M.P. Martens, and W.D. Russell. 2003. Measuring anxiety in athletics: The Revised Competitive State Anxiety Inventory-2. *Journal of Sport & Exercise Psychology* 25:519-533.

Craft, L., M. Magyar, B. Becker, and D. Feltz. 2003. The relationship between the Competitive State Anxiety Inventory-2 and sport performance: A meta-analysis. *Journal of Sport & Exercise Psychology* 25:44-65.

Craig, C.L., C. Cameron, and C. Tudor-Locke. 2013. CANPLAY pedometer normative reference data for 21,271 children and 12,956 adolescents. *Medicine & Science in Sports & Exercise* 45:123-129.

Cronbach, L.J. 1949. *Essentials of psychological testing*. New York: Harper & Row.

Cronbach, L.J., and P.E. Meehl. 1955. Construct validity in psychological tests. *Psychological Bulletin* 52(4):281-302. doi:10.1037/h0040957

Culver, D., W. Gilbert, and P. Trudel. 2003. A decade of qualitative research in sport psychology journals 1990-1999. *Sport Psychologist* 17:1-15.

Cureton, K.J., and G.L. Warren. 1990. Criterion-referenced standards for youth health-related fitness tests: A tutorial. *Research Quarterly for Exercise and Sport* 61:7-19.

Cureton, T. 1951. *Physical fitness of champions*. Urbana, IL: University of Illinois Press.

Dale, D., G. Welk, and C. Matthews. 2002. Methods for assessing physical activity and challenges for research. In *Physical activity assessments for health-related research*, ed. G. Welk, 19-34. Champaign, IL: Human Kinetics.

Danielson, C. 1997. Designing successful performance tasks and rubrics. Audio cassette tape #297072. Recorded live at the 52nd Annual Conference ASCD, Baltimore, March 22-25, 1997. Alexandria, VA: Association for Supervision and Curriculum Development.

Dennison, B., J.H. Straus, D. Mellits, and E. Charney. 1988. Childhood physical fitness tests: Predictor of adult physical activity levels? *Pediatrics* 82:324-330.

Disch, C.F., and J.G. Disch. 1979. Predictive analysis of a battery of anthropometric and motor performance tests for classifications of male volleyball players. *Volleyball Technical Journal* 4:93-98.

Disch, J.G. 1978. The construction and analysis of a test battery related to volleyball playing capacity in females. Report No. ED 148815. Washington, DC: ERIC Clearinghouse in Teacher Education.

Disch, J.G., and S.C. Disch. 2005. Performance testing athletes. *Olympic Coach* 17(3):17-21.

Dishman, R.K., and W. Ickes. 1981. Self-motivation and adherence to therapeutic exercise. *Journal of Behavioral Medicine* 4:421-436.

Dishman, R.K., R.W. Motl, R. Saunders, G. Felton, D.S. Ward, M. Dowda, and R.R. Pate. 2004. Self-efficacy partially mediates the effect of a school-based physical-activity intervention among adolescent girls. *Preventive Medicine* 38(5):628-636.

Doyle, J., and G. Parfitt. 1996. Performance profiling and predictive validity. *Journal of Applied Sports Psychology* 8:160-170.

Driskell, J.E., C. Copper, and A. Moran. 1994. Does mental practice enhance performance? *Journal of Applied Psychology* 79:481-492.

Duda, J. 1998. *Advances in sport and exercise psychology measurement*. Morgantown, WV: Fitness Information Technology.

Duda, J.L. 1989. Relationship between task and ego orientation and the perceived purpose of sport among high school athletes. *Journal of Sport & Exercise Psychology* 11:318-335.

Dunn, J.H., J. Causgrove Dunn, and D.G. Syrotuik. 2002. Relationship between multidimensional perfectionism and goal orientations in sport. *Journal of Sport & Exercise Psychology* 24:376-395.

Durand-Bush, N., J. Salmela, and I. Green-Demers. (2001). The Ottawa Mental Skills Assessment (OMSAT-3). *The Sport Psychologist* 15:1-19.

Ebel, R. 1965. *Measuring educational achievement*. Englewood Cliffs, NJ: Prentice-Hall.

Ekelund, L., W. Haskell, J. Johnson, F. Whaley, M. Criqui, and D. Sheps. 1988. Physical fitness as a predictor of cardiovascular mortality in asymptomatic North American men. *New England Journal of Medicine* 319:1379-1384.

Ellenbrand, D.A. 1973. Gymnastics skills tests for college women. Unpublished master's thesis, Indiana University, Bloomington.

Engelman, M.E., and J.R. Morrow Jr. 1991. Reliability and skinfold correlates for traditional and modified pull-ups in children grades 3-5. *Research Quarterly for Exercise and Sport* 62:88-91.

Erickson, K.I., M.W. Voss, R.S. Prakash, C. Basak, A. Szabo, L. Chaddock, J.S. Kim, S. Heo, H. Alves, S.M. White, T.R. Wojcicki, E. Mailey, V.J. Vieira, S.A. Martin, B.D. Pence, J.A. Woods, E. McAuley, and A.F. Kramer. 2011. Exercise training increases size of hippocampus and improves memory. *Proceedings of the National Academy of Sciences of the United States of America* 108(7):3017-3022. doi:10.1073/pnas.1015950108

Eysenck, H.J., and S.B.G. Eysenck. 1968. *Eysenck Personality Inventory manual.* London: University of London Press.

Feuer, M., and K. Fulton. 1993. The many faces of performance assessment. *Phi Delta Kappan* 74:478.

Fiatarone, M.A., E.F. O'Neill, N.D. Ryan, K.M. Clements, G.R. Solares, M.E. Nelson, S.B. Roberts, J.J. Kehayias, L.A. Lipsitz, and W.J. Evans. 1994. Exercise training and nutritional supplementation for physical frailty in very elderly people. *New England Journal of Medicine* 330:1769-1775.

Fields, D.A., M.I. Goran, and M.A. McCrory. 2002. Body-composition assessment via air-displacement plethysmography in adults and children: A review. *The American Journal of Clinical Nutrition* 75(3):453-467. doi:10.1093/ajcn/75.3.453

FitzGerald, S.J., C.E. Barlow, J.B. Kampert, J.R. Morrow Jr., A.W. Jackson, and S.N. Blair. 2004. Muscular fitness and all-cause mortality: Prospective observations. *Journal of Physical Activity & Health* 1:7-18.

Fleishman, E.A. 1964. *The structure and measurement of physical fitness.* Englewood Cliffs, NJ: Prentice Hall.

Fleishman, E.A., and M.K. Quaintance. 1984. *Taxonomies of human performance: The description of human tasks.* New York: Academic Press.

Fournier, J.F., C. Calmels, N. Durand-Bush, and J.H. Salmela. 2005. Effects of a season-long PST program on gymnastic performance and on psychological skill development. *International Journal of Sport & Exercise Psychology* 3:59-78.

Fox, K., and C. Corbin. 1989. The Physical Self-Perception Profile: Development and preliminary validation. *Journal of Sport & Exercise Psychology* 11:408-430.

Freedson, P.S., and K. Miller. 2000. Objective monitoring of physical activity using motion sensors and heart rate. *Research Quarterly for Exercise and Sport* 71:21-29.

Fuchs, R.K., J.J. Bauer, and C.M. Snow. 2001. Jumping improves hip and lumbar spine bone mass in prepubescent children: A randomized controlled trial. *Journal of Bone and Mineral Research* 16:148-156.

Gael, S. 1988. *The job analysis handbook for business, industry, and government.* John Wiley & Sons.

Gary, R. 2006. Exercise self-efficacy in older women with diastolic heart failure: Results of a walking program and education intervention. *Journal of Gerontological Nursing* 32(7):31-39.

Getchell, L.H., D. Kirkendall, and G. Robbins. 1977. Prediction of maximal oxygen uptake in young adult women joggers. *Research Quarterly* 48:61-67.

Gholamnezhad, Z., M.H. Boskabady, and Z. Jahangiri. 2020. Exercise and dementia. *Advances in Experimental Medicine and Biology* 1228:303-315. doi:10.1007/978-981-15-1792-1_20

Gibbs-Smith, C., and G. Rees. 1978. *The inventions of Leonardo Da Vinci.* Oxford, UK: Phaidon Press.

Glaser, R. 1963. Instructional technology and the measurement of learning outcomes. *American Psychologist,* 18: 519-522. doi: 10.1037/h0049294

Glover, K.A., and W. Zhu. 2017. Regulated sedentary behavior in occupations. In *Sedentary behavior and health: Concepts, assessment & intervention,* ed. W. Zhu and N. Owen, 71-79. Champaign, IL: Human Kinetics.

Golding, L. 2000. *YMCA fitness testing and assessment manual.* 4th ed. Champaign, IL: Human Kinetics.

Goldsmith, W. 2005. Testing: How, why, who, what, and when (and how to make sense of it). *Olympic Coach* 17(3):13-16.

Gotwals, J.K., and J.H. Dunn. 2009. A multi-method multi-analytic approach to establishing internal construct validity evidence: The Sport Multidimensional Perfectionism Scale 2. *Measurement in Physical Education and Exercise Science* 13:71-92.

Gotwals, J.K., J.H. Dunn, J. Causgrove Dunn, and V. Gamache. 2010. Establishing validity evidence for the Sport Multidimensional Perfectionism Scale-2 in intercollegiate sport. *Psychology of Sport and Exercise* 11:423-432.

Gould, D., K. Dieffenbach, and A. Moffett. 2002. Psychological characteristics and their development in Olympic champions. *Journal of Applied Sport Psychology* 14:172-204.

Gould, D., R. Horn, and J. Spreeman. 1984. Competitive anxiety in junior elite wrestlers. *Journal of Sport Psychology* 5:58-71.

Graves, J., M. Pollock, D. Carpenter, S. Leggett, A. Jones, M. MacMillan, and M. Fulton. 1990. Quantitative assessment of full range-of-motion isometric lumbar extension strength. *Spine* 15:289-294.

Green, K.N., W.B. East, and L.D. Hensley. 1987. A golf skills test battery for college males and females. *Research Quarterly for Exercise and Sport* 58:72-76.

Greenspan, M.J., and D.L. Feltz. 1989. Psychological interventions with athletes in competitive situations: A review. *Sport Psychologist* 3:219-236.

Grissom, J.B. 2005. Physical fitness and academic achievement. *Journal of Exercise Physiology Online* 8(1):11-25.

Gronlund, N.E. 1993. *Assessment of student achievement.* 6th ed. Boston: Allyn and Bacon.

Grove, J.R. 2001. Practical screening tests for talent identification in baseball. *Applied Research in Coaching and Athletics Annual* 16:63-77.

Gucciardi, D., S. Gordon, and J. Dimmock. 2008. Toward an understanding of mental toughness in Australian football. *Journal of Applied Sport Psychology* 20:261-281.

Guskey, T., and J. Bailey. 2001. *Developing grading and reporting systems for student learning.* Thousand Oaks, CA: Corwin Press.

Hagberg, J.M., J.E. Graves, M. Limacher, D.R. Woods, S.H. Leggett, C. Cononie, J. Gruber, and M.L. Pollock. 1989. Cardiovascular responses of 70-79 year old men and women to exercise training. *Journal of Applied Physiology* 66:2589-2594.

Hall, C., D. Mack, A. Paivio, and H.A. Hausenblas. 1998. Imagery use by athletes: Development of the Sport Imagery Questionnaire. *International Journal of Sport Psychology* 29:73-89.

Hall, G., R. Hetzler, D. Perrin, and A. Weltman. 1992. Relationship of timed sit-up tests to isokinetic abdominal strength. *Research Quarterly for Exercise and Sport* 63:80-84.

Hanrahan, S.J., and S.H. Biddle. 2002. Measurement of achievement orientations: Psychometric measures, gender, and sport differences. *European Journal of Sport Science* 2(5):1-12.

Harre, D. 1982. *The principles of sports training: Introduction to the theory and methods of training.* 2nd ed. Berlin: Sportverlag.

Harris, C.D., K.B. Watson, S.A. Carlson, J.E. Fulton, and J.M. Dorn. May 3, 2013. Adult participation in aerobic and muscle-strengthening physical activities – United States, 2011. *MMWR* 62(17):326-330.

Harrison, P.W., and P.A. Bradbeer. 1982. Battery tests for the assessment of physiological fitness norms. *Athletics Coach* 16(21):6-12.

Harrow, A.J. 1972. *A taxonomy of the psychomotor domain.* New York: McKay.

Hatano, Y. 1993. Use of the pedometer for promoting daily walking exercise. *International Council for Health, Physical Education, and Recreation* 29:4-8.

Henry, F.M. 1956. Coordination and motor learning. In *59th Proceedings of the Annual College Physical Education Association,* 68-75. Washington, DC.

———. 1958. Specificity vs. generality in learning motor skills. In *61st Annual Proceedings of the College Physical Education Association,* 126-128. Washington, DC.

Hensley, L.D., ed. 1989. *Tennis for boys and girls: Skills test manual.* Reston, VA: AAHPERD.

Hensley, L.D. 1997. Alternative assessment for physical education. *Journal of Physical Education, Recreation & Dance* 68(7):19-24. doi:10.1080/07303084.1997.10604978

Hensley, L.D., and W.B. East. 1989. Testing and grading in the psychomotor domain. In *Measurement concepts in physical education and exercise science,* ed. M.J. Safrit and T.M. Wood, 297-321. Champaign, IL: Human Kinetics.

Hensley, L.D., W.B. East, and J.L. Stillwell. 1979. A racquetball skills test. *Research Quarterly* 50:114-118.

Herman, J.L., P.R. Aschbacher, and L. Winters. 1992. *A practical guide to alternative assessment.* Alexandria, VA: Association for Supervision and Curriculum Development.

Hillman, C.H., K.I. Erickson, and A.F. Kramer. 2008. Be smart, exercise your heart: Exercise effects on brain and cognition. *Nature Reviews Neuroscience* 9:58-65.

Hillman, C.H., M.B. Pontifex, L.B. Raine, D.M. Castelli, E.E. Hall, and A.F. Kramer. 2009. The effect of acute treadmill walking on cognitive control and academic achievement in preadolescent children. *Neuroscience* 159(3):1044-1054. doi:10.1016/j.neuroscience.2009.01.057

Hodgdon, J.A., and A.S. Jackson. 2000. Physical test validation for job selection. In *The process of physical fitness standards development,* ed. S.P. Constable, 139-177. Wright-Patterson Air Force Base, OH: Human Systems Information Analysis Center.

Hoffman, J. 2006. *Norms for fitness, performance, and health.* Champaign, IL: Human Kinetics.

Holbrook, E.A., T.V. Barreira, and M. Kang. 2009. Validity and reliability of Omron pedometers for prescribed and self-paced walking. *Medicine and Science in Sports and Exercise* 41(3):669-673.

Holt, N., and A. Sparkes. 2001. An ethnographic study of cohesiveness in a college soccer team over a season. *Sport Psychologist* 15:237-259.

Hopkins, D.R., J. Schick, and J.J. Plack. 1984. *Basketball for boys and girls: Skills test manual.* Reston, VA: AAHPERD.

Hoskins, R.N. 1934. The relationship of measurements of general motor capacity to the learning of specific psychomotor skills. *Research Quarterly,* 63-72.

International Association of Fire Fighters. 2007. *Candidate physical ability test.* 2nd ed. Washington, DC: International Association of Fire Fighters.

Isaac, A., D.F. Marks, and D.G. Russell. 1986. An instrument for assessing imagery of movement:

The Vividness of Movement Imagery Questionnaire (VMIQ). *Journal of Mental Imagery* 10(4):23-30.

Jackson, A.S. 2006a. The evolution and validity of health-related fitness. *Quest* 58(1):160-175. doi:10.1080/00336297.2006.10491877

Jackson, A.S. 2006b. Preemployment physical testing. In *Measurement theory and practice in kinesiology*, ed. T. Wood and W. Zhu, 315-345. Champaign, IL: Human Kinetics.

Jackson, A.S., S.N. Blair, M.T. Mahar, L.T. Wier, R.M. Ross, and J.E. Stuteville. 1990. Prediction of functional aerobic capacity without exercise testing. *Medicine & Science in Sports & Exercise* 22:863-870.

Jackson, A.S., and M. Pollock. 1978. Generalized equations for predicting body density of men. *British Journal of Nutrition* 40:497-504.

Jackson, A.S., M. Pollock, and A. Ward. 1980. Generalized equations for predicting body density of women. *Medicine & Science in Sports & Exercise* 12:175-182.

Jackson, A.W., and A. Baker. 1986. The relationship of the sit and reach test to criterion measures of hamstring and back flexibility in young females. *Research Quarterly for Exercise and Sport* 57:183-186.

Jackson, A.W., A.S. Jackson, and J. Bell. 1980. A comparison of alpha and the intraclass reliability coefficients. *Research Quarterly for Exercise and Sport* 51:568-571.

Jackson, A.W., and N. Langford. 1989. The criterion-related validity of the sit and reach test: Replication and extension of previous findings. *Research Quarterly for Exercise and Sport* 60:384-387.

Jackson, A.W., D.C. Lee, X. Sui, J.R. Morrow Jr., T.S. Church, A.L. Maslow, and S.N. Blair. 2010. Muscular strength is related to prevalence and incidence of obesity in adult men. *Obesity* 18:1988-1995.

Jackson, A.W., J.R. Morrow Jr., H.R. Bowles, S.J. FitzGerald, and S.N. Blair. 2007. Construct validity evidence for single-response items to estimate physical activity levels in large sample studies. *Research Quarterly for Exercise and Sport* 78:24-31.

Jackson, A.W., J.R. Morrow Jr., P.A. Brill, H.W. Kohl III, N.F. Gordon, and S.N. Blair. 1998. Relations of sit-up and sit-and-reach tests to low back pain in adults. *Journal of Orthopaedic & Sports Physical Therapy* 27(1):22-26.

Jackson, A.W., J. Solomon, and M. Stusek. 1992. One-mile walk test: Reliability, validity, and norms for young adults [Abstract]. *Research Quarterly for Exercise and Sport* 63:A52.

Jackson, A.W., M. Watkins, and R. Patton. 1980. A factor analysis of twelve selected maximal isotonic strength performances on the universal gym. *Medicine & Science in Sports & Exercise* 12:274-277.

Jacobs, D.R, B.E Ainsworth, T.J. Hartman, and A.S. Leon. 1993. A simultaneous evaluation of 10 commonly used physical activity questionnaires. *Medicine & Science in Sports & Exercise* 25:81-91.

Janz, K., E. Letuchy, T. Burns, S. Francis, and S. Levy. 2015. Muscle power predicts adolescent bone strength: Iowa bone development study. *Medicine & Science in Sports & Exercise* 47(10):2201-2206.

Jensen, C., and C. Hirst. 1980. *Measurement in physical education and athletics*. New York: Macmillan.

Jobs, S. 1990. Interview for "The Machine That Changed the World" series. *The Paperback Computer* 102. GBH archives, http://openvault.wgbh.org/catalog/V_AD9E0BC353B-F435E83F28DEF165D4F40

Johnson, B.L., and J. Leach. 1968. *A modification of the Bass test of dynamic balance*. Commerce, TX: East Texas State University.

Johnson, B., and J. Nelson. 1979. *Practical measurements for evaluation in physical education*. 3rd ed. Minneapolis: Burgess.

———. 1986. *Practical measurements for evaluation in physical education*. 4th ed. New York: Macmillan.

Jones, G.G., S.S. Hanton, and D.D. Connaughton. 2002. What is this thing called mental toughness? An investigation of elite sports performers. *Journal of Applied Sport Psychology* 14:205-218.

Jurca, R., A.S. Jackson, M.L. LaMonte, J.R. Morrow Jr., S.N. Blair, N.J. Wareham, W.L. Haskell, M.W. van Mechelen, T.S. Church, J.M. Jakicic, and R. Laukkanen. 2005. Assessing cardiorespiratory fitness without performing exercise testing. *American Journal of Preventive Medicine* 29:185-193.

Kalamen, J. 1968. *Measurement of maximum muscular power in man*. Doctoral thesis, Ohio State University.

Kane, M.T. 1992. An argument-based approach to validity. *Psychological Bulletin* 112(3):527-535. doi:10.1037/0033-2909.112.3.527

Kang, M. 2004. *An empirical investigation of characteristics of children's physical activity recall*. Unpublished Dissertation, University of Illinois at Urbana-Champaign, Urbana.

Kang, M., and Y. Jin. 2016a. Factorial ANOVA/MANOVA. In *An introduction to advanced statistical analyses for sport and exercise scientists*, ed. N. Ntoumanis and N. Myers, 26-40. New York: John Wiley & Sons, Inc.

———. 2016b. Repeated measures ANOVA/MANOVA. In *An introduction to advanced statistical analyses for sport and exercise scientists*, ed. N. Ntoumanis and N. Myers, 41-53. New York: John Wiley & Sons, Inc.

Kang, M., M.T. Mahar, and J.R. Morrow. 2016. Issues in the assessment of physical activity in children. *Journal of Physical Education, Recreation and Dance* 87(6):35-43.

Kang, M., and D.A. Rowe. 2015. Issues and challenges in sedentary behavior measurement. *Measurement in Physical Education and Exercise Science* 19(3):105-115.

Karasu, A. U., Batur, E. B., & Karatas, G. K. 2018. Effectiveness of Wii-based rehabilitation in stroke: A randomized controlled study. *Journal of rehabilitation medicine*, 50(5), 406–412. https://doi.org/10.2340/16501977-2331

Kelly, G.A. 1955. *The psychology of personal constructs*. New York: Norton.

Kenyon, G.S. 1968. Six scales for assessing attitude toward physical activity. *Research Quarterly* 39:566-574.

Khalil, S.F., M.S. Mohktar, and F. Ibrahim. 2014. The theory and fundamentals of bioimpedance analysis in clinical status monitoring and diagnosis of diseases. *Sensors* 14(6):10895. doi:10.3390/s140610895

Kim, Y., I. Park, and M. Kang. 2013. Convergent validity of International Physical Activity Questionnaire (IPAQ): Meta-analysis. *Public Health Nutrition* 16(3):440-452.

Kirby, R.F. 1971. A simple test of agility. *Coach and Athlete* 25(6):30-31.

Kirby, R.F., ed. 1991. *Kirby's guide to fitness and motor performance tests*. Cape Girardeau, MO: Ben Oak.

Kirk, M.F. 1997. Using portfolios to enhance student learning and assessment. *Journal of Physical Education, Recreation and Dance* 68(7):29-33.

Kline, G., J. Porcari, R. Hintermeister, P. Freedson, A. Ward, R. McCarron, J. Ross, and J. Rippe. 1987. Estimation of $\dot{V}O_2$max from a one-mile track walk, gender, age, and body weight. *Medicine & Science in Sports & Exercise* 19:253-259.

Koehler, K., and C. Drenowatz. 2017 Nov. Monitoring energy expenditure using a multi-sensor device—applications and limitations of the SenseWear armband in athletic populations. *Frontiers in Physiology* 30(8):983. doi:10.3389/fphys.2017.00983

Kollock, R.O., M. Lyons, G. Sanders, and D. Hale. 2019. The effectiveness of the functional movement screen in determining injury risk in tactical occupations. *Industrial Health* 57(4): 406-418. doi:10.2486/indhealth.2018-0086

Kramer, A.F., Hahn, S., Cohen, N.J., Banich, M.T., McAuley, E., Harrison, C.R., Chason, J., Vakil, E., Bardell, L., Boileau, R.A., and A. Colcombe. 1999. Ageing, fitness and neurocognitive function. *Nature* 400(6743):418-419. doi:10.1038/22682

Krathwohl, D.R., B.S. Bloom, and B.A. Masia. 1964. *Taxonomy of educational objectives: Handbook II: The affective domain*. New York: McKay.

Kraus, H., and R.P. Hirschland. 1953. Muscular fitness and health. *Journal of Physical Education and Recreation* 24(10):17-19.

———. 1954. Minimum muscular fitness tests in school children. *Research Quarterly* 25:178-188.

Lamb, S.E., D. Mistry, S. Alleyne, et al. 2018. Aerobic and strength training exercise programme for cognitive impairment in people with mild to moderate dementia: The DAPA RCT. *Health Technology Assessment* 22(28):1-202.

LaMonte, M.J., and B.E. Ainsworth. 2001. Quantifying energy expenditure and physical activity in the context of dose response. *Medicine & Science in Sports & Exercise* 33:S370-S378.

LaPorte, R., H. Montoye, and C. Caspersen. 1985. Assessment of physical activity in epidemiologic research: Problems and prospects. *Public Health Reports* 100:131-146.

Larson, L.A. 1941. A factor analysis of motor ability variables and tests, with test for college men. *Research Quarterly* 12:499-517.

Last, J. 1992. *Dictionary of epidemiology*. 2nd ed. New York: Oxford University.

LeBlanc, A.G.W., and I. Janssen. 2010. Difference between self-reported and accelerometer measured moderate-to-vigorous physical activity in youth. *Pediatric Exercise Science* 22:523-534.

Lee, M., W. Zhu, B. Hedrick, and B. Fernhall. 2010 May. Estimating MET values using the ratio of HR for persons with paraplegia. *Medicine & Science in Sports & Exercise* 42(5):985-990. doi:10.1249/MSS.0b013e3181c0652b

Léger, L.A., D. Mercier, C. Gadoury, and J. Lambert. 1988. The multistage 20 metre shuttle run test for aerobic fitness. *Journal of Sports Sciences* 6(2):93-101.

Leone, M., G. Lariviere, and A.S. Comtois. 2002. Discriminant analysis of anthropometric and biomotor variables among elite adolescent female athletes in four sports. *Journal of Sport Sciences* 20:443-449.

Lewis, M. 2004. *Moneyball: The art of winning an unfair game*. New York: W.W. Norton.

Li, F. 1999. The Exercise Motivation Scale: Its multifaceted structure and construct validity. *Journal of Applied Sport Psychology* 11:97-115.

Li, H., Z. Chen, and W. Zhu. 2019. Variability: Human nature and its impact on measurement and statistical analysis. *Journal of Sport and Health Science* 8(6):527-531. doi:10.1016/j.jshs.2019.06.002

Lindquist, E. F. 1942. *A first course in statistics: Their use and interpretation in education and psychology*. Cambridge, MA: Riverside Press.

Livingston, S.A., and M. Zieky. 1982. *Passing scores: A manual for setting standards of performance on educational and occupational tests*. Princeton, NJ: Educational Testing Service.

Lohman, T. 1989. Assessment of body composition in children. *Pediatric Exercise Science* 1:19-30.

Looney, M.A. 2003. Facilitate learning with a definitional grading system. *Measurement in Physical Education and Exercise Science* 7:269-275.

Lund, J., and M. Kirk. 2010. *Performance-based assessment for middle and high school physical education*. Champaign, IL: Human Kinetics.

————. 2020. *Performance-based assessment for middle and high school physical education.* 3rd ed. Champaign, IL: Human Kinetics.

Lund, J., and M.L. Veal. 2013. *Assessment-driven instruction in physical education: A standards-based approach to promoting and documenting learning.* Champaign, IL: Human Kinetics.

MacDougall, J.D., and H.A. Wenger. 1991. The purpose of physiological testing. In *Physiological testing of the high performance athlete,* 2nd ed., ed. J.D. MacDougall, H.A. Wenger, and H.J. Green, 1-5. Champaign, IL: Human Kinetics.

Madaus, G.F., and L.M. O'Dwyer. 1999. A short history of performance assessment: Lessons learned. *Phi Delta Kappan* 80(9):688-695.

Mahar, M.T., and D.A. Rowe. 2014. A brief exploration of measurement and evaluation in kinesiology. *Kinesiology Review* 3:80-91.

Mahar, M.T., D.A. Rowe, C.R. Parker, F.J. Mahar, D.M. Dawson, and J.E. Holt. 1997. Criterion-referenced and norm-referenced agreement between the mile run/walk and PACER. *Measurement in Physical Education and Exercise Science* 1:245-258.

Mahoney, M., T. Gabriel, and T. Perkins. 1987. Psychological skills and exceptional athletic performance. *Sport Psychologist* 1:181-199.

Mandolesi, L., A. Polverino, S. Montuori, F. Foti, G. Ferraioli, P. Sorrentino, and G. Sorrentino. 2018. Effects of physical exercise on cognitive functioning and wellbeing: Biological and psychological benefits. *Frontiers in Psychology* 9:509. doi:10.3389/fpsyg.2018.00509

Marcus, B.H., S.W. Banspach, R.C. Lefebvre, J.S. Rossi, R.A. Carleton, and D.B. Abrams. 1994. Using the stages of change model to increase the adoption of physical activity among community participants. *American Journal of Health Promotion* 6:424-429.

Marcus, B.H., V.C. Selby, R.S. Niaura, and J.S. Rossi. 1992. Self-efficacy and the stages of exercise behavior change. *Research Quarterly for Exercise and Sport* 63(1):60-66.

Margaria, R., P. Aghemo, and E. Rovelli. 1966. Measurement of muscular power (anaerobic) in man. *Journal of Applied Physiology* 221: 1662-1664.

Markland, D., and L. Hardy. 1993. The Exercise Motivation Inventory: Preliminary development and validity of a measure of individuals' reasons for participation in physical activity. *Personality and Individual Differences* 15:289-296.

Martens, R. 1977. *Sport competition anxiety test.* Champaign, IL: Human Kinetics.

Martens, R., R. Vealey, and D. Burton. 1990. *Competitive anxiety in sport.* Champaign, IL: Human Kinetics.

Marzano, R.J., D. Pickering, and J. McTighe. 1993. *Assessing student outcomes: Performance assessment using the dimensions of learning model.* Alexandria, VA: Association for Curriculum and Development.

Matanin, M., and D. Tannchill. 1994. Assessment and grading in physical education. *Journal of Teaching in Physical Education* 13:395-405.

Matthews, C.E., K.Y. Chen, P.S. Freedson, M.S. Suchowski, B.M. Beech, R.R. Pate, and R.P. Troiano. 2008. Amount of time spent in sedentary behaviors in the United States, 2003-2004. *American Journal of Epidemiology* 167(7):875-881.

Matthews, C.E., P.S. Freedson, J.R. Hebert, E.J. Stanek, P.A. Merriam, and I.S. Ockene. 2000. Comparing physical activity assessment methods in the Seasonal Variation of Blood Cholesterol Study. *Medicine & Science in Sports & Exercise* 32:976-984.

Mayhew, T., and J. Rothstein. 1985. Measurement of muscle performance with instruments. In *Measurement in physical therapy,* ed. J. Rothstein, 57-102. New York: Churchill Livingstone.

McAuley, E., A. Szabo, N. Gothe, and E.A. Olson. 2011. Self-efficacy: Implications for physical activity, function, and functional limitations in older adults. *American Journal of Lifestyle Medicine* 5(4):361-369. doi:10.1177/1559827610392704

McCloy, C.H. 1932. *The measurement of athletic power.* New York: Barnes.

McKenzie, T.L. 2002. Use of direct observation to assess physical activity. In *Physical activity assessments for health-related research,* ed. G.J. Welk, 179-195. Champaign, IL: Human Kinetics.

McKenzie, T.L., D.A. Cohen, A. Sehgal, S. Williamson, and D. Golinelli. 2006. System for Observing Play and Leisure Activity in Communities (SOPARC): Reliability and feasibility measures. *Journal of Physical Activity & Health* 1:S203-217.

McKenzie, T.L, S.J. Marshall, J.F. Sallis, and T.L. Conway. 2000. Leisure-time physical activity in school environments: An observational study using SOPLAY. *Preventive Medicine* 30:70-77.

McKenzie, T.L., J.F. Sallis, and P.R. Nader. 1991. SOFIT: System for Observing Fitness Instruction Time. *Journal of Teaching in Physical Education* 11:195-205.

McKenzie, T.L., J.F. Sallis, T.L. Patterson, J.P. Elder, C.C. Berry, J.W. Rupp, C.J. Atkins, M.J. Buono, and P.R. Nader. 1991. BEACHES: An observational system for assessing children's eating and physical activity behaviors and associated events. *Journal of Applied Behavior Analysis* 24:141-151.

McKenzie, T.L., and H. van der Mars. 2015. Top 10 research questions related to assessing physical activity and its contexts using systematic observation. *Research Quarterly for Exercise and Sport* 86(1):13-29. doi:10.1080/02701367.2015.991264

McNair, D.M., M. Lorr, and L.F. Droppleman. 1971. *EDITS manual for POMS.* San Diego: Educational and Industrial Testing Service.

Melograno, V.J. 2000. *Portfolio assessment for K-12 physical education.* Reston, VA: National Association for Sport and Physical Education Publications.

Merlo, C.L., S.E. Jones, S.L. Michael, T.J. Chen, S.A. Sliwa, S.H. Lee, N.D. Brener, S.M. Lee, and S. Park. 2020. Dietary and physical activity behaviors among high school students—youth risk behavior survey, United States, 2019. *MMWR Supplements* 69(1):64-76. doi:10.15585/mmwr.su6901a8

Messick, S. 1989. Validity. In *Educational measurement*, 3rd ed., ed. R. L. Linn, 13-103. New York: American Council on Education/Macmillan.

———. 1995. Standards of validity and the validity of standards in performance assessment. *Educational Measurement: Issues and Practice* 14(1):5-8.

———. 1996a. Standards-based score interpretation: Establishing valid grounds for valid inferences. In *Proceedings of the Joint Conference on Standard Setting for Large Scale Assessments, Sponsored by National Assessment Governing Board and The National Center for Education Statistics.* Washington, DC.

———. 1996b. Validity of performance assessment. In *Technical issues in large-scale performance assessment*, ed. G. Philips. Washington, DC: National Center for Educational Statistics.

Miller, M.D., and S.M. Legg. 1993. Alternative assessment in a high-stakes environment. *Educational Measurement: Issues and Practice* 12(3):9-15.

Miller, P. 1985. Assessment of joint motion. In *Measurement in physical therapy*, ed. J. Rothstein, 103-136. New York: Churchill Livingstone.

Mintah, J.K. 2003. Authentic assessment in physical education: Prevalence of use and perceived impact on students' self-concept, motivation, and skill achievement. *Measurement in Physical Education and Exercise Science* 7:161-174.

Mohr, D.R., and M.J. Haverstick. 1956. Relationship between height, jumping ability, and agility to volleyball skill. *Research Quarterly* 27(1), 74-78. doi:10.1080/10671188.1956.10612853

Montoye, H.J., H.C. Kemper, W.H.M. Saris, and R.A. Washburn. 1996. *Measuring physical activity and energy expenditure.* Champaign, IL: Human Kinetics.

Mood, D. P., J. R. Morrow, and M. B. McQueen. 2019. *Introduction to statistics in human performance: Using SPSS and R.* United Kingdom: Routledge, Taylor & Francis Group.

———. 2020. *Introduction to statistics in human performance: Using SPSS and R.* 2nd ed. New York: Routledge.

Morgan, W.P. 1980. Test of champions: The iceberg profile. *Psychology Today* (July):92-93, 97-99, 102, 108.

Morris, J.N., J.A. Heady, P.A. Raffle, C.G. Roberts, and J.W. Parks. 1953 Nov 21. Coronary heart-disease and physical activity of work. *Lancet* 262(6795):1053-1057. doi:10.1016/s0140-6736(53)90665-5

Morrow, J.R. Jr., T. Fridye, and S. Monaghen. 1986. Generalizability of the AAHPERD health-related skinfold test. *Research Quarterly for Exercise and Sport* 57:187-195.

Morrow, J.R. Jr., S.B. Going, and G.J. Welk, eds. 2011. FitnessGram development of criterion-referenced standards for aerobic capacity and body composition. *American Journal of Preventive Medicine* 41(4, Suppl. 2):S63-S144.

Morrow, J.R. Jr., A. Jackson, P. Bradley, and H. Hartung. 1986. Accuracy of measured and predicted residual lung volume on body density measurement. *Medicine & Science in Sports & Exercise* 18:647-652.

Morrow, J.R. Jr., S.B. Martin, and A.W. Jackson. 2010. Reliability and validity of the FitnessGram: Quality of teacher-collected health-related fitness surveillance data. *Research Quarterly for Exercise and Sport* 81(Suppl.):S24-S30.

Morrow, J.R. Jr., S.B. Martin, G.J. Welk, W. Zhu, and M.D. Meredith. 2010. Overview of the Texas Youth Fitness Study. *Research Quarterly for Exercise and Sport* 81(Suppl 3), S1-S5.

Morrow, J.R. Jr., W. Zhu, B.D. Franks, M.D. Meredith, and C. Spain. 2009. 1958-2008: 50 years of youth fitness tests in the United States. *Research Quarterly for Exercise and Sport* 80:1-11.

Morrow, J.R. Jr., W. Zhu, and M.T. Mahar. 2013. Physical fitness standards for children. In *FitnessGram/ActivityGram reference guide*, 4th ed. (Internet Resource), ed. S.A. Plowman and M.D. Meredith, 2-12. Dallas: Cooper Institute.

Mullan, E., D. Markland, and D.K. Ingledew. 1997. A graded conceptualisation of self-determination in the regulation of exercise behaviour: Development of a measure using confirmatory factor analytic procedures. *Personality and Individual Differences* 23:745-752.

Murphy, S.M., and D.P. Jowdy. 1992. *Imagery and mental practice.* Champaign, IL: Human Kinetics.

Murray, T.D., J.L. Walker, A.S. Jackson, J.R. Morrow Jr., J.A. Eldridge, and D.L. Rainey. 1993. Validation of a 20-minute steady-state jog as an estimate of peak oxygen uptake in adolescents. *Research Quarterly for Exercise and Sport* 64:75-82.

Muzik, O., K.T. Reilly, and V.A. Diwadkar. 2018. "Brain over body": A study on the willful regulation of autonomic function during cold exposure. *Neuroimage* 172:632-641. doi:10.1016/j.neuroimage.2018.01.067

National Association for Sport and Physical Education (NASPE). 2004. *Moving into the future: National standards for physical education.* 2nd ed. Reston, VA: NASPE.

———. 2008. *PE metrics: Assessing the national standards.* Reston, VA: NASPE.

Neilson, N.P., and F.W. Cozens. 1934. *Achievement*

scales in physical education activities for boys and girls in elementary and junior high schools. New York: Barnes.

Nelson, J.K., S.H. Yoon, and K.R. Nelson. 1991. A field test for upper body strength and endurance. *Research Quarterly for Exercise and Sport* 62:436-441.

Nelson, L.R., and M.L. Furst. 1972. An objective study of the effects of expectation on competitive performance. *Journal of Psychology* 81:69-72.

Nideffer, R.M. 1976. Test of attentional and interpersonal style. *Journal of Personality and Social Psychology* 34:394-404.

Nieman, D.C. 1995. *Fitness and sports medicine: A health-related approach.* 3rd ed. Mountain View, CA: Mayfield.

NIH consensus statement. Bioelectrical impedance analysis in body composition measurement. National Institutes of Health Technology Assessment Conference Statement. December 12-14, 1994. 1996 Nov-Dec. *Nutrition* 12(11-12):749-762. PMID: 8974099.

Nitko, A.J. 1984. Defining "criterion-referenced test." In *A guide to criterion-referenced test construction,* ed. R.A. Berk, 8-28. Baltimore: Johns Hopkins University Press.

Odom, L.R., and J.R. Morrow Jr. 2006. What is this r? A correlational approach to explaining validity, reliability, and objectivity coefficients. *Measurement in Physical Education and Exercise Science* 10:137-145.

Osness, W.H., M. Adrian, B. Clark, W. Hoeger, D. Raab, and R. Wisnell. 1990. *Functional fitness assessment for adults over 60 years (A field based assessment).* Reston, VA: AAHPERD.

Ostrow, A.C., ed. 2002. *Directory of psychological tests in the sport and exercise sciences.* 2nd ed. Morgantown, WV: Fitness Information Technology.

Paluch, A.E., K. Pettee Gabriel, J.E. Fulton, C.E. Lewis, P.J. Schreiner, B. Sternfeld, S. Sidney, J. Siddique, K.M. Whitaker, and M.R. Carnethon. 2021. Steps per day and all-cause mortality in middle-aged adults in the coronary artery risk development in young adults study. *JAMA Network Open* 4(9):e2124516. doi:10.1001/jamanetworkopen.2021.24516

Park, J.H., and M. Kang. 2014. Evaluation in sports performance. In *Social networks and the economics of sports,* ed. P.M. Pardalos and V. Zamaraev, 75-87. Cham, Switzerland: Springer International.

Pate, R. 1988. The evolving definition of physical fitness. *Quest* 40:174-179.

Pate, R., J. Ross, C. Dotson, and G. Gilbert. 1985. The new norms: A comparison with the 1980 AAHPERD norms. *Journal of Physical Education, Recreation and Dance* 56(1):70-72.

Patton, M. 1990. *Qualitative evaluation and research methods.* 2nd ed. Newbury Park, CA: Sage.

Pettee Gabriel, K.K, J.R. Morrow Jr., and A.L. Woolsey. 2012. Framework for physical activity as a complex and multidimensional behavior. *Journal of Physical Activity & Health* 9(Suppl. 1): S11-S18.

Plasqui, G., and K.R. Weserterp. 2007. Physical activity assessment with accelerometers: An evaluation against doubly labeled water. *Obesity* 15(10):2371-2379.

Playground and Recreation Association of America. 1913. *Athletic badge test for boys.* New York: Playground and Recreation Association of America.

———. 1916. *Athletic badge test for girls.* New York: Playground and Recreation Association of America.

Plowman, S.A. 1992. Criterion referenced standards for neuromuscular physical fitness tests. *Pediatric Exercise Science* 4:10-19.

Plowman, S.A., and M.D. Meredith, eds. 2013. *FitnessGram/ActivityGram reference guide.* 4th ed. Dallas: Cooper Institute.

Pollock, M.L., R.L. Bohannon, K.H. Cooper, J.J. Ayres, A. Ward, S.R. White, and A.C. Linnerud. 1976. A comparative analysis of four protocols for maximal treadmill stress testing. *American Heart Journal* 92(1):39-46.

Popham, W.J. 1978. *Criterion-referenced measurement.* Englewood Cliffs, NJ: Prentice-Hall.

———. 2003. *Test better, teach better.* Alexandria, VA: Association for Supervision and Curriculum Development.

President's Council on Physical Fitness and Sports (PCPFS). 1999. *The Presidential Physical Fitness Award program.* Washington, DC: PCPFS.

———. 1999. *President's challenge.* Washington DC: PCPFS.

Prochaska, J.O., and C.C. DiClemente. 1983. Stages and processes of self-change in smoking: Toward an integrative model of change. *Journal of Consulting and Clinical Psychology* 51:390-395.

Raghuveer, G., J. Hartz, D.R. Lubans, T. Takken, J.L. Wiltz, M. Mietus-Snyder, A.M. Perak, C. Baker-Smith, N. Pietris, and N.M. Edwards. 2020. Cardiorespiratory fitness in youth: An important marker of health: A scientific statement from the American Heart Association. *Circulation* 142:e101-e118. doi:10.1161/CIR.0000000000000866

Reid, K.F., and R.A. Fielding. 2012. Skeletal muscle power: A critical determinant of physical functioning in older adults. *Exercise and Sport Sciences Reviews* 40(1):4-12. doi:10.1097/JES.0b013e31823b5f13

Reiff, G., W. Dixon, D. Jacoby, G. Ye, C. Spain, and P. Hunsicker. 1986. *The President's Council on Physical Fitness and Sports 1985 National School Population Fitness Survey.* Washington, DC: President's Council on Physical Fitness and Sports.

Rennie, K.L., T. Rowsell, S.A. Jebb, D. Holburn, and N.J. Wareham. 2000. A combined heart rate and movement sensor: Proof of concept and preliminary testing study. *European Journal of Clinical Nutrition* 54(5):409-414.

Richardson, M.T., Ainsworth, B.E., Jacobs, D.R., and A. S. Leon. 2001. Validation of the Stanford 7-day recall to assess habitual physical activity. *Annals of epidemiology* 11(2):145-153. https://doi.org/10.1016/s1047-2797(00)00190-3

Richardson, M.T., A.S. Leon, D.R. Jacobs, B.E. Ainsworth, and R. Serfass. 1995. Ability of the Caltrac accelerometer to assess daily physical activity levels. *Journal of Cardiopulmonary Rehabilitation* 15:107-113.

Rikli, R.E. 1991. *Softball for boys and girls: Skills test manual*. Reston, VA: AAHPERD.

Rikli, R.E., and C.J. Jones. 1999a. Development and validation of a functional fitness test for community-residing older adults. *Journal of Aging and Physical Activity* 7:129-161.

———. 1999b. Functional fitness normative scores for community-residing older adults, ages 60-94. *Journal of Aging and Physical Activity* 7:162-181.

———. 2013a. Development and validation of criterion-referenced clinically relevant fitness standards for maintaining physical independence in later years. *Gerontologist* 53:255-267.

———. 2013b. *Senior fitness test manual*. 2nd ed. Champaign, IL: Human Kinetics.

Rikli, R.E, C. Petray, and T.A. Baumgartner. 1992. The reliability of distance run tests for children in grades K-4. *Research Quarterly for Exercise and Sport* 63:270-276.

Robertson, L., and H. Magnusdottir. 1987. Evaluation of criteria associated with abdominal fitness testing. *Research Quarterly for Exercise and Sport* 58:355-359.

Rodriguez, M.C. 2005. Three options are optimal for multiple-choice items: A meta-analysis of 80 years of research. *Educational Measurement* 24:3-13.

Ross, J., R. Pate, L. Delby, R. Gold, and M. Svilar. 1987. New health-related fitness norms. *Journal of Physical Education, Recreation and Dance* 58(9):66-70.

Ross, R., S.N. Blair, R. Arena, T.S. Church, J.-P. Després, B.A. Franklin, W.L. Haskell, L.A. Kaminsky, B.D. Levine, C.J. Lavie, J. Myers, J. Niebauer, R. Sallis, S.S. Sawada, X. Sui, and U. Wisløff. 2016. Importance of assessing cardiorespiratory fitness in clinical practice: A case for fitness as a clinical vital sign: A scientific statement from the American Heart Association. *Circulation* 134:e653-e699. doi:10.1161/CIR.0000000000000461.

Rotter, J.B. 1966. Generalized expectancies for internal versus external control of reinforcement. *Psychological Monographs* 80 (No. 609).

Rowley, A., D. Landers, B. Kyllo, and J. Etnier. 1995. Does the iceberg profile discriminate between successful and less successful athletes? A meta-analysis. *Journal of Sport & Exercise Psychology* 17:185-199.

Ruiz, J.R., X. Sui, F. Lobelo, J.R. Morrow Jr., A.W. Jackson, M. Sjostrom, and S.N. Blair. 2008. Association between muscular strength and mortality in men: Prospective cohort study. *British Medicine Journal* 337:a439. doi:10.1136/bmj.a439

Safrit, M. 1986. *Introduction to measurement in physical education and exercise science*. St. Louis: Mosby.

Safrit, M., L. Hooper, S. Ehlert, M. Costa, and P. Patterson. 1988. The validity generalization of distance run tests. *Canadian Journal of Sport Sciences* 13:188-196.

Safrit, M., and M. Looney. 1992. Should the punishment fit the crime? A measurement dilemma. *Research Quarterly for Exercise and Sport* 63:124-127.

Safrit, M., and C. Pemberton. 1995. *Complete guide to youth fitness testing*. Champaign, IL: Human Kinetics.

Safrit, M.J., T.A. Baumgartner, A.S. Jackson, and C.L. Stamm. 1980. Issues in setting motor performance standards. *Quest* 32:152-162.

Salkind, N.J. 2015. *Excel statistics: A quick guide*. 3rd ed. Washington, DC: Sage.

Sallis, J.F., M.J. Buono, J.J. Roby, F.G. Micale, and J.A. Nelson. 1993. Seven-day recall and other physical activity self-reports in children and adolescents. *Medicine & Science in Sports & Exercise* 25:99-108.

Sallis, J.F., and B.E. Saelens. 2000. Assessment of physical activity by self-report: Status, limitations, and future directions. *Research Quarterly for Exercise and Sport* 71(2):1-14.

Santiago, J.A., and J.R. Morrow, Jr. 2020. A study of preservice physical education teachers' content knowledge of health-related fitness. *Journal of Teaching Physical Education* 40(1):1-8.

Sarason, I.G. 1975. Test anxiety and the self-disclosing coping model. *Journal of Consulting and Clinical Psychology* 43:148-153.

Sardinha, L.B., and P.B. Júdice. 2017. Usefulness of motion sensors to estimate energy expenditure in children and adults: A narrative review of studies using DLW. *European Journal of Clinical Nutrition* 71:331-339.

Sargent, D.A. 1921. The physical test of man. *American Physical Education Review* 26(April):188-194.

Sax, G. 1980. *Principles of educational and psychology al measurement and evaluation*. Belmont, CA: Wadsworth.

Schick, J., and N.G. Berg. 1983. Indoor golf skill test for junior high boys. *Research Quarterly for Exercise and Sport* 54:75-78.

Schulz, S., K.R. Westerterp, and K. Brück. 1989. Comparison of energy expenditure by the doubly labeled water technique with energy intake, heart rate, and activity recording in man. *The American Journal of Clinical Nutrition* 49(6):1146-1154. doi:10.1093/ajcn/49.6.1146

Seaman, J., and K. DePauw. 1989. *The new adapted physical education: A developmental approach.* Mountain View, CA: Mayfield.

Seaward, B.L., ed. 2021. Physical exercise: Flushing out the stress hormones. In *Essentials of managing stress,* 5th ed., 317-325. Burlington, MA: Jones & Bartlett.

SHAPE America. 2014. *National standards & grade-level outcomes for K-12 physical education.* Champaign, IL: Human Kinetics.

SHAPE America. 2019. *PE Metrics: Assessing student performance using the National Standards & Grade-Level Outcomes for K-12 Physical Education,* 3rd ed. Champaign, IL: Human Kinetics.

Shephard, R. 1990. *Fitness in special populations.* Champaign, IL: Human Kinetics.

Shifflett, B., and B.J. Shuman. 1982. A criterion-referenced test for archery. *Research Quarterly for Exercise and Sport* 53:330-335.

Sibley, B.A., and J.L. Etnier. 2003. The relationship between physical activity and cognition in children: A meta-analysis. *Pediatric Exercise Science* 15:243-256.

Siedentop, D. 1996. Physical education and education reform: The case for sport education. In *Student learning in physical education: Applying research to enhance instruction,* ed. S.J. Silverman and C.D. Ennis, 247-267. Champaign, IL: Human Kinetics.

Sierer, S., C. Battaglini, J. Mihalik, E. Shields, and N. Tomasini. 2008. The National Football League combine: Performance differences between drafted and nondrafted players entering the 2004 and 2005 drafts. *Journal of Strength and Conditioning Research* 22:6-12.

Singer, R.N. 1968. *Motor learning and human performance.* New York: Macmillan.

Smith, F.C. 1915. *Exercise and health.* Washington, DC: Government Printing Office.

Smith, R., R. Schutz, F. Smoll, and J. Ptacek. 1995. Development and validation of a multidimensional measure of sport-specific psychological skills: The Athletic Coping Skills Inventory-28. *Journal of Sport & Exercise Psychology* 17:379-398.

Smith, R., F.L. Smoll, and B. Curtis. 1979. Coach effectiveness training: A cognitive behavioral approach to enhancing relationship skills in youth and sport coaches. *Journal of Sport Psychology* 1:59-75.

Smith, R.E., F.L. Smoll, S.P. Cumming, and J.R. Grossbard. 2006. Measurement of multidimensional sport performance anxiety in children and adults: The Sport Anxiety Scale-2. *Journal of Sport & Exercise Psychology* 28:479-501.

Smith, R.E., F.L. Smoll, and R.W. Schutz. 1990. Measurement and correlates of sport-specific cognitive and somatic trait anxiety: The Sport Anxiety Scale. *Anxiety Research* 2:263-280.

Sorrentino, R.M., and B.H. Sheppard. 1978. Effects of affiliation-related motives on swimmers in individual versus group competition: A field experiment. *Journal of Personality and Social Psychology* 36:704-714.

Sparkes, A. 1998. Validity in qualitative inquiry and the problem of criteria: Implications for sport psychology. *Sport Psychologist* 12:333-345.

Spielberger, C.D., R.L. Gorsuch, and R.F. Lushene. 1970. *Manual for the State–Trait Anxiety Inventory.* Palo Alto, CA: Consulting Psychologists Press.

Stiggins, R. 1987. Design and development of performance assessment. *Educational Measurement* 6(3):33-42.

Stone, D.B., W.R. Armstrong, D.M. Macrina, and J.W. Pankau. 1996. *Introduction to epidemiology.* Madison, WI: Brown and Benchmark.

Straling, R.D. 2002. Use of doubly labeled water and indirect calorimetry to assess physical activity. In *Physical activity assessment for health-related research,* ed. G.J. Welk, 197-209. Champaign, IL: Human Kinetics.

Strand, B., and R. Wilson. 1993. *Assessing sports skills.* Champaign, IL: Human Kinetics.

Strean, W. 1998. Possibilities for qualitative research in sport psychology. *Sport Psychologist* 12:333-345.

Stuart, M. 2003. Sources of subjective task value in sport: An examination of adolescents with high or low value for sport. *Journal of Applied Sport Psychology* 15:239-255.

Sui, X., M.J. LaMonte, and S.N. Blair. 2007. Cardiorespiratory fitness as a predictor of nonfatal cardiovascular events in asymptomatic women and men. *American Journal of Epidemiology* 165:1413-1423.

Suzuki, N., and S. Endo. 1983. A quantitative study of trunk muscle strength and fatigability in the low-back pain syndrome. *Spine* 8:69-74.

Tarter, B.C., L. Kirisci, R.E. Tarter, S. Weatherbee, V. Jamnik, E.J. McGuire, and N. Gledhill. 2009. Use of aggregate fitness indicators to predict transition into the National Hockey League. *Journal of Strength and Conditioning Research* 23:1828-1832.

Taylor, M., D. Gould, and C. Roio. 2008. Performance strategies of US Olympians in practice and competition. *High Ability Studies* 19:19-36.

Telama, R., X. Yang, L. Laasko, and J. Viikari. 1997. Physical activity in childhood and adolescence as predictor of physical activity in young adulthood. *American Journal of Preventive Medicine* 13:317-323.

Tenenbaum, G., R.C. Eklund, and A. Kamata, eds. 2012. *Measurement in sport and exercise psychology.* Champaign, IL: Human Kinetics.

Terry, P. 2000. Perspectives on mood in sport and exercise. *Journal of Applied Sport Psychology* 12:1-4.

Thelwell, R., N. Weston, and I. Greenlees. 2005. Defining and understanding mental toughness within soccer. *Journal of Applied Sport Psychology* 17:326-332.

Thomas, J.R., J.K. Nelson, and S.J. Silverman. 2015. *Research methods in physical activity.* 7th ed. Champaign, IL: Human Kinetics.

Thomas, P., S. Murphy, and L. Hardy. 1999. Test of performance strategies: Development and preliminary validation of a comprehensive measure of athletes' psychological skills. *Journal of Sport Sciences* 17:691-711.

Thompson, H.R., R.C. Johnson, K.A. Madsen, and B. Fuller. 2019. Impact of physical education litigation on fifth graders' cardio–respiratory fitness, California, 2007–2018. *American Journal of Public Health* 109(11):1557-1563. doi:10.2105/AJPH.2019.305264

Troiano, R.P., D. Berrigan, K.W. Dodd, L.C. Masse, T. Tilert, and M. McDowell. 2008. Physical activity in the United States measured by accelerometer. *Medicine & Science in Sports & Exercise* 40(1):181-188.

Trudelle-Jackson, E., A.W. Jackson, C.M. Frankowski, K.M. Long, and N.B. Meske. 1994. Interdevice reliability and validity assessment of the Nicholas Hand-Held Dynamometer. *Journal of Orthopaedic & Sports Physical Therapy* 20:302-306.

Tryon, W.W. 1991. *Applied clinical psychology: Activity measurement in psychology and medicine.* New York: Plenum Press.

Tudor-Locke, C., and D.R. Bassett Jr. 2004. How many steps/day are enough? Preliminary pedometer indices for public health. *Sports Medicine* 34:1-8.

Tudor-Locke, C., Hatano, Y., Pangrazi, R. P., & M. Kang. 2008. Re-visiting "How many steps are enough?" *Medicine and Science in Sports and Exercise* 40(7):S537-S543.

Ulrich, D.A. 2016. *Test of gross motor development,* 3rd ed. Austin, TX: Pro-Ed. www.kines.umich.edu/tgmd3

U.S. Census Bureau. 2020. 65 and older population grows rapidly as baby boomers age. www.census.gov/newsroom/press-releases/2020/65-older-population-grows.html

The U.S. Army Center for Initial Military Training. 2018. *Army combat fitness test field testing manual.* Fort Eustis, VA: The U.S. Army Center for Initial Military Training.

U.S. Department of Health and Human Services (USDHHS). 1996. *Physical activity and health: A report of the Surgeon General.* Atlanta: USDHHS, Centers for Disease Control and Prevention, National Center for Chronic Disease Prevention and Health Promotion.

———. 2004. *Bone health and osteoporosis: A report of the surgeon general.* Rockville, MD: U.S. Department of Health and Human Services, Office of the Surgeon General.

———. 2008. *Physical activity guidelines for Americans.* Washington, DC: U.S. Department of Health and Human Services.

———. 2018. *Physical activity guidelines for Americans.* 2nd ed. Washington, DC: U.S. Department of Health and Human Services.

Vandewalle, H., G. Pérès, and H. Monod. 1987. Standard anaerobic exercise tests. *Sports Medicine* 4(4):268-289. doi:10.2165/00007256-198704040-00004

Van Schoyck, R.S., and A.F. Grasha. 1981. Attentional style variations and athletic ability: The advantage of a sport-specific test. *Journal of Sport Psychology* 2:149-165.

Varga, M. 2020. *Examination of the TOPS-2 Questionnaire in able-bodied athletes and athletes with a disability.* https://scholar.uwindsor.ca/etd/8488

Vasold, K.L., A.C. Parks, D.M.L. Phelan, M.B. Pontifex, and J.M. Pivarnik. 2019, Jul 1. Reliability and validity of commercially available low-cost bioelectrical impedance analysis. *International Journal of Sport Nutrition and Exercise Metabolism* 29(4):406-410. doi:10.1123/ijsnem.2018-0283

Vealey, R.S. 1986. Conceptualization of sport-confidence and competitive orientation: Preliminary investigation and instrument development. *Journal of Sport Psychology* 8:221-246.

Velicer, W.F., and J.O. Prochaska. 1997. Introduction: The transtheoretical model. *American Journal of Health Promotion* 12:6-7.

Vema, J.P., A.S. Sajwan, and M. Debnath. 2009. A study on estimating $\dot{V}O_2$max from different techniques in field situation. *International Quarterly of Sport Science* 2:42-47.

Verducci, F.M. 1980. *Measurement concepts in physical education.* St. Louis: Mosby.

Viera, A.J., and J.M. Garrett. 2005. Understanding interobserver agreement: The kappa statistic. *Family Medicine* 37:360-363.

Vingren, J.L., A.T. Woolsey, and J.R. Morrow Jr. 2014. Assessing physical activity, fitness, and progress in older adults. In *ACSM's exercise for older adults,* ed. W.J. Chodzko-Zajko. Champaign, IL: Human Kinetics.

Warburton, D.E., N. Gledhill, and A. Quinney. 2001. Musculoskeletal fitness and health. *Canadian Journal of Applied Physiology* 26(2):217-237. doi:10.1139/h01-013

Warnecke, R.B., T.P. Johnson, N. Chavez, S. Sudman, D.P. O'Rourke, L. Lacey, and J. Horm. 1997.

Improving question wording in surveys of culturally diverse populations. *Annals of Epidemiology* 7:334-342.

Weinberg, R., J. Butt, and B. Culp. 2011. Coaches' views of mental toughness and how it is built. *International Journal of Sport and Exercise Psychology* 9:156-172.

Weinberg, R., D. Gould, and A. Jackson. 1979. Expectations and performance: An empirical test of Bandura's Self-efficacy Theory. *Journal of Sport Psychology* 1:320-331.

Weinberg, R.S., and W.W. Comar. 1994. The effectiveness of psychological interventions in competitive sport. *Sports Medicine* 18:406-418.

Weinberg, R.S., D. Gould, D. Yukelson, and A. Jackson. 1981. The effect of preexisting and manipulated self-efficacy on a competitive muscular endurance task. *Journal of Sport Psychology* 3:345-354.

Welk, G.J., ed. 2002. *Physical activity assessments for health-related research.* Champaign. IL: Human Kinetics.

Weston, A.T., R. Petosa, and R.R. Pate. 1997. Validation of an instrument for measurement of physical activity in youth. *Medicine & Science in Sports & Exercise* 29:138-143.

Widmeyer, W.N., L.R. Brawley, and A.V. Carron. 1985. *The measurement of cohesion in sport teams: The group environment questionnaire.* (Available from Sports Dynamics, 11 Ravenglass Crescent, London, ON, Canada N6G 3X7)

Wier, L.T., A.S. Jackson, G.W. Ayers, and B. Arenare. 2006. Nonexercise models for estimating $\dot{V}O_2$max with waist girth, percent fat, or BMI. *Medicine & Science in Sports & Exercise* 38:555-561.

Wiggins, G. 1989. A true test: Toward more authentic and equitable assessment. *Phi Delta Kappan* 69:703-713.

———. 1998. *Educative assessment: Designing assessments to inform and improve student performance.* San Francisco: Jossey-Bass.

Wilmore, J.H., and W.L. Haskell. 1971. Use of the heart rate-energy expenditure relationship in the individualized prescription of exercise. *The American Journal of Clinical Nutrition* 24(9): 1186-1192.

Winnick, J.P., and F.X. Short. 1999. *The Brockport physical fitness test manual.* Champaign, IL: Human Kinetics.

———. 2014. *The Brockport physical fitness test manual.* 2nd ed. Champaign, IL: Human Kinetics.

Winston, W. 2009. *Mathletics: How gamblers, managers, and sports enthusiasts use mathematics in baseball, basketball, and football.* Princeton, NJ: Princeton University Press.

Yukelson, D., R. Weinberg, and A. Jackson. 1984. A multidimensional group cohesion instrument for intercollegiate basketball teams. *Journal of Sport Psychology* 6:103-117.

Zhu, W. 1997. Alternative assessment: What, why, how. *Journal of Physical Education, Recreation and Dance* 68(7):17-18. doi:10.1080/07303084.1997.1 0604977

———. 2010. Birth, growth, and challenges of "kinesmetrics" in the USA. *Annales Kinesiologiae* 1(2):95-111.

———. 2012a. "17% at or above the 95th percentile" – What is wrong with this statement? *Journal of Sport and Health Science* 1(2):67-69.

———. 2012b. Measurement practice in sport and exercise psychology: A historical, comparative and psychometric view. In *Measurement in sport and exercise psychology,* ed. G. Tenenbaum, R. Eklund, and A. Kamata, 9-21. Champaign, IL: Human Kinetics.

———. 2013. Science and art of setting performance standards and cutoff scores in kinesiology. *Research Quarterly for Exercise and Sport* 84:456-468.

Zhu, W., and W.J. Chodzko-Zajko, eds. 2006. *Measurement issues in aging and physical activity.* Champaign, IL: Human Kinetics.

Zhu, W., and M. Lee. 2010. Invariance of wearing location of Omron-BI pedometers: A validation study. *Journal of Physical Activity and Health* 7:706-717.

Zhu, W., M.T. Mahar, M. T. G.J. Welk, S.B. Going, and K.J. Cureton. 2011. Approaches for development of criterion-referenced standards in health-related youth fitness tests. *American Journal of Preventive Medicine* 41(4S2):S68-S76.

Zhu, W., and N. Owen, eds. 2017. *Sedentary behavior and health: Concepts, assessment and intervention.* Champaign, IL: Human Kinetics.

Zhu, W., M.J. Safrit, and A.S. Cohen. 1999. *FitSmart test user manual: High school edition.* Champaign, IL: Human Kinetics.

Zourbanos, N., A. Hatzigeorgiadis, S. Chroni, Y. Theodorakis, and A. Papaioannou. 2009. Automatic Self-Talk Questionnaire for Sports (ASTQS): Development and preliminary validation of a measure identifying the structure of athletes' self-talk. *Sport Psychologist* 23:233-251.

Index

Note: The italicized *f* and *t* following page numbers refer to figures and tables, respectively.

About the Authors

James R. Morrow, Jr., PhD, is a regents professor emeritus in the department of kinesiology, health promotion, and recreation at the University of North Texas at Denton. Dr. Morrow regularly taught courses in measurement and evaluation in human performance. He has authored more than 150 articles and chapters on measurement and evaluation, physical fitness, physical activity, and computer use and has made approximately 300 professional presentations. He has also conducted significant research using the techniques presented in the text.

Dr. Morrow served as president of the National Academy of Kinesiology and as chair of the science board of the President's Council on Physical Fitness and Sports. He has received research funding from the U.S. Olympic Committee, the Centers for Disease Control and Prevention, the National Institutes of Health, and the Cooper Institute. He is a fellow of the American College of Sports Medicine (ACSM), the National Academy of Kinesiology (NAK), and the North American Society of Health, Physical Education, Recreation, Sport and Dance Professionals. He is also a research fellow of SHAPE America. Dr. Morrow has chaired the AAHPERD Measurement and Evaluation Council and is a recipient of that council's Honor Award. He has produced four fitness-testing software packages, including the AAHPERD Health-Related Physical Fitness Test, and was editor in chief of *Research Quarterly for Exercise and Sport* from 1989 to 1993. He was the founding coeditor of the *Journal of Physical Activity and Health*. He enjoys playing golf, reading, traveling, and spending time with his grandchildren.

Dale P. Mood, PhD, is a professor emeritus and former associate dean in the College of Arts and Sciences at the University of Colorado at Boulder. Dr. Mood taught measurement and evaluation, statistics, and research methods courses beginning in 1970 and has published extensively in the field, including 47 articles and 6 books. He has been a consultant to five NFL football teams and chair of the AAHPERD Measurement and Evaluation Council, and he is a former president of AAALF. He was a reviewer for numerous human movement journals. In his leisure time, Dr. Mood enjoys reading, officiating summer league swimming meets, traveling, following the activities of his 18 grandchildren, and participating in a variety of physical activities.

Weimo Zhu, PhD, is a tenured full professor in the department of kinesiology and community health at the University of Illinois at Urbana-Champaign. His major area of research is kinesmetrics (i.e., measurement and evaluation in kinesiology).

Dr. Zhu's primary research interests are the study and application of new measurement theories (e.g., item response theory) and models in the field of kinesiology. His research works have earned him international recognition. He is the editor in chief of the *Research Quarterly for Exercise and Sport* and a fellow of the National Academy of Kinesiology, American College of Sports Medicine, and Research Consortium of SHAPE America. He is a member of the FitnessGram/ActivityGram advisory committee. He is also a member of the editorial board for various academic journals and serves on the executive committees of several national and international professional organizations. Dr. Zhu was the chair of the Measurement and Evaluation Council of SHAPE America and received the M&E Lifetime Achievement Award, the highest award in measurement and evaluation, from SHAPE America in 2020.

Minsoo Kang, PhD, is a full professor in the department of health, exercise science, and recreation management at the University of Mississippi. Kang earned both his bachelor's and master's degrees from Seoul National University in South Korea and his doctorate at the University of Illinois at Urbana-Champaign. His background is in analytics (measurement, applied statistics, and evaluation) in kinesiology with emphasis in IRT, Rasch, and psychometrics. Kang's research has focused on measurement and statistical methods and their applications to assessments of physical activity and sedentary behavior. He has published more than 140 refereed journal articles, made 10 book contributions, and presented more than 200 research projects. He teaches courses on data analysis, applied statistics, research methods, meta-analysis, and measurement theory and practice in human performance. He enjoys traveling and playing badminton, golf, and tennis.

Kang is a fellow of the American College of Sports Medicine (ACSM) and a research fellow of SHAPE America. He has chaired the AAHPERD Measurement and Evaluation Council and is a recipient of that council's Honor Award. Kang received the Distinguished Research Award at Middle Tennessee State University. He has served as an associate editor of the *Research Quarterly for Exercise and Sports*, the *Journal for the Measurement of Physical Behaviour*, and *Measurement in Physical Education and Exercise Science* and is also a member of the editorial board for several journals.